ADVANCES IN FUTURE MANUFACTURING ENGINEERING

Studies in Materials Science and Mechanical Engineering

eISSN: 2333-6560
VOLUME 2

PROCEEDINGS OF THE 2014 IMSS INTERNATIONAL CONFERENCE ON FUTURE MANUFACTURING ENGINEERING (ICFME 2014) HONG KONG, 10–11 DECEMBER, 2014

Advances in Future Manufacturing Engineering

Editor

Guohui Yang
International Materials Science Society, Hong Kong, Kowloon, Hong Kong

CRC Press is an imprint of the
Taylor & Francis Group, an **informa** business

A BALKEMA BOOK

Published by:
CRC Press/Balkema
P.O. Box 447, 2300 AK Leiden, The Netherlands
e-mail: Pub.NL@taylorandfrancis.com
www.crcpress.com – www.taylorandfrancis.com

First issued in paperback 2020

© 2015 by Taylor & Francis Group, LLC
CRC Press/Balkema is an imprint of the Taylor & Francis Group, an informa business

No claim to original U.S. Government works

Typeset by V Publishing Solutions Pvt Ltd., Chennai, India

ISBN 13: 978-0-367-73825-9 (pbk)
ISBN 13: 978-1-138-02817-3 (hbk)
ISBN 13: 978-0-429-22615-1 (ebook)

Visit the Taylor & Francis Web site at
http://www.taylorandfrancis.com

and the CRC Press Web site at
http://www.crcpress.com

Table of contents

Materials processing and analysis

Data processing and information management

Multimedia applications

Biochemistry and medicine

Preface

The 2014 IMSS International Conference on Future Manufacturing Engineering (ICFME 2014) was held on December 10–11, 2014, Hong Kong.

This conference gathered academics, industry managers and experts, manufacturing engineers, university students and others interested or proficient in the field of manufacturing engineering to discuss the state of the art in the field. During this conference progresses, problems and developments in manufacturing engineering was discussed. The conference is expected to play a role in the development of future manufacturing engineering.

Manufacturing engineering is a discipline of engineering dealing with different manufacturing practices and includes research, design and development of systems, processes, machines, tools and equipments. The manufacturing engineer's primary focus is to turn raw materials into a new or updated product in the most economic, efficient and effective way possible. This field also deals with the integration of different facilities and systems for producing quality products (with optimal expenditure) by applying the principles of physics and the results of manufacturing systems studies, such as Design and Manufacturing, Materials Science and Materials Processing Technology, Computer-aid Manufacturing, Mechanics.

Manufacturing engineers develop and create physical artifacts, production processes, and technology. It is a very broad area which includes the design and development of products. The manufacturing engineering discipline has very strong overlap with mechanical engineering, industrial engineering, production engineering, electrical engineering, electronic engineering, computer science, materials management, and operations management. Manufacturing engineers' success or failure directly impacts the advancement of technology and the spread of innovation.

We have received good-quality and enthusiastic responses from people in different countries and different industries. They have submitted many great papers and ideas related to this conference. Here we extend our sincere thanks to all of the people who have responded actively and spent days to write and edit the conference papers.

Also we'd like to give the keynote speakers our appreciation for their great contribution to this conference. We thank all the delegates who attended the conference here for their sharing of suggestions, opinions and ideas in the field of future manufacturing engineering. Sincere thanks go to the sponsors and organizers of this conference who have offered us a good chance to gather, discussing and sharing information and ideas in manufacturing engineering. Last but not least we would like to express our sincere appreciation to all staff and volunteers who worked so hard to make this conference a success.

Organizing committee

KEYNOTE SPEAKER

Gerald Schaefer, *Department of Computer Science, Loughborough University, Loughborough, UK*

HONOR CHAIR

Tianharry Chang, *IEEE SYS Brunei Darussalam Chapter Past Chair, Brunei Darussalam*

GENERAL CHAIRS

Enke Wang, *International Materials Science Society, Hong Kong*
Mark Zhou, *Hong Kong Education Society, Hong Kong*

PUBLICATION CHAIR

Guohui Yang, *International Materials Science Society, Hong Kong*

ORGANIZING CHAIRS

Khine Soe Thaung, *Society on Social Implications of Technology and Engineering, Maldivers*
Tamal Dasas, *Society on Social Implications of Technology and Engineering, Maldivers*

PROGRAM CHAIR

Guohui Yang, *International Materials Science Society, Hong Kong*

INTERNATIONAL COMMITTEE

Supachai Kaewpoung, *Suranaree University of Technology, Thailand*
Kenedy Aliila Greyson, *Intelligent Systems Research Unit (ISRU), Tanzania*
Kaveenga Rasika Koswattage, *Center for Frontier Science, Chiba University, Japan*
Kolchinskiy Vladislav, *Institute of Automation and Control Processes FEB RAS, Russia*
Shih Cheng-Hung, *National Sun Yat-Sen University, Taiwan*
Jin Xie, *Institute of Scientific Computing, Hefei University, China*
Xiaoyan Liu, *University of La Verne, La Verne, USA*
Qiaohua Hu, *Northwest University For Nationalities, Gansu, China*
Qiquan Wang, *China Institute of Industrial Relations, Beijing, China*
Jun Chen, *Experimental center of Jiujiang University, Jiangxi, China*

Physics and electrical energy

Advances in Future Manufacturing Engineering – Yang (Ed.)
© 2015 Taylor & Francis Group, London, ISBN 978-1-138-02817-3

Autotransformer-fed railway power supply model using MATLAB/SIMULINK

Supachai Kaewpoung, Chaiyut Sumpavakup & Thanatchai Kulworawanichpong
School of Electrical Engineering, Institute of Engineering, Suranaree University of Technology,
Nakhon Ratchasima, Thailand

ABSTRACT: This paper presents autotransformer-fed railway power supply model by using MATLAB/SIMULINK. The purpose of this study is to characterize voltage profiles and current of trains running. The system model created in this study consists of 4 main components: i) overhead contact wire and running rails ii) traction power supply substation iii) autotransformer and iv) rolling stock. The test systems used herein are 25-kV, 50-Hz double- and single-track AC railway systems with autotransformer feeding configuration to analyze voltage profiles and current of trains running, respectively. The substation spacing distance is 30 km and autotransformers are installed at every 10-km spacing. As a result, the proposed SIMULINK model can be used to analyze the voltage profiles and current of the autotransformer-fed railway power supply system, effectively.

Keywords: Autotransformer; Voltage profile; Traction power substation; MATLAB/SIMULINK

1 INTRODUCTION

The high-speed rail is one of the facilities of transportation. Continually being developed to become an important aspect of modern railways. At present, the power supply to the train tracks to an AC power supply which can afford long distances. In the past, the DC power supply at 1.5 kV to 3 kV in 1900 and 1930 were used as a control. However, there is some difficulty in turning the motors to drive [1,2] high-speed trains operating at some higher speed than the speed of typical trains in services. Typically, high-speed rail in various forms are usually powered by electric power source through wires above locomotives. To analyze power loading of train running needs power network solving algorithms that are conventionally complicated and tedious for computer programming. It would be simpler if some graphic-based programming, like MATLAB/SIMUINK, is exploited. In the past, MATLAB/SIMUINK has been applied to various research fields as a potential tool for analysis, simulation and design. However, the use of MATLAB/SIMULINK in railways is very few. Traction power supply systems [3], especially for long-distance and high-speed trains, are focused. Autotransformer feeding configuration [4] of the traction power supply system is chosen in order to performing the analysis of voltage and current in the system.

This paper presents a model of AC railway systems equipped with autotransformers by using

MATLAB/SIMUINK to analyze the voltage profile of a tested high-speed rail transit section. Comparison between three train models, which are a constant power model (S-model), a constant impedance model (Z-model) and a constant current (I-model) is engaged. Description of AC railway power supply system is summarized in Section 2. Section 3 presents circuitry models for autotransformer feeding configuration by using MATLAB/SIMUINK. Tests and their results are shown in Section 4. The last section is a conclusion remark.

2 AC RAILWAY POWER FEEDING SYSTEM

AC electric power supply is configured to supply a wide range of different industries. The standard AC railway power supply system is preferred a single-phase power feeding system. In this single-phase feeding scheme, AC power supply is connected directly to the high voltage three-phase power through traction substation transformers to convert three-phase high-voltage electric power into single-phase 25-kV electric power. In Great Britain, the standard HV/MV interfacing power systems are typically 132/25 kV, 275/25 kV or 400/25 kV [5]. The HV-side of the traction substation transformer is connected to the utility's three-phase busbars while the LV-side busbar of the

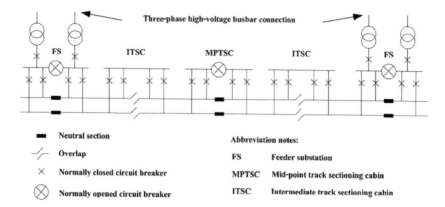

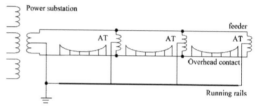

Figure 1. Typical AC railway power supply system.

▬	Neutral section
⫽	Overlap
✕	Normally closed circuit breaker
⊗	Normally opened circuit breaker

Abbreviation notes:

FS	Feeder substation
MPTSC	Mid-point track sectioning cabin
ITSC	Intermediate track sectioning cabin

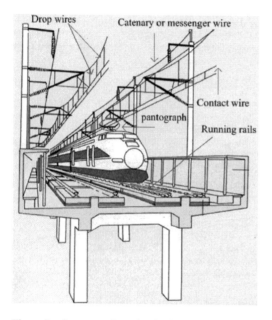

Figure 2. Structure of overhead railway power feeding system.

Figure 3. AT-fed railway power supply system.

is commonly known in the United Kingdom and Canada while messenger wire is commonly used in the United States or other European countries) [7]. The catenary or messenger wire is used in order to fix the contact wire at the minimum swing. Any running train is drawn its required power from the contact wire through the sliding contact made from graphite mounted on the roof of the power-train car, called a pantograph. All details just mentioned are shown in Fig. 2.

transformer is connected to a single-phase railway power feeding system [6] as shown in Fig. 1. The single-phase AC railway power feeding system is designed to energize electric trains. There is track-side switchgear at the middle way between two adjacent substations called MPTSC. Sometimes, it is also installed ITSC to help isolating a temporary or permanent faults in line sections. The ITSC is normally located between the substation and the MPTSC.

Structure of the AC railway power feeding system consists of overhead feeding conductors, normally called contact wire and catenary (catenary

3 MATLAB/SIMULINK MODEL OF AT-FED RAILWAY SYSTEMS

Autotransformer or AT-fed railway power systems can be described by Fig. 3. It illustrates power-feeding arrangement with 1) contact wire, 2) running rails and 3) feeder. The feeder is used for current returning purpose. It is sometimes called negative feeder. The autotransformer is installed at every one-third of the feeding length. Some might be known that current flowing into contact wire, running rails and feeder is simply computed. The current distribution in the feeding sections can be summarized as shown in Fig. 4.

4

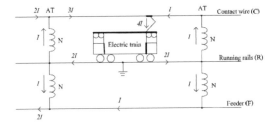

Figure 4.   Current distribution in AT-fed railway system.

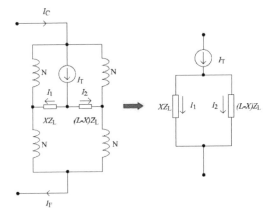

Figure 5.   Current divider model for AT feeding section.

Assume that a train consuming current IT is running between two adjacent autotransformers. By dividing the current flow into two current parts I1 and I2 as shown in Fig. 5 where ZL is impedance per length of the feeder line. The section length between two autotransformer is L. As can be seen, the current flow is depended on the train's position X, while it is shown in Fig. 5.

To model AT-fed railway power feeding system needs separate circuit for each power system component. In this paper, four major components of the AT-fed railway power system are introduced such that 1) traction power substation, 2) autotransformer, 3) overhead power line system and 4) train. Their MATLAB/SIMULINK models can be presented as shown in Fig. 6.

4   SIMULATION AND RESULTS

In this paper, two test systems are conducted. The first test system is a high-speed 25-kV, 50-Hz, double-track, AT-fed railway power system. The second test system is just a single-track AT-fed railway system. Simulation is carried out by using MATLAB/SIMULINK program with personal computer (Intel Core i3 2.26 GHz, 2 GB RAM). AC power supply substations have a distance of 60-km between two adjacent substations. This means that the feeding length of

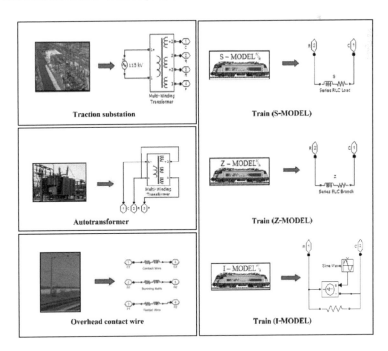

Figure 6.   MATLAB/SIMULINK model of railway system components.

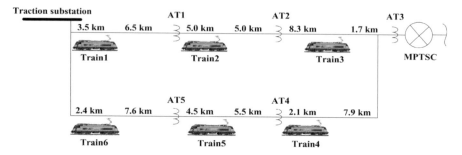

Figure 7. Test system 1.

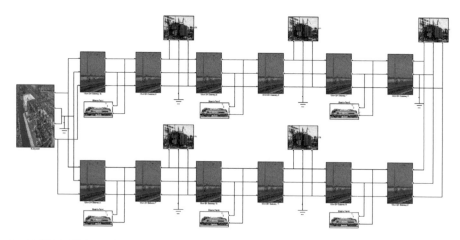

Figure 8. MATLAB/SIMULINK model of test system 1.

the test system is 30 km. Autotransformers are installed at every 10-km apart from each other. The first test system situates 6 trains running on the track, 3 trains on the up-track and 3 trains on the down-track as shown in Fig. 7. Its MATLAB/SIMULINK model can be presented in Fig. 8 to study the voltage profile and current distribution of the AT-fed railway system. This test can be detailed by using three train models (S-model, and Z-model).

Table 1 presents train model data for test system 1. It is assumed that all trains are running at 0.98 lagging power factor. Table 2 describes the voltage result at the substation and autotransformers.

The second test is a single-track system as shown in Fig. 9. Its MAT/SIMULINK model is shown in Fig. 10. It is used to prove the current distribution in the AT-fed power system as shown in Fig. 4. The test is assumed that there is only one train running in the system. The test cases are the running train located at feeding section 1 between 1) traction substation and AT1, 2) AT1 and AT2 and 3) AT2 and AT3 (MPTSC).

Table 1. Train model data for test system 1.

Positions of train (km)	S (MVA)	Z (Ω)	P.F (lagging)
Up-track	3.5 3.2066	191.02 + j38.75	0.98
	15.0 0.3236	1892.75 + j384.39	
	28.3 0.2205	2777.75 + j564.13	
Down-track	7.9 3.7015	164.47 + j33.61	
	15.5 0.3236	1892.75 + j384.39	
	27.6 3.5749	171.33 + j34.79	

Table 2. Results of test system 1.

Positions	Voltage (kV)			
	S-model	Vdrop (%)	Z-model	Vdrop (%)
Substation	25.00	0	25.00	0
AT1	24.80	0.80	24.80	0.80
AT2	24.69	1.24	24.70	1.20
AT3	24.60	1.60	24.60	1.60
AT4	24.56	1.76	24.57	1.72
AT5	24.75	1.00	24.75	1.00

Figure 9. Test system 2.

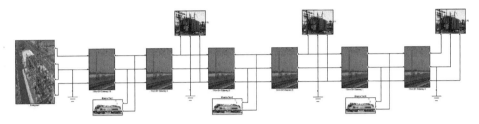

Figure 10. MATLAB/SIMULINK model of test system 2.

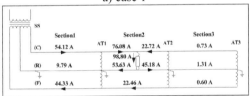

a) case 1

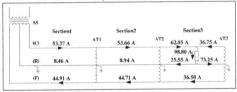

b) case 2

c) case 3

Figure 11. Current distribution of test system 2 (three test cases).

The simulation for 3 test case scenarios of the train running at the position of 3.5 km, 15.0 km and 28.3 km, respectively. For all cases, the train draws $100\angle30°$. A from the power line. The current distribution of the three test case scenarios can be viewed graphically as shown in Fig. 11. These confirm the distribution of the current flowing as presented previously in Fig. 4.

5 CONCLUSIONS

This paper presents MATLAB/SIMULINK model of AT-fed railway power system. The AT-fed railway system models consist of 1) traction substation model, 2) autotransformer model, 3) overhead power feeding line model and 4) train model. The first three models can be simply created by using RLC, voltage source and transformer blocks of MATLAB/SIMULINK program. The simulation results confirm that MATLAB/SIMULINK model can be alternative for AT-fed railway power supply analysis and simulation.

REFERENCES

[1] R.J. HILL. Electric railway traction. II. Traction drives with three-phase induction motors[J]. IET Journals & Magazines: Power Engineering Journal, June (1994), Volume: 8, Issue: 3, PP. 143–152.

[2] R.J. HILL. Electric railway traction. I. Electric traction and DC traction motor drives[J]. IET Journals & Magazines: Power Engineering Journal, February (1994), Volume: 8, Issue: 1, PP. 47–56.

[3] R.J. HILL. Electric railway traction. III. Traction power supplies[J]. IET Journals & Magazines: Power Engineering Journal, pp. 275–286, December (1994).

[4] H. BOZKAYA. A comparative assessment of 50 kV auto-transformer and 25 kV booster transformer railway electrification systems[D]. M Phil Thesis, University of Birmingham, UK, October (1987).

[5] R.W. STURLAND. "Traction power supplies", GEC ALSTHOM Transmission & Distribution Projects Ltd.

[6] R.W. WHITE, "AC supply systems and protection", Fourth Vocation School on Electric Traction Systems, IEE Power Division, April (1997).

[7] Wikipedia, Overhead line, Last edited July (2014).

[8] T. Kulworawanichpong. Optimising AC electric railway power flows with power electronic control[D]. PhD Thesis, University of Birmingham, UK, November (2003).

Advances in Future Manufacturing Engineering – Yang (Ed.)
© 2015 Taylor & Francis Group, London, ISBN 978-1-138-02817-3

Optimization of energy consumption in Electric Railways System using hit-and-run algorithm

Kenedy Aliila Greyson

Intelligent Systems Research Unit (ISRU), Department of Electronics and Telecommunications, Dar es Salaam Institute of Technology, Dar es Salaam, Tanzania

Thanatchai Kulworawanichpong

School of Electrical Engineering, Suranaree University of Technology, Muang District, Nakhon Ratchasima, Thailand

ABSTRACT: The power demand by electric train varies from a huge positive power (peak-power) in a starting phase to a negative power during braking. This power is supplied by substations along the Electrified Railway System (ERS) and the Regenerative Braking Energy (RBE) source. The kinetic energy is converted into electrical energy, reserved, and then contributes or compensates in the total energy consumed by the system. In this paper the interest is to optimize the consumed energy from the substation by utilizing the RBE. The paper discusses the possible minimum energy consumption from the grid by maximizing the RBE under the practical variable conditions, such as the maximum operational train speed, passengers ride comfort, the minimum time to cover the distance of interstations, maximum starting acceleration, and the Energy Storage Capacity (ESC). Finally, the paper presents the conclusion from the optimum solution obtained by the use of hit-and-run optimization algorithm.

Keywords: electric energy; optimization; railways; regenerative

1 INTRODUCTION

The railway transportation plays important role for public transportation in many big cities, and seems to be the best future plan for the fast growing cities. However, its energy consumption is also huge; hence the energy efficiency of railway systems should be taken into consideration during operation planning. Due to the impact of energy consumption in operating cost, it turns out that, most operators' interest is to minimize the energy consumptions. There are several techniques mentioned in several research works that include the improvement of speed profiles, charging time of the storage systems, and the torque-speed balancing. However, due to the complex relations between the train speed motion and energy consumption, which are non-linear, analytical results are scarce and available only in a simplified areas [1].

In this paper, the driving pattern which includes the starting acceleration, passengers ride comfort defined by the deceleration, the expected time to cover the interstation, and energy storage limitations are discussed. Therefore, the maximum operational train speed, energy generated from braking, and storage capacity are the concerned in subsidizing the energy drawn from the grid through substations as shown in Figure 1. The paper is based on the Sky train Sukhumvit Line (Light Green Line) in Bangkok Thailand. According to [2], the characteristics of urban and suburban type of service, initial acceleration ranges from 1.5 kph to 4 kph (0.4 m/s–1.1 m/s) and retardation ranges from 3 kph to 4 kph (0.8 m/s–1.1 m/s). For the suburban and urban train service, the train speed profile consists of the initial

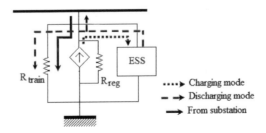

Figure 1. Current flow in the circuit during charging and discharging modes.

acceleration, coasting, and braking or retardation, with no free running period [2]. Therefore, certain state of the acceleration, coasting, and braking are used in estimating the optimum energy saving. In principle, during braking, the kinetic energy is converted into electrical energy by the use of power electronic devices. This generation is due to the opposition of the motor speed and hence the electric power is generated and can be fed back to the grid of electric network or stored for future use and eventually reduces the cost of energy cost [3][4][5]. Therefore, one of the techniques used in subsidizing the energy from substations power supply is harvesting of the Regenerative Braking Energy (RBE) [3]. The energy harvested using regenerative braking is important because it can essentially convert a train station to a micro-grid this power can be used instantly by the adjacent running trains or stored in energy storage devices as shown in Figure 1. It is assumed that, while the traction energy accounts for 60–70 percent of the total energy consumption in rail transport systems [5], about 40 percent of the traction energy consumption could be supplied by the RBE if well managed [6].

This paper presents the discussion on the optimization process by utilizing the possible maximum RBE to subsidize the energy drawn from the grid through the substations within the Electrified Railway System (ERS).

2 OPTIMIZATION PROBLEM

The energy management in electrified railways include the maximizing the regenerative braking energy and to minimize the energy consumption during the operations. The optimization involves the limits on: time to be spent in the interstations, energy storage capacity, maximum operational, initial acceleration, and the train retardation. The search of the best solution under a complicated and complex system is discussed in this section. The ultimate goal is to minimize the energy consumption from the grid. For simplicity, the minimum energy consumption problem under constrained parameters can be stated as a multi-variable optimization with inequality constraints given in (1).

$$\min f(X) = \sum_i^N f_i(v_i, a_i, \alpha_i) \qquad (1)$$

subject to:

$$t_i = f(v_i, d_i) \le T \text{ (s)}$$

$$0.4 \le a_i \le 1.1 \text{ (m/s}^2)$$

$$0.8 \le \alpha_i \le 1.1 \text{ (m/s}^2)$$

$$v_i \le 90 \text{ (kph)}$$

$$C_b \le 9 \text{ (kWh)}$$

where, v_i is the maximum operational speed of the train in the interstation i in kph, a_i is the initial acceleration in m/s^2, α_i is the deceleration in m/s^2, t_i is the time to be spent in interstation i in seconds, T is the maximum time to be spent in interstation i, d_i is the distance of the interstation i in km, C_b is the battery capacity in kWh, and N is the number of interstations in the electrified railways.

There are various mathematical programming techniques available for the solution of different types of optimization problems. However, the most practical electrified railways systems involve a number of elements with conditions. In this paper, the objective function stated in (1) is a shared objective while maintaining certain conditions. In this research, the hit-and-run optimization algorithm is used and theory behind the technique is summarized in the following section.

3 HIT-AND-RUN ALGORITHM

The hit-and-run algorithm is a random walk which has been proposed for a long time [2][7][8][9]. In this algorithm, the two hit-and-run are the hyper-sphere directions and the coordinate directions. Basically, the random direction, which is chosen with equal probability from the coordinate direction vector and their negation, is from a uniform distribution on the two directions (hyper-sphere and coordinate). The hit-and-run algorithm is as shown in [2][7][9].

For the unit ball in $\Re^n$ by $D = \{x \in R^n : \|x\| < 1\}$, and $\partial D = \{x \in R^n : \|x\| = 1\}$, where D is the d-dimensional unit sphere centered at the origin and v is the probability distribution of direction choice on ∂D. For the full dimensional region S, given by $S \in \Re^n$; with the set of constraints mentioned earlier.

Step 1: Initialize: $x_0 \in S$ and set $k = 0$;

Step 2: Generate random direction, $\theta_k \in \partial D$ according to v.

Step 3: Select λ_k from the line set $\Lambda_k = \{\lambda \in \Re : x_k + \lambda\theta_{k+1} \in S\}$ according to the density:

$$f_k(\lambda) = \frac{f(x_k + \lambda\theta_{k+1})}{\int_{\Lambda_k} f(x_k + \lambda\theta_{k+1})dr} \qquad \lambda \in \Lambda_k$$

Step 4: set $x_{k+1} = x_k + \lambda_k \theta_{k+1}$ and set $k = k+1$
Step 5: Go to step 2.

4 CASE STUDY

The electrified railway used in the simulation in this work is a BTS—Sky train Sukhumvit Line (Light Green Line) in Bangkok Thailand with 22 stations (i.e. 21 interstations), passenger mass factor of 1.4, and 21.6 km per direction. In the simulation, base parameters are assumed to be at the maximum operational speed of 80 kph, deceleration of 1 m/s², and starting acceleration of 1 m/s². The network has eight substations to provide electric power supply of a 750 V DC, third rail. The train set is made up of two motor cars and one trailer car. Motor cars carry the traction contained, the brake resistor and electrical equipment boxes. A trailer car carries the auxiliary inverter, NiCd battery, and air supply system.

5 RESULTS AND DISCUSSION

In this paper, simulations are based on the data for BTS-Sky train Sukhumvit line in Thailand shown in Table 1. For simplicity, results of two interstations are presented. The first interstation is between Mo chit and Saphan Khwai (interstation 1), a distance of 1.1 km, and the second interstation is between Saphan Khwai and Ari (interstation 2), a distance of 1.7 km. The first constraint is the time required for the train to reach and leave the station, i.e. the time to cover the interstation. For simulation purposes, the

time to cover interstation 1 and interstation 2 are 75 seconds and 110 seconds, respectively. Other constraints are as shown in (1). Test is specified at the maximum operating train speeds to range from 60 kph to 120 kph at the interval of 5 kph, maximum acceleration of 0.4 m/s² to 1.1 m/s² at the interval of 0.1 m/s², and deceleration of 0.8 m/s² to 1.1 m/s² at the interval of 0.1 m/s².

Some of the energy generated may not be stored due to the limited storage capacity, hence loss of energy. Figure 2 presents the summary of the compensated (total energy used—energy generated) in interstation 1 and 2, when unlimited ESS capacity and limited ESS capacity of 9 kWh are used. It is observed therefore that, for unlimited storage capacity, shown in Figure 2, the minimum energy compensated in interstation 1 is 6.1413 kWh, at the maximum operating speed of 80 kph, and 9.527 kWh at the maximum operating speed of 90 kph in interstation 2. At the maximum speed of 80 kph in interstation 1, and limitation

Table 1. Train model data for BTS-Sky Train Sukhumvit Line (Light Green Line).

Parameter	Value
Tare mass (tons)	107
Maximum traction effort (kN)	305
Max. operating speed (kph)	90
Friction coefficient	0.01
Efficiency (%)	79
Drag coefficients	
a	0
b	0
c	121.5

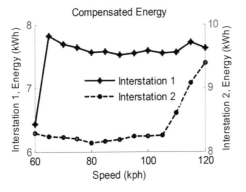
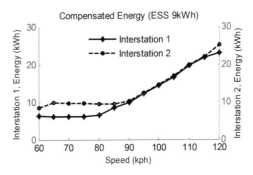

Figure 2. The energy compensated versus train speed in interstation 1 and 2, when unlimited and limited ESS are used, at given maximum operating train speeds, (maximum acceleration of 1 m/s² and retardation of 1 m/s² at both interstations).

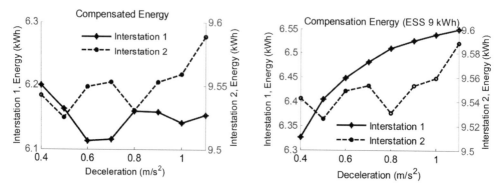

Figure 3. The energy compensated in interstation 1 and 2 versus deceleration when unlimited and limited ESS are used, (maximum acceleration of 1 m/s² and maximum operating speed of 80 m/s at both interstations).

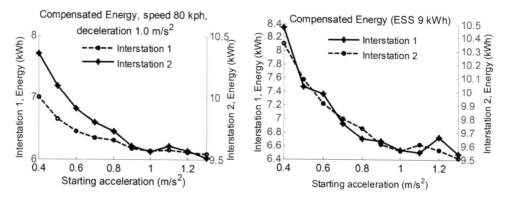

Figure 4. The energy compensated in interstation 1 and 2 versus starting acceleration when unlimited and limited ESS are used, (deceleration of 1 m/s² and maximum operating speed of 80 m/s at both interstations).

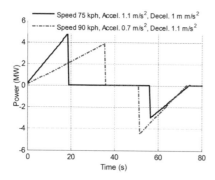

Figure 5. The comparison of train power consumption at interstation 1 at the given parameters.

on deceleration mentioned earlier, the achievable minimum compensated energy consumption was obtained at deceleration of 1 m/s² as shown in Figure 3. Figure 4 present the energy compensated when unlimited ESS capacity and limited ESS capacity are used at a given starting acceleration. The comparison shown in Figure 5 presents the two possible solutions for the problem. The energy consumption is summarized in Table 2. The solution (speed 75 kph, starting acceleration of 1.1 m/s², and deceleration of 1 m/s²), refered to as case 1, provides the better management compared to the case 2 (speed 90 kph, starting acceleration of 0.7 m/s², and deceleration of 1.1 m/s²) as shown in Table 2.

Table 2. The summary of energy consumption for interstation 1.

	Time (s)	Energy generated	Energy consumption (kWh)		
			Consumed	Unconstrained	Constrained
Case 1	74.5	7.3094	13.4782	6.1688	6.1688
Case 2	73	14.4405	19.9057	6.4652	10.9057
		Energy saved	6.4275	0.2964	4.7369

6 CONCLUSIONS

It is observed that, the effective energy management can be achieved by the effective choice of the parameters that govern the speed profile, such as starting acceleration, deceleration, and operational schedule. It can also be noted that, if well managed, the regenerative braking enegry can play the great role on the energy compensation.

ACKNOWLEDGEMENTS

Authors would like to extend their great appreciation to Suranaree University of Technology (SUT), through the research development for the support in this research work.

REFERENCES

[1] M. Larranaga, J. Anselmi, U. Ayesta, P. Jacko, A. Romo. *Optimization Techniques Applied to Railway Systems*. HAL, 2013.

[2] H.C.P. Berbee, C.G.E. Boender, A.H.G. Rinnooy Ran, C.L. Scheffer, R.L. Smith, and J. Telgen. Hit-and-run algorithms for the identification of non-redundant linear inequalities. Mathematical Programming, Vol. 37, No. 2, pp. 184–207, 1987.

[3] S. Jung, H. Lee, K. Kim, H. Jung, H. Kim, and G. Jang. A study on peak power reduction using regenerative energy in railway systems through dc subsystem interconnection [J]. Journal of Electrical Engineering & Technology, Vol. 8, No. 5, pp.1070–1077, 2013.

[4] Y. Jiang, J. Liu, W. Tian, M. Shahidehpour, M. Krishnamurthy. Energy harvesting for the electrification of railway stations: getting a charge from the regenerative braking of trains [J]. IEEE Electrification Magazine. Vol. 2, No. 3, pp. 39–48, 2014.

[5] S. Lu, P. Weston, S. Hillmansen, B. Gooi, and C. Roberts. Increasing the regenerative braking energy for railway vehicles [J]. IEEE Transactions on Intelligent Transportation Systems, accepted, 2014.

[6] S. Acikbas and M.T. Soylemez. Parameters affecting braking energy recuperation rate in dc rail transit [J]. Proceedings of ASME Conference, pp. 263–268, 2007.

[7] C.J.P. Belisle, H. Edwin Romeijn, and R.L. Smith. Hit-and-run algorithms for generating multivariate distributions [J]. Mathematics of Operations Research, Vol. 18, No. 2, pp. 255–266, 1993.

[8] R.L. Smith. Efficient Monte-Carlo procedures for generating points uniformly distributed over bounded regions [J]. Operations Res. 32, pp. 1296–1308, 1984.

[9] R.L. Smith. The hit-and-run sampler: a globally reaching markov chain sampler for generating arbitrary multivariate distribution [J]. Proceedings of the Winter Simulation Conference, pp. 260–264, 1996.

Advances in Future Manufacturing Engineering – Yang (Ed.)
© 2015 Taylor & Francis Group, London, ISBN 978-1-138-02817-3

Power dependent on supercontinuum generation in the photonic crystal fiber

Yuanfei Wei & Fuli Zhao
State Key Laboratory of Optoelectronic Materials and Technologies, Sun Yat-sen University, Guangzhou, China

ABSTRACT: We used the tunable Ti: sapphire optical parametric amplifier as the pump source and the fiber spectrometer to measure the spectroscopy of the supercontinuum generated in high nonlinear photonic crystal fiber under different input power. The PCF supercontinuum differences affected by physical mechanisms were analyzed. The experimental results showed that the spectral width is gradually widened with the increase of the incident pump pulse power. Furthermore, the width of supercontinuum finally saturated and the shape of supercontinuum was also stabilized as long as the input power reaches a certain threshold.

Keywords: photonic crystal fiber; supercontinuum; nonlinear optics

1 INTRODUCTION

Photonic crystal fiber is a new kind of optical fiber based on properties of photonic crystals. PCFs are widely used due to their several advantages, such endless single-mode, large mode area, anomalous dispersion in the visible light region and highly birefringence characteristics [1,2]. PCFs can be classified into two basic categories according to the light propagation mechanism. The first is the index-guiding PCF, we call it the TIR fiber, it is usually formed by a central solid defect region surrounded by multiple air holes in a regular triangular lattice and confines light by total internal reflection. The second one uses a perfect periodic structure exhibiting a Photonic Band-Gap (PBG) effect at the operating wavelength to guide light in a low index core region, which is also called PBG Fiber (PBGF) [3,4]. In 2000 year, the supercontinuum generation phenomenon during photonic crystal fiber experiment has been discovered and demonstrated. The research result attracted much interest of the physical mechanisms research; one significant application related research is the supercontinuum generation in the PCF.

In this paper, we use the tunable femtosecond laser as the incident light source, after the incident light coupled into the core of PCF, by changing the incident power, we recorded different spectra of supercontinuum. In order to find the relation between the femtosecond pumping and the supercontinuum spectra, we analyzed and explained the physical mechanisms which causing supercontinuum generation phenomenon.

2 THEORETICAL MODEL

In the process of the laser pulse transform in photonic crystal fibers, Supercontinuum spectra generated in PCFs of dispersion can be calculated by nonlinear Schrödinger equation.

$$\nabla^2 E - \frac{1}{c^2}\frac{\partial^2 E}{\partial t^2} = \mu_0 \frac{\partial^2 P_L}{\partial t^2} + \mu_0 \frac{\partial^2 P_{NL}}{\partial t^2} \tag{1}$$

where E is the electric field intensity, c is speed of light, μ_0 is magnetic permeability of free space, P_L is linear intensity of polarization and P_{NL} is the nonlinear intensity of polarization. When the incident pulse width is not small enough, there will be many kinds of nonlinear optical effect emerged and they cannot be ignored. In this situation, we need to describe them by the generalized nonlinear Schrödinger equation [5].

$$\frac{\partial A(z,t)}{\partial z} + \frac{\alpha}{2}A + \frac{i\beta_2}{2}\frac{\partial^2 A}{\partial T^2} - \frac{\beta_3}{6}\frac{\partial^3 A}{\partial T^3}$$
$$= i\gamma\left[|A|^2 + \frac{i}{\omega_0}\frac{\partial}{\partial T}\left(|A|^2 A\right) - T_R A\frac{\partial |A|^2}{\partial T}\right] \tag{2}$$

In the formula, z represents the transmission distance in PCF, β_2 and β_3 are coefficients of second-order dispersion and third-order dispersion, A is the amplitude of incident pulse, α is fiber loss, γ is nonlinear coefficient, ω_0 is the center frequency, T is time parameter and T_R is coefficient of the Raman Scattering [5,6].

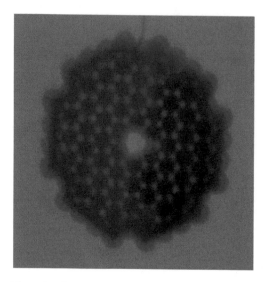

Figure 1. Cross-section of the PCF. The black regions are hollow, the red regions are pure silica.

Figure 1 is the cross-section of the PCF we used in the experiment. It is a single-mode fiber with a pure silica core surrounded by high air-filling fraction cladding. It is designed to keep single-mode transmission of the incident pulse. Meanwhile, contributions of Li Shuguang et al applied the effective index model method to simulate the equivalent model of PCFs to prove that the diameter of core and the size of the air holes both have a significant impact on the dispersion [7].

3 EXPERIMENT AND ANALYSIS

3.1 Experiment set-up

The schematic optical path diagram of the experiment is shown in Figure 1. In the experiment, these experimental instruments are decided as follows: Laser provides the incident light source. Attenuator controls the power of the incident light. Focus lens focuses the incident light. Spectrometer records the spectral profile of supercontinuum.

4 RESULTS AND ANALYSIS

There are many physical factors related to the generation of Supercontinuum (SC), such as Group Velocity Dispersion (GVD), Four Wave Mixing (FWM), Stimulated Raman Scattering (SRS), Self Phase Modulation (SPM), Cross Phase Modulation (XPM), Third-Order Dispersion (TOD), Self Steepening Effect (SSE) etc. All these physical

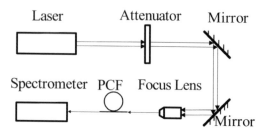

Figure 2. Schematic optical path diagram of the experiment.

phenomena have contributions to the shape and the intensity of SC [8,9].

In the experiment, the specific parameters of these experimental instruments are listed as follows:

PCF: zero dispersion wavelength $\lambda_0 = 670$ nm; Core diameter d = 1.8 um; Length of PCF L = 100 mm.

Laser: Center wavelength λ = 800 nm; Pulse Width $T_0 = 100$ fs; Repeated frequency $\omega_0 = 80$ MHz.

Spectrometer: Resolution $r_0 = 0.02$ nm; Measuring range from 200 nm to 1100 nm.

Focus lens: Magnification $M_0 = 40\times$.

We used the tunable laser to keep incident light wavelength at 710 nm to analyze how the incident power will affect the SC when incident light wavelength at none-zero dispersion region. Different normalized spectra of SC under different incident pulse are shown in Figure 3(a), (b) and (c). The average incident power delivered into the fiber was 10 mW, in the Figure 3(a), the broadening width of the generated SC ranged from 559 nm to 801 nm. According to the related references, at a low power level, GVD and SPM are the both main factors that caused the broadening of the generated SC, with the additional contribution by TOD and SRS, the shape of generated SC is asymmetrical, as is shown in Figure 3(a). As the input pump power is increased further, the bandwidth is getting wider and more peaks generated, it covered the range from 503 nm to 862 nm. We found that main components of SC are beginning blue-shifting to the short-wavelength region, nonlinear effect increased with a higher power level. At such a level, FWM effect and XPM effect begin to have a great affection on the generation of SC besides the affection of GVD, TOD, SRS [9–11].

As the input power reached up to 100 mW, main energy components of SC have been concentrated in the short-wavelength region, relative intensity of the short-wavelength components are higher than those in the long-wavelength region. Peaks get less,

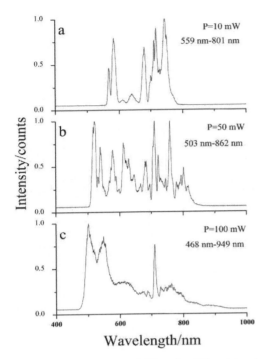

Figure 3. Spectrograms of SC under different input powers, Length of PCF L = 100 mm.

but broadening width has reached up from 468 nm to 949 nm. By the interactions of GVD, TOD, SRS, FWM, XPM and other kinds of nonlinear effect, there are more new spectral components generated, as it is shown in Figure 3(c). Along with the input power increased, we found that width of the generated SC has reached a state of saturation, which means that the spectrum would not extend any more. So, we can learn from this experiment that the width of the generated SC will increase with higher input power, but as if the input power reach a certain intensity, supercontinnum generation will be saturated and stabilized [10,11].

Figure 4 shows us the different colorful graphs with different input powers. We can see the output lights are still maintained in single-mode characteristic; they are composed by a variety of colors. With the increase of the incident power, more color appeared, from Figure 4(a) to Figure 4(c), they are becoming more and more bright and even the white light emerged.

In order to confirm our guess of the explanation to demonstrate the experiment of SC generation, we applied different length of high-nonlinear PCF, length of this PCF is 200 mm, that has the exactly same structure with the former one. We keep the input power at 10 mW, 50 mW, 100 mW. It is shown in Figure 5. With the input power increased,

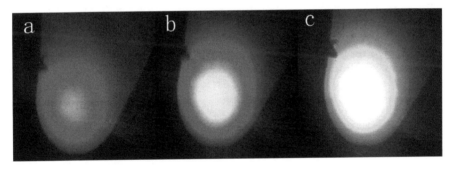

Figure 4. Different colors of superconnium from PCF under different input powers.

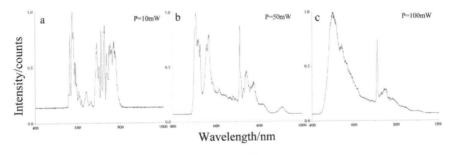

Figure 5. Spectra of SC under different input powers, L_{PCF} = 200 mm.

the broadening width of the generated SC is also increasing, the short-wavelength components are getting more and more, lots of energy is shifting to the short-wavelength region. Such a change in trend is similar to the Figure 3.

5 CONCLUSIONS

We have used the tunable femtosecond pulse laser to generate the SC in the high nonlinear PCF and demonstrated that many kinds of nonlinear effect are the main physical mechanisms caused the generation of SC. The results suggest that even incident light wavelength does not in the zero-dispersion region, we can generate the SC in the high nonlinear PCF with enough high input power; spectral width will be gradually widened with the increasing the input power, as long as it reaches a certain intensity threshold, the spectral width of SC finally saturated, the shape of SC was also stabilized. It provides a certain assistance to study the application of wavelength division multiplexing system, fiber laser system and dispersion compensation of PCF for modern optical communication system.

ACKNOWLEDGEMENTS

The authors are grateful to Y.F. Liu and X.R. Zeng for their support in the experiments. This work was supported by the National Natural Science Foundation of China (11274400, 11004208, 10574165, 11274397, 91022012 and 20973203), the Open Fund of the State Key Laboratory of High Field Laser Physics (Shanghai Institute of Optics and Fine Mechanics), the Program for Changjiang Scholars and Innovative Research Team in University (IRT13042) and Guangdong Province Project (2012B091000057, 2012B090600002).

REFERENCES

[1] J.C. Knight, T.A. Birks, P. St. J. Russell and D.M. Atkin. All silica single mode optical fiber with photonic crystal cladding [J]. Opt. Lett. Vol. 21, no. 19 (1996), pp. 1547–1549.

[2] T.A. Birks, J.C. knight and P. St. J. Russel. Endlessly single mode photonic crystal fiber [J]. Opt. Lett. Vol. 22, no. 13 (1997), pp. 961–963.

[3] Jonathan C. Knight. Photonic Crystal Fibres [J]. Nature. Vol. 424, no. 14 (2003), pp. 847–851.

[4] John M. Dudley and J. Roy Taylor. Ten years of nonlinear optics in photonic crystal fibre [J]. Nat. Photonics. Vol. 3 (2009), pp. 85–91.

[5] Li Xiao-Qing, Zhang Shu-min, Li Dan. Experimental and numerical study of supercontinuum generation in photonic crystal fiber [J]. Acta Photonica Sinca. Vol. 37, no. 9 (2008), pp. 1805–1809.

[6] J. Hu, C.R. Menyuk, L.B. Shaw. A mid-IR source with increased band width using tapered As_2S_3 chalcogenide photonic crystal fibers [J]. Opt. Commun. Vol. 293, no. 15 (2013), pp. 116–118.

[7] Li Shu-Guang, Liu Xiao-Dong, Hou Lan-Tian. The study of wave guide mode and dispersion property in photonic crystal fibers [J]. Chin. Phys. Soc. Vol. 52, no. 11 (2003), pp. 2811–2816.

[8] Wang Qing-yue, HU Ming-lie, Chai-Lu. Progress in Nonlinear Optics with Photonic Crystal Fibers [J]. Chinese J. Lasers. Vol. 33, no. 1 p(2006), pp. 57–66.

[9] Zhang Hui, Wang Yi, Chang Sheng-Jiang. Blue-shifted Spectra of Supercontinuum Generation in Photonic Crystal Fibers [J]. Acta Photonica Sinica. Vol. 39, no. 11(2010), pp. 1938–1942.

[10] Cheng Chun-Fu, Wang Xiao-Fang, Lu Bo. Nonlinear propagation and supercontinuum generation of a femtosecond pulse in photonics crystal fibers [J]. Acta Physica Sinica. Vol. 53, no. 6 (2004), pp. 1826–1830.

[11] Li Ai-Ping, Zheng Yi, Zhang Xing-Fang, Sun Qi-Bing, Li Kun. The supercontinuum generation in a photonic crystal fiber pumpedat the anomalous dispersion region [J]. Laser Technology. Vol. 32, no. 1 (2008), pp. 50–52.

Advances in Future Manufacturing Engineering – Yang (Ed.)

Determination of the interface electronic structure of L-cysteine/Ag(111) by Ultraviolet Photoelectron Spectroscopy and Photoelectron Yield Spectroscopy

KaveengaRasika Koswattage
Center for Frontier Science, Chiba University, Inage-ku, Chiba, Japan

Kinjo Hiroumi & Yasuo Nakayama
Graduate School of Advanced Integration Science, Chiba University, Inage-ku, Chiba, Japan

Hisao Ishii
Center for Frontier Science, Chiba University, Inage-ku, Chiba, Japan
Graduate School of Advanced Integration Science, Chiba University, Inage-ku, Chiba, Japan

ABSTRACT: The interface electronic structure of the sulfur-containing amino acid L-cysteine and Ag(111) was systematically elucidated by thickness dependent Ultraviolet Photoelectron Spectroscopy (UPS). In this study, L-cysteine films on silver substrate were formed by vacuum evaporation of L-cysteine. The orbital configurations at the interface in the case of L-cysteine on Ag(111) was estimated including work function, Secondary Electron Cutoff (SECO), Highest Occupied Molecular Orbital (HOMO) onset, position of an interface state, charge injection barrier, and ionization energy. In the case of SECO, the maximum shift was 0.46 eV to the higher binding energy side at the nominal thickness of 1 Å. However, from the nominal thickness of 2 Å, SECO started to shift to the lower binding energy side, and at 16 Å, the SECO shifted to a value of around 0.4 eV to the lower binding energy side to almost cancel the initial vacuum level shift. This behavior can be attributed to the weakening of the silver-sulfur bond with the increasing of L-cysteine coverage referring to the literature for L-cysteine on silver [N. B. Luque et al.: Langmuir, 28, 8084–8099 (2012)]. The Photoelectron Yield Spectroscopy (PYS) was also performed as an additional spectroscopic work, which exhibited that the work function of silver once decreased and then recovered at low coverage. This behavior can also be assigned to a weakening the interaction of L-cysteine with silver by increasing of L-cysteine coverage.

Keywords: interface electronic structure; L-cysteine; ultraviolet photoelectron spectroscopy

1 INTRODUCTION

Recently, the adsorptions of biomolecules on inorganic metallic surfaces for biocompatible surfaces are investigated for future development of bioelectronics. Amino acids are one of the simplest biomolecules and thus being received particular attention for bio-functionalization of metal surfaces. Among the amino acids, the sulfur containing L-cysteine (HS–CH2–CH(NH2)–COOH) has been specially received an attention as the most useful amino acid to attach proteins to metal surfaces for bioactive surfaces because cysteine is often located on the boarder of proteins as well as constitute the most common proteins. In the case of linking proteins and electrodes as the components of the bioelectronics devices, the L-cysteine is capable to bind on noble metal surfaces such

as gold, silver, and copper, which are commonly employed as metal electrodes in the field of bioelectronics for the construction of electronic devices such as biosensors [1–3]. In this case, the SH group of L-cysteine is known to interact strongly with above mentioned metallic surfaces and binds metallic surfaces mainly employing SH functional group and other functional groups NH_2, and COOH [4,5]. The interaction of the amino acid L-cysteine with silver surfaces is particularly interesting because L-cysteine adsorption on silver has been suggested to be stronger than on gold or copper surfaces [6] and the interactions may strongly influence the formation of novel interface states of the L-cysteine-Ag interface. For instance, Fischer et al. [6] has examined the interaction of L-cysteine with Ag(111) by combining Scanning Tunneling Microscopy

(STM), X-ray Photoelectron Spectroscopy (XPS) with Near Edge X-ray Absorption Fine Structure (NEXAFS) spectroscopy and reported that at monolayer coverage, zwitterionic L-cysteine molecules (protonated amino group and deprotonated acid group) form a dense-packed pattern on Ag(111) and being chemisorbed to the surface by the sulfur for a robust anchoring. Further, a study of Diaz Fleming et al. [7] has reported that all the three functional groups of the L-cysteine are involved in the adsorption of the zwitterionic form of L-cysteine on silver nanoparticles. Thus, such a strong interaction is found in between L-cysteine and silver surfaces by varying bonding formation and reflects a formation of novel interface electronic states by any means. These states may be located below and above the Highest Occupied Molecular Orbital (HOMO) of L-cysteine which has been suggested due to L-cysteine-metal bonding [8]. On the other hand, some DFT studies report a weakening of silver-sulfur bond with increasing coverage which may be a drawback of anchroing the biomolecules on silver surface through the L-cysteine-silver strong interaction [9].

However, research has not been sufficiently addressed for experimental investigation on interface electronic structure of L-cysteine on silver while there have been experimental and theoretical works for fundamental understanding of the interaction of L-cysteine with single-crystal silver metallic surfaces [6,9,10–12].

In the present study, we report a comprehensive clarification of L-cysteine and silver interface electronic structure by thickness dependent Ultraviolet Photoelectron Spectroscopy (UPS) using L-cysteine on Ag(111) as a model system. Further, we also report Photoelectron Yield Spectroscopy (PYS) of L-cysteine on evaporated silver surfaces in order to support the UPS analysis.

2 EXPERIMENTAL

The Ag(111) single crystal surface was cleaned by repeated cycles of sputtering utilized 1.5 keVAr$^+$ ion and subsequent annealing at around 800 K. The surface quality of the cleaned Ag(111) was verified by the low-energy electron diffraction measurements. The L-cysteine was purchased from Aldrich (Aldrich, 99% L-cysteine), which was inserted into a quartz cylindrical tube wound with a tungsten wire for heating and evaporated at a constant rate of 0.3 Å/s which was measured by a crystal monitor. Hence, the thickness values for L-cysteine coverage are nominal film thicknesses. When the desired rate of evaporation was reached, the sample was moved into position from a preparation chamber, turned to

face the evaporator for the desired time, turned upright again, and moved into the measurement chamber. Subsequently, a set of UPS measurements were performed at each step of the thickness accumulation.

The UPS measurements were performed at the beam line BL8B of the Ultraviolet Synchrotron Orbital Radiation (UVSOR) facility of the Institute for Molecular Science, Japan [13]. The UPS spectra were measured using an analyzer with a multichannel detector system (VGARUPS10). The angles of incident photons and the detected photoelectrons were 45° and 0°, respectively, relative to the substrate surface normal. The Secondary Electron Cutoffs (SECO) were measured with the sample biased at −5.00 V. All the thickness dependent UPS were performed with a photon energy of 28 eV. All the preparation steps and the measurements were performed at room temperature. To control for beam damage, adjustment for different photon energies of the intense synchrotron beam and sequential short periods of acquisition on different zones of the sample were performed. Further, for thicker films, charging accumulation was carefully monitored by comparing consecutive UPS measurements at the same experimental conditions. The PYS measurements were conducted by a home-built system [14]. UV white light from a 30 W D$_2$ lamp (Hamamatsu C1518; 30 W) is monochromatized by a grating system (Bunkoh-Keiki; M20 KV-200) filled with nitrogen gas and then introduced into the measurement vacuum chamber through a LiF window. The incident light is focused with a concave mirror to be ca. 5 mm × 2 mm at the measurement sample position. On the PYS measurements, the sample is placed at the center of a hemispherical metal shell and excitation light is introduced onto the sample from the surface normal direction through an aperture at the zenith of the hemisphere. The amount of the emitted photoelectrons is measured as a drain current (photocurrent) flowing into the sample by using a precise pico-ammeter (Keithley; 6430). A positive bias is applied to the metal shell by a battery, and the photoelectrons emitted to a wide solid angle from the sample are efficiently collected by a concentric electric field. In principle, photoelectron yield is obtained by dividing the measured photocurrent by incident photon flux. Hence, a tiny fraction of the incident light is guided into a photomultiplier to monitor the photon flux simultaneously with the measurements. In order to cut undesirable photoemission by unmonochromatized "stray" light as well as second-order light especially in low photon energy range (below 6.5 eV), quartz optical filters of various cut-off wave length are adequately inserted between the light source and grating.

3 RESULTS AND DISCUSSIONS

Figure 1 shows the results of the thickness-dependent UPS measurements. Figure 1a shows UPS spectra of L-cysteine on Ag(111) with various thicknesses and the bottom spectrum shows the UPS spectrum of the clean Ag(111). The nominal thickness value of the L-cysteine coverage is indicated by the units of Å at the right outside corner of the figure. It is clear from the figure that thin films spectra are different for thick film spectra and this may be due to the interaction of Ag(111) with L-cysteine. In the case of thin films, a clear feature is appeared at centered around 2.5 eV (dotted line). Further a partially reduced in intensity and modified in shape is observed at Ag d band region. In the case of Ag(111), Ag 4d5/2 and Ag 4d3/2 peaks are located at around 4.8 and 6.1 eV and the work function of clean Ag(111) is estimated to be 4.5 eV considering the high binding energy cutoff at 23.5 eV. In the figure, the appearance of the Fermi level of Ag(111) is gradually decreased as the depositing L-cysteine and it is completely disappeared at

64 Å thickness of L-cysteine. Thus, the 64 Å thickness of L-cysteine spectrum is assumed as only the electronic structure of L-cysteine and the feature centered at around 4.8 eV binding energy for 64 Å thickness is assigned to the HOMO of L-cysteine where the HOMO cut off is estimated to be 3.2 eV by fitting a straight line into the HOMO edge and obtain its intersect with the binding energy axis (dashed lines of HOMO peak of 64 Å thickness of L-cysteine spectrum).

Figure 1b shows the region between Fermi level of the Ag(111) substrate and the HOMO of the L-cysteine measured with a relatively small energy step. In the case of thin films, the clear spectral feature centered at around 2.5 eV(in between the Fermi level and the HOMO of L-cysteine) might be due to an interaction of a sulfur-originated state of L-cysteine HOMO interaction with Au d orbitals. According to the Renzi et al. [9] with the interpretation of Anderson–Newns model, bonding and antibonding states are possibly formed such a coupling of S 3p orbital with Ag 4d orbital. Thus, the feature centered at around 2.5 eV and the decrement of d band region can be attributed to bonding and antibonding states as coupling of S 3p orbital with Ag 4d orbital. Some works have addressed L-cysteine on Au surface and interpreted as bonding and antibonding states between sulfur and metal are located below and above the d band of metal, respectively. Thus in the UPS spectra of the L-cysteine on Ag(111), the clear feature centered at around 2.5 eV and the spectral change at d band can be assigned to S–Au antibonding and bonding states, respectively. This antibonding state which is located in between HOMO cutoff and Fermi level of Ag(111) is assigned as the interface state (or gap state) and such an interface state possibly enhances the charge transfer between L-cysteine and silver substrate. However, Luque et al. [9] has mentioned that the interpretation of S-Au antibonding and bonding states are much more complicated than simple view because the possibility of the electronic states participating in the sulfur-carbon bond interact additionally with the d states of silver. However, according to them the only situation where the bonding and antibonding states are possible if the adsorbate cysteine is "pushed" into the surface at a shorter distance than the equilibrium position.

In Figure 1a, peak shifts can be observed at around 5 eV and 9 eV (dotted lines) and the shifts are likely caused by either of the following reasons. The features at around 5 eV and 9 eV are assigned as O, C, and N atom-originated states referring the DFT study and some experiment on the valance electronic structure of L-cysteine [15]. Thus, the shifting of peaks centered at ~5 and ~9 eV may be due to an interaction of O 2 sp and N 2 sp

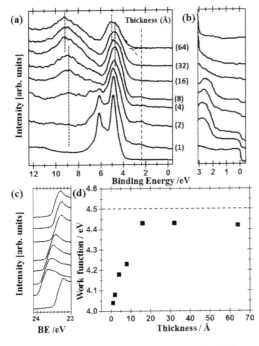

Figure 1 a. UPS spectra of L-cysteine on Ag(111) at a photon energy of 28 eV before and after each evaporation step of L-cysteine. Figure 1b. The region between the Fermi level of the Ag(111) and the HOMO of the L-cysteine measured with relatively small energy step. Figure 1c. The SECO on a magnified scale Figure 1d. The estimated work function of clean Ag(111) and L-cysteine/Ag(111) at each thickness.

orbitals with metal surfaces. This interaction may allow pushing adsorbate cysteine in to the surface at a shorter distance and consequently, satisfying the suggestion by Luque et al. [9] as the possible situation to form bonding and antibonding states due to sulfur-silver interaction. Further, it has been suggested that an appearance of the zwitterionic molecule form of L-cysteine on Ag(111) at high coverage. Thus, it may suppose as the first layer being chemisorb to the Ag(111) in neutral form and at high coverage zwitterionic form of L-cysteine are formed.

Figure 1c shows the SECO on a magnified scale and Figure 1d shows the estimated work function of clean Ag(111) and L-cysteine/Ag(111) at each thickness. The dotted line of the Figure 1d indicates the estimated value of the work function of clean Ag(111) where the shift of the SECO corresponds to the change of the work function before and after the subsequent evaporation of L-cysteine. In the case of SECO, the maximum SECO shift was 0.46 eV to a higher binding energy side and the value is simply assigned to the interface dipole between L-cysteine and the Ag(111) surface [16]. However, SECO starts to decrease at low coverage from 2 Å, and at 16 Å, the SECO shifted to a value of around 0.4 eV toward lower binding energies where at 16 Å SECO shifted back to the lower binding energy side by 0.4 eV to eventually cancel the initial vacuum level shift. This might be attributed to reduction of the amount of charge transfer cause decrease the interface dipole moment by weakening the interaction of L-cysteine and Ag(111) with the forming of over layers. Luque et al. [9] has suggested similar behavior as weakening of the silver-sulfur bond with increasing of L-cysteine coverage on silver. We performed PYS for further clarification the SECO behavior of UPS as the SECO shifts toward the low BE side at low coverage with increasing thickness as here it is attributed to weakening of the silver-sulfur bond with increasing of L-cysteine coverage on silver. The ionization potential of the chemisorbed L-cysteine estimated to be 7.24 eV by adding the work function of L-cysteine/Ag(111) at monolayer coverage as 4.04 eV and HOMO cutoff 3.2 eV.

In Figure 1, there are clear spectral changes can be observed in between nominal thickness layers of 8 Å and 16 Å, such as SECO shifting to low BE side by 0.3 eV, the peak centered at ~9 eV shifting to high BE side, and a disappearance of the peak centered at around 2.5 eV (due to interface state). The reason for the changes may be due to a formation of the second layer or more on the chemosorbed first layer as the formation of over layers can be weakening the interaction between the first layer and Ag(111) and the over layers possible to be formed in the zwitterionic form. Thus, it may

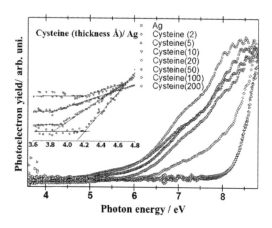

Figure 2. PYS results of L-cysteine on silver. In the figure, the inset is shown a magnification of the silver work function range in order to understand a change of the work function before and after the deposition sequence of L-cysteine.

be possible to assign the nominal thickness of 8 Å layer as around monolayer of L-cysteine. On the other hand, it can be suggested the monolayer coverage as the nominal thickness 1 Å layer because SECO reaches to the maximum value at 1 Å and the value decreases after subsequent steps.

Figure 2 shows the PYS results of L-cysteine on silver. The silver work function region is shown in the inset with a magnified scale in order to understand a change of the work function before and after the deposition sequence of L-cysteine. It is clear in the figure that the work function of Ag is decreased until 5 Å thickness of L-cysteine layer and at thickness 10 Å, it is increased again. This might due to a weakening of silver-sulfur bond with increasing the coverage, which is consistent with SECO behavior of UPS results.

4 SUMMARY

The electronic structure of the L-cysteine/Ag interface was examined using thickness dependent UPS which were involved with the estimation of the charge injection barrier from the L-cysteine HOMO to the Fermi level of Ag(111), the existence of an interface state, interface dipole and ionization potential. It was clearly observed a clear peak in between HOMO of L-cysteine and Fermi level of Ag(111) and assigned as the antibonding state due to coupling of S 3p orbital of L-cysteine with Ag 4d orbital of Ag(111). The charge injection barrier between HOMO of L-cysteine and Ag(111) was estimated to be 3.2 eV. The ionization energy of L-cysteine chemisorbed layer on

Ag(111) was estimated to be 7.24 eV. The results of SECO of UPS together with PYS was confirmed earlier suggestion by other groups as the weakening of the silver-sulfur bond with increasing of L-cysteine coverage on silver and thus anchroing the biomolecules on silver surface through the L-cysteine-silver may not be effective in the application of bioelectronics.

ACKNOWLEDGEMENTS

The UPS experiments were performed under the approval of the Joint Studies Program (25–562) of the Institute for Molecular Science, Japan. This study was also partially supported by the G-COE Program of Chiba University, Japan (Advanced School for Organic Electronics).

REFERENCES

[1] J.J. Gooding, D.B. Hibbert and W. Yang: Electrochemical Metal Ion Sensors. Exploiting Amino Acids and Peptides as Recognition Elements. Sensors, 2001, 1(3), 75–90.

[2] J. S. Lee, P. A. Ulmann, M. S. Han and C. A. Mirkin: A DNA–Gold Nanoparticle-Based Colorimetric Competition Assay for the Detection of Cysteine. Nano Lett., 8, 529–533 (2008).

[3] S. B. Lee and C. R. Martin: Anal. Chem. 73, 768–775 (2001).

[4] A.G. Brolo, P. Germain and G. Hager: Investigation of the Adsorption of l-Cysteine on a Polycrystalline Silver Electrode by Surface-Enhanced Raman Scattering (SERS) and Surface-Enhanced Second Harmonic Generation (SESHG). J. Phys. Chem. B, 106, 5982–5987(2002).

[5] M. Graff and J. Bukowska: Adsorption of Enantiomeric and Racemic Cysteine on a Silver Electrode—SERS Sensitivity to Chirality of Adsorbed Molecules. J. Phys. Chem. B, 109, 9567–9574(2005).

[6] S. Fischer et al.: l-Cysteine on Ag(111): A Combined STM and X-ray Spectroscopy Study of Anchorage and Deprotonation. J. Phys. Chem C, 116, 20356–20362 (2012).

[7] G. Diaz Fleming, J. J. Finnerty, M. Campos-Vallete, F. Célis, A. E. Aliaga, C. Fredesand R. Koch: Experimental and theoretical Raman and surface-enhanced Raman scattering study of cysteine. J. Raman Spectrosc. 40, 632–638 (2009).

[8] V. D. Renzi: Understanding the electronic properties of molecule/metal junctions: The case study of thiols on gold. Surface science, 603, 1518–1525 (2009).

[9] N. B. Luque, P. Velez, K. Potting, and E. Santos: Ab Initio Studies of the Electronic Structure of l-Cysteine Adsorbed on Ag(111). Langmuir, 28, 8084–8099 (2012).

[10] E. Santos, L.B. Avalle, R. Scurtu and H. Jones: Chem. Phys., 342, 236–244 (2007).

[11] E. Santos, L. Avalle, K. Pötting, P. Vélez and H. Jones: Experimental and theoretical studies of l-cysteine adsorbed at Ag(111) electrodes. Electrochim. Acta, 53, 6807–6817 (2008).

[12] N.B. Luque and E. Santos: Ab Initio Studies of Ag–S Bond Formation during the Adsorption of l-Cysteine on Ag(111). Langmuir, 28, 11472–11480 (2012).

[13] K. Seki et al.: A plane-grating monochromator for 2 eV ≤ hν ≤ 150 eV. Nucl. Instrum. Meth. A, 246, 264–266 (1986).

[14] Y. Nakayama, Y. Uragami, M. Yamamoto, S. Machida, H. Kinjo, K. Mase, K. R. Koswattage and H. Ishii: Determination of the highest occupied molecular orbital energy of pentacene single crystals by ultraviolet photoelectron and photoelectron yield spectroscopies. Jpn. J. Appl. Phys., vol.53, 01 AD03–01 AD03–4 (2014).

[15] M. Kamadaet al.: J. Phys. Soc. Jpn., 79, 034709-1-034709-4 (2010).

[16] Hirak Chakraborty and Munna Sarkar. Optical Spectroscopic and TEM Studies of Catanionic Micelles of CTAB/SDS and Their Interaction with a NSAID. Langmuir, 21, 3551–3558 (2004).

Advances in Future Manufacturing Engineering – Yang (Ed.)
© *2015 Taylor & Francis Group, London, ISBN 978-1-138-02817-3*

Maximum Power Point Tracking method based on gray BP neural network photovoltaic power generation system

Jing Hao & Lingquan Zeng
Northeast Dianli University, Jilin, China

ABSTRACT: In order to increase the utilization of photovoltaic energy, we should make the photovoltaic moment column all the time working on the maximum power point. Therefore, working for the maximum power point tracking for photovoltaic power generation system technology is very meaningful. In this paper, we mainly use the BP neural network and gray BP neural network to track maximum power point. During the calculation process, we simulate the photovoltaic cells characteristic of the given parameters, through the simulation of the BP neural network, we get the better parameters of the BP neural network and the tracking model of the PV photovoltaic cells and the errors of the calculation. Through Combing with gray prediction method of error correction, we draw a smaller error prediction model. The algorithm was calculated by Matlab. Result shows that the gray BP neural network method, which was based on the BP neural network method for maximum power point tracking is quicker and more accurate.

Keywords: Photovoltaic (PV) cell; illumination; Maximum Power Point Tracking (MPPT); BP neural network; grey BP neural network

1 INTRODUCTION

PV is a procedure, which directly converts solar energy into DC energy [1]. As a result, this type of energy is considered as the clean energy of power generation and at the same time, PV is the most widely used application prospects and the strongest kind of renewable energy. For these reasons, in a dozen years ago on the photovoltaic technology research, scholars generated great interests. Whether for independent power plants or power generation systems, because of its gradual increase in cell conversion efficiency and high reliability, photovoltaic power generation technologies is widely recognized by the industry [2]. Meanwhile, since the photovoltaic gradually expand the scope of application, the demand for energy depends on fossil fuels as well as pressure have got a certain amount of relief. Study shows that the PV system output voltage and current meter exhibit a non-linear relationship. From this linear relationship, we can see that at a given temperature and light conditions, PV system can be output at a maximum power point, which can be called MPPT. When the rate of change of the voltage for power is zero, that is $dP/dV = 0$, we can get the MPPT. In addition, since the output power of PV system depends on the sun light and ambient temperature, therefore, MPP will change with external environment.

To achieve the high efficiency of photovoltaic power generation system, photovoltaic power load must be considered under different weather conditions in a reasonable match. However, since the output voltage of the PV cells has nonlinear characteristics, the maximum power point tracking becomes difficult, especially when the sun light and the ambient temperature change, MPP becomes more difficult to achieve. So far, people have proposed a number of techniques, which can be implemented in MPP tracking. These technologies were different in the complexity of many aspects such as the sensor requirements, the convergence speed, the cost, the scope, and the hardware facilities requirements. In the method previously proposed [3]. The more commonly used are constant voltage method, perturbation and observation method (referred to P & O France), and admittance incremental method of which the most simple method is constant voltage method. However, when the temperature changes, the constant voltage method often cannot track the MPP, thus in theoretically the constant voltage method was often considered in theoretical and not commonly used in practice, Since the P & O is easy to implement, in practice it was more commonly used [2]. For the P & O technique, one drawback is that in the steady state, the calculated operating point will fluctuate around the maximum power point, and the energy would have been wasted. The admittance incremental method has the same parameters measurement method with P&O method, but this method is derived point of view, it ignores the impact of temperature changes. In general, in response to changing environmental conditions, in order to accurately track the MPPT, these traditional methods still have some difficulties.

This paper presents a gray neural network-based MPPT control algorithms. After calculation, we can see that compare with the previous methods, this algorithm is more accurate for MPP tracking and it is simple to implement. From the view of the nature of the algorithm itself, to some extent the algorithm alleviated the MPP and it also increase the utilization of solar energy.

2 ANALYSIS OF PV ARRAY MODEL

First of all, let's talk about the mathematical model of photovoltaic cells. From Figure 1, we can see that photovoltaic cells are current sources, which can be seen as non-linear element, it produces a current which has a positive correlation with the light. Two of resistance: R_s and R_{sh} represent the equivalent resistance of photovoltaic cells power loss. The resistance of the series branch R_s represents the loss caused by the surface of the photovoltaic cell, and the resistance of parallel branch R_{sh} represents the leakage current inside the solar cell.

PV array is connected in series or in parallel by a series of photovoltaic cells. As the sun illumination intensity and battery temperature changes, the P-V and I-V will exhibit a nonlinear characteristics. Equation (1) gives a description of the working principle of the PV cell model:

$$V_{cell} = \left(\frac{nkT_c}{q}\right)\ln\left(\frac{I_{ph}+I_0-I_{cell}}{I_0}\right) - R_s \times I_{cell} \quad (1)$$

In the formula, q is the charge amount; n is the ideality factor of photovoltaic cell which

dependents on the photovoltaic technology [3]; k is the Boltzmann constant; the I_{cell} is the output current of the battery; I_0 is the diode reverse saturation current; R_s is series resistance of the photovoltaic cell; T_c is the operating temperature of the photovoltaic cell; V_{cell} is the output voltage of the photovoltaic cell. The photocurrent I_{ph} is mainly depends on the sun light intensity and battery temperature, which is shown in formula (2):

$$I_{ph} = [I_{SC} + K_I(T_C - T_0)]G \quad (2)$$

I_{SC} is the short-circuit current when the temperature is at 25 °C, the sun shines is in the 1 kW/m, K_I is the temperature coefficient of the short circuit current; G is the light intensity of the sun, and its unit is: 1 kW/m. T_0 is the reference temperature of photovoltaic cell. From (2) it can be seen that the photovoltaic cell saturation current will change with the battery operating temperature.

$$I_S = I_{SO}(T_c/T_0)^2 \exp\left[qEg\left(\frac{1}{T_0} - \frac{1}{T_c}\right)\middle/kn\right] \quad (3)$$

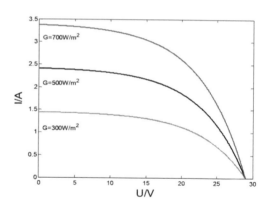

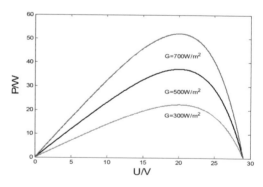

Figure 2. Output current (I)-voltage (V) and power (P)-voltage (V) characteristics for different insolation levels.

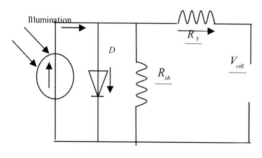

Figure 1. Simplified equivalent circuit of a solar cell.

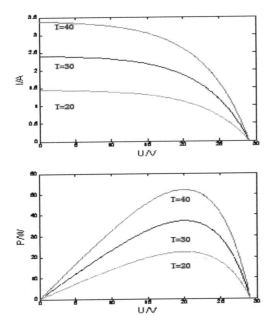

Figure 3. Output current (I)-voltage (V) and power (P)-voltage (V) characteristics for different temperatures.

In the formula (3), I_{SO} is the reverse saturation current of the battery when the battery working at the reference temperature and the light intensity is 1 kW/m. *Eg* is the total energy of the semiconductor inside the battery. We can get I_{SO} from formula (4):

$$I_{SO} = I_{SC} / \left[\exp\left(\frac{qV_{OC}}{nkN_s \times T_0} \right) \right] \qquad (4)$$

V_{OC} is the open-circuit voltage of photovoltaic cells under standard operating conditions.

N_S is the associated (parallel or series) number of cells. Since the output power of PV components as the battery temperature as well as the sun's light intensity changes, during the simulation of PV model, many factors should be considered. Here we simulate the model through the Simulink platform of Matlab. Figure 2, Figure 3 shows the I-V and P-V curve at different light intensity and temperatures.

3 THE MPPT CONTROLLER BASED ON BP NEURAL NETWORK

Maximum power point tracking is the real-time detection of the PV array output power, and use certain control algorithm to predict the maximum power of the PV array of current conditions.

By adjusting the duty cycle of power control devices to change the current equivalent impedance of the circuit, we make the PV cells work around the MPPT. This process is known as Maximum Power Point Tracking (MPPT). Since the Volt-ampere characteristics of photovoltaic cells has nonlinear characteristics, and when the outside temperature, illumination changes, Photovoltaic Maximum Power Point will also be changed, the solar cell maximum power point control is not easy to achieve.

BP neural network [4] is in an Artificial Neural Network (ANN) method commonly used in study. This method is a supervisory approach to learning, the basic principle is: the input signal through the hidden layer node to the output layer nodes, after a non-linear transformation it produce an output signal, compute error between the network output signal and the actual output signal, adjust the associations between hidden layer and the input layer, the hidden layer and output layer, make the error down along the gradient direction. After the repeated training until we determined network parameters which correspond with minimum error we get the end of training.

Back propagation [5] is an iterative process, and the calculation takes some time. If we use multi-threading technology multi-core computers to execute the algorithm, we can save a lot of computing time. The algorithm is easier to use multithreading to achieve on the computer. During the calculation, the data is assigned to each thread, and each thread could separately handle the forward and backward propagation. After each completion of iteration, all threads have to be terminated. Figure 4 shows the system to track the MPP control block diagram and Figure 5 is a configuration diagram of the algorithm.

The detailed description of BP neural network method steps are as follows:

Assuming that the network input vector is $P_k = \{a_1, a_2, ..., a_n\}$, the target vector of the network is $T_k = \{y_1, y_2, ..., y_q\}$, The intermediate layer input vector of each unit is $S_k = \{S_1, S_2, ..., S_p\}$, the output vector is $B_k = \{b_1, b_2, ..., b_p\}$. The output

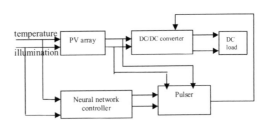

Figure 4. Block diagram of the MPPT.

27

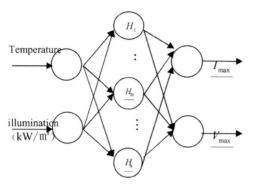

Temperature

illumination (kW/m²)

I_{max}

V_{max}

H_1

H_{i0}

H_t

Figure 5.   Back-propagation neural networks for MPPT.

layer input vector of each unit is $L_k = \{l_1, l_2, ..., l_q\}$, and the output vector is $C_k = \{c_1, c_2, ..., c_q\}$. The connection weights of input layer to the middle layer is W_{ir}, $i = 1, 2, ..., n$, $r = 1, 2, ..., p$. The connection weight of middle layer to the output layer is V_{rg}, in which $r = 1, 2, ..., p$, $g = 1, 2, ..., q$. The threshold of Intermediate layer unit is θ_r in which $r = 1, 2, ...$, path threshold of output layer unit is γ_g in which $g = 1, 2, ..., q$.

1. The W_{ir}, V_{rg}, θ_r, γ_g, were respectively endowed with the random value from -1 to 1;
2. Select the input sample and the target sample $P = (a_1^k, a_2^k, ... a_n^k)$; $T_k = (s_1^k, s_2^k, ... s_p^k)$;
3. Calculate the input value of intermediate layer: s_r: $s_r = \sum_{i=1}^{n} W_{ir} a_i - \theta_r$, $r = 1, 2, ..., p$;

   Using the transfer function to calculate the output of the intermediate layer: $b_r = f(s_r)$, $r = 1, 2, ..., p$;
4. Calculation the input value of the output layer L_g: $L_g = \sum_{r=1}^{p} V_{rg} b_r - \gamma_g$, $g = 1, 2, ..., q$,

   Using the transfer function to calculate the output value of the layer output: $C_g = f(L_g)$, $g = 1, 2, ..., q$;
5. According to target value and the calculated value actual to calculate errors of each output layer neurons generalization: $d_g^k = (\gamma_g^k - C_g) \cdot C_g \cdot (1 - C_g)$, $g = 1, 2, ..., q$;
6. According to the V_{rg}, d_g^k to calculate the errors of each middle layer of the module generalization:

$$e_r^k = \left[\sum_{g=1}^{q} d_g \cdot V_{rg} \right] b_r (1 - b_r)$$

7. According to the d_g^k and b_r to corrected for V_{rg} and θ_r: $V_{rg}(N+1) = V_{rg}(N) + \alpha \cdot d_g^k \cdot b_r$

$\gamma_g(N+1) = \gamma_g(N) + \alpha \cdot d_g^k$ $r = 1, 2, ..., p$,
$g = 1, 2, ..., q$ $0 < \alpha < 1$;

8. According to the intermediate layer generalization error e_r^k and the input value of the input layer to corrected for the connection weights:

$$W_{ir}(N+1) = W_{ir}(N) + \beta \cdot e_r^k \cdot a_i^k$$

$\theta_r(N+1) = \theta_r(N) + \beta \cdot e_r^k$ $i = 1, 2, ... n$,

$r = 1, 2, ..., p$; $0 < \beta < 1$;

9. Select the next set of learning sample trees, return to step (3) until all training of m sample group is completed. If the error E of global network is less than the pre-set minimum value, the network can be weakened. If the learning time is greater than the number of pre-set value and the global error is greater than the minimum value, the network cannot be weakened.
10. The end of the study.

Since the BP algorithm is based on the steepest descent gradient method algorithm, it has shortcomings such as easily to fall into local minima and slow convergence so that the time of BP algorithm will too long and the learning process is easy to fall into local extreme value.

This adds some difficulties for the network training. After researching the neural network and gray system theory, we found a certain similarity of the performance of information for the two theories. For the system, the output value can be approximate to a fixed value. However, due to the presence of errors, it makes the outputs fluctuate around a fixed value. In accordance with the definition of gray number in the gray system, the neural networks can be considered as a gray number. It can be seen that the neural network theory actually consists of the gray correlation theory. In this paper, the process of the forecasting respectively uses the traditional BP neural network and the gray theory, which improved BP neural network method.

Gray system model GM (m, n) is the based on the gray module concept and fitted with differential core modeling approach. The model parameter m is the order of the differential equation and n is the number of sequences involved in modeling. Based on the BP neural network above and combine with the gray prediction theory, the method is:

Assume the time series is $x^{(0)} = \{x^{(0)}(1), x^{(0)}(2), ..., x^{(0)}(n)\}$. Through the use of GM (1,1) model, we get $\{x^{\wedge(0)}(i)\}$, $i = 1, 2, n$. Define the difference between the analog values and the real value of time T is the error of time T, that is $e^{(0)}(T) = x^{(0)}(T) - x^{\wedge(0)}(T)$. Build BP neural network of which the error sequence is $\{e^{(0)}(i)\}$, $i = 1, 2, 3 ..., n$. If the prediction order is m, use the $\{e^{(0)}(i-j)\}$ to predict the error value at time i, and at the same time, use the $\{e^{(0)}(i-j)\}$, $j = 1, 2, ..., m$ as input samples of BP neural network, the $e^{(0)}(i)$ as the expectations. Using the trained BP neural network to predict the error sequence, $e^{\wedge 0(0)}$ (i), i = 1, 2, 3 ..., n. Basing on the process above, calculate the new error value $e^{\wedge (0)}(i,1)$, i = 1, 2, 3 ..., n in which $x^{\wedge (0)}(i,1) = x^{\wedge (0)}(i) + e^{\wedge (0)}(i)$.

4 RESULTS

As noted earlier, the proposed gray BP neural network prediction method was calculated as follows [6–9]. Table 1 shows that the experiment parameter used. The results show that through the algorithm we can calculate output voltage and current of the PV cells and obtain the MPP at a given temperature and illumination denoted here V_{max}, I_{max}. When the temperature and illumination change continuously, the obtained output current I_{max} and voltage V_{max} will change together, from the pulse generator shown in Figure 4, V_{max} and I_{max} can be extracted from the PV array system. Through the V_{max} and I_{max}, we can determine the switching frequency of DC/DC converter thus ensure the system working on the MPP. As can be seen through the simulation, the algorithm based on grey BP neural network MPPT shows good accuracy and faster computing speed in the temperature and illumination rapidly changing circumstances. The results were also compared with the traditional BP neural network method,

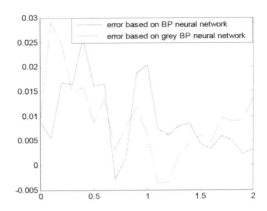

Figure 7. The percentage of the error curve of the two methods.

Figure 6 shows the working curve calculate through BP neural network MPPT controllers in the case of temperature and illumination change. Figure 7 shows the error curve of the prediction power. As can be seen from the figure, in addition to individual points, the gray BP neural network prediction error percentage is less than the traditional BP neural network prediction error percentage.

Table 1. Table of the parameters of PV cell.

Name of parameter	Value	Unit
The open circuit voltage	42.48	V
The voltage of MPP	35.28	V
The short circuit current	3.21	A
The current of MPP	2.84	A
The power of MPP	100	W

5 CONCLUSIONS

1. In this article, an approach based on BP neural network is proposed for photovoltaic maximum power point tracking. This method use the gray prediction method's advantages "accumulated generating" to weak the original data sequence and improve the regularity randomness.
2. This algorithm avoids the theoretical error of the gray prediction method, and the process is simple and easy to carry out.
3. Since combine with gray theory, the algorithm we put forward improves the accuracy of MPPT. The tracking speed is corresponding sped up. Through the algorithm we could track the maximum power point system quickly and accurately when the temperature and illumination change rapidly.

REFERENCES

[1] Zhao Zengming, Liu Jianzheng, Sun Xiaoying. Solar photovoltaic power generation and application [M]. Beijing: Science Press, 2005:20–45.
[2] Roger Messenger, Jerry Ventre. Photovoltaic System Engineering [J]. Florida, USA, CRC Press, 2005: 963–973.

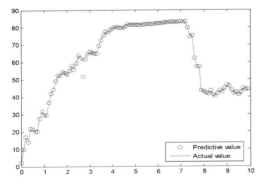

Figure 6. The working curve of the MPPT controller based on gray BP netural network.

[3] Trishan Esram, Patrick L. Chapman. Comparison of Photovoltaic Array Maximum Power Point Tracking Techniques [J]. IEEE Transactions on Energy Conversion, 2007, 439–449.

[4] Wang Song. A maximum power control system of photovoltaic cells based on neural network [J]. Journal of Shandong University (engineering science), 2004(04), pp. 45–48.

[5] Guo Liang, Chen Weirong, Jia Junbo, Han Ming. Photovoltaic cell modeling based on particle swarm optimization algorithm BP neural network. New technology of Electrical Engineering and Energy. 2011, 30(2):84–87.

[6] P.S. Revankar, W. Z. Gandhare, A.G. Thosar. Maximum Power Point Tracking for PV Systems Using MATLAB/Simulink[J], 2nd International Conference on Machine Learning and Computing, IEEE, pp. 2010: 8–11.

[7] Wang N. Chun S. Zuo K. Yukita Y. Goto K. Ichiyanagi. Research on PV model and MPPT methods in Matlab[J]. Asia Pacific Power and Energy Engineering Conferences, IEEE, 2010, pp. 1–4.

[8] Xin Wen. MATLAB neural network application design [M]. Beijing: Science press, 2001:35–72.

[9] Deng Julong. Basic methods of gray system[M]. Wuhan: Press of Huazhong University of Science and Technology, 1987:11–25.

Advances in Future Manufacturing Engineering – Yang (Ed.)
© 2015 Taylor & Francis Group, London, ISBN 978-1-138-02817-3

Research and development of high-performance informatization management and control system of power enterprises

Jia Wu, Dan Su & Wang Li
Jibei Electric Power Corporation, Beijing, China

ABSTRACT: Informatization management is an important part of modern business management. How to manage and control information projects on the whole, standardize informatization project set-up assessment system and process has become the problem remaining to solve in informatization management. This paper combines power enterprises' practice with informatization management and control theory, international leading practices and management standards and build a high-performance informatization management and control system from several dimensions such as management and control organizations, management and control process and tools, performance, etc. The purpose is to achieve the centralized management and control of key informatization projects in accord with the demand for business integration, rationality of application function, rationality of data architecture, reasonableness of security architecture, standardization and economic effectiveness.

Keywords: informatization; management and control system; power enterprises

1 INTRODUCTION

In recent years, the informatization has been developing fast. It has already played a significant role in promoting the company's overall efficiency and effectiveness. However, with the gradual deepening of informatization, the task of how to use information technology more efficiently and further enhance the information technology to promote the company's development has become an important issue. Therefore, there is an urgent need for power companies to carry out a research and development project dedicated to high-performance informatization management and control system through improvement of organizational structure and operation pattern, design and implement of informatization management and control process and tools and promotion of entire personnel informatization performance to achieve the desired strategic objectives. In addition, the international trend of enterprise information management calls for power companies to strive to go beyond the pursuit of excellence, and actively introduce advanced management concepts and tools of information technology to promote the second stage of company's information technology development, gradually catching up with international standards. Establishing a high-performance informatization management and control system is also a necessary support to building a environmental-friendly energy company, a harmonious enterprise,

a modern company and improving corporate value. This system is the indispensable precondition to the construction of modern enterprise management system, which helps to promote the company's information technology to achieve a comprehensive, coordinated and sustainable development and the realization of strategic objectives. High-performance informatization management and control system must start from aspects such as "organizational leadership, working mechanisms, operational guaranty, process control and operation supervision". Therefore, this is essential to come up with a comprehensive informatization management and control system to standardize whole company's information business processes, information management system and standards. The system combines the information management and control with business operation, manages the process of informatization stategy, plan, construction, operation, maintenance and security, rationalizes the management system, completes organizational restructuring and strengthens the functional management of information while docking the corporate-level management process, and eventually realizes the implementation of informatization management and control system. By comprehending the company's existing informatization management and control measures, introducing international and universal advanced enterprise high-performance informatization technology maturity model, updating and improving information technology procedures, systems and standards, defining company informatization competency model, we introduce the overall process of management and control system, and refine key process to reach the Four Unity of informatization management and control, that is, unified planning, unified organization, unified implementation and unified management in order to achieve "centralized lean, process transparency, process control, and information sharing" within information management.

2 THEORETICAL BACKGROUNDS AND BEST PRACTICES

This paper refers extensively to the theory of enterprise informatization management and control, and examines a number of leading international management standards and best practices including: high-performance informatization model, ITIL V3 standards, informatization capability model, informatization maturity model.

2.1 High-performance informatization model

High-performance informatization model is high-order guide for informatization job. It describes the whole lifecycle of informatization work based on the mainline of planning, construction, management and IT integrated management. Topics include: combination of information technology and business, infrastructure management and planning, project approval, service construction, service introduction, service management, IT integrated management, customer relationship management and supplier management.

2.2 ITILV3 standard

ITIL is the abbreviation for Information Technology Infrastructure Library. IT service management is the core issue of ITIL framework, which is a collaborative process used to ensure the quality of IT services through the Service Level Agreement (SLA). ITIL V2 divides the IT management activities into one management function and ten core processes. The function refers to the service desk. Ten core processes are divided into two groups, namely, Service Support and Service Delivery. The service support includes: configuration management, change management, release management, incident management and problem management. The service delivery is consisted of service-level management, financial management, sustainability management, capacity management and availability management. ITIL v3 defines five stages of the service lifecycle: Service Strategies, Service Design, Service Transition, Service Operation and Continual Service Improvement, which contains the flows of service management within the life-cycle.

2.3 Informatization capacity model

Based on the high-performance informatization model, the corporate informatization capacity contains six dimensions, coming from clear information strategy, to information systems construction, management and control, to operation and maintenance and performance management of information system, which covering the company's information technology lifecycle. These are fundamentals in building international high-performance enterprise information capacity, as well as the main points to enhance the information construction of power enterprises. The six dimensions are: informatization strategic positioning, informatization framework management, informatization value control, informatization project construction, service management and operations, personnel and performance management.

2.4 Informatization maturity model

Informatization Maturity Model is a grading analysis of the above informatization capability model. From the model in Figure 4, we can see the maturity of information management represented by the horizontal axis is divided into five grades: the highest is the fifth grade and the lowest is the first stage. The vertical axis represents the informatization capability. In practical applications, the six ability indicators can be further subdivided into detailed maturity index set. For each of the indicator set, there is a pre-set criterion. By comparing the current situation of the company's informatization with the indicators set and scoring according to criteria, company's current stage of maturity is located. This tool is mainly used to measure the current level of company informatization and provide a scientific basis for the establishment of high-performance informatization management and control system.

3 RESEARCH AND DEVELOPMENT OF HIGH-PERFORMANCE INFORMATIZATION MANAGEMENT AND CONTROL SYSTEM OF POWER ENTERPRISES—CASE STUDY OF X POWER ENTERPRISE

We use X power company as an example in this paper. Based on the theoretical basis for high-performance informatization management and control system, we analyze the current situation of informatization and build high-performance informatization management and control system in accord with the corporate practice by referring to best practices and management standards of informatization management and control system.

4 CORPORATE INFORMATIZATION MANAGEMENT AND CONTROL STATUS ANALYSIS

4.1 *Informatization organizations and operation modes*

The current department for information management includes system construction bureau, system operation bureau and integrated management bureau. The existing informatization organization architecture is in lack of complete definition for information technology staff duties. Responsibilities of employees are not clearly enough, leading to low efficiency. Besides, current organization architecture system can't fully mobilize the enthusiasm of staff and maximize the potential of information technology personnel.

4.2 *Informatization management and control process and tools*

The company carried out process design project for informatization management and control system a few years ago, but in-depth implementation is needed. The company currently does not have complete informatization planning processes. Information management center gives blueprints for the coming-years' IT development based on the abundant information provided by relevant business units. While in practice, there is a difficulty in collecting this information, which resulting in failure to give guiding planning results. Project approval of informatization projects is not proceeding at corporate level. Business units carry out their own IT project and approval based on business needs, bypassing the information management center. Information management center can't fully participate in the business sector-led informatization project construction. Business units usually conduct their own informatization construction according to their needs. Information management center can't keep abreast of the progress of the projects, costs, risks and other relevant information. There is a lack of effective program management system. Operation and maintenance of company IT system has made some achievements with a fixed operation and maintenance team. While there isn't a united operation and maintenance supporting system. The planning, project approval of IT program is mostly done by hand manner. There is no corresponding IT tools to support, resulting in difficulties in

information retrieval and non-guaranteed information accuracy. Operation and maintenance system is not supported by appropriate software. All the questions have to be submitted, distributed and handled by filling in the documents in hand manner, which isn't efficient.

4.3 *Informatization management and control performance*

The company doesn't establish the entire personnel IT performance system. Though business sectors have come to realize the importance of informatization, there isn't an appropriate performance appraisal system for informatization performance.

4.4 *Construction of high-performance informatization management and control system*

Based on the status of the company and by combing with high-performance informatization model and other best management practices, we will elaborate the construction process of high-performance informatization management and control system from the following aspects: informatization management and control organization and operations, management and control processes, tools, and entire personnel performance.

4.5 *Organizations of high-performance informatization management and control system*

"Integrated management and control" should be the core when building organization structure of high-performance informatization management and control system. Organizational reform should be done. Grassroots units establish informatization application duties, recombine informatization operation and maintenance power, build up-and-down-through information management system with characteristics such as clear responsibilities, powerful control, efficient operation, finally achieving a flat organization. There are also needs to strengthen the functions of IT departments, emphasize the match of IT strategic planning with business development strategies, enhance the company's overall planning ability in informatization projects, match the construction of IT systems with corporate IT strategies, standardize IT construction process, improve operability of IT standards and strengthen problem management in IT support process. Finally, authority separation should be considered in terms of management and control duties, including: separation of IT income assess and construction, neutrality of evaluation, separation of formulation and execution of policies, which will benefit the assessment. The main responsibilities for company informatization management and control organizations should include the following contents: 1) Formulate corporate IT management standards, policies and processes and carry out; 2) Formulate informatization construction and development blueprints, adjust and update planning according to needs, facilitate implementation of IT construction planning; 3) Set up corporate IT management technology standards, timely revise and improve corporate IT standards as required; 4) Establish and improve the company's overall IT management system, manage the informatization projects within the firm, support and cooperate in the IT system construction work; 5) Meet the actual needs of the various business units, involve in the design, development, management and support of business application system, and coordinate relationships between the various application systems; 6) In charge of corporate systems and IT construction, management, operation and maintenance task to ensure the smooth flow of information of the company system; 7) Formulate company IT facilities' operation, maintenance and transformation, manage and carry out the project plan; 8) Responsible for the management and assessment of information technology provider; 9) Responsible for the professional business training under the guide of unified training program of the firm.

4.6 *Processes, standards and tools of high-performance informatization management and control system*

Construction of high-performance informatization management and control system emphasizes on the end-to-end management and control of whole informatization life cycle. The control means include processes, procedures, systems, standards and tools.

4.7 *Processes*

Informatization management and control is intended to make information resources and ability consistent with business objectives and business needs. The informatization management and control processes provide a mutual communication channel for managers who bear the responsibility and leadership role, so as to come up with an agreed viewpoint on how and where to invest and construct informatization system and what business objectives to achieve, which developing a viable plan for the implementation of informatization construction. According to the actual situation of the company and high-performance informatization model, we build a high-performance informatization operation process framework,

which is organized into six modules, including: information security management, planning management, project approval and plans, construction management, system operation and maintenance and information resources management. From high-performance informatization management and control system process framework, we tease out the six major processes, 21 sub-processes, and clearly define the interface of the business sector and the IT management department at work. The main processes include: *Planning management.* Make clear the content and responsibility for informatization strategies and planning. The overall planning, special planning and rolling revision chain hooks up with each other, guiding the specific informatization construction. *Project approval and plan.* Center on key decision points of project approval; establish the appropriate organization, responsibilities and problems upgrading mechanisms to ensure the effective convergence of informatization construction stage. *Construction Management.* Build program management mechanism and key management roles. Based on informatization architecture planning, use means such as change management, stakeholders to manage multi-informatization projects from macro-level, ensuring the compliance with the overall objectives and interests of the company. *System operation and maintenance.* Use single point of access 81186 in the telephone service support, reasonably coordinate headquarter and primary service personnel; maintain the consistency of service workload record and service quality. *Information Security Management.* Promote safe management of the entire life cycle through the development, implementation, supervision and improvement process; enhance information security management capabilities. *Information Resource Management.* Include management processes of informatization resources such as institutions, asset, performance, knowledge and suppliers. The six major processes of operation process framework further include 21 three-level processes, 53 four-level processes.

4.8 *Standards*

According to the overall process management thinking of informatization, based on "standardized management, standardized operation" principle, we establish a comprehensive informatization standardized system, which covers regulations, standards and management system related to informatization.

4.9 *Tools*

Informatization management and control tool is the system to support informatization management and control work. An effective tool can greatly enhance the work efficiency. Companies can develop or leverage existing system to construct an integrated management and control system platform to ensure the project planning, project approval and verification proceeding within a unified system, thus improving work efficiency. Using existing ERP systems to develop informatization management and control tool not only enhance the integration of information, but also avoid difficulties in software selection and subsequent inter-system integration. The operation and maintenance system is included in the informatization management and control tools to avoid manual operation. Carry out the solidification of informatization performance indicator extraction process. Automatically generate informatization performance indicators by informatization means, promoting the development of related work.

4.10 *Entire personnel informatization performance*

Construction of informatization performance indicator system is the key point in buiding high-performance informatization management and control system. Entire personnel informatization performance system construction will measure informatization management and control level in an objective and dynamic way and enhance the informatization capability of the entire personnel. Therefore, companies need to establish informatization roles and responsibilities related to business units, design appropriate informatization performance management system, combine the company's overall strategy with the employee's personal goals, determine the employee's personal assessment index, weight and standard values and finally get the employees' work commitments. The entire personnel performance system includes indicator libraries, assessment processes, performance indicators automatic extraction system. The system establishes a complete entire personnel informatization performance assessment system in combination with RACI matrix.

5 RESULTS AND CONCLUSIONS

Building a high-performance informatization management and control system is of great significance for enhancing the power enterprise's IT work, mainly reflected in: 1) Achieve the flat organization architecture. Through combination of the business unit duty and basic level informatization position, assign personnel to right position and define responsibilities. 2) Establish high-performance management and control system. Complete end-to-end

informatization process design and solidication, focus on the whole life cycle of informatization construction, achieve lean management, process transparency, process control, build up-and-down-through information management system with powerful control and efficient operation. 3) Improve the unified informatization architecture. Construct highly integrated corporate information architecture from the overall informatization strategy to support business development and enhance business value through informatization. 4) Build an integrated support system to provide support to deepen application and improve performance, completing an "integrated management, second-level operation and maintenance, three-line support" operation and maintenance service system, achieving standardization and intensiveness of operation and maintenance process, realizing unified acceptance, unified command and unified service. 5) Establish the entire personnel informatization performance system, scientifically set up two index systems for informatization management and control and application assessment, which are used in corporate performance assessment in the third quarter and fully increase the management level, construction level, service level and application level of informatization.

ACKNOWLEDGEMENTS

This research was financially supported by the National Science Foundation.

REFERENCES

[1] M. Zuo, W.Z. Chen, R.X. Hu. Analysis of Maturity Models for Informatization and Their Comparision [J]. Chinese Journal of Management. 3(2005) 340–346.
[2] D. Madsen. Disciplinary Perspectives on Information Management [J]. Elsevier Ltd., Proceedings of the 2nd International Conference on Integrated Information (IC-ININFO 2012), Budapest, Hungary, August 30–September 3, 2012, Volume 73, 27 February 2013, Pages 534–537.
[3] S. Bahsani, A. Himi, H. Moubtakir and A. Semma. Towards a pooling of ITIL V3 and COBIT [J]. International Journal of Computer Science Issues. (2011):185–191.
[4] X. Wang, W. Yuan. Research on Corporate Informatization Maturity Model [J]. Journal of Intelligence. 10(2007) 11–14.
[5] Y. Fan. Corporate Informatization Management Strategy Framework and Maturity Model [J]. Computer Integrated Manufacturing Systems. 7(2008) 1290–1296.
[6] J.F. Bai. Research and case-study on the Evaluation Model of the Maturity of enterprise Informatization [D]. Tsinghua University. (2006).

[7] F. Yu. Construction of Corporate Informatization Management and Control System [J]. Command Information System and Technology. 2(2012) 38–43.

[8] R. Zhao, D. Guo. Exploration and Practice of Informatization Management and Control in Power Corporates [J]. Electric Power Information Technology. 5(2011) 1672–4844.

[9] W. Li. How to Break Through Corporate Financial Informatization Management and Control [J]. Shanghai State Asset Regulatory Commission. (2014).

[10] Q. Meng. Construction of Group Finance Integrated Informatization Management and Control [J]. Friends of Accounting. 17(2011) 48–50.

[11] B. Han. Application and Practice of ITIL V3 Configuration Management [J]. Silicon Valley. 12(2012), pp.125–165.

[12] C. Feng. Corporate Application of IT Service Platform Software Based on ITIL [J]. China Science and Technology Information. 9(2009) 101–102.

Advances in Future Manufacturing Engineering – Yang (Ed.)
© 2015 Taylor & Francis Group, London, ISBN 978-1-138-02817-3

Study on the integration problem of snow depth monitoring system based on laser ranging and video detection

Xiaoqing Cheng & Yong Qin
State Key Laboratory of Rail Traffic Control and Safety, Beijing Jiaotong University, Beijing, China

Zongyi Xing & Xuehui Wang
School of Mechanical Engineering, Nanjing University of Science and Technology, Nanjing, China

ABSTRACT: Nowadays, Laser ranging technology is used in most of the existing high-speed railway snow depth monitoring system in China. However, this method has small monitoring range and the false alarm problem. In order to improve the reliability and accuracy of snow depth monitoring, this paper researches and develops snow depth monitoring system based on laser ranging and video detection techniques. In this article, we study the design objectives of system, system working principle, hardware and software system design. This study achieves the cross validation by multi sensor measurement techniques to snow depth measurement data.

Keywords: snow depth monitoring system; laser ranging; video detection

1 INTRODUCTION

Natural disaster is one of the hidden dangers for the safe operation of high-speed railway. Some High-speed Railway Disaster Prevention Systems have established in France, Germany and Japan etc. Chinese high-speed railway has also exploited and applied for the Disaster Prevention System. Snow depth monitoring subsystem is an important part of this system. However, many systems is to use laser sensor for snow depth measurements, but this method of measurement is prone to cause some problems, such as false positives and the measurement error. In order to improve the accuracy and reliability of the high-speed railway snow depth monitoring system, the paper presents the research on integration technology of this system based on laser and video. This system will realize the real-time monitoring on snow condition of high-speed railway; and provides alarm, warning information for dispatching and maintenance management; and prevent or reduce the influence which caused by snow disasters on high-speed rail operating safety.

2 SITUATION ANALYSIS OF CHINA'S HIGH-SPEED RAILWAY SNOW DEPTH MONITORING SYSTEM

At present, snow depth monitoring system has been installed in Changchun-Jilin Intercity Railway, Harbin-Dalian High-speed Railway, Panjin-Yingkou High-speed Railway. The system is composed of field monitoring unit, transmission channel, data processing equipment, on-site monitoring equipment (snow depth gauge: laser sensor), dispatching and maintenance terminal. The main function of the system is to process and analyze snow depth data collected by on-site snow depth measurement equipment, to display the real-time snow depth values and send corresponding alarm information in the dispatching and maintenance terminal by comparing the snow depth values and the predefined threshold, to assist dispatchers command vehicles and maintenance personnel maintain and repair the railway line.

Laser ranging technology was used in most of the existing high-speed railway snow depth monitoring system. In the process of the use of the system, there are three problems need to be solved. Firstly, monitoring points are imperfect or their positions are not reasonable. Secondly, laser type snow depth gauge may be interfered by other factors (such as heavy fog)[1]. Thirdly, the existing system can only get the snow depth at current time, but cannot predict the change of snow depth and lack of early warning function. For the problems above, the method combing the laser sensor with video image is put forward to measure the snow depth. The laser-ranging sensor is used to measure the snow depth and the video image is used to record the snow conditions by shooting track status. The advanced video analysis technology also can predict the real-time snowfall and make an early warning for the future snow depth. Meanwhile, video image can also provide operation status of equipment at the monitoring points and avoid the interference of animals or human, the invalid measurement monitoring caused by strong winds.

3 THE GOALS OF THE SNOW DEPTH MONITORING SYSTEM DESIGN BASED ON LASER RANGING AND VIDEO DETECTION TECHNIQUES

1. Snow depth measurement based on laser ranging technique. The measurements of snow depth mainly focus on a single point and a specified number of points on the track. The system can collect real-time snow depth data and store historical data at certain time intervals.
2. Snow depth measurement based on video detection technique. The system can get a real-time snow image of specific location by a rotating PTZ and do data compression and storage automatically to analyze the snow depth condition.
3. Control and management functions. The system can realize the switch control of laser devices and video cameras as well as the rotation of PTZ and the set-up of alarm thresholds.
4. Remote monitoring and communication. Data from all regulated equipment is transmitted to the monitoring center through a wireless network. The center sends parameters set and control commands by users to each end of the monitoring unit through the network to do control and management.
5. Management platform software of monitoring center. Monitoring center software architecture is based on C/S model. And users can view snow depth information from laser, video images from cameras and snow condition estimation after data fusion through the client. The platform consists of sensor communication server, data processing section and the database.

4 THE OVERALL DESIGN OF THE SYSTEM

4.1 System components

Seeing from the entire system, there can be four layers, including the perceptual layer, the processing layer, the transport layer and the services layer. The perceptual layer is mainly composed by the laser sensor, angle sensor and the network camera, each of which is respectively for the measurement of the distance between snow surface and the datum, the installation inclination angle of the laser sensor and the video image of the monitoring station, then the collected information will be send to the processing layer, which includes a front-end processing module, used to receive the collected information, and do the data preprocessing and data fusion. The front-end processing module has some special demands about the processing performance for real-time image capturing and processing, identification of whether there is snow or not and estimation of the snow density, finally it will generate a error report, according to the measured values that passed through the accuracy inspection to eliminate errors caused by other factors except snow. The test value passed measuring accuracy will be sent to the transport layer. The transport layer contains s wireless module, with wire and wireless (3G/4G) retransmission function. This module has several RJ45 interfaces and RS232/RS485 interfaces. The wireless retransmission function needs a Telecommunication operators' billing card to send the data to the back-end services layer through the public network, or transmits within the local area network by wired means. Lastly, the services Layer, which contains the backstage data center and terminal equipment of all levels, mainly to store the uploaded data from transport layer, and distribute to the terminal display, in order to satisfied the multi-user to obtain the real-time access to snow depth measurements, video monitoring points and the image information, alarm information and error reporting.

4.2 Principle of the system

The following analysis is installation method about laser sensor and network camera, and the working principle of the system.

4.2.1 Measuring snow depth by laser

According to the characteristics of high-speed railway snow distribution, from the processing measurement data complexity and economic costs, the use of laser to measure snow depth is a better choice, laser sensors firstly measure the distance between the receiver and the measurement reference plane as a reference value, and then in subsequent measurements with the reference value for the difference, obtain the height of the reference plane and the inclined surface of the snow. Tilt sensor to measure the tilt angle of the laser probe, draw a vertical height of the reference surface and the snow surface triangular relationship that is values of snow depth. Laser sensor is installed on the power supply column in high-speed rail, installation angle of about 45°, height from the sleeper area of about 4 m. Tilt sensor to measure the inclination of laser sensors and ground, so it is generally bound with the laser probes, laser sensors collect data of deep snow every five seconds.

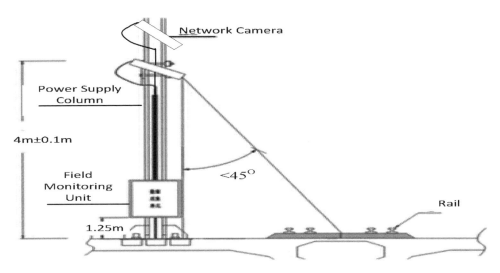

Figure 1. Installation diagram of laser sensors and cameras.

4.2.2 Video monitoring

Network cameras and laser sensors are installed on the same power supply column. Network cameras is just above (above 0.5 m) laser probe position. Observation of laser measuring is able to cover the range of Laser measuring points and the surrounding area.

Network cameras will monitor video in near real-time information collection, the compressed video stream to the front-end processing module. Front-end modules for image filtering, smoothing, contrast enhancement and other pre-processing. Video monitoring of the main features are: ① Snow depth measurement based on video. Benchmarking and testing by setting in place benchmarking calibration values to determine snow depth of snow, increase the redundancy of snow reliability. ② People and animal detection based on video. When there are persons or animals in the laser and video surveillance area, it has a chance to lead to erroneous measurements and alarms. Video detection of people and animals in the region changes to correct this kind of interference. ③ By Gaussian background modeling method of collected images to prospect and extract. The reflection properties of light according to the snow and snow models define filter rules to filter foreground pixels. We choose the most trusted snow pixels (blocks), snowflake pixels (blocks) statistics on the number. Snowflake pixel coverage ratio is a feature of snow density. Finally, the snow density is classified and we could estimate the strength of the current snowfall.

4.2.3 3G wireless communication

The system is based on telegraphic CDMA network and composed of 4 components: data capturing and compressing module, data transmission module, data recording module, and surveillance and management platform module.

4.2.4 Surveillance center

The surveillance center, as the core of the whole system, composed of surveillance center hardware system and software system, receives data of the field apparatus and executes the data analysis and decision support. Through communication module, the site acquired snow depth data and video image and then transmits them to surveillance center, which executes data management, history data savage, data processing, quick report display and image decoding to display image. After snow depth data processing and analyzing, the alarm will be given when the result go beyond the threshold.

4.3 The design of hardware system

While the snow depth surveillance system measures snow depth, the video surveillance, impeccable device protection and communication function are achieved by data analyzing methods such as pattern recognition, data fusion, stochastic filtering and intelligence prediction, as well as friendly human computer interaction interface. The principle of the system with its information flow is given in Figure 2.

4.4 Software system design

This system is a wireless Snow depth monitoring and early warning network. On the basis of the data acquisition hardware, PC software mainly implements storage of data, real-time display, data

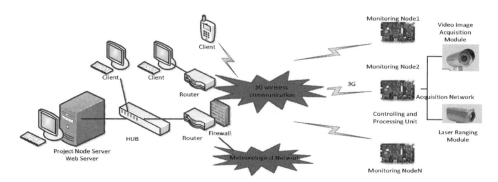

Figure 2.   Schematic diagram of the hardware system composition.

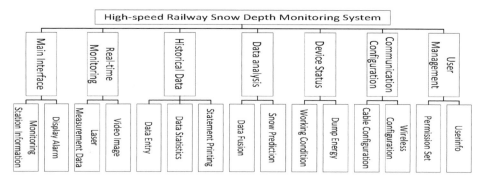

Figure 3.   Architecture of high-speed railway snow depth monitoring system.

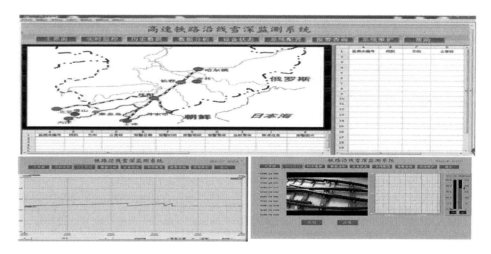

Figure 4.   System interface screenshots.

analysis and other functions. The basic architecture as shown below:

① The main interface. The main interface snow depth monitoring system for high-speed rail is the most direct reflection of snow to provide basic information for high-speed rail operation. The main interface requires concise interface design, including geographic information monitoring sites, the site displays basic information and alarm function. The alarm display is based on real depth of snow data. Also, system can acknowledge the alarm and retrieval the pictures and video information.

② Real-time monitoring. Real-time monitoring is an integrated interface for real-time acquisition of laser and video information. It is possible to get real-time snow depth data and video information. Also, it can capture and video according demand.

③ Historical Data. Mainly used for data read from the front-end outdoor equipment store card and the associated statistical analysis, including data entry, data statistics and report printing.

④ Data analysis. Snow depth data analysis is an important function of remote monitoring and early warning information system software. The main job is to analyze and process snow depth data collected by snow equipment, using data mining techniques to predict snow depth data applications. Snow forecast snow accumulation model with predictive analysis is the primary snow forecast model. The physical properties of snow, amend the conditions of the current snowfall, snow damage history information is model calculation. The model uses a combination of qualitative and quantitative analysis, computer simulation and expert the method of combined experience harmful accumulation of snow forecast analysis.

5 SUMMARY

Aiming at the requirement of complex snow depth monitoring environment of high-speed railway, application of multi sensor data fusion technology cross validation snow depth measurement data, this paper studies and designs the high reliability and multi dimension information of deep snow monitoring system. The system has been carried out laboratory test and snowfield test, the next step will be tested in railway line.

ACKNOWLEDGEMENTS

This work was supported by the Science and Technology Research and Development Program of China Railway Corporation (No. 2013T002-A-2-3). The supports are gratefully acknowledged.

REFERENCES

[1] Xiaoyu Li, Peng Zhang, Xianchun Dai, Niansheng Xi. Limit the use of monitoring systems and management optimization of high-speed rail natural disasters and foreign body invasion [J]. China Railway, 2013(10), 21–25.

[2] Zhibin Wang. HaDa Railway Passenger Dedicated snow depth monitoring system [J]. Railway Standard Design, 2012(5), 165–168.

[3] Liang Xu, Zhenhong Jia, Xizhong Tan. Video image detection and removal of snow [J]. Optoelectronics. Laser, 2007, 18(4), 478–481.

[4] Tong Wang. High-Speed Rail Disaster Prevention and Safety Monitoring System Research and Development [J]. China Railway, 2009(8), 25–28.

Advances in Future Manufacturing Engineering – Yang (Ed.)
© 2015 Taylor & Francis Group, London, ISBN 978-1-138-02817-3

Background correction algorithm for Laser Induced Breakdown Spectroscopy

Bo Zhang
*Laboratory of Networked Control Systems, Shenyang Institute of Automation, Chinese Academy of Sciences,
Shenyang, China*
University of Chinese Academy of Sciences, Beijing, P. R. China

Lanxiang Sun & Haibin Yu
*Laboratory of Networked Control Systems, Shenyang Institute of Automation, Chinese Academy of Sciences,
Shenyang, China*

ABSTRACT: Background correction is a major problem in recording the Laser Induced Breakdown Spectroscopy (LIBS), which is an important pre-processing technique used to separate signals. For this purpose, the algorithm can automatically correct varying continuum background based on a penalized least squares method. We use simulations and experiments to evaluate the validity of the algorithm.

Keywords: Laser Induced Breakdown Spectroscopy; background correction; penalized least squares

1 INTRODUCTION

Laser Induced Breakdown Spectrometry (LIBS) is one of the most promising and effective methods for directing spectral analysis of diverse origin objects [1–4]. The method is based on the use of the emission spectrum of an intensely glowing laser plasma obtained on the surface of the analyzed material. But various kinds of spectral interferences deter its use. Background interference is one of the spectral interferences. Since the LIBS background emission is strong, proper background correction is needed to obtain the analyte emission intensity [5]. The continuum background is not generated by random noise error of the measurement system, but due to relatively stable intensity of background, which is formed from continuous emission spectrum in plasma and is not removed directly during the filtering process.

Some methods that utilized experimental techniques to eliminate the background have been developed with the increasing use of LIBS recently. The two most common ways to correct continuum background are blank measurement method and off-peak background correction method [6–9]. The blank measurement method needs separate measurement of the instantaneous emission profile of a blank and the analyte emission. The background correction based on two profiles subtraction. The blank measurement method is not satisfactory if the instantaneous profiles of the blank and the sample do not match exactly [10]. Using the off-peak background correction method, the background

intensity at a wavelength adjacent to the analyte line is measured for the background estimation at the analyte wavelength by infer. This method can achieve good result sometimes, particularly in nearly linearly variable background [11–13]. It can be clearly seen from the above discussion that each of the two major background correction algorithms has its advantage and also certain shortcomings. However, they are complementary to some extent.

In current work, an automatically background correction algorithm is proposed for more general purpose background correction for LIBS based on penalized least squares background fitting. Starting with the simulated data, we use the algorithm to demonstrate the automatic optimization parameters of the algorithm for background correction with various backgrounds. Then, through application to LIBS, we demonstrate the validity of our algorithm. Finally, some conclusion and outlooks are drawn from this paper.

2 ALGORITHMS

Suppose $Y = \{y_i, i = 1, 2, ..., m\}$ is LIBS emission intensity vector, whose length is m. Suppose $Z = \{z_i, i = 1, 2, ..., m\}$ is the fitting vector, whose length is also m. The occurrence of missing signal due to measurement infeasibility or instrumentation failure is often emerging in practice. It can also be convenient to allocate exception values with a low weight, on the contrary, give a relatively high weight to high quality signal. For the balance calculation, we can introduce a vector of weights w_i corresponding to the original signal Y. The vector of weights is therefore introduced in the fidelity part z to the LIBS emission intensity y can be expressed as:

$$S = \sum_{i=1}^{m} w_i (y_i - z_i)^2 = (Y - Z)^T W (Y - Z) \qquad (1)$$

where the weights value w_i corresponding to the original signal Y, W is a diagonal matrix with w_i on its diagonal. The weights w_i can take two values depending on the relative contribution of the original signal Y and background Z at point i: $w_i = p$ if $y_i = z_i$ and $w_i = 1 - p$, otherwise. The parameter p is called the asymmetry parameter. This parameter may take values ranging from 0 to 1.

The penalization of the background estimated is expressed with the difference of order $d(\Delta^d)$, where is considered maximally penalization. They can be expressed by the squared differences between neighbors:

$$R = \sum_{i=1}^{m-1} (\Delta^d z_i)^2 = \sum_{i=2}^{m} (z_i - z_{i-1})^2, \, d = 1 \qquad (2)$$

When $d = 1$, $\Delta^d z_i$ is the derivative of the identity matrix. Second or third order difference are usually used. The formulas are given in Eqs. (3) and (4), respectively.

$$\Delta^2 z_i = \Delta(\Delta^1 z_i) = z_i - 2z_{i-1} + z_{i-2}, \, d = 2 \qquad (3)$$

$$\Delta^3 z_i = \Delta(\Delta 2 z_i) = z_i - 3z_{i-1} + 3z_{i-2} - z_{i-3}, \, d = 3 \qquad (4)$$

A balanced combination of two goals is the sum:

$$Q = S + \lambda R = W \|Y - Z\|^2 + \lambda \|\Delta^d Z\|^2 \qquad (5)$$

where λ is a user-defined parameter. The parameter λ controls the trade-off between fidelity and penalization: as $\lambda \to 0$ the solution converges to the best least-squares fitting to the data, and as $\lambda \to \infty$ the solution converges to the measurement sequence. Furthermore, the R is a monotonically increasing of function of λ, while the S is a monotonically decreasing function of λ. Large values of λ will make R term higher as the penalization effect will be larger, but at the cost of a deterioration of the fit to the data (which is the penalization concept). Thus, we have a standard sum of squares problem with penalization, where the goal is to find the series Z which minimizes Q.

We get a linear equation which can be easily solved from partial derivatives $\partial Q / \partial z = 0$:

$$(W + \lambda D^T D)Z = WY \qquad (6)$$

where D is a $(m - d) \times m$ matrix that form difference of order d.

We only have the measured signal to guide us in choosing appropriate values of p and λ in practice. Unfortunately we do not know the real background, so the optimal parameters cannot be determined accurately. We create an objective function to compare estimated background with the background using by background correction algorithm. In this paper, we write them $b_{estimated}$ and Z for short, respectively. The estimated background calculation is based on least-square fitting through the local minimum of Y.

3 SIMULATION AND EXPERIMENT

On the one hand, background correction has been done to demonstrate the feasibility of the proposed method, simulated data with the known backgrounds has been performed. On the other hand, LIBS experimental setup is prepared to test and verify the proposed method in practical applications. The simulated LIBS signal is the sum of liner or curved background, analytical signals, and random noise, which can be mathematically described as follows: $f(x) = p(x) + b(x) + n(x)$, where $f(x)$ is simulated spectrum, $p(x)$ the pure signal, $b(x)$ the liner or curved background, and $n(x)$ the random noise. The pure signal $p(x)$ is composed of four Lorentzian peaks, which can be mathematically described as follows:

$$p(x) = \frac{50}{1 + \left(\frac{x - 220}{2}\right)^2} + \frac{60}{1 + \left(\frac{x - 240}{3}\right)^2} + \frac{100}{1 + \left(\frac{x - 270}{5}\right)^2} + \frac{80}{1 + \left(\frac{x - 290}{6}\right)^2} \qquad (7)$$

Linear background $b_l(x)$ and the curved background $b_c(x)$ can be mathematically described as follows: $b_L(x) = 0.6x$, $b_c(x) = 65\sin(x/40\pi)$. Random noise is added into the simulated spectra, which can be mathematically described as follows: $n(x) = 0.1WGN(0,1)$, where $WGN(.)$ function is a white Gaussian noise generator.

The LIBS experimental setup used to obtain spectral data in this work is schematically shown in Figure 1. The laser source is a Q-switched Nd:YAG laser (CFR200 Nd:YAG from Big Sky

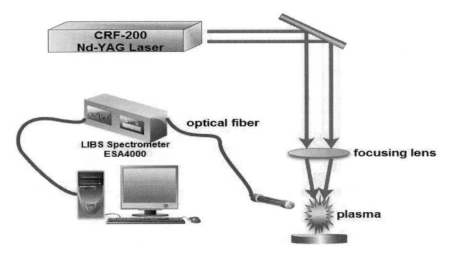

Figure 1. LIBS schematic visualization of the set-up for LIBS experiment.

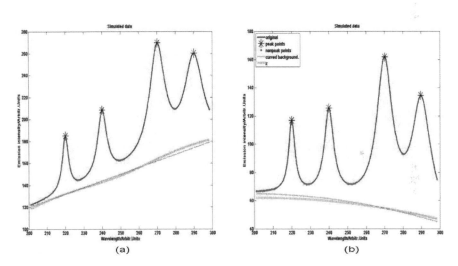

Figure 2. Results of background correction of the simulated spectra. (a) Background correction result of the linear background data. (b) Background correction result of the curved background data.

Laser) with maximum pulse energy of 200 mJ, pulse width of 12-15 ns, wavelength of 1064 nm, and repetition rate of 1–15 Hz. The sample is placed on a three-dimensional manually controlled stage, and the laser beam is focused on the sample surface by a convergent lens of 75 mm focal length. The spectrometer is ESA 4000 from LLA Instruments GmbH for analysis. In a wavelength range of 200 nm to 780 nm and with a resolution of a few picometers spectra are recorded by an ICCD camera. The samples used in experiment are some national standard copper alloy samples, and their serial number is GSB 04-2416-2008. In the measurement, the laser pulse energy is 100 mJ, the laser fluence on sample is 50 J/cm^2, and the single laser pulse frequency is 5 Hz. Because LIBS data are influenced by a variety of experimental factors in the test process, the repeatability is low. For this reason, we use the average of multi-measurement to reduce the error caused by the uneven distribution of inner components. We measured each sample for 10 times in five different positions. All experiments are under the conditions of normal atmospheric pressure and 25 degree centigrade.

4 RESULTS AND DISCUSSION

According to the results of background correction, the minimum value of f_z is 13.1068 with $k = 5$, and then we consider the optimal parameter values of (λ,p) with linear background is (3.0198e + 03, 1.9192e-05). In a similar way, the results evaluate function with curved background. We find the minimum value of f_z is 2.4688 with $k = 6$, then the optimal parameter values of (λ, p) with curved background is (1.7546e + 04, 1.8651e-06).

The background using by background correction algorithm shows better performance than estimated background in nonpeak segments, peak parts and *SBR*. When $k = 5$, Z not only improves the more fidelity in nonpeak segments than estimated background but also less deterioration character of peak parts than estimated background. Z implies better properties than estimated background in nonpeak segments and peak parts with $k = 6$.

With the purpose in finding the optimal parameters value, the evaluation function calculations of Z can be compared with the estimated background in attempt to ascertain the optimal parameters values. The wavelength range shown in Figure 4(a) is between 200 and 800 nm.

Z more approximately in nonpeak segments than estimated background, β and *SBR* of Z approach to estimated background, so we consider the optimal parameters value of (λ,p) is (8.9362,1.8239e-05) with $k = 5$. Furthermore, with the purpose in verifying the effectiveness of background correction algorithm with parameters adjustment, linear calibration of spectra (original and after background correction treatment) with six standard copper samples are studied. By processing the background correction, we use correction spectrum to do this work that subtract background from original spectrum. The calibration curves of original spectra are compared with

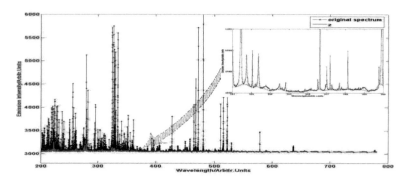

Figure 3.   Background correction algorithm with automatically optimal parameters for pair (p,λ). The wavelength range shown is between 200 and 800 nm, and partial enlarged view is on right top.

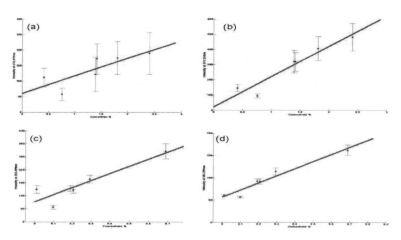

Figure 4.   Calibration curves of Pb at 373.979 nm on (a) Original LIBS spectrum and (b) Background correction LIBS spectrum; Calibration curves of Sn at 283.999 nm on (c) Original LIBS spectrum and (d) background correction LIBS spectrum.

Table 1. Characteristic of linear calibration curves.

| Element | Wavelength (nm) | Calibration curves | | | | | |
| | | Original | | | Background correction | | |
		a	b	R^2	a	b	R^2
Pb	373.979	5.6	56.0	0.8	199.3	21.87	0.9
Sn	283.999	70.7	90.6	0.9	159.0	56.1	1.0

the background correction spectra. a, b and R^2 are the slope, intercept and correlation coefficient of linear calibration curves, respectively.

5 SUMMARY

In this work, a background correction algorithm to automatically estimate the varying continuum background in LIBS is presented. From our limited study, we successfully proved the efficiency and accuracy of the background correction algorithm for simulations and experiments. The proposed algorithm is not only a valid tool for background correction in LIBS but also a useful tool in other spectral signals.

ACKNOWLEDGMENT

This work has been supported by National Natural Science Foundation of China (Grant No. 61004131) and the National High-Tech Research and Development Program of China (Grant No. 2012AA040608). The authors would like to thank LIBS project team of SIA (Shenyang Institute of Automation, Chinese Academy of Sciences) for profitable recommendations for improving method.

REFERENCES

[1] G.S. Senesi, M. Dell'Aglio, R. Gaudiuso, Heavy metal concentrations in soils as determined by Laser-Induced Breakdown Spectroscopy (LIBS), with special emphasis on chromium, Environmental Research 109(2009) 413-420.

[2] W.T.Y. Mohamed, Improved LIBS limit of detection of Be, Mg, Si, Mn, Fe and Cu in aluminum alloy samples using a portable Echelle spectrometer with ICCD camera, Opt. Laser Technol. 40 (2008) 30–38.

[3] Céline Gautier, Pascal Fichet, Denis Menut, Applications of the double-pulse Laser-Induced Breakdown Spectroscopy (LIBS) in the collinear beam geometry to the elemental analysis of different materials, Spectrochimica Acta Part B: Atomic Spectroscopy (2006) 210–219.

[4] E.H. Evans, J.A. Day, W.J. Price, C.M.M. Smith, K. Sutton and J.F. Tyson, Atomic spectrometry update. Advances in atomic emission, absorption and fluorescence and related techniques. J. Anal. At. Spectrom., 19 (2004) 775–812.

[5] E. Poussel, J.M. Mermet, Simultaneous measurements of signal and background in inductively coupled plasma atomic emission spectrometry: effects on precision, limit of detection and limit of quantitation, Spectrochim. Acta Part B 51 (1996) 75–85.

[6] P.M.J.W. Boumans, Basic concepts and characteristics of ICP-AES, in: P.W.J.M. Boumans (Ed.), Inductively Coupled Plasma Emission Spectroscopy-Part I Methodology, Instrumentation and Performance, Wiley-Interscience, New York, 1987, pp. 100–257.

[7] A.T. Zander, Spectral interference and line selection in plasma emission spectrometry, Inductively Coupled Plasmas in Analytical Atomic Spectrometry (1992), 299–339.

[8] P.M.J.W. Boumans, Line selection and spectral interference, in: P.W.J.M. Boumans (Ed.), Inductively Coupled Plasma Emission Spectroscopy—Part I Methodology, Instrumentation and Performance, Wiley-Interscience, USA, 1987, pp. 358–465.

[9] George C.-Y Chan, Wing-Tat Chan, Estimation of background continuum emission intensity of inductively coupled plasma for correction of fast changing background, Spectrochimica Acta Part B (2002), 1771–1787.

[10] R. Sing, Direct sample insertion for inductively coupled plasma spectrometry, Spectrochim. Acta Part B 54(1999) 411–441.

[11] L. Gras, M.T.C. de Loos-Vollebregt, Limiting effects on transport efficiency by matrix load in inductivity coupled plasma optical emission spectrometry with electrothermal vaporization sample introduction, Spectrochim Acta Part B 55 (2000) 37–47.

[12] S. Maestre, M.T.C. de Loos-Vollebregt, Plasma behavior during electrothermal vaporization sample introduction in inductively coupled plasma atomic emission spectrometry, Spectrochim. Acta Part B 56 (2001) 1209–1217.

[13] L. Lepine, M. Provencher, K. Thammavong, K.C. Tran, J. Hubert, Dynamic background correction by wavelength modulation in ICP atomic emission spectrometry and its application to flow injection-ICP-AES, Appl. Spectrosc. 46 (1992) 864–872.

Advances in Future Manufacturing Engineering – Yang (Ed.)
© 2015 Taylor & Francis Group, London, ISBN 978-1-138-02817-3

Research on thermal dissipation of electric bicycle lithium ion battery pack based on the microencapsulated PCMs thermal management system

Shuhua Li, Weisheng Dai, Kaijun Bi & Yinzhou Deng
Shenzhen Enter-Exit Inspection and Quarantine Bureau, Shenzhen, China

Cheng Cai
Advanced Materials Institute, Graduate School at Shenzhen, Tsinghua University, Shenzhen, China

Jian Chen
Shenzhen Enter-Exit Inspection and Quarantine Bureau, Shenzhen, China

Guoyi Tang
Advanced Materials Institute, Graduate School at Shenzhen, Tsinghua University, Shenzhen, China
Key Laboratory of Advanced Materials, School of Materials Science and Engineering, Tsinghua University,
Haidian District, Beijing, China

ABSTRACT: The microencapsulated Phase Change Materials (PCMs) are used in battery pack thermal management system to conduct the thermal dissipation experiment of ternary polymer lithium ion power battery pack. By measuring the temperature change in the process of different constant-power or constant-current discharge in different parts of the battery pack, the thermal dissipation effects of the microencapsulated PCMs thermal management system used in the battery pack are studied and compared. The experimental results show after adopting the microencapsulated PCMs thermal management system, the temperature at each point of the battery pack has been reduced to a certain extent, and the maximum reduction is up to 10 °C. In the meantime, the temperature uniformity in different parts of each battery cell has been improved significantly.

Keywords: microencapsulated PCMs; ternary lithium ion battery pack; thermal management; electric bicycle

1 INTRODUCTION

Electric bicycle enjoys great popularities by domestic and foreign customers because of its characteristics such as lightweight and environmental protection. In particular, the application of lithium ion battery with superior performance promotes its rapid development. At present, there are more than 200 million electric bicycles in China.

However, in the actual use of lithium ion battery, due to higher temperature of monomer battery (> 100 °C) [1], temperatures inside the battery module are extremely unequal. And performances between batteries do not match each other, resulting in premature failure of the battery module because of thermal runaway and even serious consequences such as spontaneous combustion and explosion. Lithium ion power battery with high performance must be equipped with the effective thermal management system, so as to ensure

proper battery working temperature, to reduce temperature difference between battery modules, to ensure that the long-term safe and stable working state of the electric power system, which is the key problem to currently restrict the development of electric vehicle power battery.

The traditional thermal management system mainly uses three methods for thermal dissipation, namely, forced ventilation, water cooling and natural convection, which have different degrees of defects in the practical application. The new lithium ion battery pack thermal management system uses Phase Change Material (PCMs) with many advantages such as latent heat of phase change (i.e., high heat storage capacity), long cycle life, non-poisonousness, harmlessness and low price, to reduce the volume of the entire battery pack, and to reduce the moving parts, actively takes intelligent control of heat in the battery pack, and need no extra energy of the battery. Therefore, PCMs is a type of thermal management material having broad prospects for application [2].

The researches on applications of PCMs in the battery thermal management are quite recently developed topics, which are still in the initial stage. According to the published studies, most researches mainly focused on paraffin wax/expanded graphite [3–7], paraffin wax/foamed aluminum [8] or paraffin saturated copper foamed [9] composite PCMs thermal management systems. However, such systems inevitably encounter the problems of volume change of PCMs during the liquid-solid phase changing and complete fusion leakage of PCMs resulting from severe overheating. In this research, microencapsulated phase-change materials synthesized using MPS (3-(trimethoxysilyl) propyl methacrylate) and VTMS (vinyltrimethoxysilane) as raw materials for hybrid shells, and n-octadecane as core materials [10], are used as PCMs for the thermal management system. The staring degradation temperature of the microencapsulated PCMs is up to 180 °C, and the shell protects the core material well. Even if PCMs in the core-shell was in the complete fusion, the problems of volume change and liquid flow would not occur. In addition, it is convenient to be encapsulated with the battery cell.

2 EXPERIMENTAL DEVICES AND SCHEME

2.1 Instrument and materials

With the ceaseless improvement of safety performance, ternary polymer lithium ion battery with high energy density is becoming the first choice of power battery of an electric bicycle. In the experiment, 48 V ternary lithium ion battery packs for the special purpose of commercially available electric bicycle is directly used. The battery pack is comprised of 65 2200 mAh, 18650 battery cells by 13S5P. Before the experiment, all battery cells have been confirmed to have consistent electrical performance through the balance test. The original protection board and charger of the battery pack are kept, under the normal condition of full charge, open circuit voltage is 54.6 V.

XBL 400-300-2000 electronic load of TDI Power is used to carry out the discharge test for the battery pack. Externally connected to 46-channel K-thermocouple, YOKOGAWA/ MX-100 91J130282 temperature recorder is used for temperature data acquisition of battery pack. 45 points are distributed respectively in the upper, middle and lower parts of 15 battery cells, as shown in Figure 1. EL-04 KA temperature and humidity chamber of Guangzhou ESPEC Environment Instruments Co., Ltd, is used to strictly monitor the test environment temperature.

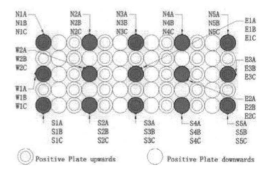

Figure 1. Schematic diagram for the temperature test point.

The microencapsulated PCMs are self synthesized in the laboratory [10], and have a melting temperature of 29.4 °C, latent heat of 132.2 kj·kg⁻¹ and n-octadecane mass fraction of 58.8%.

2.2 Establishment of the thermal management system and its working principle

Under the premise that the original structure of the battery pack basically is unchanged, the microencapsulated PCMs is filled in the gaps between battery cells and around the battery pack, and symmetrically placed in a 300 mm × 160 mm × 100 mm aluminum shell. The PCMs filling quantity is 1040.2 g.

When the batteries discharge large current, PCMs inside the shell absorbs heat out of the batteries, and then the phase change occurs. Within the polymer shell, PCMs is in the molten liquid state and stores heat, to reduce battery temperature quickly. When the battery packs stops working and is cooling, the PCMs is also cooling accordingly. The PCMs in the molten liquid is crystallized because of phase change and releases heat, which is conductive to keep battery temperature unchanged in a cold winter. The working principle of the microencapsulated PCMs is shown in Figure 2.

2.3 Experimental scheme

The original packing materials of commercially available 48 V ternary lithium ion battery pack is removed other than battery cells, and three pieces of insulation film with a width of less than 5 mm are used to support the battery pack, to make it symmetrically placed in an aluminum shell. At the temperatures of –20 °C, 0 °C, 15 °C, 22 °C and 38 °C, respectively carry out the 240 W and 500 W constant power discharge test. Then, at the temperature of 15 °C, carry out a series of tests of different constant powers and different constant

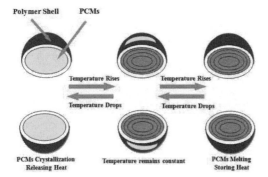

Figure 2. Working principle schematic of the microen-capsulated PCMs.

currents for the battery pack with the PCMs or non-PCMs thermal management system.

3 RESULTS AND DISCUSSION

3.1 *Characteristics of thermal dissipation of the battery pack*

3.1.1 *Discharging at different temperatures, the battery pack has obviously different absolute temperature rises*

The battery pack is fully charged and then it is naturally placed, or its temperature is forcedly controlled through the thermostatic chamber, respectively up to test temperatures of −20 °C, 0 °C, 15 °C, 22 °C or 38 °C. When the temperature at each point is stable and the difference value is less than or equal to 1 °C, respectively carry out 240 W, 500 W constant power discharge tests at the constant temperature. Temperature rise data are obtained through normalization processing of the initial temperature, and the maximum temperature rise of the battery pack is compared. The results are shown in Figure 3. Thus, it is clear that at the same temperature, absolute temperature rise of high power discharge is significantly higher; under the condition of the same discharge power, the lower the discharge temperature is, the higher the absolute temperature rise is; especially, at the temperature of −20 °C, the temperature rises sharply at the beginning of 240 W discharge, until the battery temperature is more than 5 °C, the temperature rise become stable gradually. At the temperature, however, the 500 W discharge test cannot start.

3.1.2 *The temperature rise at each part of the battery pack has no simple geometric symmetry*

Under the condition of normal discharge of low power (<500 W) or small current (<11 A),

temperature change at each symmetric position is consistent. However, when the battery pack discharges high power and large current, due to differences such as internal resistance of the battery cell, temperature difference at each symmetric position is enlarged clearly, up to more than 7 °C, as shown in Figure 4. It is predicted that with in-depth discharge of the battery pack, the difference will be enlarged. Therefore, in thermal dissipation analysis of the large power battery pack, it is inappropriate to deal with it simply according to the symmetric relation.

3.2 *Thermal dissipation effect of the PCMs thermal management system*

3.2.1 *Cooling effect of the PCMs thermal management system*

Based on the above, if the operating temperature of the battery pack is different, the absolute temperature rise is also different. For the convenience of comparison of cooling effect of the PCMs thermal management system, in all the experiments, the environmental temperature is kept at a constant 15 °C.

The battery pack with the PCMs or non-PCMs thermal management system is fully charged respectively, and the temperature at each point should be kept stable and consistent. Next, under the conditions of different constant powers and different constant currents, carry out the discharge test. The temperature measured at each highest temperature point changes with the change of discharge time, as shown in Figure 5.

It is thus evident that the PCMs thermal management system has different degrees of cooling effect. Especially it has an obvious effect in high power or large current discharge, and the decline of most temperature measuring points is more than 8 °C, as shown in Figure 6.

3.2.2 *Influence of PCMs thermal management system on the temperature uniformity*

In order to evaluate the effect of the PCMs thermal management system to improve temperature uniformity, range and coefficient of variation for evaluating the discreteness of an array in mathematic statistics are introduced. Each battery cell is taken as a unit, difference value between the highest and lowest temperature values at three points, namely, on its ends and in the middle, is taken as temperature range (Ω) of the battery cell, and mean square error of temperature values at three points is taken as temperature coefficient of variation (C_v). The battery pack with the PCMs or non-PCMs thermal management system discharges respectively in different constant currents, the difference of temperature

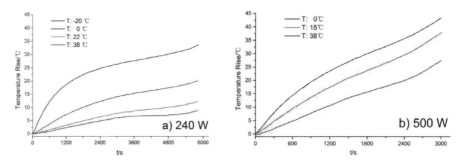

Figure 3. Maximum temperature rise of constant power discharge.

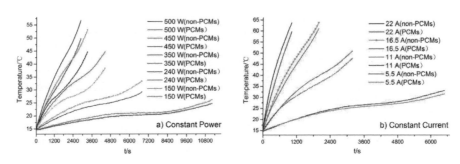

a) Four diagonal positions of the battery pack

b) Middle position between both ends of the battery pack

Figure 4. 22 A constant current discharge (15 °C).

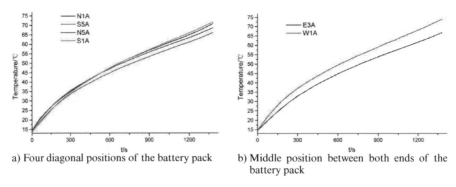

Figure 5. PCMs cooling effect in different discharge powers or current.

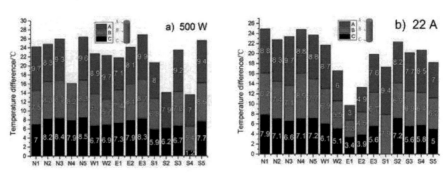

Figure 6. Difference value of temperature ($T_{non\text{-}PCMs} - T_{PCMs}$) at each point in the discharge test of the PCMs or non-PCMs thermal management system.

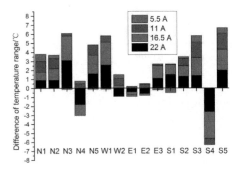 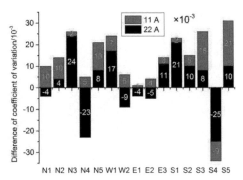

Figure 7. a) Difference of temperature range ($\Omega_{\text{non-PCMs}} - \Omega_{\text{PCMs}}$). b) Difference of temperature coefficient of variation ($C_{v \text{ non-PCMs}} - C_{v \text{ PCMs}}$).

range ($\Omega_{\text{non-PCMs}} - \Omega_{\text{PCMs}}$) of each battery cell is shown in Figure 7a, the difference of temperature coefficient of variation ($C_{v \text{ non-PCMs}} - C_{v \text{ PCMs}}$) in Figure 7b. Positive value means PCMs can improve effectively.

Thus it is obvious that except for a few battery cells (S4, N4, etc.), the PCMs thermal management system basically improves temperature uniformity, especially, in 11–16.5 A constant current discharge, the effect is obvious. In higher current discharge (such as 22 A), some battery cells may be affected by severe thermal runaway, and exceed the effect of the PCMs thermal management system under the experimental condition.

4 SUMMARY

Without changing the original structure of commercially available battery pack, microencapsulated PCMs is used for the power battery thermal dissipation management system, to effectively absorb and transfer heat released in the charge and discharge process, to reduce absolute temperature rise of the battery pack (up to 10 °C), to significantly narrow temperature range and coefficient of variation at three points, namely, on both ends and in the middle of each battery cell, and to effectively ensure normality and uniformity of working temperature of the battery pack. Those outcomes imply that the PCMs thermal management system may give more obvious effect if sufficient room between battery cells for filling in the PCMs (for example, the interval between battery cells is 5 mm [6]) can be achieved with the further improvement of encapsulation structure of the battery pack and microencapsulated PCMs.

In the research, the microencapsulated PCMs used for the battery thermal management system have a high thermal decomposition temperature

(up to 180 °C) and large specific surface area. Under the condition of phase changing, there are no problems of volume change and liquid flow for PCMs. Such advantages mentioned above make microencapsulated PCMs convenient to be encapsulated with the battery cells. Therefore, microencapsulated PCMs are expected to become a promising material in the design of novel power battery pack thermal management system.

CORRESPONDING AUTHOR

Shuhua Li, E-mail address: lish339@163.com, Tel.: 13688811339.

REFERENCES

[1] Luo Yutao & He Xiaochan: A Research on the Affecting Factors of Thermal Safety in Lithium-ion Power Battery [J]. Automotive Engineering, 2012 (Vol. 34) No. 4, p. 333–338.

[2] Siddique A, Khateeba: Thermal management of Li-ion battery with phase change material for electric scooters: experimental validation [J]. Journal of Power Sources, 142 (2005), p. 345–353.

[3] S. Al Hallaj and J.R. Selman: A Novel Thermal Management System for Electric Vehicle Batteries Using Phase-Change Material [J]. Journal of The Electrochemical Society, 147 (9), p. 3231–3236 (2000).

[4] Rami Sabbah, R. Kizilel, J.R. Selman, S. A1-Hallaj: Active (air-cooled) vs. passive (phase change material) thermal management of high power Lithium-ion packs: Limitation of temperature rise and uniformity of temperature distribution [J]. Journal of Power Sources, 182 (2008) 630–638.

[5] R. Kizilel, A. Lateef, R. Sabbah, M.M. Farid, J.R. Selman, S. Al-Hallaj: Passive Control Of Temperature Excursion And Uniformity In High-energy Li-ion Battery Packs At High Current And Ambient Temperature [J]. Journal of Power Sources, 183 (2008), p. 370–375.

[6] Riza Kizilel, Rami Sabbah, J. Robert Selman, Said Al-Hallaj: An alternative cooling system to enhance the safety of Li-ion battery packs [J]. Journal of Power Sources, 194 (2009), p. 1105–1112.

[7] Zhang Guoqing, Rao Zhonghao, Wu Zhongjie, Fu Lipeng: Experimental investigation on the heat dissipation effect of power battery pack cooled with phase change materials [J]. Chemical Industry and Englneerlng Progress, 2009 (Vol 28) No. 1, p. 23–26, 40.

[8] Siddique A. Khateeb, Shabab Amiruddin, Mohammed Farid, J. Robert Selman, Said A1-Hallaj: Thermal management of Li-ion battery with phase change material for electric scooters: experimental validation [J]. Journal of Power Sources, 142 (2005), p. 345–353.

[9] W.Q. Li, Z.G. Qu, Y.L. He, Y.B. Tao: Experimental study of a passive thermal management system for high-powered lithium ion batteries using porous metal foam saturated with phase change materials [J]. Journal of Power Sources, 255 (2014), p. 9–15.

[10] Wenhong Li, Guolin Song, Shuhua Li, Youwei Yao, Guoyi Tang: Preparation and characterization of novel MicroPCMs (microencapsulated phase-change materials) with hybrid shells via the polymerization of two alkoxy silanes [J]. Energy, 70 (2014), p. 298–306.

Advances in Future Manufacturing Engineering – Yang (Ed.)
© 2015 Taylor & Francis Group, London, ISBN 978-1-138-02817-3

The application of wavelet analysis in signal detection to vehicle Anisotropic Magnetoresistance sensor

Tao Lin
School of Automation, Chongqing University, Chongqing, China
School of Applied Electronics, Chongqing College of Electronic Engineering, Chongqing, China

Peng Wu
School of Automation, Chongqing University, Chongqing, China
Chongqing Chuanyi Automation Co. Ltd., Chongqing, China

Fengmei Gao
School of Biomedical Engineering, Xinxiang Medical University, Xinxiang, Henan, China

Linhong Wang
School of Applied Electronics, Chongqing College of Electronic Engineering, Chongqing, China

Yi Yu
School of Biomedical Engineering, Xinxiang Medical University, Xinxiang, Henan, China

ABSTRACT: To detect the signal of the vehicle Anisotropic Magnetoresistance (AMR) sensor, the Fast Fourier Transform (FFT) method cannot reflect the corresponding relation between the time domain and frequency domain, so the wavelet analysis method is proposed for the filtering precondition and the singularity fault detection. The experimental results show that the proposed method can effectively detect the fault time point of the vehicle AMR sensor signal and improve the reliability of the whole vehicle detection system based on AMR sensor.

Keywords: wavelet analysis; vehicle; anisotropic magnetoresistance sensor; signal detection

1 INTRODUCTION

The Anisotropic Magnetoresistive (AMR) sensor is a kind of magnetic field detection sensor, makes according to the anisotropic magnetoresistance effect principle. It can not only be used in testing the existence of the direct current static magnetic field, but also can determine the intensity and direction change of the magnetic field. With the support of the Permalloy (Ni-Fe) semiconductor technology, it provides better sensitivity, better cost-performance and smaller size than those traditional mechanical or other electronic magnetic field sensors [1]. The earth magnetic field is stable in a relatively wide area (about a few square kilometers), and a vehicle can be understood as a model composed of a lot of bipolar magnets with N-S polarization direction. A vehicle's static or movement can cause the disturbance of the local earth magnetic field which is as shown in Figure 1, so the AMR sensor has been widely used for detecting the presence, the operating speed and the size

of various vehicles to achieve analysis, control and management functions of the vehicles information [1, 2].

The core of the AMR sensor is usually made of Wheatstone bridge composed of 4 pieces of banded Permalloy thin film, whose bridge arm resistance

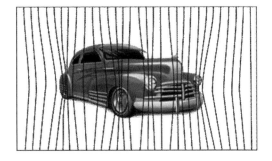

Figure 1. The disturbance of the local earth magnetic field caused by a vehicle.

values change with the external magnetic field and the corresponding voltage output signal can be provided. The AMR sensor is often used on the highway with heavy interference, because of the severe operating condition, the filtering precondition and the fault detection are crucial to the performance of the whole vehicle detection system [3].

The non-stationarity and singularity of the signal are important information reflecting the state and fault of the sensor. If the traditional FFT method and the time sequence theory are used, there are only the overal statistical average results of the signals. And it is impossible to reflect the corresponding relation between the time domain and frequency domain of the local non-stationary signal. But the wavelet analysis with the time and frequency characteristic, can reflect the signal spectrum variation with time and the frequency distribution near some time point, and is suitable for the detection of the transient abnormal information and the show of the time-frequency composition for the sensor signal [4, 5]. Therefore, the wavelet transform used to signal processing and analysis for the AMR sensor signal detection.

2 THE FILTERING PRECONDITION BASED ON THE WAVELET ANALYSIS TO THE AMR SENSOR SIGNAL

2.1 The principle of the filtering precondition based on the wavelet analysis

The signal denoising filter is essentially the process of decreasing void parts and improving value in the signal. The model of the one-dimensional signal with noise has the following general expression

$$S(k) = f(k) + \sigma \cdot e(k) \quad k = 0, 1, \ldots, n-1 \quad (1)$$

where, $S(k)$ is the signal with noise, $f(k)$ is a valuable signal, and $e(k)$ is a noise signal. This assumes that $e(k)$ is a Gaussian white noise signal $N(0, 1)$ which usually takes the form of high-frequency signal. In engineering practice, $f(k)$ may also contain many spikes or singularities, and the noise signal $e(k)$ isn't necessarily stationary noise [5, 6]. When analyzing and processing this kind of signal, the first step is the precondition to remove the noise and extract the valuable signal.

For non-stationary signal denoising, the traditional Fourior analysis method is powerless. The Fourior analysis method is to analyse the signal after converting it from the time domain to the frequency domain, and cannot give expression to the frequency variation at a given time, because the singularity at any point on the timeline can affect the whole frequency spectrum for the signal;

nevertheless, the wavelet analysis method can analyze the signal in both time field and frequency field at the same time. For a brief moment of the signal generating, the modulus maximum of the wavelet coefficient will appear and get bigger with the decomposing scale increase, reach the peak eventually. To the negative singularity of the white noise, the modulus maximum and the density of the wavelet coefficient will get smaller with the decomposing scale increase [5–7]. Using this opposite feature, the signal and noise can be separated from mixed signal by the wavelet transformation. Therefore, using the wavelet analysis method, the singularity and noise in the signal can be distinguished in order to denoise for the non-stationary signal.

2.2 The denoising method based on the wavelet analysis to the AMR sensor signal

Usually in the signal denoising process using the wavelet, the signal is first decomposed by wavelet transform. Because most of the noise signals are included in high frequency details, the wavelet coefficients obtained by decomposition can be processed through threshold value and other various forms, then the signal is reconstruct by wavelet transform in order to denoise. Thus the denoising process to the one-dimensional AMR sensor signal is divided into the following three steps [6, 7]:

Step 1. The wavelet decomposes of the one-dimensional AMR sensor signal. In order to obtain better denoising, not only appropriate wavelet function must be selected, but also the best decomposition level number N should be determined, then the signal $S(k)$ is decomposed with N levels wavelet.

Step 2. The threshold quantization of wavelet decomposes high frequency coefficients. Each high frequency coefficient from level 1 to level N is quantified using threshold value selected by soft threshold method.

Step 3. The wavelet reconstruct of the one-dimensional AMR sensor signal. According to the level N low frequency coefficient and the high frequency coefficients from level 1 to level N, the one-dimensional AMR sensor signal is reconstructed.

The normal AMR sensor signal has better continuity. In addition, the regularity of Symlets function is poor in wavelet family, so the signal is suppressed and the noise is separated out from the signal with noise. Thus, Sym8 wavelet in Symlets wavelet family is selected to achieve the level 5 filtering and denoising of the Honeywell HMC102 AMR sensor signal gathered. The wavelet decompose algorithm

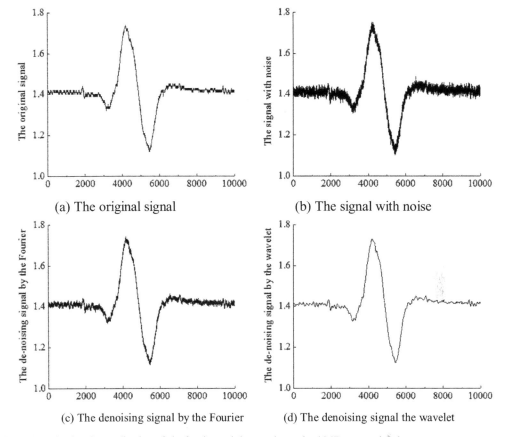

(a) The original signal

(b) The signal with noise

(c) The denoising signal by the Fourier

(d) The denoising signal the wavelet

Figure 2.　The denoise application of the fourier and the wavelet to the AMR sensor signal.

adopted is Mallat decompose algorithm, whose algorithm formula is

$$\begin{cases} C_{j+1,k} = \sum_m h(m-2k)\, C_{j,m} \\ D_{j+1,k} = \sum_m g(m-2k)\, D_{j,m} \end{cases} \quad (2)$$

where h, g are the scaling coefficient and the wavelet coefficient respectively; $C_{j+1,k}$, $D_{j+1,k}$ are smoothing coefficient and detail coefficient at level $j+1$.

The wavelet decompose high frequency coefficient is quantified by Heursure optimum forecasting variable soft threshold method, and the Honeywell HMC102 AMR sensor signal is reconstructed. Mallat reconstruction algorithm formula is

$$C_{j-1,k} = \sum_n \left[h_{k-2n} C_{j,n} + g_{k-2n} D_{j,n} \right] \quad (3)$$

As can be seen from the Figure 2, the original signal is recovered mainly from the Honeywell HMC102 AMR sensor signal with noise, and the interference caused noise is also eliminated.

3　THE FAULT DETECT BASED ON THE WAVELET ANALYSIS TO THE AMR SENSOR SIGNAL

3.1　The principle of the singularity signal detect based on the wavelet analysis

The theoretical basis of singularity signal detection based on the wavelet analysis is the relationship between the wavelet analysis and the signal singularity. The signal singularity is represented by the Lipschitz index. The singular point of the signal function is detected from the modulus maximum of the detail signals by the wavelet transformation, and the modulus maximum appears at the point where the signal singularity lies [7]. In the practical application, the fault detection is accomplished according to the extreme point of the wavelet transformation.

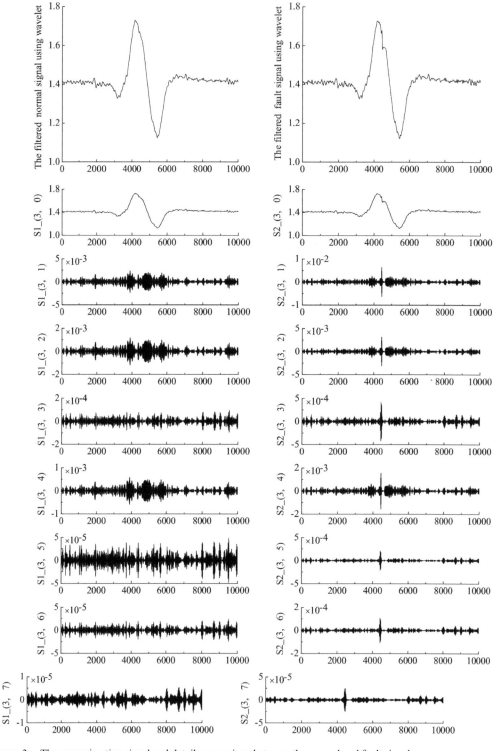

Figure 3.　The approximation signal and detail comparison between the normal and fault signal.

Definition 1. Let n is non-negative, $n < \alpha \leq n+1$, if there are two constants A and h_0 ($A > 0$, $h_0 > 0$), and a polynomial of degree n $p_n(h)$, when $h \leq h_0$,

$$| f(x_0 + h) - p_n(h) | \leq A | h |^\alpha \quad (4)$$

If above equation is true to x_0, $x_0 + h \in (a,b)$, then $f(x)$ is consistent Lipschitz α in (a,b). If $f(x)$ is not Lipschitz 1 at x_0, then $f(x)$ is singularity at x_0. The upper bound of α is Lipschitz index of $f(x)$ at x_0. If $f(x)$ is discontinuous but bounded at x_0, then its Lipschitz index is zero. The bigger Lipschitz index is and the smoother $f(x)$.

If in a certain neighborhood of x_0, there is $|Wf(s_0,x)| \leq |Wf(s_0,x_0)|$, then (s_0,x_0) is called the modulus maximum point of the wavelet transformation. If a point on a curve in two dimensional phase space (s,x), then the curve is called maximum line. The relation between the wavelet modulus value on the maximum line and Lipschitz index of $f(x)$ at x_0 is $|Wf(s,x)| \leq ks^\alpha$, k is a positive integer. When $s \to 0$, $x \to x_0$ is on the maximum line [7–10].

As a result of those conclusions mentioned above, if the maximum line in two-dimensional phase space can be obtained, according to its modulus change rate and the progress of time axis coordinate, x_0 and α of corresponding singularity point can be determined.

3.2 *The fault detect method based on the wavelet analysis to the AMR sensor*

When the response tracking performance of the AMR sensor is poor, its signal will appear discontinuous, that is singularity. Using the wavelet analysis, the singularity point and the singularity characteristic of the AMR sensor can be captured in order to take corresponding measures and improve the reliability of the AMR sensor system.

The normal and fault signal of Honeywell HMC102 AMR sensor are decomposed level 3 respectively using Db1 wavelet in Daubechies wavelet family. The approximation signal and detail comparison in level 3 are shown in Figure 3. As can be seen from Figure 3, the errors between 4000 and 5000 caused from the AMR sensor fault are clearly presented.

4 SUMMARY

The FFT signal detect method cannot reflect the corresponding relation between the time domain and frequency domain, the wavelet analysis method was used for the filtering precondition and the singularity fault detection of the vehicle AMR sensor signal with noise. Through detail decomposition, the singularity fault of the vehicle AMR sensor signal was located accurately and the reliability of the vehicle AMR sensor system was improved. If the decomposition level is further increased and the more suitable wavelet function is selected, the effect of the denoising and the singularity fault detection is more significant.

ACKNOWLEDGMENT

This research was partially supported by the Natural Science Foundation of China under the contract number 61305147 and the Natural Science Foundation of Chongqing under the contract number cstc2012jjA10129.

REFERENCES

[1] Hans. Hauser, Günther Stangl and Johann Hochreiter: High-performance magnetoresistive sensors. Sensors and Actuators A: Physical, Vol. 81 (2000), p. 27–31.

[2] Q. Chen, H.P. Tao and Zh. M. Wang: Design of Wireless Vehicle Detector Based on Magnetoresistive Sensor. Transducer and Microsystem Technologies. Vol. 30 (2011), p. 82–85.

[3] Information on http://masters.donntu.edu.ua /2007/ kita/gerus/ library/amr.pdf.

[4] D.P. Jena, Sudarsan. Sahoo and S.N. Panigrahi: Gear fault diagnosis using active noise cancellation and adaptive wavelet transform. Measurement. Vol. 47 (2014), p. 356–372.

[5] P.K. Kankar, Satish C. Sharma and S.P. Harsha: Fault diagnosis of ball bearings using continuous wavelet transform. Applied Soft Computing. Vol. 11 (2011), p. 2300–2312.

[6] L.T. Nie and K. Zhang: The Denoising Method to One-dimensional Noisy Signal Based on Wavelet Analysis. Journal of Sichuan Ordnance. Vol. 32 (2011), p. 86–88.

[7] Y.J. Li: Application of wavelet analysis to fault diagnosis for machinery. Electronic Measurement Technology. Vol. 30 (2007), p. 42–44.

[8] D.F. Zhang: The wavelet analysis and engineering application (National Defence Industry Press, China 2008).

[9] W.L. Jiang, Sh.Q. Zhang and Y.Q. Wang: Chaos and Wavelet Based Fault information Diagnosis (National Defence Industry Press, China 2005).

[10] Ingrid Daubechies: Ten Lectures on Wavelets. (Society for Industrial and Applied Mathematics, America 1992).

Advances in Future Manufacturing Engineering – Yang (Ed.)

A fault diagnosis model of power transformer using Imperialistic Competitive Algorithm and Support Vector Machine

Yiyi Zhang
Guangxi Key Laboratory of Power System Optimization and Energy Technology, Guangxi University, Nanning, Guangxi, China

Zhihua Zhong
Archives of Guangxi University, Nanning, Guangxi, China

Hanbo Zheng
State Grid Henan Electric Power Research Institute, Zhengzhou, Henan, China

Chao Tang
College of Engineering and Technology, Southwest University, Chongqing, China

Yan Sun
Guangxi Power Grid Power Dispatching and Control Center, Nanning, Guangxi, China

ABSTRACT: A fault diagnosis model using an Imperialistic Competitive Algorithm (ICA) and SVM integrated approach was established in the paper. ICA was applied to search and obtain the parameters of SVM. Then, cross validation and normalizations were applied in the training processes of SVM, and the trained SVM model with the selected parameters was established to conduct fault diagnosis. The results demonstrate that the proposed approach promotes the fault diagnosis accuracy of transformers.

Keywords: power transformer; fault diagnosis; Support Vector Machine (SVM); Imperialistic Competitive Algorithm (ICA); DGA

1 INTRODUCTION

Power transformers are considered to be the most essential equipment in a power system. A fault in a transformer can result in not only substantial repair cost but also power interruptions to thousands of customers. Therefore, it is essential to assess the working condition and conduct fault diagnosis of power transformers[1].

In the past few years, Dissolved Gas Analysis (DGA) is a common fault diagnosis practice. When an oil-filled power transformer suffers from electrical or thermal stress, detectable gasses including hydrogen (H_2), methane (CH_4), acetylene (C_2H_2), ethylene (C_2H_4) and ethane (C_2H_6) are emitted from the transformer[2]. Many approaches such as IEEE standard[3], International Electrotechnical Commission (IEC) Method 60599[4], Duvals Triangle[5], are based on the ratios of the gases diagnosis. The main drawback of these approaches is that the observed gas ratios do not match all the fault types[6]. As an effective method to resist the problems of over-fitting and local optimization,

the SVM related methods are wildly applied in the fault diagnosis problem. Generally, the classification accuracy of the SVM is largely determined by the two important parameters, punish coefficient C and kernel parameter γ. In fact, the two parameters of SVM are usually uncertain, therefore they should be optimized by an effective optimization algorithm. Some methods such as clonal selection algorithms and GA, were used to optimize the two parameters. However, the optimized result of the two parameters is sometimes in a suboptimal condition and hard to obtain the satisfactory classification result[7]. Therefore, how to obtain an effective optimization algorithm to determine the two parameters becomes a hot topic in the SVM related research.

ICA, firstly introduced by Esmaeil Atashpaz Gargari in 2007[8], is an effective optimization algorithm because of its ability of the efficient local search control and good convergence solution to the global optimum. Therefore, in this paper, a multiclass ICA-SVM integrated approach was proposed to find out whether the approach is an effective method to fault diagnosis of power transformers.

The organization of this paper is established as follows. The brief introduction of transformer fault diagnosis is presented in the first Section. The processes of ICA are briefly described in Section 2. The integrated ICA-SVM approach is developed to solve clustering and optimization problems in Sections 3. The proposed algorithm is applied to fault diagnosis of power transformers in Sections 4. The results conclusions are given in Section 5.

2 IMPERIALIST COMPETITIVE ALGORITHM

The ICA optimizes the objective function through the concept of imperialistic competition. It was first introduced by Atashpaz and Lucas in the year of 2007, and recently it has been used as an effective algorithm to solve continuous-optimization problems [9]-[10].

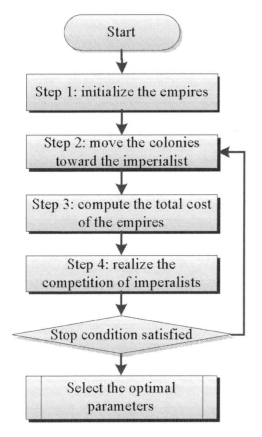

Figure 1. Flowchart of optimal parameters selection with ICA.

ICA starts with an initial random population called countries. The countries are the feasible solutions in the search space. Some of the powerful countries in the population are selected to be the imperialists, and the rest countries form the colonies of these imperialists. A imperialist and its colonies are called an empire. In ICA, N initial countries are expressed by an array of dimension N and defined as $[p_1, p_2 \dots p_n]$. The cost of the ith country s_i is calculated by the cost function F with the variables $(p_1, p_2 \dots p_n)$:

$$s_i = F(country_i) = F(p_{i1}, p_{i2}, ..., p_{iN}) \qquad (1)$$

In the paper, some powerful countries (the countries with minimum cost), N_{imp} countries, were chosen to be the imperialists from the N initial countries. The remaining countries are colonies that belongs to the imperialists. The goal of ICA processes is to obtain the most powerful country with minimum cost. The flowchart of parameter optimization using ICA, as shown in Figure 1, is described in the following steps. The details of the ICA can be seen in [11].

3 CLASSIFICATION USING MULTICLASS SVM INTEGRATED ICA

As ICA is an effective approach to seek for the best solution of an optimal problem, ICA is applied to select the best parameters of SVM (C and γ) for obtaining the best classification accuracy.

To form the optimal objective function, the traditional SVM (linear—two classification) should first be replaced by a nonlinear-multiple classification SVM. Two steps should be done as follows.
1. To be the nonlinear SVM

Assuming that $\{(x_1, y_1) \dots (x_i, y_i) \dots (x_l, y_l)\}$ is a training set (including input data $x_i \in \mathbf{R}^n$ and output $y_i \in \mathbf{R}$) and a class label $y_i \in \{-1, +1\}$ is established by input x_i. If the constraints satisfy the following formula (2), the training data can be considered as a separable set [12].

$$\begin{cases} \omega^T \varphi(x_i) + b \le -1, & \text{if } y_i = -1 \\ \omega^T \varphi(x_i) + b \ge +1, & \text{if } y_i = +1 \end{cases} \qquad (2)$$

The optimal hyperplane is described by the vector ω with the bias b. $\varphi(x_i)$ is defined as the nonlinear mapping.

Finally, the nonlinear two-class classifier can be expressed as (3).

$$y(x) = \text{sign}\left[\sum_{i=1}^{l} \alpha_i y_i K(x, x_i) + b\right] \qquad (3)$$

The One Against One (OAO) scheme is applied in this paper because the OAO combination scheme is considered as a more suitable method than others[11].

With fewer parameters to set, but excellent overall performance, the RBF is an effective option for the kernel function [12][13]. Therefore, an RBF kernel function applied in this paper is shown as follow:

$$K(x_i, x_j) = \exp(-\gamma \| x_i - x_j \|^2), \gamma > 0 \qquad (4)$$

where γ represents a parameter inversely proportion to the width of the RBF and γ is a free parameter in the RBF kernel function. In addition two major parameters, C and γ need to be predefined appropriately.

The lowest value of target function is obtained, and the best parameters C and γ can be calculated by the inverse function of the Uncertain Function (UF).

$$\mathrm{Target} = -\frac{1}{m} \sum_{i=1}^{m} \left(\frac{p_T^i}{p^i} \times 100\% \right) = UF(c, \gamma)$$
$$= s_i = F(country_i) = F(p_{i1}, p_{i2}) \qquad (5)$$

where, m represents the number of folds in the cross validation. p^i is the number of validated subsets, and p^i_T is the number of correctly classified samples in this subset. UF is the uncertain function.

Since the minimal target value will outperform other parameters, the goal is to minimize the value of the target function which is usually related to the accuracy of the classification.

In the end, the optimal parameters are determined by the processes of ICA. The steps can be summarized as follows [8]:

Step 1: Initialize the ICA parameters of the empires, and generate the target function for ICA problem.

Step 2: Move the colonies toward their relevant imperialist.

Step 3: Exchange positions between the imperialist and the colony.

Step 4: Calculate the cost of the empires.

Step 5: Realize the competition of the ICA, and get the information of the remaining empire.

Step 6: Terminate the algorithm if the selected condition is satisfied and select the optimal parameters.

4 CLASSIFICATION AND VERIFICATION

Generally, the fault diagnosis of power transformer is considered as a clustering problem which is suitable for applying the proposed algorithm. To obtain more fault types for modeling, this study collected transformer dissolved gas data from several electric power companies. After that, the multiclass SVM integrated ICA model was applied to fault diagnosis of transformers.

The samples were then randomly divided into two parts including training sample set (243 samples) and testing sample set (27 samples). A set of five commonly used fault-related gases, i.e. H_2, CH_4, C_2H_2, C_2H_4 and C_2H_6, were chosen as the input of the classifier. All input data were normalized in the range of [0 1] before training to improve the generalization performance of SVM.

There is only one imperialist at the 16th generation. The unique imperialist and its colonies formed a unique empire. In that case, the parameters selection processes could be considered as "excellent". The cost of the imperialist is the minimum value of the target function. The value of the target function was −83.95. The elapsed time of fault diagnosis is about 17.46 seconds. Finally, the best parameters obtained from ICA algorithm for the multiclass SVM classifier are $C = 62.3245$, $\gamma = 0.0187$, respectively.

With the selected parameters, a training of the multiclass SVM classifier was applied to the fault diagnosis of the transformer. 243 samples of transformers were divided into five classes by actually condition including low-energy discharge (LE-D, class 1), high-energy discharge (HE-D, class 2), thermal fault of low and medium temperature (LM-T, class 3), thermal fault of high temperature (HT, class 4), and normal condition (Normal, class 5). After cross validation (10-fold) and nominalization, the classification accuracy of training samples hit 90.535% (220/243). In addition, 27 test samples are employed to test the trained model and only the 13th sample and the 27th sample are misclassified and the classification accuracy of test samples arrives at a high rate of 92.5926% (25/27).

5 CONCLUSIONS

A one-against-one multiclass SVM model integrated ICA for transformer fault diagnosis was presented in this paper. The SVM integrated ICA method had achieved the best performance in the transformer fault diagnosis in both classification accuracy and generalization performance. Consequently, the proposed method is a reliable tool for further fault analysis of power transformer and also an effective approach for pattern recognition problems.

ACKNOWLEDGEMENT

The project is funded by "Power Transformer Reliability and Economical Life Assessment and

Its Maintenance and Decommissioning Strategy Research", "Study of Power Flow Control Form of Electromagnetic Ring Circuit in Guangxi Grid" project (K-GX2013–104), the Fundamental Research Funds for the Central Universities (Grant No. XDJK2014B031), and "The National Natural Science Foundation of China" (Project for Young Scientists Fund, Grant No. 51107103).

REFERENCES

[1] Ma, Hui; Saha, Tapan K.; Ekanayake, Chandima: "Statistical Learning Techniques and Their Applications for Condition Assessment of Power Transformer", *IEEE Transactions on Dielectrics and Electrical Insulation*, Vol. 19, No. 2, pp. 481–489 (2012).

[2] James, R.E., Su, Q.: "Condition assessment of high voltage insulation in power system equipment", *The Institution of Engineering and Technology, London (2008)*.

[3] IEEE, "IEEE guide for the interpretation of gases generated in oil-immersed transformers", *IEEE Std. C57.104*, (2008).

[4] IEC, "Guide to the interpretation of dissolved and free gases analysis", *IEC Std. 60599*, (1999).

[5] Duval, M., Pablo A.: "Interpretation of oil in gas analysis using new IEC publication 60599 and IEC TC 10 databases". *IEEE Electrical Insulation Magazine*. Vol. 17, No. 2, pp. 31–41 (2001).

[6] CIGRE working group 09 of study committee 12. "Life time evaluation of transformer". *Electra*. No. 150, pp. 39–51 (1993).

[7] Shintemirov, A., Tang, W.H., Wu, Q.H.: "Power transformer fault classification based on dissolved gas analysis by implementing bootstrap and genetic programming", *IEEE Trans. Syst. Man Cybern. C, Appl. Rev.,* Vol. 39, No. 1, pp. 69–79 (2009).

[8] Bazi, Y., Melgani, F.: "Toward an optimal SVM classification system for hyperspectral remote sensing images", *IEEE Trans. Geos. Remote Sens.,* Vol. 44, No. 11, pp. 3374–3385 (2006).

[9] Gargari, E.A. Lucas, C.: "Imperialist Competitive Algorithm: An Algorithm for Optimization Inspired by Imperialistic Competition", *IEEE Congress on Evolutionary Computation (CEC 2007),* pp. 4661–4667 (2007).

[10] Lucas, C., Nasiri-Gheidari, Z. and Tootoonchian, F. "Application of an imperialist competitive algorithm to the design of a linear induction motor", Energy Conversion and Management, pp. 1408–1411 (2010).

[11] Moghimi, H.M., Vahidi, B.: "A solution to the unit commitment problem using imperialistic competition algorithm", *IEEE Trans on Power Syst.,* Vol. 27, No. 1, pp. 117–124 (2012).

[12] Hsu, C.W., Lin, C.J.: "A comparison of methods for multiclass support vector machines", *IEEE Trans. Neural Netw.,* Vol. 13, No. 2, pp. 415–425 (2002).

[13] Melgani, F., Bazi, Y.: "Classification of electrocardiogram signals with support vector machines and particle swarm optimization", *IEEE Transactions on Information Technology in Biomedicine,* No. 5, pp. 667–677 (2012).

Advances in Future Manufacturing Engineering – Yang (Ed.)
© 2015 Taylor & Francis Group, London, ISBN 978-1-138-02817-3

Evaluating the efficiency of Renewable Energy technologies by DEA and TOPSIS method with interval data

Xiao-guang Qi & Bo Guo
College of Information System and Management, National University of Defense Technology, Changsha, China

ABSTRACT: Renewable Energy (RE) technology is an important solution to the large atmospheric pollution caused by the fossil fuel combustion. Efficiency evaluation would be an important issue during the development of RE technologies. In this paper, we use a comprehensive methodology for the efficiency evaluation of RE technologies which is combined with the DEA and TOPSIS method. The efficiency of RE technologies is first evaluated by the cross-efficiency DEA model and a cross-efficiency matrix between these RE technologies can be got. Then we propose a modified TOPSIS method for the aggregation of these cross-efficiency scores. Interval data is also considered in our proposed methodology which would be necessary to represent the imprecise information during the efficiency evaluation. A numerical example is provided to examine the validity and effectiveness of our proposed method.

Keywords: renewable energy technology; efficiency; data envelopment analysis; TOPSIS; interval data

1 INTRODUCTION

The development of Renewable Energy (RE) technology is an important solution to the large atmospheric pollution caused by the fossil fuel combustion [1]. During the development of RE technologies, efficiency would be a key managerial concept for the decision makers [1]. Because the development potential of a RE technology is related with both its performance and also its development costs. Therefore efficiency evaluation of RE technologies would be necessary for the technology selection and development.

Data Envelopment Analysis (DEA) [2] is an effective method of evaluating the efficiency of Decision Making Units (DMUs) with multiple inputs and outputs. Mukherjee [3] proposed several alternative DEA models for evaluating the energy use efficiency in the manufacturing sector of US. Sueyoshi et al. [4] proposed a DEA based approach and three efficiency measures to evaluate the performance of US coal-fired power plants. More recently, Nouri et al. [5] used DEA approach in the energy efficiency analysis in the vegetable oil industry of Iran and a multi-stage DEA model was proposed. Vlontzos et al. [6] also proposed a DEA approach for estimating both the agricultural energy and environmental efficiency of the EU member state countries. Similarly, Wang et al. [7] proposed a directional meta-frontier DEA approach for evaluating the

energy efficiency and energy saving potential in China.

In this paper, we propose a comprehensive methodology for the efficiency evaluation of RE technologies which is combined with the DEA and TOPSIS method. The efficiency of RE technologies is first calculated by using the cross-efficiency DEA model and a cross-efficiency score matrix can be got between DMUs. Then we propose a modified TOPSIS method to aggregate these cross-efficiency scores together as the final evaluation measure. Interval data is also considered in our proposed methodology. Interval data is necessary when it was difficult to get the exact data of RE technologies.

The rest of this paper is organized as follows: in section 2, some preliminaries are introduced; in section 3, our proposed methodology is formulated in detail; in section 4, a numerical example is provided to examine the validity and effectiveness of our proposed method; finally, in section 5, we give the conclusions.

2 PRELIMINARY

2.1 Data Envelopment Analysis

In a DEA problem, it is supposed that there are n DMUs (such as the Renewable Energy technologies) which consume m inputs to produce s outputs, and the vectors $x_j = [x_{1j}, x_{2j}, ..., x_{mj}]^T$ and $y_j = [y_{1j}, y_{2j}, ..., y_{sj}]^T$ are used to denote the inputs and outputs of $DMU_j (j = 1, 2, ..., n)$. The basic efficiency model for $DMU_{j_0} (j_0 = 1, 2, ..., n)$ is named as CCR model as follows [1]:

$$\text{Max } h_{j_0} = (u_{j_0})^T y_{j_0}$$

s.t.

$$(u_{j_0})^T y_j - (v_{j_0})^T x_j \leq 0, j = 1, 2, ..., n \quad (1)$$

$$(v_{j_0})^T x_{j_0} = 1$$

$$u_{j_0}, v_{j_0} \geq 0$$

In which $u_{j_0} = [u_{1 j_0}, u_{2 j_0}, ..., u_{s j_0}]^T$ and $v_{j_0} = [v_{1 j_0}, v_{2 j_0}, ..., v_{s j_0}]^T$ are the optimal weights assigned to the inputs and outputs respectively, and h_{j_0} is the efficiency score of DMU_{j_0}.

2.2 Cross-efficiency evaluation

Cross-efficiency evaluation is an important extension of the DEA method [8]. The cross-efficiency evaluation is based on the concept of peer-evaluation in which the target DMU is evaluated not only by the favorable optimal weights of itself but also by the favorable optimal weights of the other DMUs. After model (1) has been calculated for every DMU, there would be a set of optimal weighs for each DMU and the cross-efficiency of $DMU_{j_0} (j_0 = 1, 2, ..., n)$ using the optimal weights of $DMU_{j_1} (j_1 = 1, 2, ..., n$ is defined as follows:

$$k_{j_0 j_1} = \frac{(u_{j_1})^T y_{j_0}}{(v_{j_1})^T x_{j_0}} \quad (2)$$

After the cross-efficiency scores has been got for each DMU, there would be a cross-efficiency matrix and different methods can be used to aggregate these cross-efficiency scores into a single efficiency score as the final efficiency measure. In this paper, we propose a modified TOPSIS method as the aggregation method and the details of our modified TOPSIS method will be formulated in the following sections.

3 PROPOSED METHODOLOGY

3.1 Cross-efficiency evaluation with interval data

Traditional DEA models are mostly based on an assumption that all the inputs and output data are exact values. However it is sometimes difficult to get exact values in practice. Therefore interval data is introduced into DEA method to represent the imprecise information during the efficiency evaluation. Interval data is usually defined by the lower bound and the upper bound as follows:

$$x = [x^L, x^U] \quad (3)$$

Considering the interval data in DEA method, the interval inputs and interval outputs are denoted by $x_j = [x_j^L, x_j^u]$ and $y_j = [y_j^L, y_j^U]$ respectively. Wu et al. [9] proposed a pair of cross-efficiency models considering the interval inputs and outputs as follows:

$$\text{Max } k_{j_0 j_1}^L = (u_{j_1})^T y_{j_0}^L$$

s.t.

$$(u_{j_1})^T y_j^U - (v_{j_1})^T x_j^L \leq 0, j = 1, 2, ..., n$$

$$(u_{j_1})^T y_{j_1}^L - h_{j_1}^L (v_{j_1})^T x_{j_1}^U = 0 \quad (4)$$

$$(v_{j_1})^T x_{j_0}^U = 1$$

$$u_{j_0}, v_{j_0} \geq 0$$

$$\text{Max } k_{j_0 j_1}^U = (u_{j_1})^T y_{j_0}^U$$

s.t.

$$(u_{j_1})^T y_j^U - (v_{j_1})^T x_j^L \leq 0, j = 1, 2, ..., n$$

$$(u_{j_1})^T y_{j_1}^U - h_{j_1}^U (v_{j_1})^T x_{j_1}^L = 0 \quad (5)$$

$$(v_{j_1})^T x_{j_0}^L = 1$$

$$u_{j_0}, v_{j_0} \geq 0$$

in which $h_{j_1}^L$ and $h_{j_1}^U$ are calculated by the following two models:

$$\text{Max } h_{j_1}^L = (u_{j_1})^T y_{j_1}^L$$

s.t.

$$(u_{j_1})^T y_j^U - (v_{j_1})^T x_j^L \leq 0, j = 1, 2, ..., n \quad (6)$$

$$(v_{j_1})^T x_{j_1}^U = 1$$

$$u_{j_0}, v_{j_0} \geq 0$$

$$\text{Max } h_{j_1}^U = (u_{j_1})^T y_{j_1}^U$$

s.t.

$$(u_{j_1})^T y_j^U - (v_{j_1})^T x_j^L \leq 0, j = 1, 2, ..., n \quad (7)$$

$$(v_{j_1})^T x_{j_1}^L = 1$$

$$u_{j_0}, v_{j_0} \geq 0$$

3.2 A modified TOPSIS method

After solving model (4) and model (5) for every DMU, there would be a cross-efficiency matrix

K in which the cross-efficiency is also interval, as follows:

$$K = \begin{bmatrix} \left[k_{11}^L, k_{11}^U\right] & \left[k_{12}^L, k_{12}^U\right] & \cdots & \left[k_{1n}^L, k_{1n}^U\right] \\ \left[k_{21}^L, k_{21}^U\right] & \left[k_{22}^L, k_{22}^U\right] & \cdots & \left[k_{2n}^L, k_{2n}^U\right] \\ \vdots & \vdots & & \vdots \\ \left[k_{n1}^L, k_{n1}^U\right] & \left[k_{n2}^L, k_{n2}^U\right] & \cdots & \left[k_{nn}^L, k_{nn}^U\right] \end{bmatrix} \quad (8)$$

The TOPSIS method was first proposed by Hwang et al. [10] as an effective method of ranking multiple alternatives with multi-objectives. The main idea of TOPSIS method is ranking the alternatives by calculating their distances to the positive ideal solution and to the negative ideal solution. For a certain alternative, if its distance to the positive ideal solution is the shortest and its distance to the negative ideal solution is the longest, then it would be the best alternative [10].

In this section, we introduce a TOPSIS method different from [9] and [11] for the aggregation of interval cross-efficiency scores as the following steps:

Step 1: Calculating the normalized decision matrix E. As follows:

$$E = \begin{bmatrix} \left[e_{11}^L, e_{11}^U\right] & \left[e_{12}^L, e_{12}^U\right] & \cdots & \left[e_{1n}^L, e_{1n}^U\right] \\ \left[e_{21}^L, e_{21}^U\right] & \left[e_{22}^L, e_{22}^U\right] & \cdots & \left[e_{2n}^L, e_{2n}^U\right] \\ \vdots & \vdots & & \vdots \\ \left[e_{n1}^L, e_{n1}^U\right] & \left[e_{n2}^L, e_{n2}^U\right] & \cdots & \left[e_{nn}^L, e_{nn}^U\right] \end{bmatrix} \quad (9)$$

in which

$$\begin{cases} e_{ij}^L = \dfrac{k_{ij}^L}{\sqrt{\sum\limits_{a=1}^{n}\left(k_{aj}^L\right)^2 + \sum\limits_{a=1}^{n}\left(k_{aj}^U\right)^2}}, & i, j = 1, 2, ..., n \\[4ex] e_{ij}^U = \dfrac{k_{ij}^U}{\sqrt{\sum\limits_{a=1}^{n}\left(k_{aj}^L\right)^2 + \sum\limits_{a=1}^{n}\left(k_{aj}^U\right)^2}}, & i, j = 1, 2, ..., n \end{cases} \quad (10)$$

Step 2: Calculating the weighted normalized decision matrix. For the sake of simplicity, we suppose the cross-efficiency between different DMUs have the same weight. Therefore the normalized decision matrix E can be directly used in the following step.

Step 3: Determining the negative ideal solution and positive ideal solution. The negative ideal solution $A^N = \left\{e_1^N, e_2^N, ..., e_n^N\right\}$ and the positive ideal solution $A^P = \left\{e_1^P, e_2^P, ..., e_n^P\right\}$ are determined as follows:

$$\begin{cases} e_j^N = \min\{e_{1j}^L, e_{2j}^L, ..., e_{nj}^L\} \\ e_j^P = \max\{e_{1j}^U, e_{2j}^U, ..., e_{nj}^U\} \end{cases}, \quad j = 1, 2, ..., n \quad (11)$$

Step 4: Calculating the Euclidean distances. The distance from the positive ideal solution and the distance from the negative ideal solution are defined as follows:

$$\begin{cases} d_i^{LN} = \sqrt{\sum\limits_{j=1}^{n}\left(e_{ij}^L - e_j^N\right)^2}, & d_i^{LP} = \sqrt{\sum\limits_{j=1}^{n}\left(e_{ij}^L - e_j^P\right)^2} \\[3ex] d_i^{UN} = \sqrt{\sum\limits_{j=1}^{n}\left(e_{ij}^U - e_j^N\right)^2}, & d_i^{UP} = \sqrt{\sum\limits_{j=1}^{n}\left(e_{ij}^U - e_j^P\right)^2} \end{cases} \quad (12)$$

Step 5: Calculating the relative distances. The relative distances of alternative to the ideal solution would also be interval and our modified relative distance is defined as follows:

$$D_i^L = \frac{d_i^{LN}}{d_i^{LN} + d_i^{LP}}, \quad i = 1, 2, ..., n \quad (13)$$

$$D_i^U = \frac{d_i^{UN}}{d_i^{UN} + d_i^{UP}}, \quad i = 1, 2, ..., n \quad (14)$$

Step 6: Ranking the alternatives based on the relative distances. An alternative with bigger relative distance would be a better alternative.

4 NUMERICAL EXAMPLE

In this section, a numerical example is provided to examine the validity and effectiveness of our proposed method. This example is partly selected from [1] in which 6 RE technologies are considered as the DMUs with the following inputs and outputs:

Input 1(I1): Investment ratio (Euro $\times 10^3$/Kw);
Input 2(I2): Implement period (years);

Table 1. Input and output data of 6 RE technologies.

DMU	Input			Output			
	I1	I2	I3	O1	O2	O3	O4
1	0.937	1	1.47	0.5	2.35	[15, 20]	1.93
2	1.500	1	1.51	2.5	2.35	[15, 20]	9.65
3	0.700	1.5	1.45	0.5	3.10	[20, 25]	0.47
4	0.601	2	0.70	2.0	2.00	[20, 25]	0.26
5	1.803	2	4.20	5.0	2.59	[25, 30]	0.48
6	1.803	1	7.11	0.5	7.50	[15, 20]	2.52

Table 2. Interval cross-efficiency scores of 6 RE technologies.

DMU	DMU (Whose weights are used)					
	1	2	3	4	5	6
1	[0.9481, 1]	[0.9481, 1]	[0.9481, 1]	[0.8552, 1]	[0.7381, 0.9912]	[0.9481, 1]
2	[1, 1]	[1, 1]	[1, 1]	[1, 1]	[1, 1]	[1, 1]
3	[1, 1]	[1, 1]	[1, 1]	[1, 1]	[0.8932, 1]	[1, 1]
4	[0.6583, 1]	[1, 1]	[1, 1]	[1, 1]	[1, 1]	[1, 1]
5	[0.4215, 0.9861]	[1, 1]	[0.8454, 1]	[1, 1]	[1, 1]	[1, 1]
6	[1, 1]	[1, 1]	[1, 1]	[1, 1]	[1, 1]	[1, 1]

Table 3. Aggregation results of 6 RE technologies.

DMU	Interval efficiency	Mid-points	Half-width	Ranking
1	[0.6383, 0.9874]	0.8128	0.1745	4
2	[1.0000, 1.0000]	1	0	1
3	[0.8606, 1.0000]	0.9303	0.0697	3
4	[0.5428, 1.0000]	0.7714	0.2286	5
5	[0.3307, 0.9793]	0.6550	0.3243	6
6	[1.0000, 1.0000]	1	0	1

Input 3(I3): Operating and maintenance costs (Euro $\times 10^3$/Kwh);

Output 1(O1): Power (MW $\times 10^3$);

Output 2(O2): Operating hours (hours $\times 10^3$/ year);

Output 3(O3): Useful life (years);

Output 4(O4): Tons of CO_2 avoided (t$CO_2 \times 10^6$/ year).

The modified input and output data is provided in Table 1 in which output 3 is extended into the interval data case. This would be reasonable in practice because the useful life would be imprecise according to the development of RE technologies.

By the application of cross-efficiency model with interval data, the interval cross-efficiency scores are provided in Table 2.

By using our proposed TOPSIS method, the aggregation results of these interval cross-efficiency scores are shown in Table 3. These six RE technologies are ranked by the mid-points of the interval efficiency scores. And the RE technology 2 and 6 are evaluated as the most efficient technologies and followed with technology 3 and 1.

5 CONCLUSIONS

Renewable Energy technology is an important solution to the large atmospheric pollution caused from fossil fuel combustion [1]. Efficiency evaluation would be necessary for the development and selection of RE technologies. In this paper, we proposed a comprehensive methodology for evaluating the efficiency of RE technologies which is combined with the DEA and TOPSIS method. We proposed a modified TOPSIS method for the aggregation of the cross-efficiency scores calculated by the cross-efficiency DEA model. Interval data is also considered in our proposed methodology which would be necessary in practice when the information of RE technology is imprecise. The validity and effectiveness of our proposed methodology have been examined by the application into a real data example.

REFERENCES

[1] J. Ramon. A multi criteria data envelopment analysis model to evaluate the efficiency of the Renewable Energy technologies. Renewable Energy, vol. 36, 2742–2746, (2011).

[2] A. Charnes, W.W. Cooper, E. Rhodes. Measuring the efficiency of decision making units. European Journal of Operational Research, vol. 2, 429–444, (1978).

[3] K. Mukherjee. Energy use efficiency in US. manufacturing: A nonparametric analysis. Energy Economics, vol. 30, 76–96, (2008).

[4] T. Sueyoshi, M. Goto, T. Ueno. Performance analysis of US coal-fired power plants by measuring three DEA efficiencies. Energy Policy, vol. 38, 1675–1688, (2010).

[5] J. Nouri, F. Lotfi, H. Borgheipour, F. Atabi, S. Sadeghzadeh, Z. Moghaddas. An analysis of implementation of energy efficiency measures in the vegetable oil industry of Iran: a data envelopment analysis approach. Journal of Cleaner Production, vol. 52, 84–93, (2013).

[6] G. Vlontzos, S. Niavis, B. Manos. A DEA approach for estimating the agricultural energy and environmental efficiency of EU countries. Renewable and Sustainable Energy Reviews, vol. 40, 91–96, (2014).

[7] Q. Wang, P. Zhou, Z. Zhao, N. Shen. Energy efficiency and energy saving potential in China: A directional meta-frontier DEA approach. Sustainability, vol. 6, 5476–5492, (2014).

[8] T.R. Sexton, R.H. Silkman, A.J. Hogan. Data envelopment analysis: critique and extensions. InR H. Silkman (Ed.), Measuring efficiency: An assessment of data envelopment analysis. San Francisco, CA: Jossey-Bass, (1986), pp. 73–104.

[9] J. Wu, J. Sun, M. Song, L. Liang. A ranking method for DMUs with interval data based on DEA cross-efficiency evaluation and TOPSIS. J. Syst. Sci. Syst. Eng., vol. 22, 191–201, (2013).

[10] C.L. Hwang, K. Yoon. Multi-objective decision making methods and application. A state-of-the-art study[M]. Springer-Verlag, New York, (1981).

[11] G.R. Jahanshahloo, M. Khodabakhshi, F.H. Lotfi, M.R. Goudarzi. A cross-efficiency model based on super-efficiency for ranking units through the TOPSIS approach and its extension to the interval case. Mathematical and Computer Modelling, vol. 53, 1946–1955, 2011.

Advances in Future Manufacturing Engineering – Yang (Ed.)
© *2015 Taylor & Francis Group, London, ISBN 978-1-138-02817-3*

A new equipment of induction power supply at high voltage

Haokun Guo
Department of Electronic Information Engineering, Jiangyin Polytechnic College, Jiangyin, Jiangsu, China

ABSTRACT: The auxiliary devices must to be installed on high voltage transmission lines to improve capability of condition monitoring. Power supply for these devices is very difficult to provide. For lack of solar cell and induction power supply, a new kind of rechargeable induction power supply at high voltage side is designed. Working principle and circuit of module are given. Data of device performance test is obtained after the experiment. Experiment shows that: the device is correct and effective, and it can provide security, ability, and uninterrupted power.

Keywords: transmission line; induction power supply; working principle; experimental verification

1 INTRODUCTION

With the development of the smart grid, equipment for monitoring transmission line operating state is required to install on the high-voltage transmission lines. The equipment is long-term operating with high-pressure environments, and the installation location of equipment is also very demanding, so the power is very difficult to provide.

Currently, the power supply mode of equipment is including solar cells and induction power supply at high voltage. There are many disadvantages of using solar cells, such as the high cost, large size, and limited installation position. Inductive power supply is based on the electromagnetic induction principle and technology of magnetic saturation, through the Current Transformer (CT) sensing high side current to obtain electricity. But magnetic saturation technique leads to existence of a higher AC output voltage spikes, the voltage spikes can damage inside device of the power supply. And when there is a power failure or a current shortage, the current cannot be stable.

This paper presents a rational induction power supply at high voltage. And the power supply is running with lithium battery. In this way, it can still work when there is a power failure or a current shortage.

2 OVERALL STRUCTURE OF THE DEVICE

The device is composed of take power module, power-conditioning module, intelligent protection module, and lithium battery charging management circuit. The overall structure of the device is shown in Figure 1. The take power module includes bus bar, current transformer cores, secondary coils, sampling resistor, and SPDT relay. The power-conditioning module includes rectification circuit, tip discharge circuit, ripple filter capacitor, and voltage stabilizing circuit. The intelligent protection module includes single chip module, Radio Frequency (RF) module and super capacitor.

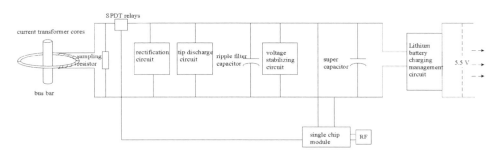

Figure 1. Schematic diagram of the device.

As shown in Figure 1, the current transformer core, secondary coils, and sampling resistor remove power in the bus bar based on electromagnetic induction theory [1–2]. Finally, the security, stability, and uninterrupted power are available through the rectification circuit, tip discharge circuit, ripple filter capacitor, and voltage stabilizing circuit. The operating temperature of tip discharge circuit is monitored by the MCU of intelligent protection module, and the temperature threshold is set in advance. After operating temperature is exceeded the threshold, MCU control module disconnects the power supply circuit by SPDT relay. After disconnecting the power supply circuit, MCU control module is powered by an internal super capacitor.

3 OVERALL STRUCTURE OF THE CORE MODULE

3.1 *Take power module*

The circuit of take power module is shown in Figure 2. In order to ensure power supply can

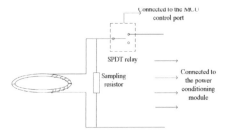

Figure 2. Circuit of take power module.

be started when the current of bus bar is 50 A. We chose 50/60 Hz steel sheet as the material of current transformer core. The factor of lamination is 10, and the radius of magnetic circuit is 9 mm. The secondary coil is wound around a total of 180 turns by the wire of 0.21 mm. Sampling resistor is connected in parallel to the output terminal of the secondary coil, for setting at a suitable voltage point of inductive alternating electromotive. Inductive power is supplied to the load equipment when SPDT relay is closed.

3.2 *Power conditioning module*

The circuit of power conditioning module is shown in Figure 3. AC voltage across the sampling resistor becomes pulsating DC voltage by the rectifier circuit (W10G). Transient fiction diode D3 is connected in parallel circuit in order to protect power module without the surge. There is great spike effect in pulsating DC voltage because of the impact of magnetic saturation. And the peak effect can destroy load equipment. Diode D1 is turned on when pulsating DC voltage peak is higher than the sum of the voltage drop of the regulator D1 and current limiting resistor R4 (5.1 KΩ, 0.5 w). Then, the power tube T1 is into a conducting state as the voltage protection of power supply. And the energy of pulsating DC peak is transferred to the bleeder resistor R1 (50 Ω, 10 w). Resistor R3 (100 KΩ, 0.5 w) is the gate current limiting resistor of T1. In this case, voltage ripple can be filtered by the filter capacitor C1 (1000 μF), and smooth DC voltage is obtained. The smooth DC voltage is regulated through voltage regulator chip

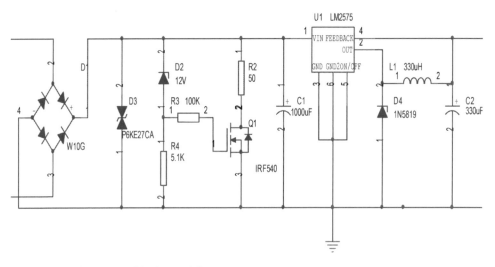

Figure 3. Circuit of power conditioning module.

(LM2575 chip), inductor L1 (330H), and diode D4. Capacitor C2 (330 µF) is the filter capacitor of regulator circuit.

3.3 Intelligent protection module

The operating temperature of tip discharge circuit is monitored by the MCU of intelligent protection module, and the temperature threshold is set in advance. After operating temperature is exceeded the threshold, MCU control module disconnects the power supply circuit by SPDT relay.

After disconnecting the power supply circuit, MCU control module is powered by an internal super capacitor. Specific circuit of RF radio module is shown in Figure 4.

3.4 Lithium battery charging management circuit

Lithium battery charging management circuit is shown in Figure 5. Power supply can supply power to electrical equipment and charge the lithium battery when the voltage is sufficient. Power supply must supply power to electrical

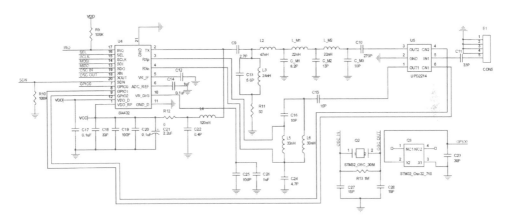

Figure 4. Circuit of RF module.

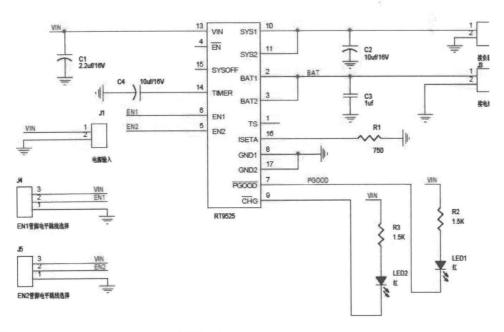

Figure 5. Charge management circuit of lithium battery.

equipment of priority when the voltage is low. If the voltage is very low, lithium battery can supply power to electrical equipment. So it can still work when there is a power failure or a current shortage.

4 EXPERIMENTAL VERIFICATION

Experiment of the induction power supply is completed in order to verify the accuracy and stability of the device. We know that the normal operating current of power supply is 50 A to 1500 A without load equipment. The results are shown in Table 1.

In Table 1, the normal operating current of power supply is 50 A to 1500 A, maximum temperature of power supply is 42°C, and the ambient temperature is 24°C. After running for some time, the lithium has been charged, then it can supply normally when the current bus is less than 50 A. Experiment results show that: this induction power supply is correct and effective.

Table 1. Data of device performance test.

Bus current/A	Output of the power supply/V	Temperature of the power supply/°C
20	0	24
50	5.5	24
100	5.4	24
1000	5.5	35
1500	5.4	42
20	5.5	24

5 CONCLUSION

A new equipment of induction power supply at high voltage is designed in order to solve that the power supply cannot work properly when there is a power failure or a current shortage. The device is composed of take power module, power-conditioning module, intelligent protection module, and lithium battery charging management circuit. Experiment of the induction power supply is completed, and the results show that: this induction power supply is correct and effective.

ACKNOWLEDGMENT

Project is supported by Jiangyin Polytechnic College of China (14-E-DZ-39).

CORRESPONDING AUTHOR

Haokun Guo, ghk1222@126.com, sixface@126.com, 15852661745.

REFERENCES

[1] Xianghong Cao, Chao Yang, Hua Zhang. *The Designment of A Inductive Current Power Supply Applied on Wireless Temperature Measurement System for High Voltage Electrical Equipment*: J. Science Technology and Engineering. Vol. 13(18) (2013), p. 5334–5338. In Chinese.
[2] Rongzeng Zheng. *Explain of Electromagnetic Induction Theory*: J. Journal of Higher Continuing Education. Vol. (2) (1994), p. 51–55. In Chinese.

Advances in Future Manufacturing Engineering – Yang (Ed.)
© 2015 Taylor & Francis Group, London, ISBN 978-1-138-02817-3

Analysis on the changing tendency of China's energy structure

Jie Yao, Yunxiu Chen & Jinshan Li
School of Economy and Management, Northeast Dianli University, Jilin, China

ABSTRACT: Since the reform and opening up, China's economy has achieved rapid growth and attracted worldwide attention. However, China's high economic growth is accompanied with the rapid growth of energy consumption. Since 1992, the total volume of China's energy consumption has exceeded that of energy production, with the disparity being expanded year by year. The energy shortage has become the bottleneck of China's economic growth. This article has carried out a thorough analysis on the changes of China's total energy production and consumption, the structure adjustment and industry distribution of China's energy consumption since 1990 by using the method of multivariate statistical analysis. Studies have shown that China's energy consumption is mainly based on coal resource and accompanied by the diversified development of multiple resources on the whole. The second industry still occupies the dominant position in China's energy consumption, and the first industry's energy consumption ratio has been decreasing while at the same time that of the second and tertiary industry on the rise. Therefore, facing with the soaring of energy prices and the reducing of global non-renewable energy resources, measures should be taken to adjust the industry structure and speed up the transformation of the pattern of economic development today to realize the sustainable development of China's economy.

Keywords: energy production; energy consumption; structure adjustment; industrial distribution

1 INTRODUCTION

Since the reform and opening up, China's economy has achieved rapid growth and attracted worldwide attention with an average annual growth rate of 9.1%, which is far higher than the average level of the world's major developed countries. However, China's high economic growth is accompanied with the rapid growth of energy consumption. According to statistics, the volume of China's energy consumption has been increased from 54.11 million tons of standard coal in 1953 to 3.61732 billion tons of standard coal in 2012, showing a sustaining growth trend as a whole. At the same time, the total volume of China's energy consumption has exceeded that of energy production since 1992, with the disparity being expanded year by year. The energy shortage has become the bottleneck of China's economic growth.

Whether the future stock of energy can support the sustainable development of China's society and economy or not, and what kinds of regularities and characteristics are presented between energy consumption and economic growth are all-important practical problems worthy of further study.

2 ANALYSIS ON THE CHANGING TENDENCY OF CHINA'S TOTAL ENERGY PRODUCTION AND CONSUMPTION

Overview. New China was founded more than 60 years ago. Since then, China has made great achievements in energy development, forming a new pattern of energy supply with coal as the main part, power as the center, and the overall development of oil, gas and renewable energy. The total self-sufficiency rate of energy has been remained over 90%. Since the reform and opening up, the total volume of China's energy production and consumption has both showed yearly growth tendency. Since 2002, China's national economy has developed into a new stage. The investment in fixed asset has increased rapidly. There has been an increasing tendency in the proportion of heavy industry. Because of the rapid expansion of high-energy-intensive industries, such as iron, steel, building materials, aluminum and so on, the energy consumption has increased dramatically.

In order to meet the demand of dramatic increase in energy consumption, the total volume of energy production has also begun to increase. In recent years, China's economy has exhibited a poor status in which the development of national economy is supported by high-energy consumption. The growth rate of energy consumption has also been significantly higher than that of energy production, leading to the emergence of energy disparity. By the end of 2012, the total volume of energy production in China has been 3.31848 billion tons, but still had a disparity of 298.84 million tons of standard coal.

Analysis on the changing tendency of China's total energy production. Through correlation analysis, we can find that the total volume of China's energy production and consumption has showed a good correlative relationship with time series T. So this paper first did regression analysis on the total volume of energy production with T in order to find its growth trend. The data selected is from 1990 to 1990. And we have used the classical growth model, which is log—linear model. By using the method of Ordinary Least Squares (OLS), a regression model between LOG(EP) and time series T has been established, as shown below.

$$LOG(EP) = 11.42396 + 0.055609*T. \quad (1)$$

The decision coefficient of the model is 0.947, F statistic value is 378.220, and the significance level is 0.00. By using Eviews software to make forecast, the forecast evaluation index shows that the prediction accuracy of model 1 is very high. Therefore, this regression model can be said to have practical significance, and we can use it to predict the total volume of China's energy production in recent years. The model also shows that the annual growth rate of China's total energy production has been increasing linearly from 1990 to 2012, with an annual growth rate of more than 5.56%, which is showed by the coefficient of time series T. Thus, we can see that China's total energy production has maintained sustainable growth momentum since 1990.

Analysis on the changing tendency of China's total energy consumption. As mentioned above, the total volume of China's energy consumption also has a strong correlative relationship with time series T. So we process the time series data of China's total energy consumption in the same way as that of China's energy production. By using the method of Ordinary Least Squares (OLS), a regression model between LOG(EC) and time series T has been established as follows.

$$LOG(EC) = 11.41934 + 0.061286*T. \quad (2)$$

The decision coefficient of the model is 0.959, F statistic value is 493.393, and the significance level is 0.00. By using Eviews software to make

forecast, the forecast evaluation index shows that the prediction accuracy of model 2 is also very high. Therefore, this regression model can be said to have practical significance, and we can use it to predict the total volume of China's energy consumption in recent years. The model also shows that the annual growth rate of China's total energy consumption has been increasing linearly from 1990 to 2012, with an annual growth rate of more than 6.12%, which is showed by the coefficient of time series T. By comparing the coefficients of time series T in model 1 and model 2, we can find that the total volume of China's energy consumption has been growing faster than that of energy production.

With the rapid economic and social development, China has been in great demand for energy in recent years. And the speed of energy supply can't keep up with that of energy consumption. Therefore, the problem of energy shortage has increasingly become an important factor that limits the economic development of China.

3 ANALYSIS ON THE STRUCTURE ADJUSTMENT OF CHINA'S ENERGY CONSUMPTION

From the point of energy consumption structure, coal is still dominant in China's total energy consumption. From 1978 to 2012, the average ratio of coal consumption to total energy consumption is 71.5%. At the early stage of reform and opening up, because of the rapid development of economy and extensive economy pattern with high input, high consumption and low output, China's coal consumption has ever been more than 76% of the total energy consumption. With the development of oil and gas industry and hydropower industry in China, the proportion of coal consumption has declined, which was 66.6% in 2012. But China is still in the stage of heavy industrialization, so is in strong demand for coal resources. In order to meet the demand of economic development, China has been continuing to mine coal resources, so that the proportion of coal production to the total energy production is on the rise, which was 76.5% in 2009.

On the whole, the proportion of Oil consumption has showed a rising trend. From 1978 to 2012, the average ratio of oil consumption to total energy consumption is about 19.4%. Compared with the other energy, the proportion of China's oil production has exhibited a stable declining tendency, which was 23.7% in 1978 and then fell to 8.9% in 2012. In 1978, the proportion of China's natural gas consumption is 3.2%. The rate has remained to be between 2% and 3% during a very long time.

With the operation of the central Asian gas pipeline and the west part of the second line of West to east gas pipeline, and also with the fast development of Liquefied Natural Gas (LNG) project, the production and consumption of China's natural gas has been grown at the speed of two digits in recent years. At the same time, the proportion of China's natural gas production and consumption to that of the total energy also rises sharply, which has been 4.3% and 4.3% respectively by the end of 2012.

In recent years, China's power industry has achieved leapfrog development with the acceleration of structural adjustment and industry upgrading. The renewable energy represented by hydropower, nuclear power and wind power has developed very fast. And the proportion of their production to national total has presented a steady growth trend. Overall, China's energy consumption is mainly based on coal resource and accompanied by the diversified development of multiple resources.

4 ANALYSIS ON THE INDUSTRIAL DISTRIBUTION OF CHINA'S ENERGY CONSUMPTION

From the ratio of each sector's energy consumption to national total (as shown in Table 1), we can notice that the second industry has occupied the dominant position in China's energy consumption. From 2006 to 2011, the ratio of the second industry's energy consumption to national total has been about 73%. And the energy consumption of manufacturing sector has accounted for the most part of the second industry. Since the reform and opening up, China has made great achievements in manufacturing sector. According to statistics, the output of China's manufacturing sector has been the first in the world since 2010. However, due to long-term high-energy consumption and high pollution, the energy consumption and emission of China's unit GDP in manufacturing sector is obviously worse than that of developed countries. In addition to the second industry, the proportion of the third industry's energy consumption to the total has also accounted for about 25%, with life consumption sector accounting for 10.75% and transportation, warehousing, postal service sectors accounting for 8.2% of nation total respectively.

The energy consumption of the tertiary industry has a lot to do with the living standard of residents. On the one hand, the year-by-year increasing of energy consumption shows that the residents' living standard has been improved a lot. On the other hand, it also shows that the investment in the construction of residents' living infrastructure has continuously been increased in recent years, which has in turn improved the public service level. Compared with that of the second and the tertiary industry, the ratio of the first industry's energy consumption to national total has declined which was 3.14% in 2006 and dropped to 1.94% in 2011. In recent years, lots of rural labors have transferred to cities, leading to the reducing of the first industry's labor force, and the shrinking of the cultivated land in many areas, and finally leading to the declining of the first industry's energy consumption.

The decreasing of the first industry's energy consumption ratio and the increasing of the second and the tertiary industry's reflects the adjustment of China's industry structure to some extent.

Table 1. Ratio of each sectors' energy consumption to national total 2006–2011(%).

Sector	2006	2007	2008	2009	2010	2011
Agriculture, forestry, animal husbandry and fishery	3.41	3.10	2.06	2.04	1.99	1.94
Industry	71.12	71.60	71.81	71.48	71.12	70.82
Mining	5.39	5.29	5.85	5.73	5.66	5.75
Manufacturing	58.09	58.82	59.05	58.89	58.01	57.59
Production and distribution of electricity, gas and water	7.64	7.49	6.91	6.85	7.45	7.47
Construction	1.51	1.52	1.31	1.49	1.92	1.69
Traffic, transport, storage and post	7.55	7.77	7.86	7.73	8.02	8.20
Wholesale and retail trades, hotels and catering services	2.24	2.24	1.97	2.09	2.10	2.24
Other sectors	3.87	3.67	4.04	4.14	4.21	4.36
Living consumption	10.31	10.09	10.94	11.04	10.64	10.75

Data sources: Assembled and computed according to "China statistical yearbook 2008–2013".

5 SUMMARY

Through the above analysis on the changes of China's total energy production and consumption, on the adjustment of China's energy structure and on the industrial distribution of China's energy consumption, this paper draws the following conclusions. Overall, China's energy consumption is mainly based on coal resource and accompanied by the diversified development of multiple resources. According to the ratio of each sector's energy consumption to national total, we can notice that the second industry has still occupied the dominant position in China's energy consumption. From the point of the changing tendency, the first industry's energy consumption ratio has been decreasing, while that of the second and tertiary industry has been on the rise. Therefore, facing with the soaring of energy prices and the reducing of global non-renewable energy resources, measures should be taken to adjust the industry structure and speed up the transformation of the pattern of economic development today to realize the sustainable development of China's economy.

ACKNOWLEDGMENTS

The research work was supported by the scientific research promotion project of Northeast Dianli University—Soft Science, Humanities and Social Sciences Special Assistance Scheme under Grant No. 201311.

CORRESPONDING AUTHOR

Jie Yao, leeyooab@163.com, Mobile phone 86-15844223069.

REFERENCES

[1] Jian-hua Yin, Zhao-hua Wang, Empirical study on the relationship between China's energy consumption and economic growth-Based on the data from 1953 to 2008, Journal of scientific research management, 2011(7), p. 121.
[2] Xiao-ping He, The energy saving potential and influencing factors of China's industry, Journal of Financial Research, 2011(10), p. 34–46.
[3] Dong-sheng Zhang, The present situation and prospects of new energy industry development in China, Economics and Management Strategies Journal, 2012(01), p. 66–77.
[4] Guo-feng Guo, Yan-peng Wang, Analysis on the saving potential and saving target of China's energy during the twelfth five-year period, Quantitative Economics in the 21st Century, 2013(13), p. 354–370.

Advances in Future Manufacturing Engineering – Yang (Ed.)
© 2015 Taylor & Francis Group, London, ISBN 978-1-138-02817-3

Study on characteristics of grounding impedance of large grounding grid

Bingping Xie & Yunlong Wang

Huizhou Power Supply Bureau, Guangdong Power Grid Corporation, Huizhou, China

ABSTRACT: As the large grounding grid occupies large area, the potential difference on the grounding grid is obvious, and the proportion of inductive components in the grounding impedance is larger. Then the design and the operation of the grounding grid are different from the small grounding grid. The paper studies the characteristics of resistive components and inductive components in the grounding impedance of the large grounding grid. The factors affecting the grounding impedance and the features of the large grounding grid are discussed. The results can provide a technical reference for design, operation and maintenance of the large grounding grid.

Keywords: grounding; grounding impedance; inductive reactance; soil resistivity; substation

1 INTRODUCTION

The grounding system is the basic guarantee of safe and reliable operation of the grounding system [1]. Along with continuous increases of voltage levels, the transmission capacity becomes larger and larger, the short-circuit current of the substation also increases correspondingly, so that the requirements for the ultra-high (extra-high) voltage substation are higher. On the other hand, due to inadequate land and more and more difficult selection of substation site, many substations are built in areas with higher soil resistivity, so that a greater challenge is brought to the design of the large grounding grid. The grounding grid is paid more attention to by design, construction and operation departments.

As the grounding grid of the large substation occupies large area, before the fault current flows into ground, it flows through the grounding conductor for a long distance. At the same time, a large number of steel with very high permeability and then high inductive components in the longitudinal impedance is used as grounding materials in China, so that the industrial frequency grounding resistance of the large grounding grid, which is called the grounding resistance in both the IEEE standard and the China's latest grounding standard, is inductive [1]. In the traditional analysis, as only the grounding resistance is concerned, the grounding impedance is rarely studied, and the grounding impedance characteristics of the large grounding grid and the factors affecting the grounding impedance are penuriously studied.

Along with sustainable rapid increase of China's power industry, many 500 kV, 750 kV and 1000 kV substations will be constructed, so it is necessary to develop the relevant study work.

Based on the electromagnetic field numerical calculating theory, the paper studies the characteristics of the grounding impedance of the large grounding grid and the factors affecting the grounding impedance by using the simulation analysis method and defines the features of the large grounding grid, thereby providing a technical reference for design, operation and maintenance of the large grounding grid.

2 CALCULATION METHOD

From the view of whether the potential drop on the grounding body is considered or not, the numerical method of grounding analysis can be divided into an equipotential method and an unequal potential method. In the equipotential method, the grounding body is considered as the ideal conductor, and potentials at all positions of the grounding body are equal. The equipotential method is suitable for the analysis of the small grounding grid or the situations under which the grounding material conductivity is good, and the soil resistivity is higher, and the result obtained by the equipotential method only is the grounding resistance caused by the distribution of soil resistivity, in which the grounding impedance cannot be contained [2, 3]. As the resistivity, the magnetic conductivity and the like in the grounding body are considered in the unequal potential method, the potentials at all positions are unequal [4–6]. The unequal potential method is suitable for the analysis of the large grounding grid or other situations under which the grounding material conductivity is worse, the magnetic conductivity is higher, and the grounding impedance calculated by using the unequal potential method contains inductive components. So the calculated result is more effective. The paper uses the developed idiostatic method and the unequal potential method grounding calculation method [6] for the grounding resistance of the grounding grid, the characteristics of the grounding impedance of the large grounding grid and the factors affecting the grounding impedance. In the following analysis, the result calculated by the equipotential method is called the grounding resistance, and the result calculated by the unequal potential method is not called the grounding impedance.

3 EFFECT OF GROUNDING GRID AREA

Along with the increase of grounding grid area, the total grounding impedance decreases, and the proportion of inductive reactance in the grounding resistance increases. Therefore, first of all, the effect of grounding grid area on the characteristics of the grounding impedance is analyzed. The frequency takes 50 Hz, the soil resistivity takes 10 Ω, the burial depth of the grounding body takes. 8 m, the radius takes 0.01 m, the current is injected from the center of the steel, the grounding impedance results calculated by using the equipotential method and the unequal potential method are analyzed when the grounding grids of which the areas are 50 m × 50 m, 100 m × 100 m, 200 m × 200 m, 300 m × 300 m, 480 m × 480 m are 10 m arranged at equal intervals of 10 m, and the unequal potential method is used for analyzing the real part and the imaginary part. The calculated results are shown in Table 1.

The analysis is performed according to the calculated results shown above. In the calculation, the

difference between the equipotential method and the unequal potential method lies in that the unequal potential method considers the inductance and resistance effects of the grounding grid conductor in the calculation; and in the equipotential method, the whole grounding grid is considered as the equipotential body, and the inductance and the conductor resistance can be neglected. Therefore, the results calculated by using the unequal potential method are more actual, and the equipotential body can be used for the analysis of the small grounding grid (in the small grounding grid, the effect of the inductance is less). The amplitude values calculated by the unequal potential method and the equipotential method can be compared in Table 2.

The above results are analyzed. Under the situation of power frequency fault, the smaller the grounding grid area is, the smaller the error calculated when the grounding grid is considered as the equipotential body is. At the moment, the grounding grid can be considered as the small grounding grid, the whole grounding grid is considered as the equipotential body, and the resistance and the inductance existing in the conductor are not considered.

When the grounding grid area is 200 m × 200 m or below, if the error calculated by using the equipotential method and the unequal potential method is less than 5%, the results calculated by the both can be considered to be the same. When the grounding grid area is enlarged to be 300 m × 300 m, the algorithm error is about 10%; and when the grounding grid area reaches 480 m × 480 m, the algorithm error is about 20%. As the engineering construction may bring along 5–10% of error to the grounding resistance, if the algorithm error exceeds 10%, the grounding grid is considered as the large grounding grid. At the moment, the resistance and the inductance of the grounding grid conductor have resulted in big enough potential difference existing on the grounding grid and the grounding grid area has also resulted in the non-ignorable inductance existing on the grounding grid at the same time;

Table 1. Grounding impedance results calculated by using equipotential method and unequal potential method under grounding grids with different areas.

| Area | Results calculated by using equipotential method (Ω) | | | Results calculated by using unequal potential method (Ω) |
	Amplitude value	Radian	Complex number	
50 × 50	0.9392	0.0086	0.9392 + 0.0080i	0.9388
100 × 100	0.4633	0.0215	0.4632 + 0.0100i	0.4569
200 × 200	0.2343	0.0521	0.2340 + 0.0122i	0.2241
300 × 300	0.1671	0.1080	0.1661 + 0.0180i	0.1509
480 × 480	0.1152	0.2213	0.1124 + 0.0252i	0.0943

Table 2. Comparison on deviation of results calculated by using unequal potential method and equipotential method under grounding grids with different areas.

Area	Power frequency impedance calculated by using unequal potential method $Z1(\Omega)$	Power frequency impedance calculated by using equipotential method $R2(\Omega)$	Deviation of values calculated by using equipotential method and unequal potential method $(Z1-R2)/Z1 *100\%$
50×50	0.9392	0.9388	0.04%
100×100	0.4633	0.4569	1.38%
200×200	0.2343	0.2241	4.35%
300×300	0.1671	0.1509	9.69%
480×480	0.1152	0.0943	18.14%

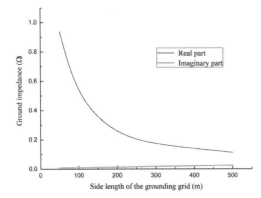

Figure 1. Relation between side length of grounding grid and real part and imaginary part of grounding impedance.

and if the inductance is neglected, the algorithm error exceeds the usual construction error.

The results calculated by using the unequal potential method are divided into the real part and the imaginary part, wherein the real part is the resistance part of the grounding impedance of the grounding grid; and the imaginary part is the inductance part of the grounding impedance of the grounding grid. Both the real part and the imaginary part are shown in Figure 1.

Observed from Figure 1, along with the increase of the grounding grid area, the resistive part of the grounding impedance of the grounding grid continuously decreases, and the inductive part of the grounding grid continuously increases. Under the situation that the grounding grid area is 300 m × 300 m, the inductive part is 10.9% of the resistive part; and under the situation that the grounding grid area is 480 m × 480 m, the inductive part is 22.4% of resistive part. At the moment, the inductance characteristics of the grounding grid must be considered.

The calculated results from this section show that when the soil resistivity is about 100 Ω, and

the outer contour side length of the rectangular grounding grid is 300 m, the grounding grid can be considered as the large grounding grid, and its voltage level is 500 kV. In the construction of the substation, the grounding grid with a voltage level of 500 kV, 750 kV and 1000 kV needs to be calculated according to the model of the large grounding grid. In fact, not only the grounding grid area, but also the soil resistivity can affect the proportion of inductive impedance in the grounding resistance, which is discussed below.

4 EFFECT OF SOIL RESISTIVITY

In fact, not only the grounding grid area, but also the soil resistivity can affect the inductive components and the resistive components of the grounding impedance. The frequency takes 50 Hz, the grounding grid area takes 300 m × 300 m, the burial depth of the grounding body takes 0.8 m, the radius takes 0.01 m, the current is injected from the center of the steel, the grounding impedance results calculated by using the equipotential method and the unequal potential method are analyzed when the soil resistivity is 10 Ω, 50 Ω, 100 Ω, 500 Ω, 1000 Ω and 2000 Ω, and the unequal potential method is used for analyzing the real part and the imaginary part. The calculated results are shown in Table 3.

The above calculated results show that when the soil resistivity increases, the speed of the resistive part of the grounding impedance (the real part in the results calculated by using the unequal potential method) in direct proportion to the soil resistivity quickly increases, and the speed of the inductive part in the grounding impedance (the imaginary part) in direct proportion to the soil resistivity slowly increases. The inductive part in the grounding impedance is mainly related to the grounding grid area.

The above calculated results show that the soil resistivity mainly affects the resistive part (the real part) of the grounding impedance and slightly

Table 3. Grounding impedance results calculated by using equipotential method and unequal potential method under different soil resistivity.

Soil resistivity	Results calculated by using idiostatic method (Ω)			Results calculated by using unequal potential method (Ω)
	Amplitude value	Radian	Complex number	
10	0.0351	0.4905	0.0309 + 0.0165i	0.0151
50	0.0927	0.1932	0.0910 + 0.0178i	0.0755
100	0.1671	0.1080	0.1661 + 0.0180i	0.1510
500	0.7672	0.0237	0.7670 + 0.0182i	0.7549
1000	1.5183	0.0120	1.5182 + 0.0182i	1.5097
2000	3.0205	0.0060	3.0205 + 0.0182i	3.0195

affects the inductive part. When the soil resistivity is lower, the inductive part of the grounding impedance is hard to neglect compared with the resistive part, and the side of the smallest frame of the large grounding grid model is shorter. For example, when the soil resistivity is 100 Ω, and the large grounding grid model has an area of more than 300 m × 300 m, the inductive impedance occupies about 10% of total impedance; and when the soil resistivity is 1000 Ω, the inductive impedance occupies about 1% of total impedance. Therefore, if 10% of algorithm error is exceeded or about 10% of inductive impedance occupied in the total impedance is exceeded, the grounding grid is considered as the large grounding grid; and the large grounding grid is also closely related to the soil resistivity. The lower the soil resistivity is, the smaller the large grounding grid area is, and the higher the soil resistivity is, the larger the large grounding grid area is, so that the complexity of large grounding definition is increased. In fact, if the soil resistivity is very high, either the rock or the gravel is considered. And at the moment, the grounding grid cannot be buried in, and the fine soil or other backfill materials with low resistivity should be backfilled. Therefore, the grounding grid is usually buried in the soil layer with the soil resistivity being within hundreds of omegas; and through contrasting the above analysis results, it is reasonable that the grounding grid with an outer contour side length being more than 300 m is considered as the large grounding grid. As a result, the grounding grid with a voltage level of above 500 kV is considered as the large grounding grid.

5 EFFECT OF GROUNDING MATERIALS

The characteristics of the grounding impedance of the grounding grid made of either steel or copper can be affected. The frequency takes 50 Hz, the grounding grid area takes 300 m × 300 m, the burial depth of the grounding body takes 0.8 m, the radius takes 0.01 m, the current is injected from the center, the grounding impedance results calculated by using the equipotential method and the unequal potential method when the radius of the steel and the radius of the copper are 0.01 m and 0.02 m respectively are analyzed when the soil resistivity is 10 Ω in the Section IV, and the unequal potential method is used for analyzing the real part and the imaginary part.

Obviously, a minimum difference exists between the real part and the imaginary part of the grounding impedance when the soil resistivity is 10 Ω. Under this situation, the unequal potential method is used for calculating the grounding impedance to obtain Table 4.

The calculated results show that under the situation that the whole grounding grid is considered as the equipotential body, the radius of the grounding grid conductor slightly affects the calculated results of the grounding impedance of the grounding grid. Under the correspondingly real calculation situation (unequal potential method), along with the increase of the grounding grid radius, the grounding impedance more obviously decreases.

When the radius of the grounding grid conductor made of either steel or copper increases, the resistive part and the inductive part of the grounding impedance of the grounding grid simultaneously and consistently decrease; and the decrease range of the inductive part is slightly larger than that of the resistive part.

The calculated results show that under the situation of calculation by using the equipotential method, the grounding impedance of the grounding grid is not related to the conductor radius. Therefore, under the situation of larger conductor radius and smaller conductor magnetic conductivity, the difference between the radius and the magnetic conductivity is smaller.

Table 4. Calculated results of grounding impedance of conductors made of steel and copper under different grounding conductor radiuses.

| Radius (m) | Steel | | Copper | |
	Unequal potential method (Ω)	Equipotential method (Ω)	Unequal potential method (Ω)	Equipotential method (Ω)
10	0.0351 (0.0309 + 0.0165i)	0.0151	0.01597 (0.0156 + 0.0036i)	0.0151
50	0.0248 (0.0227 + 0.0010i)	0.0150	0.01544 (1.5106 + 0.3177i)	0.0150

Table 5. Calculated results of grounding impedance of grounding grid at different fault current injection positions.

Position	Real part (Ω)	Imaginary part (Ω)	Amplitude value (Ω)	Radian
Angular point	0.0603	0.0398	0.0723	0.5834
Midpoint of one side	0.0415	0.0250	0.0484	0.5422
Center of grounding grid	0.0309	0.0165	0.0351	0.4905

6 EFFECT OF CURRENT-INTO-GROUND POSITION OF GROUNDING GRID

Due to the existence of the longitudinal impedance of the grounding body, if the current-into-ground position is different, the grounding impedance is different. The frequency takes 50 Hz, the grounding grid area takes 300 m × 300 m, the burial depth of the grounding body takes 0.8 m, the radius takes 0.01 m, the grounding impedance results calculated by using the unequal potential method under different current injection positions when the steel radius is 0.01 m are analyzed when the minimum soil resistivity of the grounding impedance is 10 Ω, and the unequal potential method is used for analyzing the real part and the imaginary part. When 1 A current flows into ground, the distribution of contact voltage and step voltage of grounding grid diagonal line is calculated.

In the Section IV, a minimum difference exists between the real part and the imaginary part of the grounding impedance when the soil resistivity is 10 Ω. The model is established according to the above assumptions, the short-circuit current is respectively injected at the angular point, the midpoint of one side and the center of the grounding grid respectively, and the calculated results of the corresponding grounding impedance are shown in Table 5.

The calculated results show that along with the movement of the calculated position towards the center position of the grounding grid, the amplitude value of the grounding impedance obtained by calculation decreases. Meanwhile, the real part and the imaginary part of the grounding impedance can monotonously decrease. Therefore, the closer the distance between the fault current injection position and the center of the grounding grid is, the better the effect is, so that the grounding impedance of the grounding grid can be better reduced.

7 CONCLUSIONS

Along with the increase of grounding grid area, the resistive part of the grounding impedance of the grounding grid continuously decreases, and the inductive part of the grounding impedance of the grounding grid gradually increases. If value that the inductive part is 10% of the resistive part is exceeded, the grounding grid is considered as the large grounding grid; and when the grounding grid area is larger than above 300 m × 300 m, the grounding grid can be considered as the large grounding grid. At the moment, the resistance and the impedance of the grounding grid conductor have resulted in big enough potential difference existing on the grounding grid, and the grounding grid area has also resulted in the non-ignorable inductance existing on the grounding grid. As usual, the voltage level of the grounding grid with an area of 300 μ × 300 m in the substation is 500 kV. Therefore, the grounding grid with a voltage level of 500 kV or above in the substation in the engineering is considered as the large grounding grid.

The ratio of the resistive part to the inductive part changes along with the soil resistivity and the grounding materials, and the lower the soil resistivity is, the higher the magnetic conductivity is, and the larger the proportion of the inductive part is. In the large grounding grid, the current-into-ground position has a greater effect on the grounding impedance; and the closer the distance between the current-into-ground position and the center of the grounding grid is, the smaller the grounding impedance is.

REFERENCES

[1] IEEE Guide for Safety of AC Substation Groundings, IEEE Std. 80–2000, Jan. 2000.

[2] Takehiko Takahashi, Taro Kawase, "Calculation of earth resistance for a deep-driven rod in a multi-layer earth structure", IEEE Transactions on Power Delivery, vol. 6, pp. 608–614, Apr. 1991.

[3] A.P.S. Melipoulos, "An Advanced Computer Model for Grounding System Analysis," IEEE Transaction on Power Delivery, vol. 8, pp. 13–23, Jan. 1993.

[4] F.P. Dawalibi, "Electromagnetic fields generated by overhead and buried short conductors, part 2—ground conductor," IEEE Transaction on Power Delivery, vol. 1, pp. 112–119, Oct. 1986.

[5] Jiansheng Yuan, Huina Yang, Liping. Zhang, "Simulation of Substation Grounding Grids with Unequal-Potential," IEEE Trans. on Magn., vol. 36, pp. 1468–1471, July 2000.

[6] A.F. Otero, J. Cidras, J.L. del Alamo, "Frequency-dependent grounding system calaulation by means of a conventional nodal analysis technique," IEEE Transaction on Power Delivery, vol. 14, pp. 873–878, July 1999.

Advances in Future Manufacturing Engineering – Yang (Ed.)
© 2015 Taylor & Francis Group, London, ISBN 978-1-138-02817-3

The research on active distribution network reconfiguration method based on OPF

Bin Shu & Kai Zhang
Beijing Electric Eco-Tec Research Institute, Beijing, China

Xiaoxi Gong
SGCC Beijing Electric Power Company, Beijing, China

Qihe Lou
Beijing Electric Eco-Tec Research Institute, Beijing, China

ABSTRACT: Distribution network reconfiguration is influenced by the output of distributed generations in electricity procurement schedule; the network topology of the reconfigured distribution network result in loss of the power loss, and then affect the electricity procurement schedule. The article takes into account both solutions of minimum purchase cost and comprehensive optimum power loss cost. Based on the bi-level programming theory: A OPF model is built in the first layer of planning aiming at minimizing the purchase cost, solve this model by means of interior point method and a electricity purchasing strategy can be obtained, then pass the node power to the second layer of panning; a distribution network reconfiguration model is built in the second layer of planning aiming at minimizing the power loss, solve this model by means of genetic algorithm and the network topology can be obtained, pass the network topology to the first layer of panning and a electricity purchasing strategy is acquired again, repeatedly iterate until the precision is satisfied, the electricity purchasing strategy and the network topology obtained is the most economical. Finally, an example demonstrates the effectiveness of the proposed method.

Keywords: active distribution network; distribution network reconfiguration; bi-level programming

1 INTRODUCTION

Active distribution network technology could solve the challenge of applying large-scale intermittent renewable energy, improving the efficiency of green power, optimizing the structure of primary energy, etc. Active distribution network is a distribution network with the ability of combination controlling various DG, controllable loads, storage devices, demand side management devices, etc. Its aim is to improve renewable energy accommodated capability of the distribution network, raise the utilization ratio of the distribution network assets, delay the investment and update of the distribution network, and improve the quality and the reliability of power services.

There are various DG in active distribution network: gas turbines, photovoltaic power generation, wind power generation, etc. The generation scheduling of the distribution network is arranged based on power balance constraints and branch overload constraints in current network structure.

Rationalize the purchasing share from transmission grid and DG aiming at minimizing power purchase cost. Distribution network reconfiguration is to get a network topology with minimum power loss. The aim of these two is all to optimize the economic indicators, and they are mutual influenced. In recent years, due to increased DG, people began to study distribution network reconfiguration considering DG. There are three situations when dealing with DG in distribution network reconfiguration: (1) the process of DG is restricted to treating DG as "negative" load[1][2], and it couldn't reflect the operation advantage of DG; (2) optimize the output of DG only for minimizing the power loss in distribution network reconfiguration[3][4][5], and generation cost isn't considered; (3) build a stochastic power flow model considering renewable energy like wind power, and build a reconfiguration model aiming at minimizing expected value of power loss[6][7]; (4) defining minimum generation cost as objective function[8], optimizing capacitor switching and network topology without considering power loss, and a OPF purchasing strategy considering DG is established.

The research did not fully consider the joint optimization of power purchase cost and power loss in the presence of DG, so the solution isn't the optimal solution. In this article, a OPF model is built in the first layer of planning aiming at minimizing the purchase cost, solve this model and a electricity purchasing strategy can be obtained, then pass the node power to the second layer of panning; a distribution network reconfiguration model is built in the second layer of planning aiming at minimizing the power loss, solve this model and the network topology can be obtained, pass the network topology to the first layer of panning and a electricity purchasing strategy is acquired again, repeatedly iterate, the integrated electricity purchasing strategy and the network topology obtained is the most economical.

2 THE FIRST LAYER OF PLANNING MODEL

2.1 The objective function of OPF

Draw up generation scheduling in active distribution network, dispatch DG and purchase electricity energy from transmission grid, and the expression of generation cost appears as shown below:

$$\min f_1 = c_{sysV} P_{sys} + \sum_{i=1}^{n} c_{DGVi} p_{DGi} \qquad (1)$$

In the formula above: f_1—generation cost; c_{sysV}—the price of electricity purchased from the grid in a certain period; p_{sys}—the active power purchased from the grid in a certain period; n—node number of the distribution network; c_{DGVi}—the electricity price in the i node; p_{DGi}—power generated by DG in the i node;

2.2 The constraints

1. The constraints of power flow equations:

$$\begin{cases} P_i = U_i \sum_{j=1}^{n} U_j (G_{ij} \cos \delta_{ij} + B_{ij} \sin \delta_{ij}) \\ Q_i = U_i \sum_{j=1}^{n} U_j (G_{ij} \sin \delta_{ij} - B_{ij} \cos \delta_{ij}) \end{cases} \qquad (2)$$

In the formula above: P_i, Q_i—the active power and reactive power respectively injected from i node; U_i, U_j—the respective voltage of i node and j node; G_{ij}, B_{ij}, δ_{ij}—the conductance, susceptance and phase angle difference of i node and j node.

2. The voltage constraints

$$U_{i\min} \leq U_i \leq U_{i\max} \qquad (3)$$

In the formula above: $U_{i\max}$, $U_{i\min}$—the upper limit and lower limit of i node voltage effective value.

3. The branch power constraints

$$S_j \leq S_{j\max} \qquad (4)$$

In the formula above: S_j—the actual power flow of j branch; $S_{j\max}$—the maximum power flow allowed for the j branch.

4. The constraints of power purchased from the transmission grid

$$p_{sys} \leq p_{sys\max} \qquad (5)$$

In the formula above: p_{sys}—the actual active power purchased from the transmission grid in a certain period; $p_{sys\max}$— the maximum active power purchased from the transmission grid in a certain period.

5. The constraints of power generated by DG

$$p_{DGi} \leq p_{DGi\max} \qquad (6)$$

In the formula above: p_{DGi}—the actual active power generated by DG in a certain period; $P_{DGi\max}$—the rated installed capacity of DG in i node.

3 THE SECOND LAYER OF PLANNING MODEL

3.1 The objective function of distribution network reconfiguration

There are multiple objectives in distribution network reconfiguration, here we take the minimum power loss as the object, the formula is:

$$\min f_2 = \sum_{j=1}^{b} k_j r_j \frac{P_{ij}^2 + Q_{ij}^2}{U_j^2} \qquad (7)$$

In the formula above: f_2—the power loss obtained through power flow calculation; b—the branch number; k_j—the state variable of j switch, it's a 0–1 discrete magnitude, 0 represent off and 1 represent on;

r_j—the resistance of j branch; P_{ij}, Q_{ij}—the active power flow and reactive power flow at the end of j branch; U_j—the node voltage at the end of j branch.

3.2 *The constraints of distribution network reconfiguration*

1. The constraints of power flow equations:

$$
\begin{cases}
P_i = U_i \sum_{j=1}^{n} U_j \left(G_{ij} \cos \delta_{ij} + B_{ij} \sin \delta_{ij} \right) \\
Q_i = U_i \sum_{j=1}^{n} U_j \left(G_{ij} \sin \delta_{ij} - B_{ij} \cos \delta_{ij} \right)
\end{cases}
\tag{8}
$$

In the formula above: P_i, Q_i—the active power and reactive power respectively injected from i node; U_i, U_j—the respective voltage of i node and j node; G_{ij}, B_{ij}, δ_{ij}—the conductance, susceptance and phase angle difference of i node and j node.

2. The voltage constraints

$$
U_{i\min} \leq U_i \leq U_{i\max}
\tag{9}
$$

In the formula above: $U_{i\max}$, $U_{i\min}$—the upper limit and lower limit of i node voltage effective value.

3. The branch power constraints

$$
S_j \leq S_{j\max}
\tag{10}
$$

In the formula above: S_j—the actual apparent power flow of j branch; $S_{j\max}$—the maximum apparent power flow allowed for branch j.

4. The constraints of network topology

The reconfigured distribution network is in radial structure, and looped network doesn't exist. The network is connected, and it doesn't contain islands and isolated points.

Figure 1 is a bi-level programming diagram. The power from DG and transmission grid can be obtained through the first layer of planning. Passing the two variables to the second layer of planning, then solving the problem of distribution network reconfiguration with the help of genetic algorithm, and a distribution network topological structure with minimum power loss can be obtained. In such structure changed power loss lead to the change of power supplied by transmission grid, so the primary electricity procurement schedule isn't optimum. The structure can cause some branches overload with the primary electricity procurement schedule, so it need power flow calculation once again. In the end, pass the calculated results of the second layer of planning—the network topology to the first layer.

4 THE SOLVING PROCEDURE

1. **Known conditions.** The known date need by OPF: (1) node class: the node types (load nodes, generator nodes, DG nodes), the power into the nodes (active load, reactive load); (2) branch class: the terminal nodes number, the branch

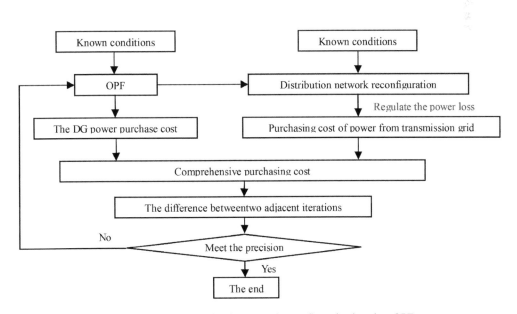

Figure 1. The solving flow chart on active distribution network reconfiguration based on OPF.

types (transformer, transmission line), the impedance, the admittance, the transformation ratio; (3) economic class: the transmission grid generating cost curve, the DG generating cost curve; (4) constraints: the power balance constraints, the node voltage constraints, the branch power constraints, the generator power constraints; (5) control variables of the calculation: the accuracy of the iterative computations, the iterations, the population of the genetic algorithm, the iterations of the genetic algorithm, the probability of crossover and mutation, etc.

2. **The calculation of OPF.** In current topological structure, the power purchased from transmission grid and DG can be obtained by calculating the power flow that meet the power balance constraints, the node voltage constraints, the branch power constraints and the generator power constraints with the method of interior point algorithm, then the power into the nodes can be obtained.

3. **Distribution network reconfiguration.** On the base of the power into the nodes in the second step, a network topological solution with minimum power loss can be obtained through reconfiguring the distribution network with the method of genetic algorithm, which meets the power balance constraints, the node voltage constraints and the branch power constraints.

4. **Judgement.** Calculate the sum of the power purchase cost and the power loss cost, iterate twice, if the accuracy is less than the demand, stop, otherwise return to the second step to continue the iteration.

5 ANALYSIS OF EXAMPLES

Figure 2 shows a distribution network with three feeders, the reference system capacity is 100 MVA, the reference voltage is 23 kV, the total network load is 28.7+j5.9 MVA.

Solution 1: the network isn't reconfigured, the opened line number is 14,15,16, the DG at node 4 and 13 are dispatched at the moment.

Solution 2: the DG at node 4 and 13 are dispatched, and the network is reconfigured.

The optimal power flow problem is carried out without consideration for distribution network reconfiguration in solution 1, the total cost is 19,830 yuan, and the optimal power flow problem is carried out with consideration for distribution network reconfiguration in solution 2, the total cost is 19,820 yuan, obviously the cost is decreased.

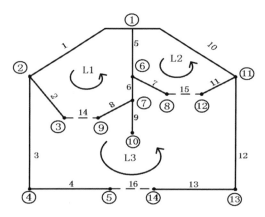

Figure 2.   IEEE three feeder system.

Table 1.   Power data that is known.

| Node | Type | Active power | | The variable cost [yuan/MWh] |
		The lower limit [MW]	The upper limit [MW]	
1	Transmission grid	–	28	584
4	Gas turbine	0	2	810
13	Internal combustion engine	0	2	710

Table 2.   Results of the optimal operation without taking into account the line power constraints.

Solution	The opened line number	Power loss [MW]	All-in cost [10,000 yuan]
Solution 1	14, 15, 16	0.4111	1.983
Solution 2	8, 7, 16	0.395	1.982

Although the decrease is little, the long-standing cost is considerable owing to the cost calculated above is instantaneous cost.

6 CONCLUSIONS AND PROSPECTS

This paper proposed an active distribution network reconfiguration method based on OPF, which solve the problem with bi-level programming theory. The method could decrease the total power purchase cost of the power grid and improve the operating economy of the system. Better power quality and economic benefit can be obtained with this method coordinating with the switch of reactive-load compensation equipment. This method could play a significant role in the planning and operation of active distribution network and smart grid.

ABOUT THE AUTHOR

ShuBin work in Beijing Electric Eco-Tec Research Institute.
e:13801100270@163.com. m:+8613801100270. t:+86–010–63678562. f:+86–010–63371500
a:Beijing Xichengqu Guang-an-men-che-zhan-xi-jie 15#.p:100055.

REFERENCES

[1] Wang Xiuli, Wu Hongxiao, Bie Chaohong, etc. Distribution System Reliability and Network Reconfiguration [J]. Electric Power, 2001, 34(9): 40–43.

[2] Wang Jiajia, Lv Lin, Liu Junyong, etc. Reconfiguration of Distribution Network Containing Distribution Generation Units Based on Improved Layered Forward-Backward Sweep Method [J]. Power System Technology, 2010, 34(9): 60–64.

[3] Cui Jinlan, Liu Tianqi, Li Xingyuan. Network reconfiguration at the distribution system with distributed generation [J]. Power System Protection and Control, Power System Protection and Control, 2008, 36(15): 37–40, 49.

[4] Rugthaicharoencheep N. Sirisumrannukul S. Feeder reconfiguration with dispatchable distributed generators in distribution system by tabu search [C]. Universities Power Engineering Conference (UPEC),2009, Proceedings of the 44th International, 2009: 1–5.

[5] Mojdehi M.N. Kazemi A. Barati M. Network operation and reconfiguration to maximize social welfare with distributed generations [J]. EUROCON, IEEE, 2009: 552–557.

[5] Rugthaicharoencheep N. Sirisumrannukul S. Optimal feeder reconfiguration with distributed generators in distribution system by fuzzy multiobjective and Tabu search [C]. Sustainable Power Generation and Supply, 2009, SUPERGEN'09, International Conference on, 2009:1–7.

[6] Chen Guang, Dai Pan, Zhou Hao, etc. Distribution System Reconfiguration Considering Distributed Generators and Plug-in Electric Vehicles [J]. Power System Technology, 2013, 37(1): 82–88.

[7] Shirmohammadi D, Hong H.W. Reconfiguration of electric distribution networks for resistive line loss reduction [J]. IEEE Trans on PWRD, 1989, 4(2): 1492–1498.

Advances in Future Manufacturing Engineering – Yang (Ed.)
© 2015 Taylor & Francis Group, London, ISBN 978-1-138-02817-3

Melting curve of Mo under high pressure

Cui-E Hu & Zhao-Yi Zeng

College of Physics and Electronic Engineering, Chongqing Normal University, Chongqing, China
Laboratory for Shock Wave and Detonation Physics Research, Institute of Fluid Physics,
Chinese Academy of Engineering Physics, Mianyang, China

Jia-Jin Tan

Laboratory for Shock Wave and Detonation Physics Research, Institute of Fluid Physics,
Chinese Academy of Engineering Physics, Mianyang, China

ABSTRACT: The melting curve of Mo was investigated from MD simulations combining with EAM potential. We predict the thermal EOS in a large range of pressure and temperature. It is found that there is no phase transition before melting. Fitting data to Simon function, we yield the melting curve: $2900(1+P/44.659)^{0.474}$. Our melting curve confirms the conclusions from the shock-wave measurements.

Keywords: refractory metal; molecular dynamics; melting

1 INTRODUCTION

It is an intriguing problem of melting curve of Molybdenum (Mo), *i.e.* there is enormous divergence in melting curves between laser-heated Diamond Anvil Cell (DAC) experiments and Shock-Wave (SW) measurements. One possible reasons is that the melting temperature (T_m) is underestimated in DAC experiments [1], while overestimated in SW experiments [2]. On the other hand, a possible solid-solid phase transition before melting results in the large slope (dT/dP) of the melting curve [3,4]. This issue has attracted much attention in the last few years and is receiving continual attention now. For Mo, the acoustic velocity data by Hixson et al. showed that the longitudinal acoustic velocity disappeared around 390 GPa in SW experiments. And for the liquid material, only the bulk acoustic velocities can be measured. The break of acoustic velocity indicated shock melting [5]. The T_m datum estimated by Hixson et al. is as high as 10000 K. Extrapolating DAC melting data to shock melting pressure, about 5000 K of discrepancies exist. Though Errandonea corrected the SW data by considering 30% superheating [2], the revised T_m (7700 ± 1500 K at 390 GPa) is also much larger than the T_m (below 4000 K) at this pressure extrapolated from DAC experiments. For other refractory metals, the results also show the same problem [6]. In this work, we obtained the melting curve of Mo by using Molecular Dynamic Simulation (MD). It can help us understand the melting law of refractory metals.

2 COMPUTATIONAL METHOD

For the present MD simulations, three-dimensional periodic boundary conditions and solid-liquid coexistence method were applied. As Mo is stable in bcc structure at a broad pressure and temperature range, we tread bcc Mo as the simulation system. There are two methods to simulate the melting behavior, *i.e.*, the one-phase method and the two-phase (solid-liquid coexistence) method. In present MD simulations, we applied one phase method to test the validity of the potential and obtain the thermal Equation Of State (EOS). And the two-phase method is suitable to simulate the melting properties. From our tests on size dependence of melting point, we found that the difference of melting point obtained from 16000 atoms and 31250 atoms are less than 100 K in the whole range of pressure. To construct initial configuration of solid-liquid coexistence, we heated 16000 atoms up to 4000 K (far beyond the melting point) in NVE ensemble. Then we merged the liquid and solid parts in the box along z-axis. As a result, the coexistence phase (including 32000 atoms) containing a solid-liquid interfaces was constructed. This box was used as the initial configuration for further simulations. The NPT ensemble was used to achieve constant temperature and pressure. The Verlet leapfrog integration algorithm is used with a time step of 0.5 fs. The total simulation time steps were 100,000. The first 60,000 steps were used for equilibrium, and the last 40,000 steps for statistical average of properties. Zhou et al. showed the parameters of the Embedded Atom Model (EAM) potential for several metals, including Mo [7].

3 RESULTS AND DISCUSSIONS

The thermal EOS was obtained from one-phase MD method. At ambient conditions, the obtained equilibrium volume is 15.66 Å³, which agrees well with the experimental datum 15.583 Å³ [8]. The isobar of bcc Mo at 1 bar is shown in Figure 1 (a). The present MD results consist with the experimental data [9–11] and our previous work in the temperature range 298–3000 K [4]. The isotherms are shown in Figure 1 (b). The 298 K isotherm agrees well with the experimental data [12,13] and the calculated results [14]. The good agreements can also confirm the validity of the EAM potential.

To understand the discrepancy on melting data between DAC and SW experiments, we should judge whether the structural transition results in the large slope of the melting curve. Therefore, to clarify the solid-solid phase transition before melting is of crucial importance. In one phase method, besides to obtain the thermal EOS, we can also analysis the local atomic structure at the simulated conditions by using Common Neighbor Analysis (CNA) method. In this method, it is powerful enough to clearly distinguish between various local

structures, in particular fcc, hcp and bcc. Figure 2 (a) and (b) are the configurations at 3000 K and 5000 K under 150 GPa. At 3000 K, the bcc structure remains the most composition. And when the temperature increases to 5000 K, the system melts gradually. It is obvious that there is no solid-solid phase transition before melting. In all the simulated pressure and temperature range, the results show the same conclusion. The present results confirm the viewpoint concluded from reference-coexistence technique [15]. Meanwhile, in the latest SW experiments, the acoustic velocities of Mo up to 440 GPa were measured by Nguyen et al. [16] It is found that the sound speed increases linearly with the increased pressure. This newest results suggest that there is no evidence for the previous solid-solid phase transition and Mo remains in bcc phase on the Hugoniot up to the melting pressure.

In MD simulations, the planar density is a convenient tool to monitor the simulations. The planar density is defined as the number of atoms presented in a slice of the simulation box cutting parallel to the solid-liquid interface. As is shown in Figure 3, the planar density in the solid region is a periodic function, while it is simply a random number fluctuating around an average value in the liquid region. By narrowing the interval of the temperature, the T_m datum can be estimated at the fixed pressure in the simulation. From Figure 3, it

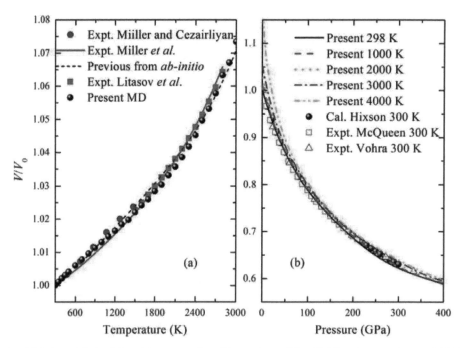

Figure 1. (a) Isobar of the bcc Mo at 1 bar, comparing with experimental data [9–11] and our previous work [4]. (b) Isotherms of the bcc Mo at different temperatures, comparing with the experiments [12,13] and previous calculations [14].

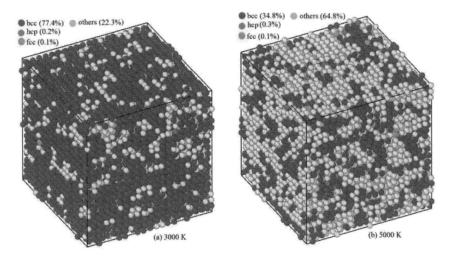

Figure 2.　The configurations of system at (a) 3000 K and (b) 5000 K under 150 GPa.

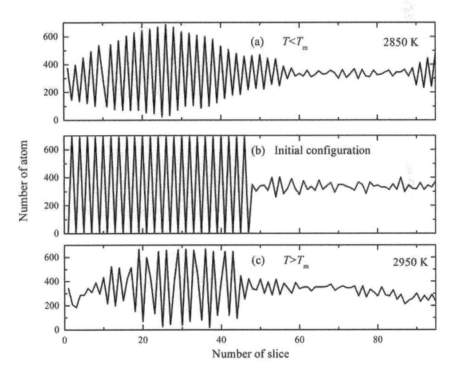

Figure 3.　Density profile for the coexistence configuration. (a) the moving of interfaces towards liquid part when $T < T_{\mathrm{m}}$; (b) the initial configuration; (c) the moving of interfaces towards solid parts when $T > T_{\mathrm{m}}$.

can be seen that at 1 bar, the T_{m} is 2900 ± 50 K. We repeated the steps at different pressures, and then we can obtain the whole melting curve under high pressure. Fitting the melting data as Simon function, it can be shown as follows:

$$T_m(P) = 2900\left(1 + \frac{P}{44.659}\right)^{0.474}$$

The dT_{m}/dP at 1 bar is about 30 K/GPa.

93

The SW experiments show the melting pressure is ~390 GPa, and the T_m datum is estimated to be ~10000 K by Hixson et al. [5]. As is known that the overestimation of the T_m exists in SW experiments, Errandonea corrected the SW datum at 390 GPa by considering 30% superheating [2]. The revised melting temperatures located at 7700 ± 1500 K, also much larger than the melting temperature (just above 4000 K) at this pressure extrapolated from DAC experiments. To compare SW melting data with DAC data directly, the more melting data at low pressure must be obtained. Generally, SW measurements can only obtain one shock T_m of Mo (~390 GPa). So the datum cannot be extrapolated to the whole pressure range. To obtain more melting data, the initial state of the sample should be changed. In the previous SW measurements, Zhang et al. obtained two shock-induced release T_m data by measuring the porous Mo [18]. The melting data are 6320 ± 682 K at 136 GPa and 6503 ± 397 K at 197 GPa. Recently, in our group, Zhao measured the melting datum of preheated Mo. The sample was preheated up to 973 K, and the release T_m is 5695 ± 129 K at 176 GPa [19]. As is shown in Figure 4, it is obvious that our MD results agree with all the SW data. Comparing with other theoretical data, our results are in good agreement with that obtained

by Cazorla et al. [17], but lower than that from Belonoshko et al. [3]. The possible reason for the melting data produced by Belonoshko et al. higher than the others is the size effects and the simulation method. The system simulated by Belonoshko et al. is associated with insufficiently large computational cells.

Below 150 GPa, the melting data in present work also much higher than the previously reported anomalously low melting curve in DAC experiments. The main reason is due to the limitation of the DAC measurements. By Laser-Heated DAC experiments, Dewaele et al. obtained the high melting curve of Ta up to 120 GPa by using X-Ray Diffraction (XRD) as the primary technique [1]. Dewaele et al. considered that two effects are identified to alter the melting determination of refractory metals in DAC experiments. Firstly, the strong chemical reactivity of the sample with the pressure transmitting media and with carbon diffusing out from the surface of the anvils is observed. Secondly, pyrometry measurements can be distorted when the pressure medium melts. So, the previous low melting curves of the refractory metals may be incorrect. The new melting data of Ta is also lower than that from SW experiments and MD simulations, but they are comparable. Remarkably, by using MD simulations, Wu et al. reported a shear-induced, partially

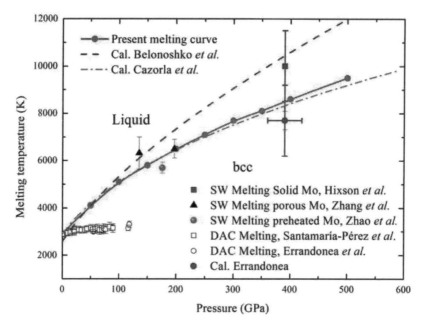

Figure 4.   Melting curve of Mo. Present MD data (solid circles and line); AIMD simulations of Belonoshko et al. (dashed line) [3]; AIMD results of Cazorla et al. (dashe-dotted line) [17]; Considering 30% superheating of the SW data by Errandonea (solid circle with error bar) [2]; Experimental SW (solid squares [5], triangles [18] and sphere [19] with error bar) and DAC (open circles [20] and squares [6] with error bars) measurements.

disordered viscous plastic flow from bcc Ta under heating before it melted into the liquid [21]. Perhaps the melting state in DAC experiments is a one-dimensional structure analogous to a liquid crystal or glass. It is worthy to note that this transition is fully consistent with the previously reported anomalously low-temperature melting curve. This mechanism is not specific to Ta and is expected to hold more generally for other refractory metals. The melting of refractory metals under high pressure is still a challenging topic, and needs more experiments and theoretical simulations to explore.

4 CONCLUSIONS

To explore the melting law of refractory transition metal at high pressure is a longstanding problem. In this letter, we investigate the melting curve of Mo by using classical MD simulations combining with the EAM potential. Based on reproducing the available experimental data, we predict the thermal equation of state in a large range of pressure and temperature. The results from Common Neighbor Analysis show that the bcc Mo is the stable phase, which indicated no solid-solid phase transition before melting. Fitting our T_m data to the well-known Simon form, we yield the melting curve: $2900(1+P/44.659)^{0.474}$. The MD results confirm the conclusions from the SW measurements.

ACKNOWLEDGMENT

The authors acknowledge the support by the supported by the National Natural Science Foundation of China (Grant nos. 11304408, 11347019) and China Postdoctoral Science Foundation (Grant nos. 2014M552380, 2014M552541XB, 2014M562336).

REFERENCES

[1] A. Dewaele, M. Mezouar, N. Guignot, P. Loubeyre: High melting points of tantalum in a laser-heated diamond anvil cell, Phys. Rev. Lett. Vol. 104 (2010), p. 255701.

[2] D. Errandonea: Improving the understanding of the melting behaviour of Mo, Ta, and W at extreme pressures, Physica B Vol. 357 (2005), p. 356–364.

[3] A.B. Belonoshko, L. Burakovsky, S.P. Chen, B. Johansson, A.S. Mikhaylushkin, D.L. Preston, S.I. Simak, D.C. Swift: Molybdenum at high pressure and temperature: melting from another solid phase, Phys. Rev. Lett. Vol. 100 (2008), p. 135701.

[4] Z.Y. Zeng, C.E. Hu, L.C. Cai, X.R. Chen, F.Q. Jing: Lattice Dynamics and Thermodynamics of Molybdenum from First-Principles Calculations, J. Phys. Chem. B Vol. 114 (2010), p. 298–310.

[5] R.S. Hixson, D.A. Boness, J.W. Shaner, J.A. Moriarty: Acoustic velocities and phase transitions in molybdenum under strong shock compression, Phys. Rev. Lett. Vol. 62 (1989), p. 637–640.

[6] D. Errandonea, B. Schwager, R. Ditz, C. Gessmann, R. Boehler, M. Ross: Systematics of transition-metal melting, Phys. Rev. B Vol. 63 (2001), p. 132104.

[7] X.W. Zhou, H.N.G. Wadley, R.A. Johnson, D.J. Larson, N. Tabat, A. Cerezo, A.K. Petford-long, G.D.W. Smith, P.H. Clifton, R.L. Martens, T.F. Kelly: Atomic scale structure of sputtered metal multilayers, Acta mater. Vol. 49 (2001), p. 4005–4015.

[8] L. Ming, M.H. Manghnani: Isothermal compression of bcc transition metals to 100 kbar, J. Appl. Phys. Vol. 49 (1978), p. 208.

[9] A.P. Miiller, A. Cezairliyan: Thermal Expansion of Molybdenum in the Range 1500–2800 K by a Transient Interferometric Technique, Int. J. Thermophys Vol. 6 (1985), p. 695–704.

[10] G.H. Miller, T.J. Ahrens, E.M. Stplper: The equation of state of molybdenum at 1400°C, J. Appl. Phys. Vol. 63 (1988), p. 4469–4475.

[11] K.D. Litasov, P.I. Dorogokupets, E. Ohtani, Y. Fei, A. Shatskiy, I.S. Sharygin, P.N. Gavryushkin, S.V. Rashchenko, Y.V. Seryotkin, Y. Higo, K. Funakoshi, A.D. Chanyshev, S.S. Lobanov: Thermal equation of state and thermodynamic properties of molybdenum at high pressures, J. Appl. Phys. Vol. 113 (2013), p. 093507.

[12] R.G. McQueen, S.P. Marsh, J.W. Taylor, J.N. Fritz, W.J. Carter: High—Velocity Impact Phenomena, (Academic Press, New York, 1970).

[13] Y.K. Vohra, A.L. Ruoff: Static compression of metals Mo, Pb, and Pt to 272 GPa: comparison With shock data, Phys. Rev. B Vol. 42 (1990), p. 8651–8654.

[14] R.S. Hixson, J.N. Fritz: Shock compression of tungsten and molybdenum, J. Appl. Phys. Vol. 71 (1992), p. 1721–1728.

[15] C. Cazorla, D. Alfè, M.J. Gillan: Constraints on the phase diagram of molybdenum from first-principles free-energy calculations, Phys. Rev. B Vol. 85 (2012), p. 064113.

[16] J.H. Nguyen, M.C. Akin, Ricky Chau, D.E. Fratanduono, W.P. Ambrose, O.V. Fat'yanov, P.D. Asimow, N.C. Holmes: Molybdenum sound velocity and shear modulus softening under shock compression, Phys. Rev. B Vol. 89 (2014), p. 174109.

[17] C. Cazorla, M.J. Gillan, S.T.e. al.: Ab initio melting curve of molybdenum by the phase coexistence method, J. Chem. Phys. Vol. 126 (2007), p. 194502.

[18] X.-L. Zhang, L.-C. Cai, J. Chen, J.-A. Xu, F.-Q. Jing: Melting Behaviour of Mo by Shock Wave Experiment, Chin. Phys. Lett. Vol. 25 (2008), p. 2969–2972.

[19] K. Zhao, L.-C. Cai: (2014), personal communication (unpublished data).

[20] D. Santamaría-Pérez, M. Ross, D. Errandonea, G.D. Mukherjee, M. Mezouar, R. Boehler: X-ray diffraction measurements of Mo melting to 119 GPa and the high pressure phase diagram, J. Chem. Phys. Vol. 130 (2009), p. 124509.

[21] C.J. Wu, P. Söderlind, J.N. Glosli, J.E. Klepeis: Shear-induced anisotropic plastic flow from body-centred-cubic tantalum before melting, Nature Mater. Vol. 8 (2009), p. 223–228.

Advances in Future Manufacturing Engineering – Yang (Ed.)
© 2015 Taylor & Francis Group, London, ISBN 978-1-138-02817-3

The study and application of power supply reliability forecasting based on wavelet neural network

Lie-luan Yang, Xiang-dong Xu, Zhi-yuan Liu, Qian Wang & Jian-yong Gao
State Grid Gansu Business Operation Monitoring Center, Lanzhou City, Gansu Province, China

ABSTRACT: Power supply reliability is the ability of sustainable supply of power supply system for its users. Traditional power supply reliability assessment highly depends on accurate grid structure and equipment reliability parameters, which makes it difficult to apply to large complex distribution network. Therefore, this paper proposes a novel power supply reliability forecasting method based on wavelet neural network power which integrates the time-frequency localization characteristics of wavelet transform as well as the self-learning and self-organizing ability of neural network. Indices that are both highly correlative to power supply reliability and represent the basic features of power grid are picked up as input variables of wavelet neural network such as Total User Number, Total Length of Cable Line, Average Interruption Times of Customer and Average Interruption Hours of Customer. The empirical study of 10 kV urban power supply reliability forecasting indicates that the developed forecasting method based on wavelet neural network is feasible and effective.

Keywords: power supply reliability; wavelet neural network; indices forecasting; urban power network

1 INTRODUCTION

Nowadays electricity, used widely in all walks of life, plays an increasingly important role in social development and economic growth. Social production and living are inseparable from secure, high-quality, and efficient power supply. Specifically, power enterprises tend to be customer-oriented, in order to better understand and meet customers' demand; "customer satisfaction" is regarded as the ultimate goal and the pursuit of improving service levels never stops [1]. Naturally, the power outage has a detrimental influence on the normal production, lowers living quality and directly causes user dissatisfaction. Therefore, it has become an inevitable trend that power supply reliability evaluation and predictive analytics are integrated into power supply system management.

Power supply reliability refers to the ability of power supply system to provide continuous power for its users, which is a comprehensivereflection of the grid structure, technical equipment and management level. Accurately predicting power supply reliability has very important significance. According to thorough literature research, the traditional reliability assessment methods are mainly listed as follows: Failure Mode and Effect Analysis, Minimal Path Method considering the branch line protection, isolation switch and circuit breaker impact [2], Minimum-cut set applied for

distribution system with rings [3], the reliability evaluation algorithm for complex distribution system based on minimal path method as well as the equivalent method, minimum fault zone algorithm integrated with the tree traversal method, reliability evaluation method for distribution system interval staking the element parameter uncertainty, feeders partition and load transfer restrictions into account, load points reliability computing and system weaknesses identification based on four-layer Bayesian network and grid block technique and graph theory traversal algorithm.

Based on previous research, it can be concluded that the traditional power supply reliability forecasting methods usually require detailed distribution network structure and a lot of accurate equipment reliability parameters, such as cable failure rate, mean time to repair each line, transformer failure rate, contact switching time and the number of fuses, etc. Therefore, the traditional methods are only applied to regional power system with simple structure and complete equipment reliability parameters. When it comes to large-scale grid with complex structure and missing equipment reliability parameters, researchers have proposed novel supply reliability prediction algorithms based on statistical methods which greatly reduce the dependence on the device parameters, and have broad application prospects. Song et al. [4] proposed a reliability prediction method for complex state grid based on BP neural network. Firstly, the important characteristics of variables that affect supply reliability indicators were determined, including the maximum load, the average length of overhead lines, line segments switch sets the average number of lines the average number of contact switches and other units; secondly, the neural network was trained using historical data, and then was used for prediction in objective urban grid reliability index.

This paper summarizes previous work and wavelet neural network is firstly applied for power supply reliability prediction. Meanwhile, the relevant factors characterizing basic properties of power grid are considered, such as the total number of users, the total length of the cable line, the Average Interruption Hours of Customer as well as the Average Interruption Times of Customer.

2 RELATED METHODOLOGY

In this paper, based on wavelet neural network, a novel kind of power supply reliability forecasting method is constructed. So, to begin with, the relevant methodology including wavelet theory and wavelet neural network is introduced.

2.1 Wavelet theory

Wavelet transform [5] is a time-frequency localization analysis method, changing both time and frequency window and the wavelet transform of the square integrable function $x(t)$ is defined as:

$$f_x(a, \tau) = \frac{1}{\sqrt{a}} \int_{-\infty}^{+\infty} x(t) \varphi^* \left(\frac{t - \tau}{a} \right) dt, \ a > 0. \tag{1}$$

where $\varphi^*(t)$ is the complex conjugate of wavelet base function $\varphi(t)$; τ is the location parameter, which determines the transformation results in the time domain; a is known as the scale parameter which changes the bandwidth of the filter, and thus determines the frequency information. From equation (1), it can be seen that the wavelet transform, as a filter on the signal, is the inner product of the original signal and wavelet function.

2.2 Wavelet neural network

Wavelet neural network [6] is based on Back Propagation (BP) neural network topology and regards the wavelet function as the hidden layer transfer function. Signals propagate forwards and simultaneously errors propagate backwards. Wavelet neural network combines the time-frequency localization features of wavelet transform with the self-learning and self-organizing capacity of traditional neural

network, which has been successfully applied to fault detection, pattern recognition, intelligent control and many other areas.

3-layer wavelet neural network comprises an input layer, a hidden layer and an output layer. In the hidden layer, wavelet function is regarded as the excitation function. During the training process of neural network, parameters are constantly revised to self-adaptively adjust the shape of wavelet function, and thus the network predicted output keeps approaching the desired output.

Let $x_1, x_2, \ldots x_k$ to be the input parameters of wavelet neural network. According to the algorithm of wavelet neural network, based on the sequence of the input signal, the output of the hidden layer is calculated as

$$h(j) = h_j \left(\frac{\sum_{i=1}^{k} \omega_{ij} x_i - b_j}{a_j} \right), j = 1, 2, \ldots, l. \tag{2}$$

where: l is the number of nodes in the hidden layer; $h(j)$ is the output value for the j-th hidden layer node; ω_{ij} is the connection weight between input layer and hidden layer; h_j denotes wavelet function with location parameter b_j and scale parameter a_j. The final output layer value of wavelet neural network is calculated as

$$y(k) = \sum_{j=1}^{l} \omega_{jk} h(j), k = 1, 2, \ldots, m. \tag{3}$$

where: m is the number of nodes in the output layer; $y(k)$ is the output value for the k-th output layer node; ω_{jk} is the connection weight between hidden layer and output layer.

The weights can be modified using a gradient correction method, so that the predicted output of wavelet neural network keeps approximating the desired output. The gradient correction process is described below:

1. Calculate the network prediction error

$$e = \sum_{k=1}^{m} (yn(k) - y(k)). \tag{4}$$

where: $yn(k)$ is the desired output; and $y(k)$ represents the wavelet neural network prediction output.

2. According to the prediction error e, both the wavelet neural network weights and wavelet function parameters are corrected as follows:

$$\omega_{n,k}^{j+1} = \omega_{n,k}^{j} - \eta \frac{\partial e}{\partial \omega_{n,k}^{j}}. \tag{5}$$

$$a_k^{i+1} = a_k^i - \eta \frac{\partial e}{\partial a_k^i}. \qquad (6)$$

$$b_k^{i+1} = b_k^i - \eta \frac{\partial e}{\partial b_k^i}. \qquad (7)$$

where η is the learning rate.

The basic steps of wavelet neural network modeling include network initialization, sample classification, forecasting output, weight amendments and termination conditions. Once the training process of wavelet neural network is completed, the network can be used for next step forecasting.

3 SIMULATION RESULTS AND DISCUSSION

In this section, empirical research is conducted to verify the feasibility and effectiveness of the proposed power supply reliability prediction method. To begin with, input parameters are selected which are closely related to power supply reliability rate and can characterize the basic properties of the power grid. In the second place, the structure of the wavelet neural network is determined. Then the well-trained neural network is employed to complete power supply reliability rate prediction.

3.1 Index selection

The power supply reliability evaluation system in China takes the average of the mean value management approach, based on power customers, and regulates interruption frequency, power outage duration and range as the basic concerned statistical factors. Among these, the most commonly used indicators are defined as follows.

Average Interruption Hours of Customer (AICH-1) refers to the average interruption hours during the statistical period, whose calculation formula is:

$$AIHC\text{-}1 = \frac{\sum t_i \times n_i}{N}. \qquad (8)$$

where: t_i is the interruption hours each time, n_i is the number of users for each outage and N is the total number of power users.

Average Interruption Times of Customer (AITH-1) refers to the average number of times during the statistical period and it is a reflection of interruption frequency regarding the power supply system, which is defined as:

$$AITH\text{-}1 = \frac{\sum n_i}{N}. \qquad (9)$$

Reliability on Service in total (RS-1) is the ratio of hours with reliable power supply and total hours of the statistical period. As a percentage, it reflects the reliability of the power supply system and is regarded as the core index in this paper whose formula is:

$$RS\text{-}1 = (1 - ATHC\text{-}1/T) \times 100\%. \qquad (10)$$

where T is the statistical period.

In order to characterize the basic features of the grid and avoid data which are difficult to collect, we elect the following two types of parameters as input parameters of wavelet neural network: the total number of power users and the total length of the cable line.

Table 1 shows totally five kinds of parameters [7, 8] of urban 10 kV grid from 2005 to 2013 in China index.

3.2 Prediction model

The goal of the numerical example is to forecast the power supply reliability in 2013 by means of wavelet neural network. The forecasting process can be demonstrated as follows:

Step 1. The structural design of wavelet neural network. With a wide range of applications,

Table 1. 10 kV city power supply reliability and correlative factors in China.

Year	User number	Cable line length (km)	RS-1 (%)	AITH-1	AICH-1
2005	894363	100353.00	99.7660	4.180	20.4910
2006	967354	142303.49	99.8494	3.046	13.1911
2007	1057056	168902.70	99.8817	2.434	10.3603
2008	1113966	201639.00	99.8626	2.715	12.0705
2009	1277298	220822.00	99.8960	1.998	9.1110
2010	1385020	274284.00	99.9230	1.735	6.7220
2011	1412383	266299.41	99.9200	1.220	7.0100
2012	1774323	314592.37	99.9485	1.100	4.5300
2013	1866400	352799.67	99.9582	0.733	3.6600

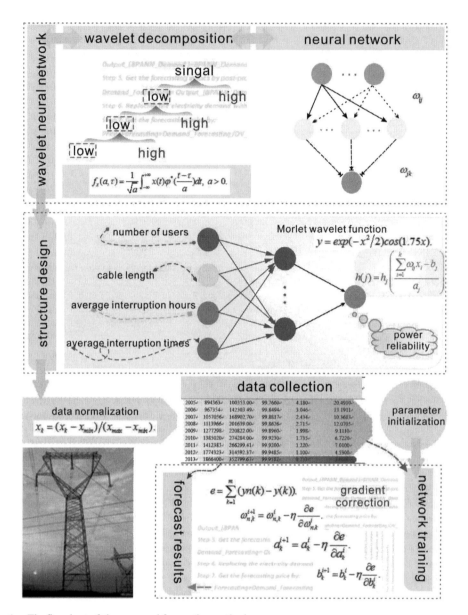

Figure 1.　The flowchart of the proposed forecasting method.

three-layer neural network is chosen. The number of the input and output layer nodes is 4 and 1 respectively. And according to the empirical formula, the number of the hidden layer nodes is determined as 6.

Step 2. The selection of wavelet function. After extensive testing, we finally selected Morlet wavelet neural networks as the excitation function, whose mathematical expressions is

$$y = exp(-x^2/2)cos(1.75x). \tag{11}$$

Step 3. The sample data normalization. In order to eliminate the adverse effects of magnitude difference on the neural network learning process, firstly sample data are normalized, which means:

$$x_k = (x_k - x_{min})/(x_{max} - x_{min}). \tag{12}$$

Step 4. The weights and parameters initialization. Then the sample is divided into training and test sets.

100

Step 5. By means of the well-trained wavelet neural network, the predicted power supply reliability of China's urban 10 kV grid is 99.9152%, 0.0376% lower than the real value 99.9528%. Therefore, the proposed power wavelet neural network method is feasible and effective for power reliability prediction. The flowchart of the proposed forecasting method is shown in Figure 1.

4 CONCLUSIONS

For the simulation example, it can be concluded that wavelet neural network is feasible for power system reliability rate forecast. Compared with traditional methods, forecast method based on wavelet neural networks relies less on equipment reliability parameters and is more suitable for large-scale grid with complex structure. In this paper, only a small part of related data is available, which affects the final prediction accuracy to a certain extent. However, the predicted results of the study can still reflect the effectiveness of the wavelet neural network prediction method.

In respect to determine the input parameters of wavelet neural network, several variables, which are closely related to the power supply reliability rate and are capable of characterizing the basic grid properties, are selected. This outcome has important reference value for subsequent research.

CORRESPONDING AUTHOR

Name: Zhi-yuan Liu, Email: liuzy_lz@126.com, Mobile phone: +86 13519660661.

REFERENCES

[1] M. Zeng, Z. Liang, Y. Ju, K. Tian. Evaluation on the Service Strategy for Power Supplying Enterprises Based on AHP, Technoeconomics & Management Research. 2(2010) 8–11 (in Chinese).

[2] R.L. Chen, K. Allen, R. Billinton. Value-Based Distribution Reliability Assessment and Planning, IEEE Transactionson Power Delivery. 10(1995) 421–429.

[3] X.L. Wang, S. Luo, S.Y. Xie, X. Wang, Y.L. Zhang. Reliability evaluation of distribution systems with meshed network based on the minimum-cut set, Power System Protection and Control. 39(2011)52–58 (in Chinese).

[4] Y.T. Song, J.L. Wu, D. Peng et al. A BP Neural Network Based Method to Predict Power Supply Reliability of Urban Power Network, Power System Technology. 32(2008)56–59 (in Chinese).

[5] G.J. Osórioa, J.C.O. Matiasa, J.P.S. Catalão. Short-term wind power forecasting using adaptive neuro-fuzzy inference system combined with evolutionary particle swarm optimization, wavelet transform and mutual information, Renewable Energy. 75(2015) 301–307.

[6] H. Chitsaza, N. Amjadyb, H. Zareipour. Wind power forecast using wavelet neural network trained by improved Clonal selection algorithm, Energy Conversion and Management. 89(2015) 588–598.

[7] Information on http://www.chinaer.org/list.aspx?m = 20100424125434250110.

[8] Information on http://chinaer.cec.org.cn/zhibiaofabu/linianzhibiao/.

Advances in Future Manufacturing Engineering – Yang (Ed.)
© *2015 Taylor & Francis Group, London, ISBN 978-1-138-02817-3*

Improved time series business expansion combination forecasting model

Xiang-dong Xu, Zhi-yuan Liu, Jing Wang, Long Li & Xi Song
State Grid Gansu Business Operation Monitoring Center, Lanzhou City, Gansu Province, China

ABSTRACT: According to the indicators, including industry users' capacity average utilization, industry-assembling capacity, industry users' electricity consumption and industry average capacity utilization and so on, the power consumption of whole society can be calculated with business expansion, given the characteristics of different industries of electricity. And the whole society electricity consumption of Gansu province will be forecasted consequently.

The time series theory will be taken into consideration to construct the exponential smoothing forecasting model, the first-order coefficient adaptive forecasting model and the second-order coefficient adaptive forecasting model. And the combination method is to improve the accuracy of business expansion, which combines two single forecasting models together to get the combination forecasting values. Furthermore, the PSO is also employed to optimize the parameter of each single model to get better performance. An optimal forecasting model will be chosen by comparing the forecasting error of each model.

Keywords: time series; combination model; PSO neural network

1 INTRODUCTION

Power load prediction has attracted a great deal of attention from both the practice and academia. The load forecasting has a definitive impact on the daily operations of a power utility [1]. It is used for various purposes, such as price and income elasticity, energy transfer scheduling, unit commitment and load dispatch. With the emergence of load management strategies, the load prediction has played a broader role in utility operations [2,3]. The aim of load forecasting is to make the best use of electric energy and relieve the conflict between supply and demand. Inaccurate forecast of power load will leads to a great deal of loss for power companies. Therefore, the development of an accurate, fast, simple and robust load forecasting model is important to electric utilities and its customers.

Since the early 1990s, the process of deregulation and the introduction of competitive markets have been reshaping the landscape of the traditionally monopolistic and government-controlled power sectors. In many countries worldwide, electricity is now traded under market rules using spot and derivative contracts. However, electricity is a very special commodity. It is economically non-storable, and power system stability requires a constant balance between production and consumption. At the same time, electricity demand depends on weather (temperature, wind speed, precipitation, etc.) and the intensity of business and everyday activities (on-peak vs. off-peak hours, weekdays vs. weekends, holidays and near-holidays, etc.).

With the reform of electric power system and the development of computer technology, construction of the power working process the for the must take the market and customer service as the center, to facilitate the customers for the purpose, optimize business processes, innovative service mode, strengthen supervision ability, improve the level of the enterprise decision-making and management. Through various ways to improve service quality, to develop electric power consumer market, to create better social and economic benefits for power supply enterprises, become the core concept of electric power marketing management information system.

This paper is organized as follows. Section 2 outlines the basic Exponential Smoothing forecasting model and Adaptive Exponential Smoothing forecasting model. Section 3 discusses the proposed forecasting model. We report the experiments and evaluation results on Gansu Province data set. Section 4 concludes this paper.

2 BASIC FORECASTING MODELS

2.1 Exponential smoothing

Exponential smoothing is prominent among the various techniques that can help to extrapolate past date into future trends [4,5]. It is a kind of method popular not only in production forecasting but in both medium-term and short-term economic development tendency forecasting. It takes advantage of whole data with different weights according to time index, which is exponential smoothing usually based on the premise that the level of time series should fluctuate about a constant level or change slowly over time [6]. Under such a premise, the travel time series $y(t)$ can be described by

$$y(t) = \beta(t) + \varepsilon(t). \tag{1}$$

where, $\beta(t)$ takes a constant at time t and may change slowly over time, $\varepsilon(t)$ is a random variable and is used to describe the effect of stochastic fluctuation.

Applied exponential smoothing forecast the expanding new capacity in the monthly industry and the average user capacity utilization rate for the next year.

Determine the parameter α to satisfy $0 < \alpha < 1$ and the initial value $s_0 = x_1$.

Therefore, in the exponential smoothing series $s_t = \alpha x_t + (1-\alpha)s_{t-1}$, the forecasting value x_{t+1} at time $t+1$ can be replaced with the smoothing value s_t at time t.

Consequently, we get:

$$\hat{x}_{t+1} = \alpha x_t + (1-\alpha)\hat{x}_{t-1}. \tag{2}$$

2.2   Adaptive exponential smoothing

Sometimes, it is necessary to change the smoothing parameter α used in exponential smoothing when the rate at which β changes over time changes. This suggests that an adaptive smoothing parameter would produce improved forecasts.

Adaptive exponential smoothing method is the further development of exponential smoothing, the traditional static smoothing parameter in solving practical problems tend to produce bigger errors. As data changes, the original value may no longer applicative. First order adaptive exponential smoothing considers to revise the value of α constantly according to changing conditions to make better forecasting results [6,7].

Replace α with α_t, the formula (2) is transformed into:

$$\hat{x}_{t+1} = \alpha_t x_t + (1-\alpha_t)\hat{x}_{t-1}. \tag{3}$$

Easily, α_t should be increase if the forecasting errors are larger than a certain value (for example, all the e_t are positive or negative), which means the larger the forecasting errors are, the larger α_t is to strengthen the modification magnitude.

Introduce two error signals: $E(t)$ and $M(t)$. Define another constant $\beta, 0 < \beta < 1$, and exert Exponential Weighted Average on the first t times $e_K (K = 1, 2, ..., t)$:

$$E_t = \beta e_t + \beta(1-\beta)e_{t-1} + \cdots \beta(1-\beta)^{t-1}e_1. \tag{4}$$

The absolute value of $E(t)$, $|E_t|$ can display the forecasting error.

We can easily get:

$$E_t = \beta e_t + (1-\beta)E_{t-1}. \tag{5}$$

In order to make α_t satisfies $0 < \alpha_t < 1$, similarly, we get the recursion formula

$$M_t = \beta|e_t| + (1-\beta)M_{t-1}. \tag{6}$$

Therefore, set $\alpha_t = |E_t|/M_t$ to satisfy the requirements.

2.3   Combination model

Combination forecasting is one of the best ways to improve the prediction accuracy, since neither linear nor nonlinear model is a universal model that is suitable for all situations. Since it is difficult to completely know the characteristics of the time series data in a real problem, a combination strategy that has both linear and nonlinear modeling abilities is a good alternative for forecasting the seasonal time series data [8]. Both the linear and nonlinear models have different unique strength to capture data characteristics in linear or nonlinear domains, so the combination models proposed in this study is composed of the linear component and the nonlinear component. Thus, the combination models can model linear and nonlinear patterns with improved overall forecasting performance.

The precision of combination model prediction mainly depends on the weight of each single forecasting model, and here are two basic forms: equal weights and unequal weights. The principle and using method of these two forms are exactly the same, ranging from right combination of combination forecast method is used to more accurate results.

2.4   Particle swarm optimization algorithm

We consider the parameters of the prediction model using the method of artificial intelligence [10]. Although the practical application of exponential smoothing is effective, but considering it is difficult to artificially select optimal parameters to achieve the smallest prediction error is insufficient, so this project will be through the use of artificial intelligence algorithm for parameter optimization to solve the problem.

Particle swarm optimization algorithm is a kind of random optimization algorithm based on swarm intelligence. The algorithm simulation bird flying foraging behavior, cluster by bird collective collaboration between groups [11,12].

Assume the searching space with D dimension, and the number of the particle is n. The information of particle I can be represented by a vector with D dimensions—the location is $X_i = (x_{i1}, x_{i2}, ..., x_{il})^T$ and the speed is $V_i = (v_{i1}, v_{i2}, ..., v_{il})^T$ and so is other items. Therefore, the update formula of location and speed is:

$$v_{id}^{k+1} = wv_{id}^k + c_1 rand_1^k (pbest_{id}^k - x_{id}^k)$$
$$+ c_2 rand_2^k (gbest_d^k - x_{id}^k). \quad (7)$$

$$x_{id}^{k+1} = x_{id}^k + v_{id}^{k+1}, 1 \le i \le n, 1 \le d \le D. \quad (8)$$

where, w is *inertia weight*, v_{id}^k is the speed of particle i in the d dimension kth recursion, x_{id}^k the present location of particle i in the d dimension kth recursion, $pbest_{id}$ is the coordinate of extreme point of particle i in the d dimension kth recursion; c_1, c_2 is acceleration coefficient (or learning factor), $rand_1()$ and $rand_2()$ are two random functions ranging in [0, 1].

3 EXPERIMENTS AND DESIGN

Due to the increase in quantity sold mainly comes from the increase of industrial customers and commercial electricity power consumption. Because of these two types of users implement discriminatory prices pricing policy, bring electricity and electricity price double high sell electricity income increase.

3.1 *Data set*

On the basis of the eight major industry business expanding new capacity, is derived for industry expansion referring to the efficiency and increasing the income of electricity and sale volumes of electricity.

3.2 *Evaluation indicators*

Prediction accuracy is an important criterion for evaluating a forecasting algorithm. We use three kinds of evaluation metrics to evaluate the prediction accuracy: the accuracy of forecasting a single point, the overall accuracy of forecasting multiple points, and the accuracy of prediction trend.

A usual way to examine a single point forecasting accuracy of a model is to evaluate the Prediction Error (PE) for the ith year by comparing the actual sequence $\{x^0(i)\}_{i=1}^n$ to the forecasting sequence $\{\hat{x}^0(i)\}_{i=1}^n$.

$$PE(i)(\%) = \left| \frac{x^0(i) - \hat{x}^0(i)}{x^0(i)} \right|, \quad i = 1, 2, ..., n. \quad (9)$$

Mean Absolute Percentage Error (MAPE)

$$MAPE = \frac{1}{n} \sum_{i=1}^n \left| \frac{\hat{y}_i - y_i}{y_i} \times 100 \right| \quad (10)$$

Root of Mean Squared Error (RMSE)

$$RMSE = \sqrt{\frac{1}{n} \sum_{i=1}^n (\hat{y}_i - y_i)^2} \quad (11)$$

4 EXPERIMENTS AND RESULTS

Apply the Exponential Smoothing (ES), First order Adaptive Exponential Smoothing (F_AES), Second order Adaptive Exponential Smoothing (S_AES) and their combination model, the Exponential Smoothing-First order Adaptive Exponential Smoothing (ES-F_AES), the Exponential Smoothing-Second order Adaptive Exponential Smoothing (ES-S_AES) and First order Adaptive Exponential Smoothing-Second order Adaptive Exponential Smoothing (F_AES-S_AES) to the electricity data originated from the business expanding data in Gansu Province.

From Table 1, the best forecasting model among these six forecasting models is the combination model of the Exponential Smoothing and Second order Adaptive Exponential Smoothing, which is ES-S_AES.

Table1. The MAPE and RMSE of each model.

	ES	F_AES	S_AES	ES-F_AES	ES-S_AES	F_AES-S_AES
MAPE	4.45%	4.61%	4.64%	4.53%	4.27%	4.35%
RMSE	4.3283	4.2593	4.2641	4.2091	4.0152	4.108

5 CONCLUSIONS

The purpose of this paper is to evaluate the performance of existing forecasting models based on electricity sale data originated from the business expanding data. The main contribution of this paper is twofold. First, an introduction of how to get electricity sale data from the business expanding data is proposed. Second, a combination model is proposed to overcome the drawbacks of previous exponential smoothing and adaptive exponential smoothing model. The results of experiments show the forecasting performance of ES-S_AES is superior to other models according to MAPE and RMSE.

However, each model seems to have its own strength and weakness, and have a sort of typical or optimal forecasting horizon; no single model is expected to be the best in all aspects. Our future study will be laid on a versatile forecasting model that combines the existing forecasting models and has good performance on all spans of forecasting horizons and electrical conditions.

CORRESPONDING AUTHOR

Name: Zhi-yuan Liu, Email: liuzy_lz@126.com, Mobile phone: +86 13519660661.

REFERENCES

[1] Jianzhou Wang, Wenjin Zhu, Wenyu Zhang, Donghuai Sun, A trend fixed on firstly and seasonal adjustment model combined with the epsilon-SVR for short-term forecasting of electricity demand, J. Energy Policy 37 (2009) 4901–4909.

[2] Jianzhou Wang, Suling Zhu, Wenyu Zhang, Haiyan Lu, Combined modeling for electric load forecasting with adaptive particle swarm optimization, J. Energy. (2010) 35 1671–1678.

[3] Bretschnerder S. Estimating forecast variance with exponential smoothing Some new results, J. International Journal of Forecasting. (1986) 2(3) 349–55.

[4] Billah B, King ML, Snyder RD. Exponential smoothing model selection for forecasting, J. International Journal of Forecasting. (2006) 22(2) 239–47.

[5] Taylor JW. Exponential smoothing with a damped multiplicative trend, J. International Journal of Forecasting. (2003) 19 715–25.

[6] Billah B, King ML, Snyder RD, Koehler AB. Exponential smoothing model selection for forecasting, J. International Journal of Forecasting. (2006) 22 239–47.

[7] Yu Zhenming, GuoYajun, Yang Zheng, et al. Combination forecast based on experts' rational expectation method and its application, J. Journal of Northeastern University: Natural Science. (2010) 31(6) 902–905

[8] Fang-Mei Tseng, Hsiao-Cheng Yu, Gwo-Hsiung-Tzeng. Combining neural network model with seasonal time series ARIMA model, J. Technological Forecasting & Social Change. (2002) 69 71–87.

[9] Uraikul V, Chan CW, Tontiwachuthikul P. Artificial intelligence for monitoring and supervisory control of process systems, J. Engineering Applications of Artificial Intelligence. (2007) 20(2) 115–31.

[10] Hedayat A, Davilu H, Barfrosh AA, Sepanloo K. Optimization of the core configuration design using a hybrid artificial intelligence algorithm for research reactors, J. Nuclear Engineering and Design. (2009) 239(12) 2786–99.

[11] Kennedy J, Eberhart R. Particle swarm optimization, J. Proceedings of IEEE International Conference on Neural Network. (1995) 1942–48.

[12] Navaiertpron T, Afzulpurkar NV. Optimization of tile manufacturing process using particle swarm optimization, J. Swarm and Evolutionary Computation. (2011) 1(2) 97–109.

Advances in Future Manufacturing Engineering – Yang (Ed.)
© 2015 Taylor & Francis Group, London, ISBN 978-1-138-02817-3

Experimental studies on pipe stress testing based on inverse magnetostriction effect

Shixiong Yuan, Haimin Guo & Rui Deng
Key Laboratory of Exploration Technologies for Oil and Gas Resources, Ministry of Education,
Yangtze University, Wuhan, China
School of Geophysics and Oil Resources, Yangtze University, Wuhan, China

ABSTRACT: With magnetic anisotropy technique based on inverse magnetostriction effect, we make experimental studies on stress behavior of the pipe. Exert pressure on the pipe on test machine, and detect the magnetic signals under the load action with special measuring set, we find the way to seek the best frequency and sensitivity of the detection probe. When one point of the pipe has force, we detect the vertical and annular magnetic response through dividing the pipe into 72 measurement points, then analyze the circular force, the results indicate that there is good correlation between the magnetic signals distribution and the stress distribution.

Keywords: inverse magnetostriction; magnetic probe; non-destructive testing; stress and strain

1 INTRODUCTION

Under complicated load, the service life of ferromagnetic material pipe column will reduce. In order to monitor the pipe column's operation condition and prolong its service life, the characteristic parameters should also be tested other than necessary corrosion protection. The stress distribution is one of the important parameters, and the engineers attach great importance to its research method.

The earliest stress testing method is Barkhausen Effective Stress Testing Method which was proposed by H. Barkhausen (1919) and based on noise signals caused by discontinuous magnetic domain and magnetic domain walls. This method is the most widely used method which is greatly affected by the material structure and impurity. Based on the phenomenon of metal magnetic memory, Chen Yuling (2004) and other people made experimental study on the test of stress concentration. This testing method is also a new method which cannot be tested quantitatively. Based on the technology theory and experiments offluxoid stress testing of magnetomechanical effect, XiongErgang (2007) and other people made study. Based on inverse magnetostriction effect, the engineers propose a new stress testing method now. Yao Shixuan (2007) and other people made theoretical deduction and experimental study on the mechanism of magnetostrictive sensor, which made this stress testing method feasible. Based on the magnetic anisotropy testing technique of inverse magnetostrictive effect, this paper makes an experimental study on the condition of pipe column under load.

2 METHOD AND PRINCIPLE

Magnetostriction refers to a phenomenon that the ferromagnet volume or shape will change correspondingly when its magnetization state is changed. Magnetostriction phenomenon is reversible, namely, the magnetization will also change when the ferromagnet deformation occurs. This is inverse magnetostrictive effect or piezomagnetic effect, materials are magnetic isotropic when the ferromagnet is in unstressed state and materials are anisotropic when the ferromagnet is in stress state (Wang Wei, 2005; Wen Xiqin, 2002; Yao Shixuan, 2007). So the change of stress or strain will cause the change of magnetic permeability and reluctance of the ferromagnetic materials. The change of permeability can reflect the average degree of change of stress or strain on the material surface or in certain depth of the material. When constant magnetomotive force is provided to the magnetic probe, the change of permeability in magnetic circuit will cause flux change. The average level and distribution of stress and strain on the specimen surface or in a certain depth can be measured by measuring the induced electromotive force change of induction coil (Wang Wei, 2005).

According to the effect of stress on magnetic properties and reference to the results of relevant researchers, this paper presents a magnetics model which is about the relation between the stress and the change of relative permeability of ferromagnetic material. The relation is

$$\Delta\mu = (2k/B^2)\,\sigma\mu^2 \tag{1}$$

$\Delta\mu$ is permeability variation, k is magnetostrictive coefficient, B is magnetic induction intensity, σ is stress, μ is permeability. Therefore, the change of stress can be analyzed through studying and analyzing the change of magnetic flux.

3 EXPERIMENTAL MATERIAL AND METHOD

3.1 Magnetic signal receiving device

For the same size of different ferromagnetic materials, different mechanical properties (such as stress transformation, hardness, strength, structure, heat treatment state), different surfaces and inner defects, the current values induced by them in the magnetic field are different.

A number of experiments show that the values induced by them in the magnetic field has a corresponding relation which can be found out by this nondestructive tester through a large number of pre-measurements. The measurement results are compared and showed with unit of magnetic flux so that the test of different performance parameters of the measured pieces with same specification size can be conducted.

3.2 Experimental preparation process

The testing pipe column should be smooth steel pipe column without any damage. The experiment should be conducted after the pipe column is laid flat for a period of time so that the effect caused by material processing can be eliminated to the maximum extent. The dimension of experimental column: length is 610 mm, outer diameter is 272 mm, inner diameter is 263 mm, and thickness is 9 mm. As shown in Figure 1 and Figure 2, 8 points of the pipe column are measured in horizontal direction and 9 points are measured at intervals of 50 mm in vertical direction.

In order to get the magnetic response of the whole cylinder, we divide the pipe column into 72 points to conduct measurement in circumferential direction and get the magnetic response distribution of the whole pipe column through interpolation.

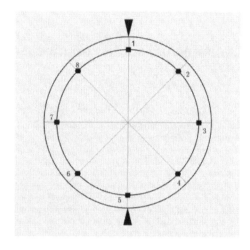

Figure 1.   Cross section graph of pipe model.

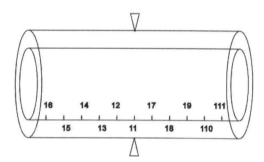

Figure 2.   Vertical section graph of pipe model.

4 RESULTS AND DISCUSSION

4.1 Sensitivity and frequency

In general, lower the detecting frequency, deeper the detection, and vice versa. Because of the uncertainty of the measured pieces' state, the selection of detecting frequency should depend on the actual situation of the measured pieces. The method of frequency selecting: in the pipe column without load, adjusting compensation knob to make magnetic flux near "0", then testing the pipe column with load. The value of magnetic flux should not exceed the upper limit value and lower limit value of the instrument when external force is the greatest. If exceeding, "0" needs to be moved up or down. Sensitivity should be reduced if it still doesn't enter the visual range.

A proper detecting frequency and sensitivity should be selected when any workpiece of some specifications is tested for the first time. When it has been selected, this instrument state can be directly used.

With the increase of frequency, the magnetic flux will increase, and there is a linear relationship between them within a certain range. As shown in Figure 3, the abscissa is frequency, the ordinate is magnetic flux, and the distance between the two lines' corresponding points becomes larger and larger with the increase of abscissa.

4.2 *The relation between magnetic and external force*

Putting the probe in the place of 0 mm and 0-degree angle, adjusting external force, can measure the response of magnetic signal. As can be seen from Figure 4, the magnetic signal becomes large with the increase of external force, and there is a linear relationship between magnetic and external force. There is also a linear relationship between stress and permeability.

Putting the probe in the place of 150 mm and 0 degree angle, namely 150 mm away from the pressure point. The magnetic signals are unlikely to change before the external force reaches 20 KN. However, the magnetic signals change greatly after the external force reaches 20 KN.

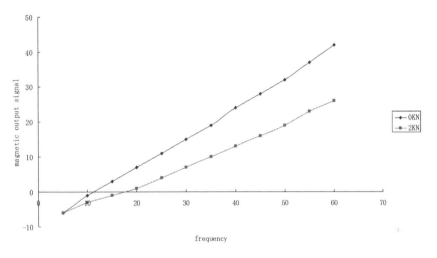

Figure 3.   The relation between magnetic signal and frequency.

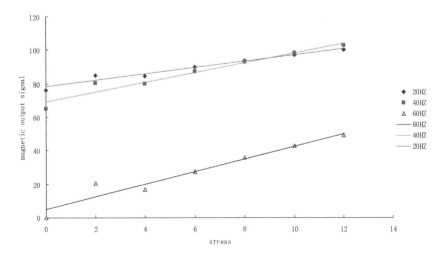

Figure 4.   The relation between magnetic signal and external force in the place of 0 mm.

109

From the distribution of magnetic field in Figure 5, we can know that the magnetic field value increases with the increase of load and has a redistribution of magnetic field relative to the zero load.

4.3 *The magnetic signals changes with measuring angles*

Figure 6 shows the distribution of different magnetic signals from 0 degree angle to 180 degree angle in the place of 0 mm, which is completely symmetrical to the distribution from 180° to 360°, and they have the same change law.

In the process of measurement, the circumferential measurement of pipe column is conducted under various load conditions. The load range is from 0 to 30 KN, and the load progressively increases by 5 KN. Figure 6 shows the curve of magnetic output and measurement angle when the pipe load is 15 KN. The curves are the same under other load conditions, which won't be shown here.

As can be seen from the figure, the relation between magnetic output and angle is similar to the sine function. The magnetic signal is strongest at 0 degree angle and begins to decrease from 45 degree angle and 315 degree angle. It is the weakest at 90 degree angle and 270 degree angle

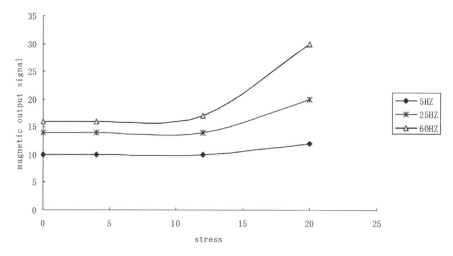

Figure 5.   The relation of magnetic signal and external force in the place of 150 mm.

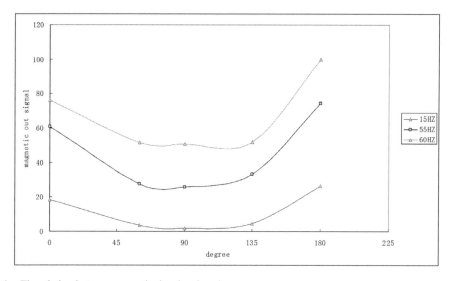

Figure 6.   The relation between magnetic signal and angle.

110

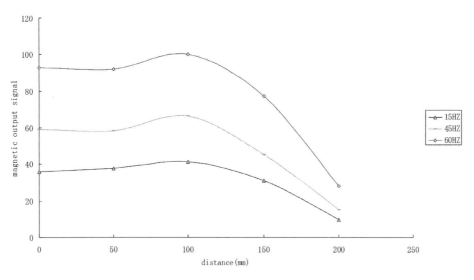

Figure 7.   The relation between magnetic signal and distance.

and the strongest at 180 degree angle. The stress distribution is the sum of two opposite cosine functions, which is the same as our measurement result regardless of positive and negative values.

4.4   *The magnetic signal changes with the distance*

As can be seen from Figure 7, the farther from the pressure point is, the smaller the magnetic flux. The reason why the values from 0 to 100 mm don't change is that the pressure is a linear pressure. The three curves refers to the magnetic signals under different frequencies.

5   SUMMARY

Based on inverse magnetostriction effect, we make systematical experimental studies on the flux response when the pipe column is under load and we also study the relation between magnetic flux and angle, the relation between magnetic flux and distance. The farther from the point of force application, the smaller the magnetic flux. Unfolding the pipe column, we can get the distribution of magnetic flux and find that it has some correspondence with the external force. After analyzing the magnetic flux on the circumference, we can find whether there is a stress concentration in some area of the pipe column, which is valuable for us to know the property of pipe material and monitor pipe condition.

ACKNOWLEDGEMENTS

This work is supported by National Natural Science Foundation of China (Grant No. 41474115 and Grant No. 41474116) and open fund of Key Laboratory of Exploration Technologies for Oil and Gas Resources (Yangtze University), Ministry of Education (NO. K2013-02).

REFERENCES

[1] Abuku S. Magnetic studies of Residual Stress in Iron and Steel induced by Uniaxial Deformation [J]. Japanese Journal of Applied physics, 1977, 16(7):1161–1170.
[2] Langman R. Measurement of the mechanical stress in mild steel by means of rotation of magnetic field strength [J]. NDT & E International, 1981, 14(5): 255–262.
[3] Langman R. Measurement of the mechanical stress in mild steel by means of rotation of magnetic field strength-part 2: Biaxial stress [J]. NDT & E International, 1982, 15(2): 91–97.
[4] Langman R. Measurement of the mechanical stress in mild steel by means of rotation of magnetic field strength-Part 3: Practical applications [J]. NDT & E International, 1983, 15(2): 59–65.
[5] Deng Rui, MengFanshun. Influence Factors of Marine Riser Stress Detection Using Magnetic Anisotropy [A], Proceedings of the Twentieth (2010) International Offshore and Polar Engineering Conference, ISOPE2010, Vol 2, pp. 37–40.

Advances in Future Manufacturing Engineering – Yang (Ed.)
© 2015 Taylor & Francis Group, London, ISBN 978-1-138-02817-3

Research on synthesis and electrochemical properties of cathode materials for lithium ion batteries based on LiFePO$_4$

Lingju Meng, Junfeng Gao, Zhaokun Xuan, Jinyu Liu, Yan Wang,
Liansheng Jiao & Hongli Chen
Department of Chemistry, Hebei Normal University for Nationalities, Chengde, China

ABSTRACT: This paper makes use of various synthetic methods and analysis techniques to make a systematic research on lithium iron phosphate from material preparation, structure characterization of electrochemical performance, and prepare electrode material with good performance. The paper aims at the two fatal disadvantages of LiFePO$_4$ cathode material properties that are low electron conductivity and low lithium ion diffusion rate. Take the material particles, particle refinement surface deposition of carbon conductive layer and Mg^{2+} ion doping and other measures to modify and explore, in order to improve the electrochemical performance of LiFePO$_4$ cathode material. Electrochemical performance tests show that the initial discharge cathode materials for lithium ion LiFePO$_4$ synthesis by microwave can reach 118 mAh$\cdot$g^{-1} ratio capacity. Preparation of the material is high nearly 20% compared with the traditional high temperature solid state method.

Keywords: lithium cathode materials; LiFePO$_4$/C composite; grain refinement; microwave synthesis

1   INTRODUCTION

Entering the twenty-first Century, the relationship between the electrochemical energies are drawing more and more attention. Lithium ion battery is a new type with environmentally friendly new energy, is rapidly entering into human society in various fields, getting the maximum utilization, and has largely been able to affect people's daily life. Lithium ion battery has high energy density, high output voltage, long cycle life, fast charging and discharging can be safe. Compared with lithium metal batteries, lithium ion batteries have short circuit resistance, impact resistance, and shockproof, anti-overcharge and over discharge characteristics. And lithium ion batteries do not contain cadmium, lead, mercury and other harmful metals, will not cause pollution to the environment. It is the most development potential third generation cell with environmental protection. Lithium ion battery has no memory, can charge or discharge at any time repeatedly. At the same time, the lithium ion battery has excellent low-temperature discharge performance. In view of the excellent performance of lithium ion battery, lithium ion has become a research focus of new environmental protection energy. While the performance of cathode materials for lithium ion directly determines the quality of the performance of lithium ion battery.

From the development of the positive and negative electrode materials of view, at present what restrict the lithium ion battery energy density ascension are the cathode materials. Therefore, the research and development of new cathode materials system replace the current widely used LiCoO$_2$ cathode materials, introduce higher energy, cheaper prices. A new generation of lithium ion battery is safe and reliable, and has the important application value and practical significance. LiFePO$_4$ is expected to become one of the new generation of lithium ion battery cathode materials. It will be extended to the electric vehicle, the material of choice for a large industrial battery in the field of pumped storage power station, military weapons such as high capacity and high power.

2   STUDY ON THE SYNTHESIS OF LiFePO4$_4$ MATERIALS

The conventional heating is an external heat source by radiation or convection mode, from the outside to the inside or by means of heat conduction bottom heating; microwave penetration ability can even deeper into the sample interior in a short period of time, so that the heating of the sample center temperature rises rapidly, the whole sample almost at the same time is evenly heated; thereby greatly shorten the heating time, the whole experimental sintering process need only ten minutes. Compared with the traditional solid heat for several hours, microwave synthesis has the advantages of short reaction time, sensitive reaction, high efficiency and low energy consumption[1–4].

In addition, because the strength of material for microwave absorption ability mainly depends on the consumption coefficient of the material, consumption coefficient is the ratio of the dielectric loss factor and dielectric constant. The dielectric constant of measurable substance can prevent microwave penetrate its ability, capacity and loss coefficient reflects the material dissipation of microwave. At certain frequencies, consumption coefficient of the sample is greater, it has stronger the ability for absorbing microwave[5], along with prolonging the radiation time of microwave system temperature rise quickly, 2 min can reach 197 DEG, 4 min can reach 394 DEG, 8 min can achieve 732 DEG, reaction is complete. Microwave heating efficiency is particularly high. The time is too long will cause overheating and burning sample.

2.1 *The experimental process*

The process of the microwave synthesis method is as shown in Figure 1. The main raw materials are Lithium acetate, Ferrous oxalate, Two ammonium hydrogen phosphate. According to the stoichiometric ratio of target compounds (molar ratio) Li:Fe:P = 1:1:1 weighing lithium acetate, ferrous oxalate, ammonium hydrogen phosphate two chemical reagent, used alcohol as dispersant to disperse in agate mortar constantly stirring, broken, to mix evenly; the mixture is then placed in the magnetic boat into the tube heating type heating furnace, temperature control is 65 DEG C, the drying time is 4 h, in the dry process it needs to pass into the protective gas (such as nitrogen or argon) protection, gas flow rate is about 40 L/h. After cooling the sample aliquots into N parts, and adding a small amount (about 2%) of iron, the material and the hard alloy ball milling tank are arranged according to the mass ratio of 1:10 hard alloy, ball milling in star type ball mill: speed is 300 rad/min, time is 3 H; after ball milling powder

in the pressure under the pressure of 98 Mpa for mechanical compression. Rated frequency of a microwave oven is 2.45 GHz, power is 700 W, heating time is different, to find out the optimal heating time according to the result of the measurement.

2.2 *Assembly of the battery and its test of electrochemical performance*

The preparation of the electrode charge discharge and cyclic performance testing: according to the positive electrode active material, a conductive acetylene black: Polytetrafluoroethylene (PTFE) = 75:20:5 ratio mixing, mechanical milling in 3 h and drying at 120 DEG in 6h to prepare cathode material. Taking lithium as the electrode, using Merck company's LP30 as the electrolyte (the composition of mass ratio of EC:DMC = 1 to 1,1 mol/ LLiPF6), electrode diaphragm for polypropylene micro pore membrane (Celgard2400), battery assembly in argon glove box; using the LAND charge and discharge system testing battery charge and discharge performance and cycle performance. The charge and discharge interval is 2.8 ~ 4.2 V, current density 0.1 mA/cm^{-2} (about 0.05 C).

3 CHARACTERIZATION OF THE PHYSICAL AND CHEMICAL PROPERTIES OF MATERIALS

3.1 *Phase characterization*

Figure 2 shows P1 (6 min), P2 (9 min), P3 (10 min), P4 (12 min) XRD spectra deriving from different time by microwave heating, comparing with XRD diffraction spectrum can be found. Lithium iron phosphate compounds have four kinds of products in P2, P3 contains only pure phase. P1 microwave heating products due to lack of time, there has been a lot of impurity phase peak, almost no lithium iron phosphate phase. And the diffraction peak is not sharp; strength is very low, indicating that this reaction process is not really. Although products have P4 high intensity, the diffraction peak is sharp, FWHM is relatively narrow, but there obvious impurity phase (peak of Fe$_2$O$_3$ diffraction peaks in Figure 2 cross star calibration),

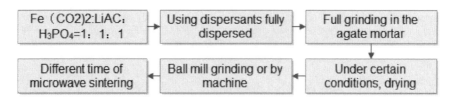

Figure 1. Process chart of the procedure of microwave synthesizing.

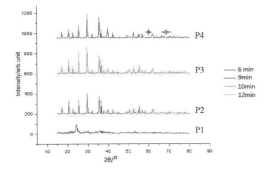

Figure 2. XRD pattern of sample P1, P2, P3 and P4.

Table 1. Structural parameter of several synthetics and standard sample.

Type	a/nm	b/nm	c/nm	c/a	V/nm³
P1	–	–	–	–	–
P2	0.6010	1.030	0.4689	0.7802	0.2903
P3	0.6011	1.034	0.4697	0.7814	0.2919
P4	0.6013	1.036	0.4707	0.7828	0.2932
Standard sample	0.6008	1.033	0.4693	0.7811	0.2914

which indicates that the reaction temperature has gone too far, too high cause Fe^{2+} ion is oxidized to Fe^{3+} ions, resulting in the XRD spectrum appeared obvious Fe_2O_3 impurity phase peak chart. Figure 2 XRD spectra also show that, although the product P2, P3 phase is pure, P2 diffraction peak intensity than P3, indicating its reaction has not yet reached the perfect condition. Visible, microwave heating time is too long or too short will not ferrous phosphate lithium cathode materials for synthesis of relatively pure, and also proves that the optimization time is 10 min of microwave synthesis.

3.2 The determination of crystal parameters

Table 1 lists completely the crystal parameters of $LiFePO_4$ crystal cell parameters and experimental preparation of the three samples, from Table 1 can be found that only the samples closest to P3. The longer microwave sintering time, changed in different degree of crystal field parameters. The crystal parameters of P2 are small, the crystal parameter P4 is too large; P4 compared with P2, the parameter of P4 increased 0.05%, C parameters increase 0.3%. Although there are some differences for P3, the little difference can be neglected, and the standard phase parameters agree. So you can think of the time of microwave heating of sample P3 is the ideal time for sintering microwave synthesis of

lithium ion battery anode material lithium iron phosphate $LiFePO_4$.

3.3 The morphology characterization of lithium iron phosphate

Using Scanning Electron Microscopy (SEM) observe samples of P1, P2, P3, P4 by microwave heating, the results are shown in Figure 3. Samples P1 does not have enough sintering time, crystallization has not yet happened; crystalline character of the sample P2 is incomplete, the existence has coarse particles; samples of P3 crystallize fully from the particle surface morphology, angular, smooth surface, grain full, particle size distribution in the 1 μm; sample dispersion of P4 is good, but the particles grew significantly. The particle becomes extremely thick, it will make the lithium ion diffusion distance in the charge discharge cycle variable length, and is not conducive to the lithium ion embedded in the process of charge and discharge and prolapse.

3.4 B.E.T analysis

The test results are shown in Table 2. From the table we can see that, with the increase of microwave irradiation time, the specific surface area of the samples is also changed. For the samples P1, the specific surface area that reaction has

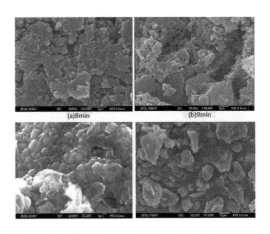

Figure 3. SEM images of all samples irradiated by microwave for different period of time.

Table 2. B.E.T results of different samples.

Type	P1	P2	P3	P4
B.E.T surface area (m²g⁻¹)	53.245	21.3548	17.5468	2.5614

not yet been carried out, the specific surface area decreases for the sample P2, continuing to increase the time of microwave irradiation. Specific surface area of synthesis P4 sample is only 2.5614 $m^2 \cdot g^{-1}$ after 12 min; on the other hand, it also reflects the change of sample particle size.

3.5 *The electrochemical performance test*

Figure 4 is the electrochemical performance test curve that synthesized cathode materials in a different time and under the microwave irradiation. The initial discharge capacity curve shows that the current density is 0.05 C in the process of charge and discharge, synthetic material not capacity curve by microwave irradiation in 6 min, because the irradiation time is too short. The reaction temperature is too low and cause the reaction has not yet been; microwave irradiation after 10 min, the synthesis of positive electrode material of initial discharge capacity can reach 118 $mAh \cdot g^{-1}$, but the time of microwave irradiation at 12 min. The synthesis of the initial discharge specific capacity is only 100 $mAh \cdot g^{-1}$, which is due to microwave sintering time is too long, resulting in high temperature. So that the material grain size is too large, the diffusion rate of lithium ion becomes small, also appeared the new phase, resulting in the electrochemical properties of the materials quickly attenuating.

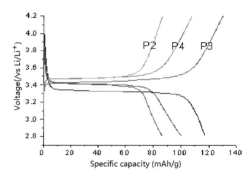

Figure 4. The charging/discharging profiles of different samples.

4 CONCLUSIONS

1. Microwave method can prepare $LiFePO_4$ compounds with good performance in a short time, low energy consumption condition. Under these experimental conditions, the optimum microwave heating time is 10 min.

2. Microwave heating of the sample with 10 min can reach 118 $mAh \cdot g^{-1}$ in the first discharge specific capacity. The reason is that under these conditions, $LiFePO_4$ crystalline preparation is complete. Powder particle size, thus partially overcomes the unfavorable factors of $LiFePO_4$ lithium ion diffusion rate are low in the charge discharge cycle in. So, microwave synthesis method can refine the grain size and improve the electrochemical performance of $LiFePO_4$.

ACKNOWLEDGEMENT

Hebei province science and technology support project 13214408D.

REFERENCES

[1] Shi Zhicong, Li Chen, Yang Yong et al. Research on lithium iron phosphate positive electrode material of electrochemical performance of novel [J]. Electrochemistry, 2003, 9 (2): 9 ~ 14.

[2] Li Haiying, Zhai Xiujing, Fu Yan et al. Microwave synthesis and structure characterization of $LiCoO_2$ cathode materials for lithium ion batteries [J]. Journal of Molecular Sciences, 2002, 18 (4): 199 ~ 203.

[3] Zhai Xiujing, Sun Xiaoping, Tian Yanwen et al. Research on lithium iron phosphate positive electrode material of electrochemical performance of novel [J]. Journal of Molecular Sciences, 2002, 18 (4): 199 ~ 203.

[4] Lu Junbiao, Zhang Zhongtai, Tang Zilong et al. A new type of cathode material—$LiFePO_4$ for lithium ion batteries [J]. Rare metal materials and engineering, 2004, 33 (7): 679 ~ 683.

[5] Zheng Honghe, Shi Lei, Zhao Yang et al. Nano cathode materials for lithium ion batteries [J]. Chemical bulletin, 2005 (8): 591 ~ 600.

Materials processing and analysis

Advances in Future Manufacturing Engineering – Yang (Ed.)
© 2015 Taylor & Francis Group, London, ISBN 978-1-138-02817-3

Structural and mechanical features of formation of high-speed surface layers of flame sprayed NiAl

P.O. Rusinov & Zh.M. Blednova
Krasnodar, Russia

ABSTRACT: The optimum process conditions in NiAl-surface modification of steel 1045 involve high-speed gas-flame spraying of mechanically activated NiAl powder. The control processing parameters that control the structural state of the material at the stage of the surface-modified layer and the subsequent combined treatment are established, which influence the properties of the shape memory effect of the surface layer of NiAl. We achieve high cycle fatigue resistance, and increased wear of the layers of alloys with the shape memory effect based on NiAl steels, in combination with giving them new functional properties.

Keywords: intermetallics compound; shape memory effect; nikelid aluminum; martensite; austenite; endurance limit; wear rate; mechanical activation

1 INTRODUCTION

Various materials with a Shape Memory Effect (SME) have been successfully realized in modern units and structures. One of the areas of applied research in the field of semi-finished products and technologies using alloys with the shape memory effect is their creation on the basis of detachable joints and parts [1]. Due to the effects of power generation and the stress relaxation layer of the surface-modified alloy, thermomechanical memory, which is only a small fraction of the total mass, may be provided with a new functionality of parts and structures. A widely known intermetallics compound is based on nickel aluminum NiAl, which has a high-temperature shape memory effect (the temperature of martensitic transformations in alloys with shape memory effect NiAl can reach 1000 K) and the formation of several different variants of martensite with different structures [2]. Martensitic transformation in NiAl alloys is shown with nickel content above 60 at %. This martensitic transformation is characterized by a complete and high reversibility of the induced deformation, a small temperature hysteresis, the shape memory effect and superelasticity [4–15].

The aim of this work is to study the possibility of designing the structure of the surface layers of alloys of nickel-aluminum with a high-speed flame spraying of mechanically activated powder for functional and mechanical properties, and to provide cost-effective functional materials and components on this basis; investigated are the influence of subsequent processing, including heat treatment and plastic deformation of the surface, and the properties of the reversible deformation of the surface modified layer of alloy with a shape memory effect.

2 THE EXPERIMENTAL PROCEDURE

The formation of the surface layers of the shape memory was carried out by applying massive layers of NiAl steel 1045. We used the technology of high-speed flame spraying of mechanically activated powder brand PN80YU20. Deposition was performed on cylindrical NiAl (Ø 10 mm steel 1045) samples, with the total thickness of the layers varying 0.5 mm.

XRD analysis was performed on a diffractometer Shimadzu XRD—7000 Cu-K$_\alpha$ radiation (λ = 1.54051) with an Ni-filter. Chemical analysis of the steel alloy NiAl was performed with an ARC-MET 8000 optical emission metals and alloys analyzer. An ultrahigh resolution JSM-7500F scanning electron microscope was used to investigate the structure and the chemical analysis of the NiAl alloy. Evaporation and heating of the analyzed image was carried out using arc and spark discharges, and then the spectra were decrypted

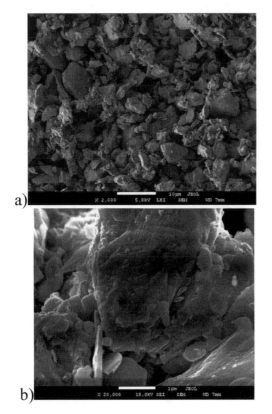

a)

b)

Figure 1. PN80YU20 powder crushed and mechanically activated in an attritor for 1 h, ×2000—a); ×20000—b).

the substrate, a pre-shot blasted steel surface was used with subsequent etching of 15% nitric acid solution. As the combustible gas mixture, methane and oxygen were used, and the carrier gas for the powder was argon. High-speed flame spraying was carried out at an angle of inclination of the burner of 40–90°. The surface layers of NiAl were gradually subjected to full Thermomechanical Cycle (TMC) treatment, including mechanical, thermal and combined Thermomechanical Processing (TMP). Figure 2 shows the results of an

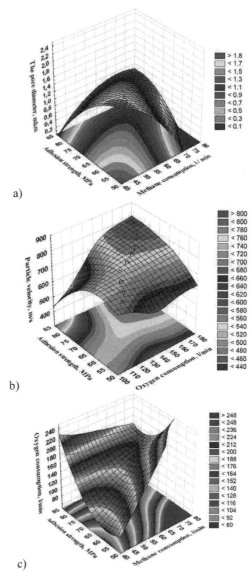

a)

b)

c)

Figure 2. Influence of technological factors on the strength of adhesion of the coating to the base NiAl, a–c).

and the intensity of the spectral lines of the corresponding elements was assessed.

Cycle fatigue of the cylindrical steel 1045 samples was examined on an MIE-6000 installation. A quantitative evaluation of the wear resistance was carried out gravimetrically using a WA-33 analytical balance.

NiAl for high-speed flame spraying of mechanically powder was activated in a high-speed planetary ball mill (attritor) steel drum (paddle) and steel balls ∅ 6 mm. The stirrer speed was 900 min⁻¹, and the weight ratio of balls to powder weight load was 20:1.

The powder particles after PN80YU20 mechanoactivation are flat disks ranging in size from 1 to 10 mm (Fig. 1).

3 RESULTS AND DISCUSSION

High-speed flame spraying was carried out in a vacuum system (RF patent number 2502829). In order to improve the adhesive strength of the coating to experiment to assess the impact of the high-speed flame spraying parameters on the performance properties of the coating, processed using the program Statistica 10.0.

After deposition and machining to size, annealing samples: steel 1045+ NiAl were held in a vacuum at T = 1073K for 2 h. To determine the surface layers' functionally necessary mechanical properties, they were subjected to the thermomechanical surface treatment method of Plastic Deformation (PPD) in ambient conditions, and at high temperatures in several stages. Run-cylindrical specimens of steel 1045 with NiAl coating were carried out under conditions of high (175–235 °C) temperatures, with the following running roller parameters: contact load P = 3.7 kN, roll diameter d1—Ø 50 mm, roller width b1–8 mm, running speed v = 95·10^{-3} m/s, traverse—S = 0.07 mm/rev. Control thermomechanical return samples were subjected to a combined treatment of the thermomechanical cycle, produced after heating to the temperature required for reverse martensitic transformation.

As a result, the high-speed flame spraying of mechanically activated nanocrystalline powders formed a coating with a minimum content of 1% of the pore, increasing the adhesion strength of the coating to the substrate.

The results of X-ray analysis showed that at room temperature the initial phase state of the NiAl layer after high-speed flame spraying of mechanically activated powder is a B2 austenite phase with a cubic lattice, and L1$_0$ martensite phase with tetragonal lattice.

Studies have shown that the structure of the deposited layers of NiAl alloy has extremely weak etched conventional reagents, largely due to strong grain refinement as a result of high-speed heating, rapid cooling, and significant deformation (Fig. 3), which provides special structural effects. In the NiAl coating it is possible to observe an inhomogeneous structure with a grain size of 80–190 nm (Fig. 3). The microhardness of the NiAl-layer varies from H$_\mu$ = 6.8–8.9 GPa.

To create plug connections, managed by thermomechanical recovery, you need to implement a reversible shape memory effect. In this paper, to guide the spontaneous reversible shape memory effect in order to manage the complex and structure-sensitive functional and mechanical properties of the NiAl, we used the combined method of treatment of the surface layer of the alloy with the shape memory effect in a single cycle of heat treatment and plastic deformation. In the thermal treatment, there is a stabilization of the structure at a possible relaxation of internal residual stresses after high-speed flame spraying. With increasing annealing temperature, memory alloys are activated and provide a process of stress relaxation

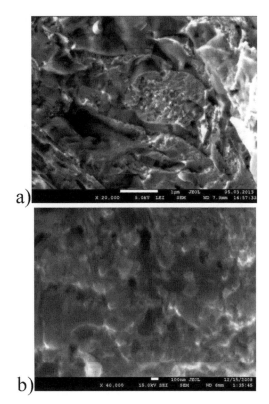

Figure 3. Nanostructured NiAl coating obtained by high-speed flame spraying of mechanically activated powder: ×20000—a); ×40000—b).

and elimination of defects. Such treatment is a necessary step in the formation of a high-quality technology of nickel-aluminum coating that provides an effective increase of the complex functional and mechanical properties.

The properties of reversible plasticity of the surface-modified layer of an alloy with the shape memory effect are the basis for cost-effective functional materials. To achieve the desired levels of developing reactive stresses, shape recovery and reversible deformation, the necessary complex processing was carried out, including surface plastic deformation of the surface layer $\delta_{NiAl} = 70$–80 mm in the temperature range of direct martensitic transformation M$_s$–M$_f$. The effective layer of an NiAl alloy in the free state recovers induced deformation when heated in the temperature range of the reverse martensitic transformation A$_s$–A$_f$.

After annealing, followed by plastic deformation of the NiAl alloy layer with the shape memory effect, there is a high density of crystal defects of the B2 austenite phase to form fine austenite (grain size 150–240 nm) (Fig. 4), which leads to an increase in the microhardness (partially withdrawn annealing)

Figure 4. Microstructure of NiAl alloy layer after thermomechanical treatment, ×50,000.

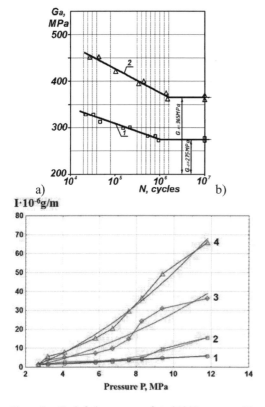

a)   b)

Figure 5. Cycle fatigue curves of steel 1045: no cover (1), after the surface modification of the alloy with shape memory effect NiAl (2)—a); dependence of the wear rate of the pressure I drive F: sliding speed drive 0.5 m/s—1; 1 m/s—2; 1.5 m/s—3; 2 m/s—4—b).

and the cycle life of all of the samples. A full cycle of processing, including high-speed flame spraying of mechanically activated NiAl powder, a annealing, and then surface-plastic deformation, leads to the formation of a homogeneous nanocrystalline

structure, improving the durability and functional and mechanical features of the alloy with shape memory effect. Thermomechanical processing layers of alloy with shape memory effect NiAl in several stages yields a combination of high hardness and fatigue resistance multicyclic loading with stable functional characteristics of the shape memory effect.

We carried out tests on the high-cycle bending fatigue with the rotation of the samples from the steel 1045 surface modified alloys with shape memory effect based on NiAl and showed a significant increase in longevity (Fig. 5a). The highest value of the amplitude of the alternating voltage at which the samples are not destroyed by the base number of cycles (endurance limit, σ_{-1}) for uncoated steel 1045 is 275 MPa, and after surface modification of the alloy with shape memory effect NiAl—365 MPa, that is, an increase of $\approx 33\%$.

Figure 5b shows the results of the wear test on specimens of steel 1045 with a surface layer of nanostructured NiAl with a thickness of 0.5 mm, obtained by the high-speed flame spraying scheme disk drive. The tests were conducted under conditions of dry friction at various disk sliding speeds. In the process of loading, the temperature was measured at the point of contact. The experimental data was obtained using a Statistica v10.0 application package in the SPSS environment. Analysis of the relationship shows (Fig. 5b) that at high speeds the test steel 1045 showed virtually no break-in period, and at low speeds it is weakly expressed. Increase in the wear resistance of the surface modified by a layer of NiAl is 2–2.5 times that of steel.

4 SUMMARY

On the basis of analysis of experimental data, the optimal modes of the surface modification process with NiAl of steel 1045 by high-speed flame spraying of mechanically activated powder are determined. The result is a nanocrystalline structure, improving the quality of the reinforcement cover with the shape memory effect. Thanks to the mechanical activation of powder with a shape memory effect, there is an increase in the density of the coating (far less than 1%), as well as an increase in the strength of adhesion of the coating to the substrate by up to 85 MPa. We established the control processing parameters that control the structural state of the material at the stage of the surface-modified layer and the subsequent combined treatment, which allows the properties that influence the shape memory effect of the surface layer of NiAl. It was established experimentally that after a high-speed flame spraying with mechanically activated shape memory NiAl powder, the cyclic durability under high-cycle fatigue is increased by ~ 33%. The wear resistance of steel 1045 after the deposition and annealing of the NiAl layer increased by 2 times, and after their treatment in the thermomechanical cycle, by 2.5 times. Thus, we achieved high cycle fatigue resistance, increased wear of the layers of alloys with the shape memory effect based on NiAl steels, in combination with giving them new functional properties.

ACKNOWLEDGMENTS

This work was supported by a grant of the President of the Russian Federation and the Ministry of Education and Science of the Russian Federation within the framework of the project in 2416 (2014g.).

REFERENCES

[1] Likhachev V.A. Materials with shape memory effect. NIIH SPSU T.1. 1997. p. 424.

[2] Kositsyn S.V., Valiullin A.I., Kataeva N.V., Kositsyna I.I. Investigation of Microcrystalline NiAl-Based Alloys with High-Temperature Thermoelastic Martensitic Transformation: I. Resistometry of the Ni-Al and Ni-Al-X (X = Co, Si, or Cr) Alloys// Physics of Metals and Metallography. 2006. V. 102, No 4, P. 391–405.

[3] Khachin V.N., Pushin V.G., Kondratyev V.V. Nikelid titanium. The structure and properties. Nauka, Moscow. 1992. p. 161.

[4] Kim H.Y., Miyazaki S. Martensite transformation behavior in Ni-Al and Ni-Al-Re melt-spun ribbons // Scripta Mater. 2004. V.50. p. 237–241.

[5] Potapov P.L., Ochin P., Pon S.J., Schryvers D. Nanoscale inhomogencities in melt-spun Ni-Al // Acta Mater. 2000. V.48. P. 3833–3845.

[6] George E.P., Liuy C.T., Horton J.A., Sparks C.J., Kao M., Kunsmann H., King T. Characterization, Processing, and Alloy Design of NiAl-Based Shape Memory Alloys // Materials characterization. 1994. V. 32. P. 139–160.

[7] Otsuka K., Kakeshita T. Science and technology of shape-memory alloys: new developments // mrs bulletin. 2002. V.2. p. 91–98.

[8] Kositsyn S.V., Valiullin A.I., Kataeva N.V., Kositsyna I.I. Investigation of Microcrystalline NiAl-Based Alloys with High-Temperature Thermoelastic Martensitic Transformation: II. Construction of Isothermal Diagrams of Decomposition of a Supersaturated b Solid Solution of $Ni_{65}Al_{35}$ and $Ni_{56}Al_{34}Co_{10}$ Alloys // Physics of Metals and Metallography. 2006. V. 102. No 4. pp. 406–420.

[9] Potapov P.L., Song S.Y., Udovenko V.A., Prokoshkin S.D. X-ray Study of Phase Transformations in Martensitic Ni-Al Alloys // Metallurgical and materials transactions A. 1997. V. 28A. No 5. p. 1133.

[10] Kainuma R., Ohtani H., Ishida K. Effect of Alloying Elements on Martensitic Transformation in the Binary NiAl(β) Phase Alloys // Metallurgical and materials transactions A. 1996. V. 27A. No 9. p. 2445.

[11] Zh.M. Blednova, P.O. Rusinov. Formation of nanostructured surface layers by plasma spraying the mechanoactivated powders of alloys with Shape Memory Effect. Nanotechnologies in Russia V. 5, No 3–4, (2010).

[12] Zh.M. Blednova, P.O. Rusinov. Formation of nanostructured surface layers from Materials with shape Memory effect TiNiCu in conditions. Materials Science Forum. Vols. 738–739 (2013). pp. 512–517.

[13] Z.M. Blednova, P.O. Rusinov, M.A. Stepanenko. Superficial Modifying by SME materials In Engineering Appendices. Materials Science Forum. Vols. 738–739 (2013). pp. 595–600.

[14] Zh.M. Blednova, P.O. Rusinov. Mechanical and Tribological Properties of the Composition "Steel-nanostructured Surface Layer of a Material with Shape Memory Effect Based TiNiCu". Applied Mechanics and Materials. Vols. 592–594 (2014). pp. 1325–1330.

[15] Z.M. Blednova, P.O. Rusinov, M.A. Stepanenko. Influence of Superficial Modification of Steels by Materials with Effect of Memory of the Form on Wear-fatigue Characteristics at Frictional-cyclic Loading. Advanced Materials Research. Vols. 915–916 (2014). pp. 509–514.

Advances in Future Manufacturing Engineering – Yang (Ed.)
© 2015 Taylor & Francis Group, London, ISBN 978-1-138-02817-3

Analysis of factors affecting thermal conductivity of low concentration silver nanofluids prepared by a single step chemical reduction method

Mingzheng Zhou*
State Nuclear Power Technology Research and Develop Center, SNPTRD, Beijing, China

Guodong Xia
Key Laboratory of Enhanced Heat Transfer and Energy Conservation, Ministry of Education,
College of Environmental and Energy Engineering, Beijing University of Technology, Beijing, China

Chunlai Tian
State Nuclear Power Technology Research and Develop Center, SNPTRD, Beijing, China

ABSTRACT: For the purpose of this study silver-water nanofluids are prepared by a single-step chemical reduction method via an Ultrasound-Assisted Membrane Reaction (UAMR), using a hollow fiber membrane (about 30 nm) to control the reaction process. The work investigates the effects on the thermal conductivity of the nanofluids of key factors, such as surfactant (Polyvinylpyrrolidone, PVP) concentration and volume fraction of silver nanoparticles. The results indicate that the variation of thermal conductivity of silver nanofluids with concentration is similar to that of the PVP solution without nanoparticles. Compared with pure water, the thermal conductivity enhancement reaches 6%, with a volume fraction of silver of 0.012%.

Keywords: silver nanofluid; single-step method; thermal conductivity; PVP

1 INTRODUCTION

Nanofluids have recently gained significant interest because of their properties of enhanced thermal conductivity and little pipe wall abrasion, which can be obtained by the incorporation into the base liquid of thermally conductive particulate solids such as metals or metal oxides [1–4].

The preparation of nanofluids is a basic key step for the use of nanoparticles to improve the thermal conductivity of fluids. Nanofluids have often been synthesized either by a two step approach that first generates the nanoparticles and subsequently disperses them into the base fluid [5] or by a single-step physical method that simultaneously makes and disperses the nanoparticles into the base fluid[6]. Compared with the two-steps, the single-step method, which combines the preparation of nanoparticles and the preparation of nanofluids, can avoid the processes of drying, storage, transportation, and re-dispersion of nanoparticles. Eastman [6] has used a single-step physical method to prepare nanofluids, in which Cu vapor was directly condensed into nanoparticles by contact with a

flowing low-vapor-pressure liquid (ethylene glycol), but this method appears to be cost ineffective and has the limitation that only low vapor pressure fluids are compatible with the process. In an attempt to synthesize high-quality nanofluids with controllable microstructures, a single-step chemical solution method has been recently developed.

Kumar [7] presented a one-step method for the preparation of stable, non-agglomerated copper nanofluids by reducing copper sulphate pentahydrate, with sodium hypophosphite as reducing agent and ethylene glycol as base fluid by means of conventional heating.

The chemical reduction method emphasizes the control of the contacting area of reactants and the reaction rate, thereby controlling the size and structure of nanoparticles. In the present study, we introduce a hollow fiber membrane (about 30 nm) to control the reaction process.

Silver nanoparticles, which have high thermal conductivity and excellent chemical and physical stability, are considered a good choice for nanofluids. Patel [8] investigated the thermal conductivities of silver nanofluids with a volume fraction of 0.001%, and concluded that the particle size obviously affects the thermal conductivity of nanofluids. Sharma [9] synthesized silver

*Corresponding author: Tel: +86-010-56681679 Fax: +86-010-56681000 Email: zhoumingzheng@snptrd.com

nanofluids using a single-step chemical reduction method and got silver nanofluid in ethylene glycol, finding that the thermal conductivity of silver nanofluids increased to 10, 16, and 18% as the amount of silver particles in the nanofluid were 1000, 5000, and 10,000 ppm, respectively.

In the present study, silver-water nanofluids are prepared via an Ultrasound-assisted Membrane Reaction (UAMR) in a single-step chemical reduction method. A hollow fiber membrane (about 30 nm) is used to control the reaction process. $NaBH_4$ is used to reduce silver ions in the aqueous solution, while Poly-vinylpyrrolidone (PVP) is used as stabilizer and surfactant. The key parameters for heat transfer fluids, thermal conductivity and viscosity are systematically investigated. Thermal conductivities are measured by the Transient Plane Source (TPS) technique and the measurement of viscosities is performed using a torsional oscillation viscometer. The effects of PVP concentration, temperature and weight fraction of silver nanoparticles on the thermal conductivity and viscosity are investigated experimentally.

2 EXPERIMENTAL

2.1 *Preparation of nanofluids*

The silver-water nanofluids used in this paper are prepared via an Ultrasound-assisted Membrane Reaction (UAMR) using a single-step method. The $NaBH_4$ is used to reduce silver ions in aqueous solution, while PVP is used as stabilizer and surfactant [10]. Two kinds of PVP with different molecular weights (PVP-1 = 10000, PVP-2 = 30000) are used in this work.

The experimental apparatus is shown in Figure 1 and the specific preparation process are as follows: First dissolving 6.0 g PVP into 200 ml deionized water and 0.53 g $NaBH_4$ into 50 mL deionized water, respectively. Then after mixing the two solutions in the glass reactor and putting the reactor in generator of ultrasonic wave. A 50 mL 3.2 g/L $AgNO_3$ solution is pumped into the pipeline of hollow fiber membrane with a flow rate of 1.5 ml/min. The $AgNO_3$ solution diffuses into the glass reactor via the microporous of hollow fiber membrane (< = 0.03~0.3μm), and reacts with the $NaBH_4$ solution in the glass reactor at a certain stirring speed and ultrasonic frequency.

After the chemical reaction, the ultrasound and stirring continues for 1 hour to obtain steady water-based Ag nanofluids. Different concentrations of water-based Ag nanofluids can be obtained by changing the amount of reactants.

Compared with the "two-step" preparation, the one step method can effectively avoid the problem of hard agglomeration, caused at the drying, collection and storage stages. No significant sediment is found when the nanofluids keep standing for three months.

As shown in Figure 2, the shape of the most silver particles is spheroidal, and almost no aggregation occurs. Through the microporous of the hollow fiber membrane, the droplets of $AgNO_3$ solution diffuse uniformly into the glass reactor. Under a certain stirring speed and ultrasonic frequency, the silver particles generate quickly and leave the reaction area. During this process through electrostatic, hydrogen bonding and charge transfer interactions, PVP polymer molecules are effectively packaged on the silver particles to reduce the

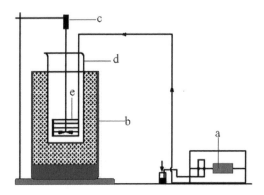

Figure 1. Schematic diagram of preparing Ag nanoparticles. a. Column piston pump; b. generator of ultrasonic wave; c. mechanical stirrer; d. reaction kettle made of glass; e. hollow fiber film and subassembly; f. storage tank of metal salt solution.

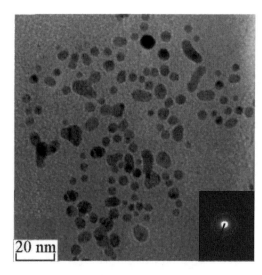

Figure 2. TEM images of silver nanoparticles.

126

agglomeration of silver nanoparticles and achieve a stabilization effect. The range of particle size of Ag particles is 3 to 9 nm, with an average size of about 4.8 nm.

3 MEASUREMENT OF THE THERMAL CONDUCTIVITY OF FLUIDS

The thermal conductivities of nanofluids are measured in this paper by the thermal constants analyzer 2500S (from Hot Disk Inc., Sweden). It utilizes a sensor element in the shape of a double spiral which acts both as a heat source for increasing the temperature of the sample and a resistance thermometer for recording the time-dependent temperature increase of the heat source itself.

This probe has been specifically designed to maximize the heating area and length scale for a given sample size, maintaining close agreement between the physical model and real experimental geometry.

The test of thermal conductivity is performed by recording the voltage variations over the TPS element while its temperature is slightly raised (≈ 1 K) by a pulsed electrical current (≈ 300 mA). A constant electric power supplied to the sensor results in an increase in temperature $\Delta T(\tau)$ which is directly related to the variation in the sensor resistance $R(t)$ by the equation:

$$R(t) = R_0(1 + \Omega \Delta T(\tau)) \tag{1}$$

where, Ω is the temperature coefficient of the resistance (TCR); R_0 is the resistance of TPS-element in the beginning of the recording (initial resistance); τ is the variable on the time of electrification and be defined as:

$$\tau = \sqrt{\frac{t\kappa}{a^2}} \tag{2}$$

where, t is the measuring time; κ is the thermal diffusivity of the sample; a is the radius of the sensor.

According to Fourier Law of heat conduction, if no natural convection of a fluid occurs, $\Delta T(\tau)$ can be calculated as:

$$\Delta T(\tau) = \frac{P}{\pi^{3/2} aK} D(\tau) \tag{3}$$

where,

$$D(\tau) = [m(m+1)]^{-2}$$
$$\times \int_0^\tau \frac{d\sigma}{\sigma^2} \left\{ \sum_{l=1}^m l \left[\sum_{k=1}^m k \times \exp\left(\frac{-(l^2+k^2)}{4\sigma^2 m^2} \right) E\left(\frac{lk}{2\sigma^2 m^2} \right) \right] \right\} \tag{4}$$

P is the total output power; K is the thermal conductivity of the sample and E is a modified Bessel function. The exact evaluation of the integral equation of the bifilar spiral pattern of the sensor does not seem an easy task. However, as has been pointed by Suleiman [11], the approximation of the solution by concentric rings did not affect the results obtained, especially for values of τ less than 3, $D(\tau)$ is the theoretical expression of the time dependent increase, which describes the conducting pattern of the disk shaped sensor, assuming the disk consists of a number m of concentric ring sources. By fitting the experimental data to the straight line given by Eq. (3), the thermal conductivity can be obtained by calculating the value of the slope for the fitting line $P(\pi^{3/2} aK)$.

The experimental setup is schematically shown in Figure 3. The setup includes a thermal constants analyzer, a stainless container and a constant temperature bath. The probe of the thermal constants analyzer is immersed vertically in the test liquid, which fills the stainless steel container. The stainless steel container has a high thermal conductivity and can effectively prevent liquid convection.

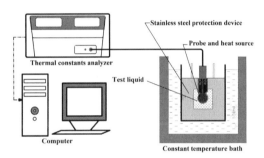

Figure 3. Schematic diagram of the experimental setup.

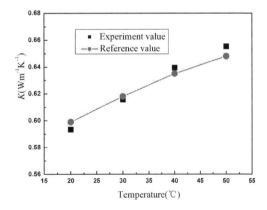

Figure 4. Validation of the thermal constant analyzer with deionized water.

127

3.1 Measurement procedure and accuracy verification

To calculate the mean value of the experimental data, the experiments are repeated 10 times for thermal conductivity. The thermal conductivity of the deionized water are firstly measured at temperatures ranging from 20°C to 50°C, to validate the thermal constants analyzer. The results with the deionized water will also be used as basis for comparison with the results on nanofluids. As shown in Figure 4, the results show good agreement between the measured values and the reference values with an uncertainty of the measurement within 1.2%.

4 RESULTS AND DISCUSSION

4.1 Effect of surfactant concentration on thermal conductivity

The stability of nanofluids has a good corresponding relation with thermal conductivity, with a better dispersion stability behavior of the nanofluids at higher thermal conductivity values [12]. The addition of PVP improves the stability of the nanofluids significantly. However, there may be negative effects on the thermal conductivity with the addition of surfactant. When adding 5 wt.% PVP, the thermal conductivity of EG-surfactant mixture lowers by about 2% compared to that of pure EG[13].

Figure 5 represents the thermal conductivity ratio of PVP-1 solution and silver nanofluids with variations of PVP-1 concentration. As shown in Figure 5, the thermal conductivity ratio of PVP solution decreases rapidly reaching the lowest value at 0.1wt%, and then rises gradually with increasing concentrations. The variation of the thermal conductivity ratio from 0.4 wt% to 1.2 wt% is

quite small. The ratio at 0.8 wt% is slightly higher compared with other concentrations. However the value is only 75% of water. When the concentration of PVP is too low, the silver nanoparticles can not be kept stable and aggregate quickly, due to the high surface energy of nanoparticles. The smallest concentration of PVP used in present work is 0.4 wt%. The trend of thermal conductivity of silver nanofluids is similar to the PVP solution. A similar phenomenon was found by Li [14], when they investigated the variation of the thermal conductivity of Cu-H2O nanofluids using Sodium Dodecyl Benzene Sulfonate (SDBS) as surfactant.

4.2 Effect of nanoparticle concentration on thermal conductivity compared with the base fluid and other formulae

Figure 6 depicts the thermal conductivity ratios of the nanofluids as a function of the volume fraction of silver nanoparticles at room temperature. Thermal conductivity ratio increases quickly with the volume fraction of silver nanoparticles, especially at higher volume fractions. Substantial increases in thermal conductivity are observed for the measured nanofluids, with thermal conductivity ratio of 1.15 at 0.012 vol% of nanoparticles.

The Maxwell equation fails to predict this enhancement of thermal conductivity. The predictions are much lower than experimental data. The thermal conductivity ratios obtained by Kumar [7] for Au 4 nm nanoparticles in toluene are also presented in Figure 7. Comparing the two situations, both nanoparticles have nearly the same large size, the temperature of silver nanofluids is a little lower than Au nanofluids and the base fluid PVP solution has higher thermal conductivity than that of toluene. The thermal conductivity enhancement increases with growing temperatures [15] and with

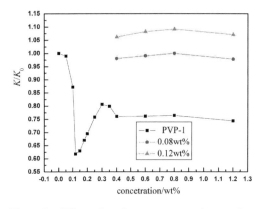

Figure 5. Effects of surfactant concentration on thermal conductivity ratio at room temperature.

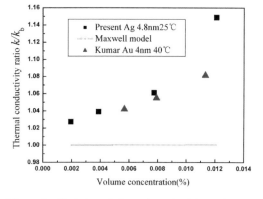

Figure 6. Variation of thermal conductivity ratio with volume fraction of silver nanoparticles.

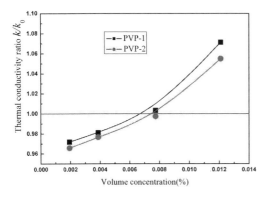

Figure 7. Variation of thermal conductivity ratio of nanofluid and pure water with volume fraction of silver nanoparticles.

decreasing values of thermal conductivity of the base fluid [16]. However, the thermal conductivity ratio of nanofluids with 0.0077 vol% sliver nanoparticle is 1.061, while the ratio of 0.0079 vol% Au nanofluids is only 1.054. That is because that the thermal conductivity of silver is an order of magnitude higher than gold. This indicates that the kind of nanoparticles may affect significantly the thermal conductivity of nanofluids. It confirms that there are many parameters influencing the thermal conductivity enhancement process of nanofluids and their combined effects should be considered.

4.3 *Effect of nanoparticle concentration on thermal conductivity*

The addition of surfactants can affect the size and structure of nanoparticles in the process of reaction and improve the stability of nanoparticles after the reaction. However, they also affect the thermal conductivity of nanofluids. Different characteristics of surfactants may contribute to different effects.

The thermal conductivity ratios of silver nanofluids and pure water are shown in Figure 7, varying with the volume fraction of silver particles at room temperature. Because of the passive effect of PVP, the thermal conductivities of nanofluids with low silver volume fraction (0.002% and 0.004%) are smaller than that of pure water, and as displayed in Figure 8 the ratios are below 1.0. The thermal conductivity ratio increases quickly with the fraction of silver nanoparticles. The ratios are equal or a little higher than that of pure water, when the volume fraction of silver particles increases to 0.008%. Once the volume fraction of silver reaches 0.012%, the thermal conductivities of nanofluids are higher than that of pure water. The ratios are higher than

1.06. Comparing the two surfactants, the thermal conductivity ratios of nanofluids synthesized with PVP-1 are higher than that with PVP-2, at the same volume fraction of silver nanoparticles. This is because that PVP-2 has a higher aggregation and thereby a longer molecular chain. The longer chain will produce more effect on thermal conductivity, when the surfactant molecules are absorbed on the surface of silver nanoparticles.

5 CONCLUSIONS

As described above, the thermal conductivities are measured by the Transient Plane Source (TPS) technique. The effects of PVP concentration, temperature and weight fraction of silver nanoparticles on the thermal conductivity are investigated. Key conclusions can be summarized as follows:

The variation of thermal conductivity of silver nanofluids with concentration is similar to that of PVP solutions without nanoparticles. The thermal conductivity ratio increases quickly with the volume fraction of silver nanoparticles, especially at higher volume fractions. A higher thermal conductivity of nanoparticles is positive for conductivity enhancement of nanofluids. Comparing with pure water, the thermal conductivity enhancement reaches 6%, at a 0.012% volume fraction of silver.

ACKNOWLEDGEMENTS

The work is supported by the National Natural Science Foundation of China (51176002), the National Basic Research Program of China (2011CB710704), the Research Fund for the Doctoral Program of Higher Education (20111103110009) and the Staff Independent Innovation Fund of SNPTC (State Nuclear Power Technology Company, China) (SNP-KJ-CX-2013-10).

NOMENCLATURE

a	radius of the sensor, mm
A	effective area of surfactant hydrophilic group, nm^2
b, c, d	constant
E	modified Bessel function.
K	thermal conductivity of solution, $W \cdot m^{-1} \cdot K^{-1}$
K_b	thermal conductivity of base fluid, $W \cdot m^{-1} \cdot K^{-1}$
K_0	thermal conductivity of water, $W \cdot m^{-1} \cdot K^{-1}$
l_0	length of surfactant hydrocarbon chain, nm
M	mass of the liquid sample, kg
P	total output power; W
r	inner radius of the vessel; m
R_0	initial resistance of sensor, Ω

t   measuring time, s

V   volume of surfactant hydrocarbon core, nm^3

Greek symbol

δ logarithmic damping decrement

τ characteristic time, s

κ thermal diffusivity, $m^2 \cdot s^{-1}$

ρ density of sample, $kg \cdot m^{-3}$

Ω temperature coefficient of resistance, K

Subscript

1   empty vessel.

REFERENCES

[1] H. Xie, J. Wang, T. Xi, Y. Liu. Thermal conductivity of suspensions containing nanosized SiC particles [J]. International Journal of Thermophysics. 23 (2002) 571–580.

[2] S.H. Kim, S.R. Choi, D. Kim. Thermal conductivity of metal-oxide nanofluids: particle size dependence and effect of laser irradiation [J]. Journal of Heat Transfer. 129 (2007) 298.

[3] G.D. Xia, M.Z. Zhou, L.J. Zhou. Heat transfer of sub-merged jet impingement on pin-fin heat sinks with silver nanofluids [J]. Journal of the Chemical Industry and Engineering Society of China. 62 (2011) 916–921.

[4] E. Abu-Nada, A.J. Chamkha. Effect of nanofluid variable properties on natural convection in enclosures filled with a CuO-EG-Water nanofluid [J]. International Journal of Thermal Sciences. 49 (2010) 2339–2352.

[5] M. Chopkar, S. Kumar, D.R. Bhandari, P.K. Das, I. Manna. Development and characterization of Al2Cu and Ag2 Al nanoparticle dispersed water and ethylene glycol based nanofluid [J]. Materials Science and Engineering: B. 139 (2007) 141–148.

[6] J.A. Eastman, S.U.S. Choi, S. Li, W. Yu, L.J. Thompson. Anomalously increased effective thermal conductivities of ethylene glycol-based nanofluids containing copper nanoparticles [J]. Applied Physics Letters. 78 (2001) 718–720.

[7] S.A. Kumar, K.S. Meenakshi, B.R.V. Narashimhan, S. Srikanth, G. Arthanareeswaran. Synthesis and characterization of copper nanofluid by a novel one-step method [J]. Materials Chemistry and Physics. 113 (2009) 57–62.

[8] H.E. Patel, S.K. Das, T. Sundararajan, A.S. Nair, B. George, T. Pradeep. Thermal conductivities of naked and monolayer protected metal nanoparticle based nanofluids: Manifestation of anomalous enhancement and chemical effects [J]. Applied Physics Letters. 83 (2003) 2931.

[9] P. Sharma, I.H. Baek, T. Cho, S. Park, K.B. Lee. Enhancement of thermal conductivity of ethylene glycol based silver nanofluids [J]. Elsevier: Powder Technology. Volume 208, Issue 1, 10 March 2011, pp.7–19.

[10] Z.Y. Wang, X. Guan, H. He. Preparation of Ag Nanoparticles by Ultrasound-assisted membrane Reaction [J]. Chemical Journal of Chinese Universities. 28 (2007) 1756–1758.

[11] B.M. Suleiman, S.E. Gustafsson, L. Börjesson. A practical cryogenic resistive sensor for thermal conductivity measurements [J]. Sensors and Actuators A: Physical. 57 (1996) 15–19.

[12] X. Wang, D. Zhu. Investigation of pH and SDBS on enhancement of thermal conductivity in nanofluids [J]. Chemical Physics Letters. 470 (2009) 107–111.

[13] W. Yu, H. Xie, L. Chen, Y. Li. Investigation on the thermal transport properties of ethylene glycol-based nanofluids containing copper nanoparticles [J]. Powder Technology. 197 (2010) 218–221.

[14] X.F. Li, D.S. Zhu, X.J. Wang, N. Wang, J.W. Gao, H. Li. Thermal conductivity enhancement dependent pH and chemical surfactant for Cu-H2O nanofluids [J]. Thermochimica Acta. 469 (2008) 98–103.

[15] S.K. Das, N. Putra, P. Thiesen, W. Roetzel. Temperature dependence of thermal conductivity enhancement for nanofluids V Journal of Heat Transfer. 125 (2003) 567.

[16] H. Xie, M. Fujii, X. Zhang. Effect of interfacial nanolayer on the effective thermal conductivity of nanoparticle-fluid mixture [J]. International Journal of Heat and Mass Transfer. 48 (2005) 2926–2932.

Advances in Future Manufacturing Engineering – Yang (Ed.)
© 2015 Taylor & Francis Group, London, ISBN 978-1-138-02817-3

The thermodynamic characteristics of silica film calculated via Molecular Dynamics method

Cheng Chen, Hong Fang & Zhilin Xia
School of Materials Science and Engineering, Wuhan University of Technology, Wuhan, Hubei, China

ABSTRACT: This paper has calculated some thermodynamic characteristics of silica bulk and film by using the molecular dynamics method. The effect of temperature and film thickness on these thermodynamic characteristics has been investigated.

Keywords: silica film; thermodynamic characteristics; molecular dynamics method

1 INTRODUCTION

As for laser induced damage of optical films, researchers have put forward a series of damage mechanisms, including avalanche ionization, multi-phonon absorption, defect absorption, thermal explosion and thermal-mechanical coupling [1, 2, 3]. In order to analyze the mechanism and process of the laser induced damage of films, it is necessary to know the thermodynamic characteristics of the used materials, and the affection of temperature and film thickness on these characteristics. In the present paper, Molecular Dynamics (MD) method has been adopted [4] to calculate the thermodynamic characteristics of silica. The influence of films' thickness and temperature on the thermodynamic performances of films has been studied.

2 SIMULATION MODEL

SiO_2 crystal is a three-dimensional networked cubic crystal, and it is made up of silicon atoms and oxygen atoms in the ratio of 1:2. The used simulation model of the block materials was cubic unit, and the periodic boundary conditions were applied in all directions. Figure 1 showed the selected model of the film materials. Periodic boundaries were set with y and z direction, and vacuum layer was set with x direction. Morse potential function was adopted to calculate the interaction between two atoms: $V(r) - D_e = D_e(e^{-2a(r-r^e)} - 2e^{-a(r-r^e)})$. Here, r is the distance between two atoms; r_e is the balance distance between two atoms; D_e is the depth of the potential well; α is the width of the potential well. The used parameters are listed in Table 1.

In simulation, NTV canonical ensemble is used and the cutoff is 9Å [5].

3 RESULTS AND DISCUSSION

3.1 *Heat conductivity of SiO₂*

The heat conductivities as functions of temperature are shown in Figure 2(a). At the simulation temperature about 100 K, the heat conductivity of the block materials and the film with thickness of 4.2 nm is 7.5 $W \cdot m^{-1} \cdot K^{-1}$ and 0.8 $W \cdot m^{-1} \cdot K^{-1}$ respectively. As the temperature increases, the heat conductivity gradually decreases. When the simulation temperature reached to 600 K, these two kinds of heat conductivity are reduced to 2.8 $W \cdot m^{-1} \cdot K^{-1}$ and 0.5 $W \cdot m^{-1} \cdot K^{-1}$ respectively. The heat conductivity trends to be stable with the temperature continuing increasing. It also can be seen from Figure 2(a) that the heat conductivity of film materials is less than that of the block materials by one order of magnitude.

The heat transmission of SiO_2 crystal is dominated by phonons. The average phonon number in crystals can be gained through the Bose statistics: $\bar{N} = 1 / e^{\hbar\omega/k_B T} - 1$, $\hbar$ stands for the Planck constant; ω stands for frequency of phonon; k_B is the Boltzmann constant; and T is temperature. The number of phonons increases as the temperature rises. As a result, the mean free path of phonon and the heat conductivity of materials are reduced. As the temperature continues to increase, the phonon number increase slowly, consequently, the heat conductivity tends to be stable.

When the film thickness is smaller than the mean free path of phonons, the film thickness will

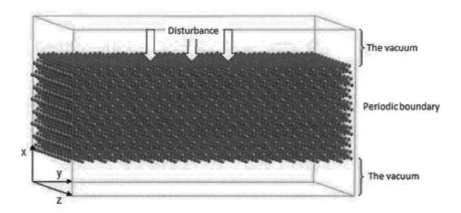

Figure 1.   MD simulation model.

Table 1.   Parameters of the potential function [6].

Bond	m	D/[gÅ²/fs²]	α[1/Å]	r[Å]
Si–Si	2.0000000e+000	3.1979232e-026	2.6518000e+000	1.6280000e+000
O–O	2.0000000e+000	3.7286166e-028	1.3731000e+000	3.7910000e+000
Si–O	2.0000000e+000	1.2328852e-028	2.0446000e+000	3.7598000e+000

Charge: $q(Si) = +1.3$, $q(O) = -0.65$.

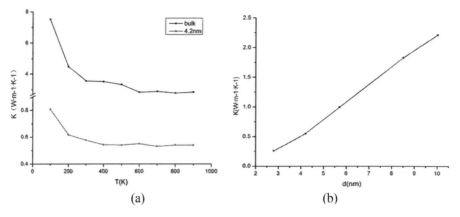

(a)   (b)

Figure 2.   (a) The heat conductivity of the SiO_2 block and film with thickness of 4.2 nm; (b) At 300 K, the heat conductivity of the films with different thicknesses.

restrict the mean free path of phonons. The mean free path of phonons is approximate to the film thickness, and it is far smaller than the mean free path of the block state. As a result, the conductivity of film materials is far smaller than that of block materials. As the Figure 2(b) shows, when the films thickness is thinner than 10 nm, the heat conductivity of films has approximately linear relation with the films' thickness.

3.2   *Specific heat of SiO₂*

The specific heat at constant volume as functions of temperature is also shown in the Figure 3(a).

At the simulation temperature about 100 K, the specific heat at constant volume of SiO_2 block materials and the film with thickness of 4.2 nm is 0.55 $J \cdot K^{-1} \cdot g^{-1}$ and 0.41 $J \cdot K^{-1} \cdot g^{-1}$ respectively. These data increased with the gradual change of temperature. When the temperature comes to 900 K, the specific heat at constant volume of the block materials and the film with thickness of 4.2 nm, increases to 0.72 $J \cdot K^{-1} \cdot g^{-1}$ and 0.58 $J \cdot K^{-1} \cdot g^{-1}$ respectively. Figure 3(b) shows the change of the specific heat at constant volume with different thickness of film.

According to the Dulong-Petit law, the specific heat at constant volume as below:
$$C_v = \frac{\partial E}{\partial T} = 9Nk_B \left(\frac{T}{\theta_D} \right)^3 \int_0^{x_D} \frac{x^4 e^x}{(e^x - 1)^2} dx, \theta_D \text{ stands for}$$
Debye temperature, $x = \beta \hbar \omega, x_D = \beta \hbar \omega_D = \theta_D / T$.

When the temperature is higher than Debye temperature, or x << 1, $C_v \approx 9Nk_B \left(\frac{T}{\theta_D} \right)^3 \int_0^{x_D} x^2 dx = 3Nk_B$, the specific heat tends to be stable and the experimental result can well coincide with the Dulong-Petit law. When the temperature is lower than Debye temperature, or x >> 1, the specific heat is proportion to the temperature T^3. With the temperature increasing, the result still coincides with the Dulong-Petit law.

The specific heat at constant volume of the film's alteration with different thickness can be seen from the Figure 3(b). When the thickness of the film is greater than 3 nm, the specific heat increases with the thickness of the film's increasing. The result is consistent with the research results in ref. 7.

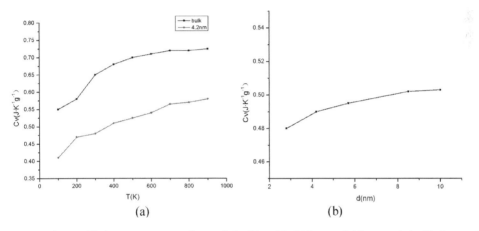

Figure 3. (a) The specific heat at constant volume of the film with thickness of 4.2 nm and the block materials; (b) At 300 K, the specific heat at constant volume of the film with different thicknesses.

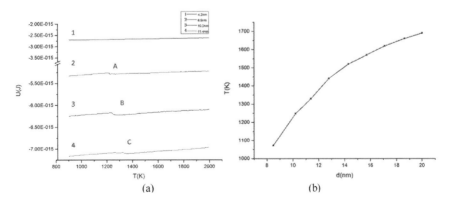

Figure 4. (a) The energy change of SiO_2 films with different thicknesses as temperature rises; (b) the melting temperature of SiO_2 films with different thicknesses.

3.3 Melting temperature of the films

The simulation results of the energy evolutions of the SiO_2 film with different thickness as temperature rises have been shown in Figure 4(a). Different from the energy evolution curve of silica film with thickness of 4.2 nm, there are some dramatic changes (the point of 'A' 'B' 'C' in the fig) on the energy evolution curves of films with thickness of 8.5 mm, 10.2 nm and 11.4 nm. These dramatic changes suggest that the film is melting. After this, the energy evolution curves regain the stability, which suggests the film melting has ended and the materials have transformed into the liquid state.

The characteristic of the crystal materials is that they have a specific melting temperature. Compared the energy curve of films thickness with 4.2 nm and 8.5 nm, it can be deduced that the film belongs to the non-crystalline structure when the film thickness is smaller. Thus, there is no specific temperature in the whole process. The critical thickness is between 4.2 mm to 8.5 mm.

It can be seen from Figure 4(b) that the thicker the silica film is, the higher the melting temperature are. The slope of the curve decreases gradually, which suggests that the size effect becomes weak with the increasing of thickness. The formula for calculating the melting temperature of the nano-film can be given as follow [8]: $T_{mp} = T_{mb}\left[1 - \left(\frac{2}{3}\right)\frac{D}{d}\right]$, D stands for the atomic diameter, d is the thickness and l is the length factor. T_{mp} stands for the melting temperature of the nanocrystalline, T_{mb} stands for the melting temperature of corresponding block materials. It can be seen that, when the film thickness increases, T_{mp} increases.

4 CONCLUSIONS

Theoretical models have been established to interpret some thermodynamics characteristics of the silica film. The result reflect that the heat conductivity of the silica film is significantly smaller than that of bulk materials by one order of magnitude, and with the thickness increasing, the heat conductivity of the silica film is in an approximate linear relationship with the films' thickness within the simulated film thickness. The simulation result shows significant size effect. Compared with the bulk materials, the specific heat at constant volume of the film is less than that of the bulk materials with the temperature increasing, and the specific heat at constant volume of the film elevates as the thickness increases. There is no a specific melting point when the thickness of the silica film is comparatively small, the films' state undergoes a transformation from the non-crystal to the crystal. The critical value of the thickness transformed is between 4.2 nm to 8.5 nm.

ACKNOWLEDGMENTS

The authors gratefully acknowledge the financial support of the National Science Foundation of China (Grant U1430121), the Open Research Fund of Key Laboratory of Material for High Power Lasers,—Chinese Academy of Sciences.

REFERENCES

[1] L.B. Freund, S. Suresh. Thin film materials: stress, defect formation and surface evolution [M]. English: Cambridge University Press, 2003: 769.

[2] M.R. Kozlowski, I.M. Thomas, J.H. Campbell, and F. Rainer. *High-Poweroptical coatings for a megajooule class ICF laser* [J]. Thin Film for Optical Systems, SPIE, 1993, 1782:105–121.

[3] Yuan-an Zhao. *Pulsed laser on the damage mechanism of optical thin film and testing technology research* [D]. Shanghai: Chinese Academy of Sciences Shanghai Light Electric Machinery Research Institute Dissertation confidential, 2005.

[4] Hua Liu, Chunyan Li, Jianchao Chen and Xiaofeng Yang. *Molecular dynamics simulation of the status and its application in materials science* [J]. Materials Review, 2011, 21(4):5–8.

[5] E. Demiralp, T. Cagin, W.A. Goddard III. *Morse stretch potential charge equilibrium force field for ceramics: application to the quartz-stishovite phase transition and to silica glass* [J]. Physics Review Letters, 1999, 82(8):1708.

[6] A. Takada, P. Richet, C.R.A. Catlow, G.D. Price. *Molecular dynamics simulations of vitreous silica structures.* Journal Non-Crystal Solids, 2004, (345): 224–229.

[7] Prasher R. S. and Phelan P. E.. *Size effects on the thermodynamic properties of the thin sold films.* 1998.

[8] Akhter J.I., Jin Z.H., Lu K. *Super heating in confined Pb (110) films* [J]. Journal of Physics Condense Matter, 2001, 13(35):7969–7975.

Advances in Future Manufacturing Engineering – Yang (Ed.)
© 2015 Taylor & Francis Group, London, ISBN 978-1-138-02817-3

A new composite nanostructure of Au@SiO$_2$/AAO

Hong Fang, Cheng Chen & ZhiLin Xia
School of Materials Science and Engineering, Wuhan University of Technology, Wuhan, Hubei, China

ABSTRACT: This paper puts forward a new type of medium structure. The composite structure of medium is supported by porous alumina film, with gold nanorods enveloped by mesoporous silica dispersed disorderly in the template by electrophoretic deposition. Through Scanning Electron Microscope (SEM), we can see that spherical Au @ SiO$_2$ can be evenly dispersed in the hole of AAO. This pure inorganic composite structure is stable and will have a great application prospect in new optical storage.

Keywords: AAO; gold nanorods; silica; electrophoretic deposition

1 INTRODUCTION

Scientists predicted that "The photon technology will lead to an industrial revolution in excess than electronic technology caused, and this revolution will bring a huge shock to the industry and society." The combination of nanophotonics with advanced technology, such as information technology, green energy, the early detection and treatment of cancer, cell engineering and water purification, indicates the photonic technology become one of the most important forces to change the world. Moreover, due to the surface plasmon resonance of metal nanostructures, the spectrum of nanorods presents optical properties nonlinear, such as electromagnetic properties, Fano resonance, etc. AAO template also has its special properties. For example, assembling all kinds of metal nano-materials in the hole of the AAO template, it can be used as a decorative film material by taking advantage of the different absorption of light band. And it also can be used as capacitor by its different dielectric properties of composite membrane. Due to their special features of the gold nanorods and porous alumina template, this innovation of the compound medium structure with abnormal optical property will thoroughly change the traditional disc structure, as in the butterfly effect. Ultimately, it will realize the property of ultrahigh density, super fast, long life, safe storage [1], and promote the rapid development of various fields.

2 EXPERIMENTAL DETAILS

2.1 The fabrication of AAO

A high purity (99.99%) aluminum foil with 0.1 mm thickness was used. After pretreatment, two-step anodization was performed using an electrochemical cell equipped with a cooling stage at a temperature of 0–1°C [2]. The first anodization step was performed under 120V for 6–8 h in 0.3M H$_3$PO$_4$. Afterward, the formed porous oxide film was chemically removed by a mixture of 6 wt% of H$_3$PO$_4$ and 1.8% chromic acid for a minimum of 3–6 h at 75°C. The second anodization step was performed by cyclic anodization in the same condition with the first anodization.

2.2 The fabrication of Au@SiO$_2$

Gold nanorods having aspect ratios of three to five were prepared using a wet-chemical method similar to the method described in ref. 3. The nanorods were then transferred to a silica matrix by a sol–gel method [4]. A solution consisting of 5 g of 2-propanol and 50 mg of ammonia solution (25% in water) was added to a solution of 60 µl ethanol, 110 ul tetramethylorthosilicate (TMOS) and 27 µl of distilled water. After 10 min, the aqueous nanorods were added to the mixture in a 1:1 volume ratio.

2.3 Assembling Au@SiO$_2$ with AAO

Firstly, three drops of the Au @ SiO$_2$ sol add into 10 ml deionized water in the sink. After ultrasonic dispersion, assemble Au@SiO$_2$ with AAO by electrophoretic deposition [5,6] at the voltage of 120 V, where there AAO template act as the anode. About 5 minutes later, terminate the process and dry it in the air under room temperature environment.

3 RESULTS AND ANALYSIS

Figure 1 shows Scanning Electron Microscope (SEM) images of porous alumina membranes

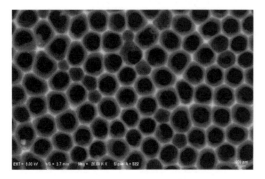

Figure 1. SEM micrographs of anodic porous alumina prepared under a constant voltage condition of 120 V in a 0.3M H₃PO₄ solution.

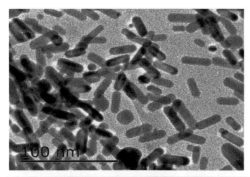

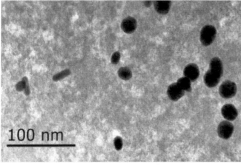

Figure 2. (a) is the Transmission Electron Microscopy (TEM) image of nanorods with an aspect ratio of four. Figure 2(b) TEM images of silica-coated gold NRs (silica thickness: 40 nm, aspect ratio: 4).

fabricated by the two step anodic oxidization at 120 V. All samples exhibit hexagonal pore arrays, and the pore diameters are about 300 nm. In this way, we can prepare the porous alumina membranes with the larger pore diameter at higher voltage.

In Figure 2 (a), the GNRs have the uniform morphology with 45 nm in length and 16 nm in diameter. As to Figure 2 (b), Transmission Electron

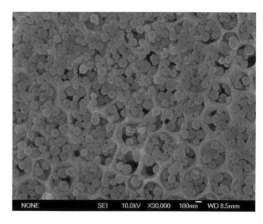

Figure 3. SEM micrographs of anodic porous alumina prepared under a constant voltage condition of 120 V in a 0.3M H₃PO₄ solution.

Microscopy (TEM) was conducted on thick silica-coated GNRs (aspect ratio: about 4, silica shell thickness: ca. 40 nm). The method to prepare core-shell structure is simple and finally the size of integrity structure is about 120 nm.

The microstructure of the medium is observed by means of SEM, as Figure 3 shows, and it can be seen that Au @ SiO₂ evenly dispersed in the AAO template hole. Only a small amount of hole is incompletely filled and it can be continuous improvement by increasing the voltage or extension of the time. The biggest benefit of the assembling and what we need to apply is that the Au @ SiO₂ fixed in the hole without any agglomeration. It is also observed that the pore size of the AAO is about 380 nm, which is much bigger than that prepared before. So the reason may be attributed to the electrophoretic deposition. As the AAO act as anode, it probably occur further oxidation and the average pore diameter increased.

4 CONCLUSION

This composite structure in this paper is stable enough to preserve longer than a century. The most advantage is that it not only realized high density but also avoid the common agglomerate phenomena of the metal nanoparticles, which is indeed great important to the application such as optical storage and photocatalysis.

ACKNOWLEDGMENTS

The authors gratefully acknowledge the financial support of the National Science Foundation of China (Grant U1430121), the Open Research Fund of Key Laboratory of Material for High Power Lasers, — Chinese Academy of Sciences.

REFERENCES

[1] Tanaka, T.; Kawata, S. *Three-dimensionalmulti-layered fluorescent optical disk*. In Technical Digest Int. Symp. Opt. Mem. (2007).

[2] Jessensky, O.; Müller, F.; Gösele, U. *Self-Organized Formation of Hexagonal Pore Arrays in Anodic Alumina* [J]. Appl. Phys. Lett. 72, 1173–1175 (1998).

[3] N.R. Jana, L. Gearheart, C.J. Murphy, J. Phys. Chem. B 2001, 105, 4065–4067.

[4] James W.M. Chon, Craig Bullen, Peter Zijlstra, and Min Gu. *Spectral Encoding on Gold Nanorods Doped in a Silica Sol–Gel Matrix and Its Application to High-Density Optical Data Storage* [J]. Adv. Funct. Mater. 17, 875–880 (2007).

[5] D.J. Lee, S.S. Yim, K.S. Kim, S.H. Kim, and K.B. Kim. *Formation of Ru nanotubes by atomic layer deposition onto an anodized aluminum oxide template, Electrochem* [J]. Solid-State Lett., vol. 11, no. 6, K61–K63 (2008).

[6] Kai Kamada, Haruto Fukuda, Keita Maehara, Yukiko Yoshida, Masumi Nakai, Shunji Hasuo, and Yasumichi Matsumoto. *Insertion of SiO₂ Nanoparticles into Pores of Anodized Aluminum by Electrophoretic Deposition in Aqueous System* [J]. Electrochemical and Solid-State Letters, 7 (8), B25–B28 (2004).

Advances in Future Manufacturing Engineering – Yang (Ed.)
© 2015 Taylor & Francis Group, London, ISBN 978-1-138-02817-3

Express measurement of optical material refractive index with using laser beam profiler

Vladislav Kolchinskiy
Institute of Automation and Control Processes, FEB RAS, Vladivostok, Russia
School of Natural Sciences, Far-Eastern Federal University, Vladivostok, Russia

Ikai Lo & Cheng-Hung Shih
Department of Physics, Department of Materials and Optoelectronic Science, Center for Nanoscience and Nanotechnology, National Sun Yat-Sen University, Kaohsiung, Taiwan

Yuri Kulchin & Roman Romashko
Institute of Automation and Control Processes, FEB RAS, Vladivostok, Russia
School of Natural Sciences, Far-Eastern Federal University, Vladivostok, Russia

ABSTRACT: New method for express measuring of optical material refractive index for plane-parallel samples or optical elements is proposed and studied. The method is based on using laser beam profiler for measurement of laser beam displacement caused by its refraction in a sample of known geometry. The developed method was used to measure the refractive index of gallium nitride in VIS spectral range at wavelengths 470, 561 and 633 nm. The technique provides a refractive index measurement with accuracy 10^{-3}. Simplicity of the measurement procedures allows one to perform it in express way.

Keywords: refractive index; laser beam profiler; gallium nitride

1 INTRODUCTION

The refractive index is one of the most important parameters of optical material. Currently a number of techniques for refractive index measurement are available. Among them goniometric, refractometric, and ellipsometric methods are commonly used [1]. They provide measurement accuracy up to 10^{-4}–10^{-5} but require fulfillment of specific conditions and/or using relatively complex equipment (goniometers, elipsometers, etc). However, same practical applications do not require very high accuracy in refractive index measurement but demand performing such measurements in expressway. In this paper, we propose new method for determining the refractive index of the optical material using laser beam profiler.

2 PRINCIPLE OF MEASUREMENT

Diagram illustrating the principle of measuring the refractive index of the plane-parallel sample using the laser beam profiler is shown in Figure 1. Thin laser beam is incident at angle α to the input facet of the plane-parallel sample. The refracted beam is shifted by the value h due to refraction. Herewith this beam should be much thinner than the expected shift h. Rotation of the sample leads to a change of the incidence angle α and, respectively, to change of the beam displacement h. Being measured α and h are used to calculate the refractive index n of the material according to the equation derived from Snell's law:

$$n = \frac{\sin\alpha}{\sin\left(a\tan\left(\tan\alpha - \frac{h}{d\cos\alpha}\right)\right)} \tag{1}$$

where d—thickness of the sample.

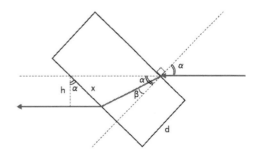

Figure 1. Propagation of light beamina plane-parallelplate.

The peculiarity of the proposed approach is that sample should be of known geometry. It is clear that in this case all relations of geometrical optics can be easily applied for calculation of refractive index from experimentally measured parameters of light beam propagation in a material. One of the most simple and widely used geometry is a plate with two commonly parallel facets. This geometry is implemented in practice in many types of optical elements (plates, wafers, prisms, etc.) and naturally provided for crystals [2]. Eq. (1) can be easily modified if sample has another geometry.

3 EXPERIMENT

Scheme of experimental setup for measuring the refractive index of plane-parallel sample using a laser beam profiler is shown in Figure 2. Here laser diodes generating at wavelengths 470 and 561 nm, and He-Ne laser generating at wavelength 632.8 nm were used as the light sources. In case of using laser diodes, collimated light beam was formed by means of the lenses. In addition, in order to remove distortions in the beam profile the latter was transmitted through the spatial filter is a combination of a plano-convex lens and diaphragm. Optical isolator at the output of the laser was used for the stability of laser operation, which provides light transmission in one direction almost without losses, and in the other (opposite) direction with high attenuation.

The investigated sample was placed in a vertical position on a precision rotary table which provide changing the incident angle α with accuracy 0.25 degrees. Coordinates of the laser beam spot center was determined by a laser beam profiler NewPort LBP-4-USB with accuracy 8 microns. Shift of laser beam caused by refraction in the sample was calculated from new position of spot center for every new incidence angle.

The proposed approach was applied for obtaining the refractive index of the undoped sample of gallium nitride GaN [3] having dimensions 10×10 mm^2 and a thickness of $d = 150$ mm. The sample was grown at the National Sun Yat-Sen University.

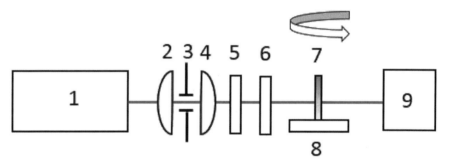

Figure 2. Scheme of experimental setup: 1—laser; 2,4—plano-convex lenses, 3—diaphragm; 5—polarizer; 6 –λ/4 wave plate; 7—sample; 8—precision rotatory stage; 9—laser beam profiler.

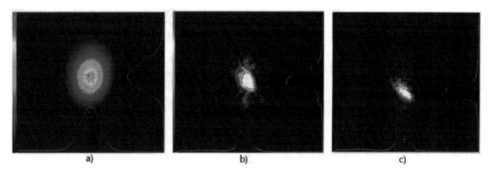

Figure 3. Profiles of the laserbeam (a) incident on sample, (b) after passing through the sample rotated axially by 1 deg, (c) after passing through the sample rotated axially by 35 deg.

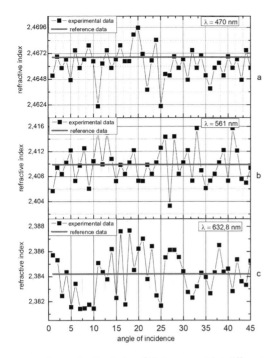

Figure 4. Refractive index of GaN measured at different angles of incidence for three wavelengths: (a)—470 nm, (b)—561 nm, (c)—632.8 nm.

Table 1. The values of the refractive index of GaN.

λ [nm]	$n_{experimental} \pm \Delta n$	n*
470	2,4677 ± 0,0026	2,4668
561	2,4095 ± 0,0055	2,4098
633	2,3844 ± 0,0043	2,3842

*in work [4].

The angle of incidence was changed in the range from 0 to 45 degrees; the greater rotation lead to a significant distortion of the profile of the laser beam. Figure 3 shows the profiles of the laser beam incident on the sample (Fig. 3a), and profile of this beam after it passing through the sample at an incidence angle of 1 deg (Fig. 3b) and 35 deg (Fig. 3c).

As will be seen below, the distortion of the beam at higher angles of incidence (up to 45 deg) does not lead to a noticeable increase in measuring errors.

Figure 4 shows dependencies of GaN refractive index on incidence angle (in the range of 1 to 45) obtained at three wavelengths. The refractive index at each wavelength in the figure is shown in comparison with reference data [4]. The Table 1 summarizes the average values of the refractive index of gallium nitride obtained experimentally at three wavelengths in comparison with reference data.

As seen from the presented data the experimentally obtained refractive indexes are coincide with reference data with accuracy to a measurement error Δn which is not exceed $5{,}5 \times 10^{-3}$.

4 CONCLUSIONS

Thus, in this paper we propose and approbated a method of determining the refractive index using a laser beam profiler. The refractive index of plane-parallel gallium nitride sample was measured at wavelengths of 470, 561 and 633 nm. The experimental values coincide with the reference data with accuracy to 10^{-3}. This method of measuring the refractive index distinguishes simplicity of installation schemes and low cost. Method can be used for express-measurement of refractive index of optical elements.

ACKNOWLEDGMENTS

The research was funded by Russian Scientific Foundation (project #14-12-01122).

REFERENCES

[1] I.S. Grigoriev and E.Z. Meilikhov, Handbook of Physical Quantities (Energoatomizdat, Moscow, 1991; CRC Press, Boca Raton, Florida, United States, 1997).
[2] A. Pimpinelli, and J. Villain. Physics of Crystal Growth [M]. Cambridge University Press, Cambridge, 1998, PP. 1–400.
[3] A.N. Turkin: Components and Technology. Vol. 5 (2011) 176–1180.
[4] A.S. Barker Jr., and M. Ilegems: Phys. Rev. B. Vol. 7. (1973) 743.

Advances in Future Manufacturing Engineering – Yang (Ed.)
© 2015 Taylor & Francis Group, London, ISBN 978-1-138-02817-3

Flow cessation mechanism of β phase solidification of titanium aluminium alloy with high niobium content

Ziqi Gong, Lihua Chai, Ziyong Chen, Feng Zhou, Xiao Hou & Guifang Yang
College of Materials Science and Engineering, Beijing University of Technology, Beijing, China

ABSTRACT: Spiral sample of Ti-45 Al-8 Nb–0.2 W-0.25Cr alloy was used to test the flow cessation mechanism by microstructure observation. It turned out that the morphology and size of columnar crystal and the equiaxial crystal changed a lot with the increase of flow length; flow cessation mechanism of the alloy showed that, besides end congestion causes, the main reason to stop the alloy melt flow is the abundant growth of columnar crystal in the root, which causes to form the transcrystalline structure and finally gives rise to the "necking", leading to flow cessation.

Keywords: high Nb-TiAl; flow cessation mechanism; microstructure; flow model

1 INTRODUCTION

The flow cessation mechanism of alloy melt has always been an issue for the foundry workers, and it has experienced a long time to research the stopping flowing mechanism. Many casting scholars have their own theoretical basis and experimental basis. The American M.C. Flemings scholar [1] took Al-Sn alloy as the research object and put forward two flow cessation mechanisms: The narrow crystallization temperature range and wide crystallization temperature range to the flow cessation mechanism of alloy melt. Russian scholars [2] made macrostructure observation of the pure titanium melt spiral sample, drew the flow cessation model and made the flow cessation mechanism of the narrow crystallization temperature range alloy melt that similar to M. C. Flemings'. China's Zhou Yao and others [3] drew the flow cessation mechanism of overheating pure metals, eutectic alloys and wide solidification interval alloy with the studying of Sn-Pn alloy. Ye Rongmao [4] took diluted Al-Si alloys as the research object and obtained the flow cessation mechanism of alloy melts with different purity. Qing chun Li [5] summed the flow cessation mechanism of alloy melts up in two: the neck blockage flow cessation mechanism and the end blockage flow cessation mechanism. It can be seen from the above study about the flow cessation mechanism of alloy melt, the flow cessation mechanism of various alloy melt are different. While there have been no reports on the flow cessation mechanism of Ti-Al based alloys melt.

TiAl based alloys are considered to be the most promising lightweight high-temperature structural materials due to its low density, high specific strength, excellent high temperature mechanical properties and oxidation resistance and other characteristics, and it has been the focus of research development [6–10]. In recent years, the high-niobium TiAl intermetallic compound rising between domestic and foreign has a good comprehensive performance. Furthermore, many studies have shown that the introduction of β-phase stabilizing elements, such as W, Cr, Mn, V, etc., can improve the β phase content of the alloy, thereby improving the high temperature deformation resistance of the alloy [11]. Meanwhile, the using of β-phase solidification path can available to obtain fine small-cast TiAl alloy tissue. However, TiAl based alloys are more sensitive to technological parameter during casting process; the filling is more difficult and poor surface quality of castings.

This article through the casting spiral flow performance experiments of the multi-alloy β-phase solidification high-NbTiAl alloy to observe the macroscopic tissue of melt in overall flow path, derive the flow cessation mechanism of alloy, and establish the flow cessation mechanism of TiAl based alloys. It has important significance to improve the filling capacity and the formed member's surface quality of high-Ti based alloy melt.

2 EXPERIMENTAL PROCEDURE

Ti-45 Al-8 Nb-0.2 W-0.25Cr alloy is adopted in the experiment of high Nb-TiAl alloy. In the experiment, the argon gas under high vacuum is used to protect electrode melting of self-consuming, the spiral shaped casting sample as shown in Figure 1(a). To ensure the stability and the

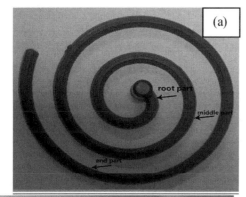

(a)

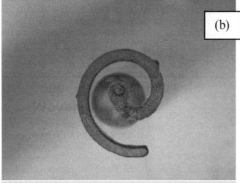

(b)

Figure 1. The spiral sample (a) and the casting spiral sample (b).

Table 1. The casting process parameters.

The name of parameter	The numerical value
Each combination alloy quality/g	40
The sample length/mm	500
The initial temperature of cast/°C	200
The melting power/KW	50

comparison of experimental results, the pouring process, the distance between the mold runner of samples and the metal casting speed will remain the same. And alloy was cast three times cyclically, in which the casting process parameters is shown in Table 1.

The sample was observed by longitudinally cutting the spiral shaped alloy, was then mechanically polished using the water-proof abrasive paper of P80 to P2000 and surface corrosion treatment employed the etchant of 90 ml distilled water + 5 ml HNO_3 + 5 ml HF. In the end, microstructure observation of the sample was carried out by the optical OLYMPUS-TH3 microscope.

3 RESULTS AND DISCUSSION

3.1 Microstructure characterization

Spiral flow of the alloy specimen shown in Figure 1(b). Sample filling length is 107 mm, and the sample front-end is convex.

Figures 2, 3, and 4 were the root, middle and end of sample macrostructure, the direction of arrow represents the flow direction of the sample filling. The graphic size is helical length of the sample.

As shown in Figure 2, the root of samples is made by the accumulation of columnar crystal and the equiaxial crystal zone, and along the flowing direction increases gradually. In contrast, the columnar crystal growth of the mould wall is perpendicular to the direction of the flow. The columnar crystal growth gradually close to the blocking zone with the increase of flow length, the organization of the columnar crystal then turn into the rich

Figure 2. The root region of spiral sample.

Figure 3. The middle mixed region of spiral sample.

Figure 4. The end region of spiral sample.

and thick, both the mould wall sides of columnar crystals almost together, it is certain that the blocking of alloy melt stopped the melt flowing in here.

Figure 3 presents the middle mixing area of the columnar and equiaxial crystal in the spiral specimens; it is found that the organization of the columnar crystal significantly smaller than the organization in Figure 2, and the columnar crystal organization of mixed zone decreases gradually with the increase of flow length. However, the arrangement of the columnar crystals is very neat, the clamp grain of between columnar crystals is very uniform, the grain structure is small and compact, and there is no significant change in grain size.

The end area of the spiral-shaped specimen appears as the thin layer of columnar crystals in Figure 4, it be further observed that the center organization gradually small and the columnar crystal along the flow direction of fine grain zone disappear completely when the spiral line further extends to the mould wall and the columnar crystal of the mould wall turn into a thin layer. At the moment, the columnar crystal disappear completely, the grain size of samples is completely fine, the grain organization is very dense and occupy the whole sample, especially the grain of the arced end zone is the most dense.

3.2 Filling capacity calculation

Assumptions: (1) the surface contact temperature between the mould and the liquid metal is constant in the flowing process of liquid metal; (2) the mean velocity of liquid metal is constant in the process of filling; (3) the physical properties value do not changing with temperature; (4) the liquid metal above liquidus is considered to Newton's body, and the yield stress of metal linearly change with temperature in the two phase region of solid and liquid. The filling process is shown in the figure below.

3.2.1 The filling length of liquid metal temperature dropped to the liquidus temperature T_L

First of all, differential equations are listed as follows according to the energy conservation principle.

$$\frac{\partial^2 T}{\partial X^2} - \frac{V}{\alpha}\frac{\partial T}{\partial X} - \frac{hP}{\lambda F}(T - T_0) = \frac{1}{2}\frac{\partial T}{\partial t} \qquad (1)$$

boundary conditions: $x = 0$, $T = T_j$; $t = \infty$, $T = T_0$.
In the formula:
λ—the thermal conductivity of liquid metal, W/(m·°C)

α—the thermal diffusivity of liquid metal, m²/s
P—the circumference of the filling pipe, m
F—the sectional area of the filling pipe, m²
h—the heat transfer coefficient of metal and mould, W/(m²·°C)
V—the mean flow rate of liquid metal in the filling process, m/s
T_0—the mould temperature, °C
T—the liquid metal temperature at random time, °C
t—the time, s
T_j—the casting temperature, °C.
Solve differential equations and substitute into boundary conditions:

$$T = (T_j - T_0)e^{\left(\frac{V}{2\alpha} - \sqrt{\left(\frac{V}{2\alpha}\right)^2 + \frac{hP}{\lambda F}}\right)x} + T_0 \qquad (2)$$

When the temperature of the liquid metal T fell to the liquidus temperature, the flowing length of the metal liquid is the filling length of X_L, so the formula (2) $T = T_j$, and then,

$$X_L = \frac{\ln\dfrac{T_j - T_0}{T_L - T_0}}{\sqrt{\left(\dfrac{V}{2\alpha}\right)^2 + \dfrac{hP}{\lambda F}} - \dfrac{V}{2\alpha}} \qquad (3)$$

3.2.2 The filling length of liquid metal temperature dropped to the flow stop from the liquidus temperature T_K

Due to the presence of solid phase, the physical properties value of metal change with temperature, so an asterisk is added to distinguish it. According to the energy conservation principle, the same differential equation with the formula (1) is listed, substitute into $X = X_L$, $T = T_L$:

$$T^* = (T_j - T_0)^{\frac{\beta^*}{\beta}} \times (T_L - T_0)^{1 - \frac{\beta^*}{\beta}e^{-\beta^*}} + T_0 \qquad (4)$$

In the formula: $\beta = \sqrt{\left(\dfrac{V}{2\alpha}\right)^2 + \dfrac{hP}{\lambda F}} - \dfrac{V}{2\alpha}$

$$\beta^* = \sqrt{\left(\frac{V}{2\alpha^*}\right)^2 + \frac{hP}{\lambda^* F}} - \frac{V}{2\alpha^*}$$

$$\alpha^* = \frac{\lambda^*}{C^*\rho^*}$$

$$C^* = C + \frac{L}{T_L - T_S}$$

L—the latent heat of crystallization, J/kg
T—the solidus temperature, °C

145

ρ^*—the average density of solid and liquid, kg/m³

C^*—the average specific heat capacity of solid and liquid, J/(kg·°C)

λ^*—the average thermal conductivity of solid and liquid, W/(m·°C)

C—The liquid phase specific heat capacity, J/(kg·°C).

For convenience, $\rho^* = \rho$, $\lambda^* = \lambda$, and lower temperature will increase the yield stress τ_S which block the flow; assume that varying pattern is linear $\tau_S = K(T_L - T^*)$, substitute equation (4) into:

$$\tau_S = K\left[T_L - (T_j - T_0)^{\frac{\beta'}{\beta}} \cdot (T_L - T_0)^{1-\frac{\beta'}{\beta}} e^{-\beta'} - T_0\right] \quad (5)$$

In the type, K is the changes of yield stress caused by unit temperature changes, N/(m²·°C).

This can be calculated block the flow of forces FZ. In crystal growth type wall caused by the narrow channel size boundary of a cylindrical body, the drag for:

$$F_Z = 2\pi R \int_{x_L}^{x^*} \tau_S dx \quad (6)$$

substitute (3), (5) into (6), and approximately $X^* - X_L$ as a smaller value, so,

$$e^{-\beta'(X^*-X_L)} \approx 1 - \beta'(X^* - X_L) + \frac{\beta'^2}{2}(X^* - X_L)^2,$$

then$\beta' = \sqrt{\left(\frac{V}{2\alpha^*}\right)^2 + \frac{hP}{\lambda^* F}} - \frac{V}{2\alpha^*}$, obtain from(3)

$$e^{-\beta' x_L} = \left(\frac{T_L - T_0}{T - T_0}\right)^{\frac{\beta'}{\beta}},$$

and so,

$$F_Z = \pi R K \beta' \left(x^* - x_L\right)^2 (T_L - T_0) \quad (7)$$

because of $d\beta'/d\alpha^* > 0$, while $\alpha^* = \lambda^*/C^*\rho^*$, the value of α^* is reduced by the effect of latent heat. The R of the formula (7) remains unclear, but it can be determined by the solidification heat transfer law, there have both the columnar crystal and the equiaxial crystal in the most narrow junction area. Under ignoring the heat transfer condition along the flow direction, the length of the columnar crystal was calculated. Because the temperature is T_L here, according to the heat transfer law the following equation be listed:

$$Ph(T_L - T_0) = \frac{F\rho_S L}{r}\frac{dR'}{dt}$$

in the formula: R'—the narrow crystal length, m

r—the loss coefficient of dendrite

$$\rho_S - \frac{\text{crystal volume of not wash away}}{\text{crystal volume}}$$

Make $W = (Ph(T_L - T_0)/F\rho_S L) \cdot r$ as solidification rate, so $R' = Wt$, assume flow velocity V as a constant, then the time of stop the flow is $t = (x^* - x_L/V)$, therefore:

$$R' = \frac{W}{V}(x^* - x_L) \quad (8)$$

Make R^* as the pipe size, so $R = R^* - R' = R^* - W/V \, (X^* - X_L)$, substitute this formula into (7):

$$F_Z = \pi K \beta' \left[R^* - \frac{W}{V}(x^* - x_L)\right](x^* - x_L)^2 \cdot (T_L - T_0) \quad (9)$$

And by the force balance equation: $\pi R^2 \overline{P} = F_Z$, therefore:

$$\overline{P}\left[R^* - \frac{W}{V}(x^* - x_L)\right] = K\beta'(x^* - x_L)^2 \cdot (T_L - T_0)$$

In the formula:$\overline{P}$—the effective pressure, P_a

Solution of the equation:

$$x^* = \frac{\left[-\dfrac{\overline{P}W}{V} + \sqrt{\left(\dfrac{\overline{P}W}{V}\right)^2 + 4K\beta'\overline{\beta}(T_L - T_0)R^*}\right]}{2K\beta'(T_L - T_0)} + X_L$$

Substitute into (3):

$$X^* = \frac{\left[-\dfrac{\overline{P}W}{V} + \sqrt{\left(\dfrac{\overline{P}W}{V}\right)^2 + 4K\beta'\overline{\beta}(T_L - T_0)R^*}\right]}{2K\beta'(T_L - T_0)}$$

$$+ \frac{1}{\beta}\ln\frac{T_j - T_0}{T_L - T_0} \quad (10)$$

This is an alloy filling length expression; the alloy is a Bingham body in the two phase of solid and liquid, so the expression is universal on the alloy of a certain range crystallization temperature. Considering two extreme cases:

1. For alloy with wide crystallization temperature range, $r=0$, the value of K is large, then:

$$X^* = \left[\frac{R \cdot \overline{P}}{K\beta^*(T_L - T_0)}\right]^{1/2} + \frac{1}{\beta}\ln\frac{T_j - T_0}{T_L - T_0} \quad (11)$$

2. For alloy with narrow crystallization temperature range or pure metal, $r = 1$ (No dendrites fall off), $K = 0$ (flow stops mainly caused by the connection of the columnar crystal), the limit calculated based on formula (10):

$$X^* = \frac{R^*V}{W} + \frac{1}{\sqrt{\left(\frac{V}{2\alpha}\right)^2 + \frac{h\rho}{\lambda F} - \frac{V}{2\alpha}}} \cdot \ln\frac{T_j - T_0}{T_L - T_0} \quad (12)$$

In this case, $P = 500$ mm, $F = 19.625$ mm^2, $T_0 = 25°C$, T_j closely approximates 1800°C, the rest of the parameters can be obtained through chemical manual query. The alloys solidify is in a narrow crystallization temperature range, the result calculated on formula (12) is 112 mm, close to the experimental data 107 mm.

3.3 Cessation mechanism of flow

Diagrammatic drawings of cessation mechanism of flow are shown in Figure 5, the cessation of the flow can be divided into the following several stages:

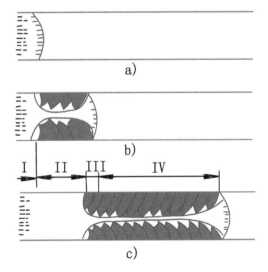

Figure 5. Diagrammatic drawings of cessation mechanism of flow.

a. The pure liquid flow beyond the superheat (5a).
b. The terminal of the cold wall solidified (5b).
c. The metal liquid behind flow in heated pipes, cooling intensity decreased (5c).

Due to fluid flow through area I, in which there is a certain degree of superheat, the solidified shell will be remelted, turning into the zone II. Thus, this area solidified first is eventually melted. Section III is part of solid phase area retained by the melting, in the end overheat of the liquid metal will be run out. In zone IV, The liquid phase and solid phase share the same temperature—crystallization temperature. Clogging often happens in its neighborhood because in the starting position crystallization starts earlier and crystallization is also relatively earlier beyond the section. The suction caused by front liquid metal solidification shall help to form trumpet-shaped shrinkage cavity. As the melt flows, the decrease of the composition of supercooling is conducive to the growth of the columnar crystal in zone I and II, the columnar crystal would joint in the center of the casting by growing and extending.

The solidification end looks like convex protrusion, showing that it is end block cessation, which shows a narrow crystallization temperature range; root shows the necking flowcessation mechanism, namely the wide-crystallization-temperature-range cessation mechanism of flow temperature range. So it can be concluded that high Nb-TiAl alloy cessation is not only about the end clogging, the main reason to stop the flow is that the columnar crystal is rich in the root, and the side post crystal touches each other, that forms a transgranular organization leading to necking stop.

While the alloy melt goes into the mould, there is largest temperature difference between the melt and the mould, so a thin layer of columnar crystal will come out beyond the flow channel walls. As subsequent liquid metal continuously flow into gradually heated mould, the chilling effect is reduced, the columnar crystal crust becomes more and more thin, and liquid metal heat makes crust continuously fall off. In addition, because of the casting chilling effect, columnar crystal crust also grows up; this makes a speed difference between the growing speed and the melting speed. Namely

$$\Delta V = V_{growing} - V_{melting}$$

$V_{growing}$ is growing speed (m/s), $V_{melting}$ is the melting speed (m/s).

While $\Delta V > 0$, The crust of the columnar crystal gets thicker, and finally form the necking stop; while $\Delta V < 0$, Columnar crystal crust is thinner and thinner, crust remelting disappears along with the stream flow to the end of the mould.

In the process of the alloy melt flow cessation, crystal growing rate and melting rate alternately play a leading role. At the beginning, temperature difference is big between the liquid metal and the mold, crystal grows up faster than that of crystal melting. With the increase of mould temperature, heat dissipation of solid metal slows, but the heat spread to crystal remains the same, thus crystal melting speed begins to play a key role. But with the loss of the liquid metal temperature, the capacity of the crystal melting reduces, alternately crystal growing speed will gradually play a leading role, the crystal grows up faster than crystal melts, the flow of liquid metal in the wall and thin crystal layer scour to form a "necking". The causes of necking is due to the enrichment of solute discharge on the crystal side. As for Ti-45 Al-8 Nb-0.2 W-0.25Cr, it contains a so high content of solutes that it is easier to enrich to form necking, "Necking" parts under the flow of scour and temperature fluctuation fuse or fall off, flowing to the terminal of the mould along with the liquid. because of the large temperature gradient in the helix root, it is easy to maintain a relatively narrow constitutional supercooling area, narrow constitutional supercooling which is the best condition for the growth of the columnar crystal will help columnar crystal stretch into the center of the casting where two parts of columnar crystal meet, forming a particularly "transgranular" structure.

With the increase of flowing distance, the length of the columnar crystal begins to shorten. The lower temperature of liquid metal creates favorable conditions for the stability of the existed washed crystal parts, and indirectly hinders the growth of the columnar crystal. In the spiral end of chilling area, equiaxed grains are fine and close. Clumps of crystal and completely elimination of metal overheating limit the growth of columnar crystal, shown in Figure 4. Therefore, a large number of crystal shards accumulates where columnar crystal disappeared, limiting the crystal growing completely and becoming a dense equiaxed grains zone.

To sum up, cessation mechanism of flow of high Nb-TiAl alloy should be mainly the backend necking plugging mechanism, the front-end plugging mechanism is complementary. The spiral sample can be divided into three zones (see Figs. 2, 3, and 4), that is transcrystalline structure in helical roots, the middle area of helix is mostly washed-down crystal and the coexisted columnar crystal, the back is a dense equiaxed fine grains area. End flow stop is mainly due to the backend transgranular blocking so that formed to be convex shape.

4 SUMMARY

1. For the alloy Ti-45 Al-8 Nb-0.2 W-0.25Cr,the terminal of the spiral sample is convex, the flowing length of the sample is 107 mm.

2. The abundant and coarse columnar crystals beyond the cavity wall help to form transcrystalline structure; Along the central flowing direction, the columnar crystal organization reduced gradually and slowly, the arrangement of columnar crystals is very neat, between them are the uniform and fine grains; To end the columnar crystals disappear completely in the terminal; fine grains occupying the center run through the whole sample.

3. The Flow Cessation Mechanism of high Nb-TiAl alloy is mainly caused by the abundant columnar crystals in the root, which will give rise to transcrystalline structure so as to form the so-called "necking", and finally stop the flow.

ACKNOWLEDGMENT

This work is financially supported by The National Natural Science Fund (contract No.: 51301005), which is gratefully acknowledged.

CORRESPONDING AUTHOR

Ziqi Gong, Email: ziqi1029@163.com, Mobile phone: +86-13366169211.

REFERENCES

[1] M.C. Flemings, E.F. Niiyama and H.F. Taylor: AFS Trans Vol. 69 (1961), p. 625.
[2] Intitute of metal research, Chinese academy of sciences, in: Titanium Alloy Casting Performance Abroad, chapter, 20, Shanghai institute of science and technology information (1974).
[3] Z. Zhang, Y.H. Zhou: Journal of Northwestern Polytechnical University Vol.1 (1965), p. 29.
[4] R.M. Ye, L.G. Jiang, Z.L. Jiang and H.G. Wang: Metal Science and Technology Vol. 2 (1983), p. 41.
[5] Q.C. Li, in: Cast Forming Theory Foundation, Mechanical Industry Publishers (1980).
[6] Z.Q. Gong, Z.Y. Chen, L.H. Chai, Z.L. Xiang and Z.R. Nie: Acta Metallurgica Sinica Vol. 49, (2013), p. 1369.
[7] G.H. Cao, G.J. Shen, J.M. Liu:Scripta Materialia Vol. 49 (2003), p. 797.
[8] H.M. Yang, Y.Q. Su, L.S. Luo: Rare Metal Materials and Engineering Vol. 40, (2011), p. 1.
[9] L. Cheng, H. Chang, B. Tang: Rare Metal Materials and Engineering Vol. 43, (2014), p. 36.
[10] Z.Y. Chen, Z.Q. Gong, L.H. Chai, Z.L. Xiang and Z.R. Nie: Materials Science and Technology Vol. 29, (2013), p. 937.
[11] Y.Y. Chen, F. Yang, F.T. Kong: Journal of Rare Earth Vol. 29, (2011), p. 114.

Advances in Future Manufacturing Engineering – Yang (Ed.)
© 2015 Taylor & Francis Group, London, ISBN 978-1-138-02817-3

The luminescence property of Rhodamine B grafted in different structured channels of mesoporous silicate

Meilu Wang, Yuzhen Li, Wei Xu & Zhihui Huang
College of Environmental Science and Engineering, Taiyuan University of Technology,
Yingze District, Taiyuan, China

ABSTRACT: Rhodamine B has been grafted in different structured channels of mesoporous silicate materials (MCM-41, SBA-15). The products mesostructure were characterized by small angle X-Ray Diffraction (XRD) and N_2 adsorption-desorption measurement (BET). The morphology of composite materials is that the mesopores maintained were got from Transmission Electron Microcopy (TEM). Fourier Transform infrared spectrometry (FT-IR) shows that the Rhodamine B has been introduced into mesoporous. Finally, the luminescent property was obtained from Photoluminescence spectroscopy (PL), which indicates that the emission spectrum of RhB/APTES/MCM-41 shows more blue shift than RhB/APTES/SBA-15.

Keywords: mesoporous silica; Rhodamine B; composite materials; luminescence

1   INTRODUCTION

As one of the nano-materials, mesoporous plays an important role in many fields such as adsorption [1], catalyst [2], medicine [3] due to its particular properties including highly ordered structure, tunable particle size, larger surface area and pore volume, etc [4]. Nowadays, mesoporous materials can be used as the host of dyes. In our previous study, we synthesized the mesoporous luminescent materials that used the 1,8-Naphthalic anhydride as luminescent guest and the bimodal mesoporous materials as the host, which can cause the efficient luminescence [5]. Rhodamine B that has longer wavelength of absorption and emission, larger molar absorption coefficient, higher fluorescence quantum yield is a famous and hot organic dye [4]. It has been considered to be one of the best luminescence dyes.

In this paper, we chose the famous ordered mesoporous materials (MCM-41 and SBA-15) as the host materials and Rhodamine B as the guest dye. MCM-41 and SBA-15 are ordered mesoporous materials with larger surface area, good adsorption capacity, and high thermal stability. MCM-41 has a variable pore size in the range of 2–6 nm with smooth pore wall, but SBA-15 has a big pore size in the range of 5–10 nm with roughness pore wall [6]. The products were prepared by simple method. The structure and photoluminescence characters have been discussed.

2   EXPERIMENTAL SECTION

2.1   *Chemicals*

Hexadecyl trimethyl ammonium bromide (CTAB) and Ethyl silicate (TEOS) were supplied by Sinopharm Chemical Reagent limited Corporation. Tri-block copolymer poly(ethylene oxide)-poly(propylene oxide)-poly(ethylene oxide) (P123) was purchased from Sigma-Aldrich. APTES was got from Alfa Aesar Company. Rhodamine B was provided by Aladdin industrial Corporation. Phosphoric acid was offered by Tianjin Fengchuan Chemical Reagent Science and Technology Company Limited. Tetrahydrofuran (THF) was supplied by Tianjin Tianda Chemical Experiment Factory. Ethanol was got from Tianjin Beichen Fangzheng Reagent Factory. All the materials were all A.R. grade.

2.2   *Synthesis of composite materials*

MCM-41 was got by a simple method [7]: 3.644 g of CTAB was dissolved in 100 ml of deionized water and then 10 ml of NaOH solution (1 M) was added. After stirring for 15 min, 18.6 ml of TEOS was added drop wise. The mixture was stirred for another 2 h. The obtained gel was poured into an autoclave with Teflon lining hermetically closed and was heated at 110 °C for 48 h. The product was filtered, washed with distilled water and dried at 120 °C. To remove the surfactant, the solid was calcined at 550 °C for 6 h with a heating rate of 5 °C/min. The resulting materials were noted as MCM-41.

SBA-15 was obtained according to the literature [8]: 3.0 g P123 and 5 ml phosphate were added into 70 ml distilled water, and stirred until the P123 was dissolved totally. Then 8 ml TEOS was added quickly. After being stirred for 1 h at 40 °C, the obtained gel was poured into an autoclave with Teflon lining hermetically closed at 100 °C for 24 h. The product was filtered and dried for 3 h at 80 °C. In the end, the product was calcined at 550 °C for 6 h.

The method of preparing APTES/MCM-41-(SBA-15): 0.5 g of MCM-41 (SBA-15) was dissolved in 50 ml dried THF under stirring at 25 °C for 10 min. Then about 0.586 ml APTES was added to the mixture under continuous stirring. After stirring for 5 h, the resulting samples were filtered and dried at 80 °C for 4 h.

The process of making RhB/APTES/MCM-41-(SBA-15): 0.25 g of APTES/MCM-41 (SBA-15) was added into 50 ml ethanol solution of Rhodamine B (1×10^{-4} mol L^{-1}). After stirred for 5 h, the product was filtered, washed with ethanol several times until no color was observed in the ethanol, and then dried over night at room temperature.

3　RESULTS AND DISCUSSION

Figure 1 showed the small angel diffraction of (a) MCM-41, (b) APTES/MCM-41, (c) RhB/

APTES/MCM-41 (lift) and (a) SBA-15, (b) APTES/SBA-15, (c) RhB/APTES/SBA-15 (right). It is obvious that the characteristic (100) diffraction peak of mesoporous materials was shown in all samples at the 2θ in the range of 0.6°–8° and (110) diffraction peak was found in the right samples. The peak intensity of all samples decreased obviously after modification and grafting because the ordered structure degree has fallen caused by the random distribution of functional groups [5]. The value of d space of MCM-41 decreases from 3.805 nm to 3.763 nm after modification due to the—NH$_2$ groups were loaded on the pore surface [5]. The value turns into 3.679 nm for RhB/APTES/MCM-41 and the value of the d space decreases from 8.489 nm to 8.328 nm for SBA-15 after grafting because functional groups have been introduced into the mesostructure [7].

The TEM image of RhB/APTES/MCM-41 (left) and RhB/APTES/SBA-15 (right) were shown in Figure 2. It can be found that the well-ordered mesopore structure is maintained after grafting. As shown in the picture, the pore size of mesopore is around 2 nm for MCM-41 and about 5 nm for SBA-15. It is can be judged that the materials have a narrow pore distribution.

FT-IR spectra of sample A (a) MCM-41, (b) ATPES/MCM-41, (c) RhB/APTES/MCM-41 and sample B (a) SBA-15, (b) APTES/SBA-15, (c) RhB/APTES/SBA-15 were displayed in the Figure 3. There is a strong infrared absorption

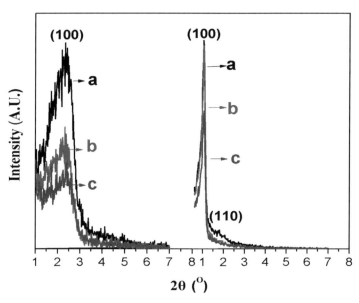

Figure 1.　XRD patterns of samples, (a) MCM-41, (b) ATPES/MCM-41, (c) RhB/APTES/MCM-41 in the left and (a) SBA-15, (b) ATPES/SBA-15, (c) RhB/APTES/SBA-15 in the right.

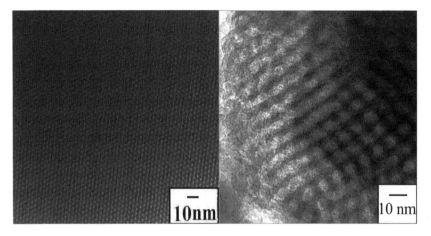

Figure 2.   TEM image of RhB/APTES/MCM-41 in the left and RhB/APTES/SBA-15 in the right.

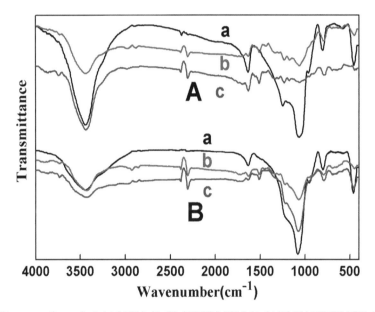

Figure 3.   FT-IR spectra of sample A (a) MCM-41, (b) ATPES/MCM-41, (c) RhB/APTES/MCM-41 and sample B (a) SBA-15, (b) APTES/SBA-15, (c) RhB/APTES/SBA-15.

band around 3400 cm^{-1} for all samples of MCM-41 series and around 3437 cm^{-1} for SBA-15 series that can be considered as the Si–OH stretching vibration [3,9]. The peaks near 1640, 804 and 460 cm^{-1} for all samples of MCM-41 series and 1637, 790 and 458 cm^{-1} for SBA-15 series come from the O–H bending vibration of water molecules [10], the symmetric Si–O stretching vibration and the Si–O–Si bending vibration [5]. The bending vibration of Si–OH is found at 968 cm^{-1} for MCM-41 (A.a) and 956 cm^{-1} for SBA-15 (B.a), but the peak becomes weaker or disappears after modification

and grafting because of the lack of Si–OH group [5]. The peaks near 1517 and 696 cm^{-1} in Figure 4A (b, c) and B (b, c) are contributed by the symmetric—NH$_2$ bending vibration [11] and the vibration of the O-Si-C bond [12]. We can judge that the mesoporous materials have been modified by APTES. Moreover, the characteristic peak of Rhodamine B is found at 1705 cm^{-1} in Figure 4A (c) and 1712 cm^{-1} in Figure 4B (c) which arises from C = O [13]. So we can draw a conclusion that Rh B has been successfully embedded in mesoporous materials.

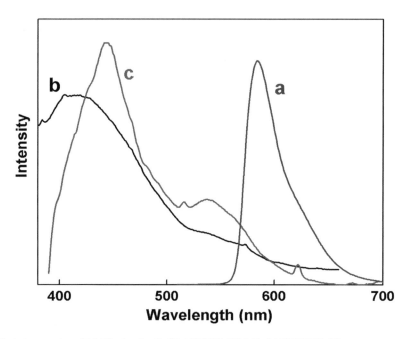

Figure 4.   Emission spectra of (a) Rhodamine B, (b) APTES/MCM-41, (c) RhB/SBA-15.

Figure 4 showed the Emission spectra of liquid Rhodamine B in ethanol with the concentration of 1×10^{-4} mol L^{-1} at 500 nm excitation wavelength, solid RhB/APTES/MCM-41 with the excitation wavelength of 350 nm (b) and solid RhB/APTES/SBA-15 with 380 nm (c). The strong characteristic band of RhB appears at 584 nm (a). When it was introduced into the channels of MCM-41 and SBA-15, the value of peak was blue-shifted to 415 nm (b) and 444 nm (c), respectively. The organic dyes are dispersed much better after grafting in the mesoporous and the solid surface may have influence on the luminescent property. The lack of aggregation of dye molecules has a strong influence on the energy increase between the ground state and the excited state [14]. The emission spectrum of RhB/APTES/MCM-41 show more blue shift than RhB/APTES/SBA-15 is because with the reducing of pore size, the degree of dispersion further rises [15]. For MCM-41 with the pore size of mesopore is around 2 nm, which owned the long and one-dimensional channels, May be only a small amount of Rhodamine B molecules were introduced into the pore of the MCM-41, the light-emitting molecules was in a decentralized state. While for SBA-15 with the pore size is about 5 nm, larger amount of Rhodamine B molecules

were adsorbed and then accumulated at the channels of the pores which is the reason why the peak of the RhB/APTES/MCM-41 is blue-shifed to that of RhB/APTES/SBA-15.

4   SUMMARY

In conclusion, we made luminescent composite materials (RhB/APTES/MCM-41 and RhB/APTES/SBA-15) by chemical bonding method. The physicochemical properties of materials were investigated by XRD, TEM, FT-IR and PL. The results show that the mesoporous were successfully modified and grafted and the mesoporous is maintained. We found that the materials that after grafting reveal a big blue-shifed.

ACKNOWLEDGEMENTS

This work was supported by the National Natural Science Foundation of China (21203136), the Project for Importing Talent of Taiyuan University of Technology (tyut-rc201120a) and the Natural Science Foundation of Shanxi (2013011040-1). E-mail address: liyuzhen123456@126.com.

CORRESPONDING AUTHOR

Dr. Yuzhen Li (Y.-Z. Li), Tel: 0086-0351-6018970,
E-mail address: liyuzhen123456@126.com.

REFERENCES

[1] D. Zhou, K. Li, J.J. Deng, X.H. Lu and Q.H. Xia. Tunable adsorptivity of mesoporous MCM-41 materials for organics and water: submitted to Materials Letters 122 (2014) 170–173.

[2] Z. Liu, S.H. Shen and L.J. Guo. Study on photocatalytic performance for hydrogen evolution over CdS/M-MCM-41 (M = Zr, Ti) composite photocatalysts under visible light illumination: submitted to International Journal of Hydrogen Energy 37 (2012) 816–821.

[3] Z. Bahrami, A. Badiei and F. Atyabi. Surface functionalization of SBA-15 nanorods for anticancer drug delivery: submitted to Chemical Engineering Research and Design 92 (2014) 1296–1303.

[4] D.L. Xu, G.W. Zhou, L. Zhang, Y.J. Li, J.Y. Zhang and Y. Zhang. Fluorescent hybrid assembled with Rhodamine B entrapped in hierarchical vesicular mesoporous silica: submitted to Powder Technology 249 (2013) 110–118.

[5] Y.Z. Li, J.H. Sun, X. Wu, L. Lin and L. Gao. Posttreatment and characterization of novel luminescent hybrid bimodal mesoporous silicas: submitted to Jounral of Solid State Chemistry 183 (2010) 1829–1834.

[6] D.P. Liu, X.Y. Quek, H.H. Adeline Wah, G.M. Zeng, Y.D. Li and Y.H. Yang. Carbon dioxide reforming of methane over nickel-grafted SBA-15 and MCM-41 catalysts: submitted to Catalysis Today 148 (2009) 243–250.

[7] L. Gao, J.H. Sun and Y.Z. Li. Functionalized bimodal mesoporous silicas as carriers for controlled aspirin delivery: submitted to Journal of Solid State Chemistry 184 (2011) 1909–1914.

[8] L. Gao, J.H. Sun, L. Zhang, J.P. Wang and B. Ren. Influence of different structured channels of mesoporous silicate on the controlled ibuprofen delivery: submitted to Materials Chemistry and Physics 135 (2012) 786–797.

[9] P.K. Tapaswi, M.S. Moorthy, S.S. Park and C.S. Ha. Fast, selective adsorption of Cu^{2+} from aqueous mixed metal ions solution using 1,4,7-triazacyclononane modified SBA-15 silica adsorbent (SBA-TACN): submitted to Jounral of Solid State Chemistry 211 (2014) 191–199.

[10] X.W. Cheng, X.J. Yu and Z.P. Xing. Synthesis and characterization of C–N–S-tridoped TiO_2 nano-crystalline photocatalyst and its photocatalytic activity for degradation of rhodamine B: submitted to Journal of Physics and Chemistry of Solids 74 (2013) 684–690.

[11] M. Guli, X.T. Li, Y. Chen, H. Ding, J.X. Li and S.L. Qiu. Encapsulation of Coumarin 151 into the mesopores of modified rodlike SBA-15: submitted to Materials Research Bulletin 45 (2010) 1–5.

[12] Y.Z. Li, J.H. Sun, L. Gao, X. Wu and B. Yang. Novel luminescent hybrid materials by covalently anchoring 2-[3-(triethoxysilyl)propyl-1H-Benz[de] isoquinoline-1,3(2H)-dione to bimodal mesoporous materials: submitted to Journal of Luminescence 132 (2012) 1076–1082.

[13] X.Q. Gao, J. He, L. Deng and H.N. Cao. Synthesis and characterization of functionalized rhodamine B-doped silica nanoparticles: submitted to Optical Materials 31 (2009) 1715–1719.

[14] G.D. Chen, L.Z. Wang, J.L. Zhang, F. Chen and M. Anpo. Photophysical properties of a naphthalimide derivative encapsulated within Si-MCM-41, Ce-MCM-41 and Al-MCM-41: submitted to Dyes and Pigments 81 (2009) 119–123.

[15] J.C. Tu, N. Li, Y. Chi, S.N. Qu, C.X. Wang, Q. Yuan, X.T. Li and S.L. Qiu. The study of photoluminescence properties of Rhodamine B encapsulated in mesoporous silica: submitted to Materials Chemistry and Physics 118 (2009) 273–276.

153

Advances in Future Manufacturing Engineering – Yang (Ed.)
© 2015 Taylor & Francis Group, London, ISBN 978-1-138-02817-3

Mechanical behavior analysis of 2.5D woven composites at elevated temperature based on progressive damage analysis method

Jian Song, Weidong Wen, Haitao Cui & Ying Xu
Nanjing University of Aeronautics and Astronautics, Jiangsu Province, P.R. China

ABSTRACT: A new parameterized finite element model of 2.5D woven composites called Full-cell model, which considers the influence of the outer structures, has been established. Based on this model, three-dimensional progressive damage analysis method was used to predict the stress-strain behaviors and progressive damage processes of 2.5D woven composites subjected to weft tension at 180°C. Additionally, the corresponding experiments were also conducted to validate the model. Compared with experimental data, the modulus and strength errors in weft are lower than 10% and the failure local and modes agree with the test results.

Keywords: 2.5D woven composites; full-cell model; mechanical property; temperature; progressive damage analysis

1 INTRODUCTION

2.5D woven composite materials have been more and more interesting in advanced aerospace industry, owing to their good comprehensive mechanics performance. Though the structure compared to the laminate composites is relatively complex, textile composites provide more balanced properties in the fabric plane and higher delamination resistance than laminates [1]. In textile composites, a relatively new class of 3D angle-interlock braided composite, called 2.5D woven composites in this paper, has been applied in aeronautics and aerospace fields, which possesses the advantages mentioned above and the braiding technology is more simple in comparison with other 3D woven composites.

Until now, the modulus prediction methods of 2.5D woven composites at Room Temperature (RT) are mainly concluded: Selective Averaging Method (SAM) [2, 3], Finite Element Method (FEM) [4–7], Bridging Model [8, 9]. Hallal and Younes [3] proposed an analytical model based on SAM, called three stages homogenization method, to predict the elastic properties of 2.5D interlock composite at RT. Dong [6] studied the stiffness, strength and damage extended issues of 2.5D braided composites based on commercial finite element analysis software ANSYS. Chou [7] proposed first the bridging model, which was used to analyses the elastic behaviors of 2D woven composites. For the strength prediction method of 2.5D woven composites at RT is focused on FEM.

Lu et al. [4, 5] studied the failure behaviors of 2.5D textile composites based on the unit cell model and the progressive damage model calculated by FEM were conducted to predict the on-axis and off-axis tensile strength of this class of material. Zheng [10] and Qiu [11] predicted the strength of 2.5D woven composites, as well as the progressive damage behaviors were also studied at RT based on FEM.

However, a number of manufacturing structural components exposed to long-term temperatures in 100–200°C range, such as aero-engine casing, required polymer matrix composite materials have the advantage of elevated temperature resistance performance. Unfortunately, due to the high-cost and difficult-to-test at elevated temperatures, experimental researches and theory models subjected to static loadings are relatively backward. Additionally, until now the effect of elevated temperatures on mechanical properties was focus on FRP [12–14], but for the textile composite materials, due to the difficulty of synthesis with heat-resistant resin, study on the mechanical characteristic is further limited [15–17]. Selezneva et al. [16] investigated the failure mechanism in off-axis 2D woven laminates at elevated temperature by experiments and found that the woven yarns began to straighten out and rotate towards the loading direction just prior to failure. Vieille and Taleb [15] studied the influence of temperature and matrix ductility on the behavior of 2D woven composites with notch and un-notched and the results revealed that the highly ductile behavior of thermoplastic laminates is quite effective to accommodate the overstresses near the hole at temperatures higher than their T_g.

At present, although several works have analyzed the mechanical properties of 2.5D textile composites by FEM, most of the models only took inner cell into consideration, which inevitable introduces error. Meanwhile, most studies are applicable to the room property prediction loaded in warp direction. Therefore, in this paper, a more reasonable model called Full-cell was firstly proposed based on the microscopic observation of 2.5D woven composites. After that, three-dimension progressive damage analysis method, which takes the influence of temperature into consideration, was applied to predict the mechanical properties of 2.5D woven composites at 180°C loaded in weft direction.

2 FULL-CELL FINITE ELEMENT MODEL

In this section, the geometric and finite element model of 2.5D woven composites will be established. The actual microstructure is illustrated in Figure 1.

From Figure 1, a rectangular and two anti-quadratic curve shapes are selected to describe the cross-sections of inner warp and weft yarns and the inner cell model can be established as follows:

① The boundary dimensions of the inner cell:

$$L_x = 10(N_f - 1) M_w, \ L_y = 10N_j/M_j. \quad (1)$$

where L_x and L_y are the longitudinal and transverse length (mm), respectively. N_f means the number of weft yarn at the same height in which $N_f = 3$. N_j means the number of warp yarn in the transverse direction, in which $N_j = 2$. M_j and M_w represent the warp and weft arranged density, in which $M_j = 10$ (tows/cm) and $M_w = 3.5$ (tows/cm).

② The cross-sectional sizes of the warp yarn:

$$A_j = T/(1000\rho P_j), \ W_{1j} = 10/M_j, \ W_{2j} = A_j/W_{1j}. \quad (2)$$

where A_j is the cross-sectional area of warp yarn (mm²). T is the linear density of yarns (g/1000m). ρ is the material density (g/cm³). P_j is the packing factor of fiber in warp yarn.

③ The cross-sectional sizes of the weft yarn and dip angle:

$$A_w = T/(1000\rho P_w), \ W_{2w} = (L_z - (N_h + 1)W_{2j})/N_h. \quad (3)$$

where A_w is the cross-sectional area of weft yarn (mm²). L_z is the height in the z direction.

In order to obtain the width of weft yarn W_{2w} and the dip angle of the straight segment of the warp yarn θ. A series of equations were given in Eq. 4~8. Furthermore, the configuration of the weft yarn is assumed quadratic curve presented as follows:

$$z = ax^2 + c \ (b = 0). \quad (4)$$

Firstly, a condition of the first-order continuous in point C (shown in Fig. 2(b)) must be ensure the smooth transition in that point where the two curves are connected.

$$z'|_{x = W1w/2} = -\tan \theta = aW_{1w}. \quad (5)$$

where θ is the dip angle. In addition, the dip angle can also be described by Eq. 6.

$$\tan \theta = (W_{2j} + W_{2w} + W_{2j} \cos \theta)/ \\ (L_{x/2} - W_{1w} - W_{2j} \sin \theta). \quad (6)$$

By Eq. 6, the dip angle can be obtained by the bisection method on the condition of the known W_{1w}. Furthermore, according to the continuity condition in point C, another Eq. 7 can be obtained:

$$(W_{2j} + W_{2w})/2 = a(W_{1w}/2)^2 + c. \quad (7)$$

Finally, the configuration of weft yarn is changed by adjusting W_{1w} to make sure that the area of weft yarn is equal to A_w.

$$4 \int_0^{W1w/2} (ax^2 + c - (W_{2j} + W_{2w})/2) \, dx = A_w. \quad (8)$$

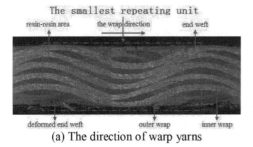

(a) The direction of warp yarns

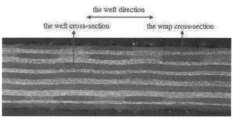

(b) The direction of weft yarns

Figure 1.   Cross-section microphotographs of bend-joint structure of 2.5D woven composites.

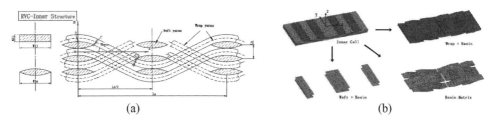

(a)   (b)

Figure 2.   (a) Illustration of the geometric relation in the inner structure; (b) Microstructure finite element model of 2.5D woven composites without outer cell.

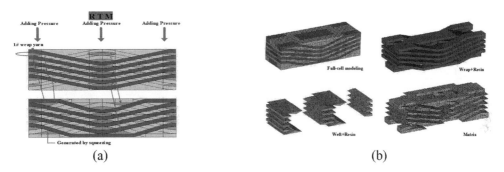

(a)   (b)

Figure 3.   (a) Molding process of 2.5D woven composites based on RTM technology; (b) Molding process of 2.5D woven composites based on RTM technology.

Therefore, the shape parameters (W_{1w}, a, c) and dip angle θ can be calculated according to simultaneous Eq. 4~8. Ultimately, the Inner-cell model shown in Figure 2(a) can be established by the above parameters, as shown in Figure 2(b).

In my view, the reason for formatting this type of complex microstructure is due to the influence of molding pressure which leads to the fact that the outer weft yarn deviate the center line of the inner weft yarn (see Fig. 3(a)). Therefore, the Full-cell model can be then established based on the inner cell model as shown in Fig. 3(b).

3 PROGRESSIVE DAMAGE APPROACH

In this part, the three-dimensional progressive damage approach, which can consider the influence of temperature is developed. The classical progressive damage approch has been successfully utilized to study the behavior of composite laminater at RT [18]. It concludes three main components: stress analysis, failure analysis and material property degradation. However, in this issue, the yarn is regarded as unidirectional composite with different fiber volume fraction. Additionally, the key point is to focus on the influence of temperature on the mechanical properties of 2.5D woven composites,

therefore, the study on the mechanical properties of components material T300–3k/QY8911-IV at various temperatures is quite important.

The theoretical mechanical models of unidirectional composites at various temperatures. The corresponding experiments have been performed for $[0]_{8/10/12}$, $[90]_{8/10/12}$ and $[\pm 45]_{2s/3s}$ at 20°C, 160°C and 200°C. The mechanical models, which take temperature and fiber volume fraction into account, are shown in Table 1.

Failure analysis. Failure mechanisms of composite structures are very complex. The three-dimensional Hashin-type criterion [19] has been developed successfully and applied to analyze damage failures of 2.5D woven composites at RT [11]. Furthermore, based on the SEM observation by this author, the modified tree-dimensional Hashin-type criterion is given as follows:

① The longitudinal failure of yarns:

$$(\sigma_{11}/X_{11}(V_f,T))^2 + (\sigma_{12}/S_{12}(V_f,T))^2 + (\sigma_{13}/S_{13}(V_f,T))^2 \geq 1. \tag{9}$$

② The transverse failure of yarns:

$$((\sigma_{22} + \sigma_{33})/Y_{22}(V_f,T))^2 + ((\sigma_{12}^2 - \sigma_2\sigma_3) /S_{23}(V_f,T))^2 + (\sigma_{12}/S_{12}(V_f,T))^2 + (\sigma_{13}/S_{13}(V_f,T))^2 \geq 1. \tag{10}$$

Table 1. The mechanical models of unidirectional composites with various temperatures and fiber volume fraction.

Mechanical properties	Models
Longitudinal tensile modulus	$E_{11} = a(V_f E_{f1} + V_m E_m)(1-k\sin((T-T_0)/(T_r-T_0)))$
Transverse tensile modulus	$E_{22} = E_{f2} Em(V_f + aV_m)/(E_m V_f + E_{f2} V_m)(1-k\sin((T-T_0)/(T_r-T_0)))$
Loading directional modulus	$E_{45} = (a\ E_{f2} Em/(E_m V_f + E_{f2} v_m) + (1-a)(E_{f1} V_f + E_m V_m))(1-k\sin((T-T_0)/(T_r-T_0)))$
Longitudinal tensile strength	$X_{11} = a(V_f X_{f1} + V_m X_m)(1-k\sin((T-T_0)/(T_r-T_0)))$
Transverse tensile strength	$Y_{22} = c(1-(a\ \sqrt{V_f} -bV_f))(1-E_m/E_{f2})X_m(1-k\sin((T-T_0)/(T_r-T_0)))$
In-plane shear strength	$S_{12} = (1 + V_f(1/a-1)b)(1-k\sin((T-T_0)/(T_r-T_0)))$
Resin matrix modulus	$E_m = E_0(1-k\sin((T-T_0)/(T_r-T_0)))$
Resin matrix strength	$S_m = S_0(1-k\sin((T-T_0)/(T_r-T_0)))$

Table 2. Dagradation method of component materials.

Mode of failure	Degradation method				
	E_1	E_2	G_{12}	G_{13}	G_{13}
Longitudinal failure in warp yarns	0.2	0.2	0.2	0.2	0.2
Transverse failure in warp yarns	1	0.4	0.4	0.4	0.4
Longitudinal failure in weft yarns	0.2	0.2	0.2	0.2	0.2
Transverse failure in weft yarns	1	0.4	0.4	0.4	0.4
Matrix failure	0.4	0.4	0.4	0.4	0.4

③ The pure resin matrix failure:

$$(\sigma_{11} - \sigma_{22})^2 + (\sigma_{22} - \sigma_{33})^2 + (\sigma_{11} - \sigma_{33})^2 + 6(\tau_{12}^2 + \tau_{13}^2 + \tau_{23}^2) \geq 2X_m(T). \tag{11}$$

where σ_{ij} (i, j = 1,2,3) are the stress components in the material coordinate system, X and S denote the related strength, respectively.

Material property degradation. The material properties will be degraded in the damage areas once damage occurs. Many researchers [20–22] have been proposed various material property degradation rules according to the experimental research. In this issue, the property discount approach is used to consider the material property degradation rule as shown in Table 2.

Analysis flow. Based on ANSYS software, a parametric three-dimensional progressive damage program was established. The flowchart of this approach is explained briefly as follows:
1. Input woven parameters and material properties at specified temperature; 2. Create the Full-cell model based on mentioned method; 3. Exert boundary conditions and displacement load to simulate; 4. Stress and

failure analysis: check out whether damage has occurred in the elements. If damage has occurred, the corresponding material property degradation can be used and failure analysis will be recalculated until no damage is checked out in this load step. Otherwise the load will be increased and the analysis of stresses will be executed in the next step. If the catastrophe happened at some load step, the program will be stopped.

4 RESULTS AND DISCUSSION

In order to verify the reasonable and accuracy of three-dimensional progressive damage approach, a serial of specimens made by RTM technics were prepared along the weft directions at 180°C. The material properties of T300–3k/QY9911-IV at RT and 2.5D woven parameters are listed in Table 3 and 4. Additionally, the mechanical properties of components at 180°C can be calculated by equations listed in Table 1.

The stress-strain curves of 2.5D woven composites under static tension in the weft direction at 180°C are shown in Figure 4(a). It is clearly seen that the stress-strain curve shows a linear behavior up to ultimate failure, which suggests that the mechanical properties is slightly influenced by the elevated temeprature environment. Furthermore, the catastrophic failure mode reflects a brittle material characteristic. Accordign to Figure 4(b), the high percentages of longitudinal damage in weft yarns indicates that the final failure of this material loaded in weft is caused by the fiber break failure in weft yarns.

Table 5 gives the comparison results between the experimental and simulation. From Table 5, it can be obtained that the modulus and strength errors are within 10%, which also demonstrates the progressive damage approach proposed by this author can predict the mechanical properties of 2.5D woven composites in weft tensile loading at 180°C.

158

Table 3. Property parameters of components materials at room temperature.

	E_{f1}/Em	E_{f2}	G_{f12}/G_m	G_{f23}	u_{f12}/u_m
T300-3K	230	40	17	4.8	0.3
QY8911-IV	4.16	–	–	–	0.34

Table 4. Woven parameters of the specimens.

Warp arranged density M_w (tows/cm)	Weft arranged density M_w (tows/cm)	Number of weft yarn at the same height N_f	Number of layers in weft direction N_h	Height L_z (mm)	Packing factor of fiber in warp yarn P_j
10	3.5	5	6	2.04	0.765

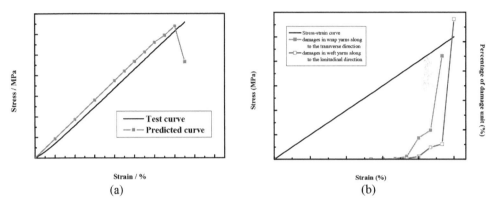

(a) (b)

Figure 4. (a) The weft tensile stress-strain curves of 2.5D woven composites at 180°C; (b) Fracture element and the percentages of damage corresponding to each failure modes.

Table 5. Comparison of predicted tensile properties and experimental results of 2.5D woven composites loaded in weft at 180°C.

	Temperature	Experiment value	Predicted value	Error
Modulus	180°C	27.185	28.068	3.248%
Strength		192.001	191.779	0.116%

The damage evolution processes of components at 180°C are shown in Figure 5. The initial damage mode is the longitudinal failure mode at the crossing point in outer-layer weft yarns and grows along the perpendicular loading direction up to the final fracture of 2.5D woven composites. In this process, the main damage mode in wrap yarns is the transverse damage observed in the contact area between adjacent warp yarns (these areas are also parallel to load direction). As the external tension increases further, the extension direction is the perpendicular loading direction. Therefore, the simulation results suggest that the finial fracture surface is perpendicular loading direction. Additionally, few damages

observed in warp yarns and matrix show that the fracture is even, and with obvious characteristic of brittle fracture.

Figure 6 illustrates the SEM photographs of fracture for 2.5D woven composites loaded in weft. It shows that the damage locations are focused on the straights of warp yarns and delamination damages are few. The fractured surface morphology of weft yarns presents quite even and the brittle characteristic (see Fig. 6). According to Figure 6a and c, the fracture local between test and predicted results are basically identical, which demonstrates the modified three-dimension progressive damage approach is well suited

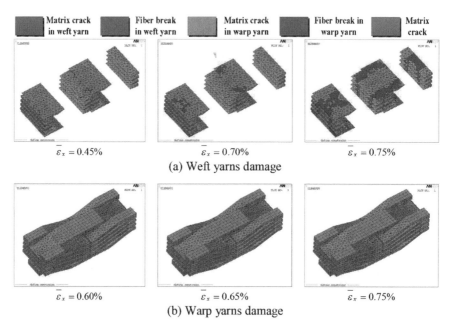

| Matrix crack in weft yarn | Fiber break in weft yarn | Matrix crack in warp yarn | Fiber break in warp yarn | Matrix crack |

$\bar{\varepsilon}_x = 0.45\%$ $\bar{\varepsilon}_x = 0.70\%$ $\bar{\varepsilon}_x = 0.75\%$

(a) Weft yarns damage

$\bar{\varepsilon}_x = 0.60\%$ $\bar{\varepsilon}_x = 0.65\%$ $\bar{\varepsilon}_x = 0.75\%$

(b) Warp yarns damage

Figure 5.   Progressive damage under static tension in warp direction at 180°C.

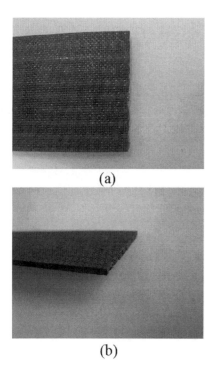

(a)

(b)

Figure 6.   Fracture micrographs of 2.5D woven composites at 180°C.

to predict the mechanical behavior of 2.5D woven composites subjected to weft loading at elevated temperatures.

5   SUMMARY

A numerical analysis has been conducted to investigate the mechanical behavior of 2.5D woven composites subjected to weft loading at elevated temperatures. The main conclusions are summarized as follows: (1) a new parameter model called Full-cell model, which takes the influence of outer structure into account, has been established; (2) The influence of temperature is introduced into three-dimensional progressive approach and Comparison with test results, the predicted errors are within 10% as well as the predicted stress-strain curve is agree with the experimental results; (3) According to the simulation results, the effect of temperature on the mechanical behavior is not apparent and the stress-strain curve shows a linear behavior up to finial failure; (4) The weft tensile tests illustrates the fracture surface is quite even and the ultimate failure mode is the fiber break in weft yarns.

ACKNOWLEDGMENT

The work was supported by Jiangsu Innovation Program for Graduate Education [grant
number KYLX_0237] and Fundamental Research
Funds for the Central Universities [grant number
NZ2012103].

CORRESPONDING AUTHOR

Jian Song, dfsongjian2006@126.com, +86
15951964648.

REFERENCES

[1] A.P. Mouritz, M.K. Bannister and P.J. Falzon,
et al. "Review of applications for advanced three-
dimensional fibre textile composites" Composites
Part A-Applied Science and Manufacturing 30
(1999) 1445–1461.
[2] B.V. Sankar, R.V. Marrey, "Analytical method for
micromechanics of textile composites" Composites
Science and Technology 57(1997) 703–713.
[3] A. Hallal, R. Younes and F. Fardoun, et al. "Improved
analytical model to predict the effective elastic properties of 2.5 D interlock woven fabrics composite"
Composite Structures 94(2012) 3009–3028.
[4] Z. Lu, Y. Zhou and Z. Yang, et al. "Multi-scale finite
element analysis of 2.5 D woven fabric composites
under on-axis and off-axis tension" Computational
Materials Science 79(2013) 485–494.
[5] Y. Zhou, Z. Lu and Z. Yang, "Progressive damage analysis and strength prediction of 2D plain
weave composites" Composites Part B-Engineering
47(2013) 220–229.
[6] W.F. Dong, J. Xiao and Y. Li, "Finite element
analysis of the tensile properties of 2.5 D braided
composites" Materials Science and Engineering
A-Structural Materials Properties Microst 457(2007)
199–204.
[7] T. Chou, T. Ishikawa, "One-dimensional micromechanical analysis of woven fabric composites"
AIAA journal 21(1983) 1714–1721.
[8] T. Ishikawa, T. Chou, "Elastic behavior of woven
hybrid composites" Journal of Composite Materials
16(1982) 2–19.
[9] T. Ishikawa, T. Chou, "Stiffness and strength behaviour of woven fabric composites" Journal of Materials Science 17(1982) 3211–3220.
[10] J. Zheng, "Research on elastic property prediction
and failure criteria of 2.5D woven composites"
Nanjing: Nanjing University of Aeronautics and
Astronautics 2008.
[11] R. Qiu, W.D. Wen and H.T. Cui, "Meso-structure of
2.5D woven composites and its strength prediction
model " Acta Materiae Compositae Sinia 31(2014)
788–796.
[12] H. Blontrock, L. Taerwe and P. Vandevelde, Fire
tests on concrete beams strengthened with fibre composite laminates Forum Vol. 2(2000), p151–161.
[13] L.M. Di, F. Piscitelli and A. Prota, et al. "Improved
mechanical properties of CFRP laminates at elevated temperatures and freeze–thaw cycling" Construction and Building Materials 31(2012) 273–283.
[14] Y. Bai, T. Keller and T Vallee T "Modeling of stiffness of FRP composites under elevated and high
temperatures" Composites Science and Technology
68(2008) 3099–3106.
[15] B. Vieille, L. Taleb "About the influence of temperature and matrix ductility on the behavior of
carbon woven-ply PPS or epoxy laminates: Notched
and unnotched laminates" Composites Science and
Technology 71(2011) 998–1007.
[16] M. Selezneva, J. Montesano and Z. Fawaz, et al.
"Microscale experimental investigation of failure
mechanisms in off-axis woven laminates at elevated
temperatures" Science and Engineering A-Structural
Materials Properties Microst 42(2011) 1756–1763.
[17] J. Montesano, Z. Fawaz and C. Poon, et al. "A
microscopic investigation of failure mechanisms in a
triaxially braided polyimide composite at room and
elevated temperatures" Materials & Design 53(2014)
1026–1036.
[18] F. Chang, K. Chang, "Post-failure analysis of bolted
composite joints in tension or shear-out mode failure"
Journal of Composite Materials 21(1987) 809–833.
[19] Z. Hashin, "Failure criteria for unidirectional
fiber composites" Journal of Applied Mechanics-
Transactions of the ASME 47(1980) 329–334.
[20] F.K. Chang, L. Lessard and J.M. Tang, "Compression
response of laminated composites containing an
open hole" SAMPE Q 19(1988) 46–51.
[21] F. Chang, L.B. Lessard, "Damage tolerance of laminated composites containing an open hole and subjected to compressive loadings. I: Analysis" Journal
of Composite Materials 25(1991) 2–43.
[22] L.B. Lessard, F. Chang, "Damage tolerance of
laminated composites containing an open hole and
subjected to compressive loadings. II: Experiment"
Journal of Composite Materials 25(1991) 44–64.

Advances in Future Manufacturing Engineering – Yang (Ed.)
© 2015 Taylor & Francis Group, London, ISBN 978-1-138-02817-3

Pd-TiO$_2$/C anodic catalysts for formic acid electrooxidation

Weifeng Xu
Department of Chemical Engineering, Harbin Institute of Petroleum, Harbin, China

Ying Gao & Bing Wu
Key Laboratory of Design and Synthesis of Functional Materials and Green Catalysis,
Colleges of Heilongjiang Province, Harbin Normal University, Harbin, China
College of Chemistry and Chemical Engineering, Harbin Normal University, Harbin, China

Lingfei Li & Hongyan Li
Department of Chemical Engineering, Harbin Institute of Petroleum, Harbin, China

ABSTRACT: Different atomic ratios of Pd-TiO$_2$/C catalysts were prepared by chemical reduction method with intermittent microwave radiation, and they were characterized by X-ray diffraction spectroscopy and energy dispersive spectroscopy. The catalysts were investigated by voltammetry and chronoamperometry. Electrochemical measurements show the Pd-TiO$_2$/C-21 catalyst, which is atomic ratios of Pd to Ti about 2:1, is significantly higher catalytic activity and also it is stability for formic acid electrooxidation than other catalysts. Thus, Pd-TiO$_2$/C-21 is one of promising candidates as anode catalyst for direct-formic acid fuel cells.

Keywords: formic acid; Pd-TiO$_2$/C catalysts; eletrocatalysis

1 INTRODUCTION

Nowadays, Direct Formic Acid Fuel Cells (DFAFCs) have attracted much attention as potential clean and mobile power sources due to its numerous advantages, such as high energy conversion efficiency, high electromotive force, limited fuel crossover and high practical power density at low temperature [1–6].

Many investigations have found that the electrocatalytic activity of Pd catalyst was better than Pt catalyst for formic acid oxidation [7–12].

However, poor stability was one of drawbacks of Pd catalyst as formic acid anode catalyst [13,14]. In order to improve the electrocatalytic performance and stability of the Pd catalyst, some metals and metal oxide were added into, such as Pd-Sn [2,15], Pd-Au [14], Pd-Ir [16], Pd-Ni [1], Pd-TiO$_2$ [17].

In this paper, we prepared Pd-TiO$_2$/C catalysts which with different atomic ratios of Pd to Ti by HCOOH reduction and microwave-assisted method with the expectation to enhance the activity and stability of the formic acid oxidation. The results demonstrated that the Pd-TiO$_2$/C catalysts activity and stability were improved when an appropriate amount of Ti was introduced.

2 EXPERIMENTAL

2.1 Chemicals and reagents

Vulcan XC-72 carbon was obtained from Cabot Company (Boston, USA). PdCl$_2$ was purchased from Sinopharm Chemical Reagent Co., Ltds. (Shanghai, China). All other reagents were analytical grade. All the aqueous solutions were prepared with triply distilled water.

2.2 Catalysts preparation

The Pd-TiO$_2$/C catalysts were prepared by chemical reduction with intermittent microwave heating method. The process of preparation as following [17]: First, to prepare TiO$_2$/C. 20 mL anhydrous ethanol was added into a beaker, then 0.2 mL tetrabutyl titanate was dropped in the beaker under constant stirring at room temperature for 30 min and solution (1) was got; 2 mL distilled water was added into another beaker, then 0.2 mL nitric acid was added into the distilled water and followed with adding of 20 mL anhydrous ethanol solution (2) was got; then solution (1) was added dropwise into solution (2) under constant stirring at room temperature, and the even transparent colloidal TiO$_2$ was obtained.

A certain amount of carbon (Vulcan XC-72R activated carbon) was added into the resulting colloidal TiO$_2$, thoroughly stirred and aged for 24 h and got slurry. The TiO$_2$/C was obtained after the slurry was filtered and dried in the vacuum condition at 80°C in nitrogen for 6 h. Second, to prepare the different atomic ratios of Pd-TiO$_2$/C catalysts, a certain amount of TiO$_2$/C which obtained above was dispersed into 15 mL ethylene glycol solvent by agitating ultrasonically for 1 h. In order to keep the different ratios of Pd to Ti, the appropriate amount of PdCl$_2$ (3.93 g·L^{-1}) was added dropwise under mechanical stirring for 1 h at 65°C. The pH of the mixture was adjusted to 8–9 by the addition of NaOH to the solution and continuously magnetic stirring for 1 h. Then, excess HCOOH was added into the mixture to reduce PdCl$_2$ to form Pd catalyst with intermittent microwave irradiation. As the following steps: The beaker was placed in the center of a domestic 2,450 MHz, 700 W microwave oven (Galanz). The microwave was operated in intermittent microwave irradiation, which is on for 20 s, and off for 50 s for two times, and then on for 15 s, and off for 60 s for six times. Then the beaker was cooled to room temperature. The slurry in the beater was filtered and washed with triply distilled water until no Cl$^-$ was detected. Finally the resulting precipitate was filtered and dried at 100°C in vacuum and the Pd-TiO$_2$/C catalysts powder was gotten.

The resultant Pd-TiO$_2$/C catalysts with the atomic ratios of Pd to Ti = 1:1, 2:1 and 3:1 were denoted as Pd-TiO$_2$/C-11, Pd-TiO$_2$/C-21 and Pd-TiO$_2$/C-31, respectively. For comparison, 20 wt% Pd/C catalyst was prepared with the same method. The Pd loadings in the catalysts are all 20% by weight.

2.3 Preparation of the electrodes

The Pd/C, Pd-TiO$_2$/C and TiO$_2$/C electrodes were prepared as follows [2,17]: A fixed amount of the catalyst was mixed with 5% Nafion solution, 20% PTFE and ethanol. Slurry was got when the mixture above was ultrasonicated for 5 min. Then the slurry was spread on a carbon paper and dried at the room temperature. The geometrical surface area of the electrode was 0.5 cm^2 and the loading of Pd was 1 mg·cm^{-2}.

2.4 Electrochemical measurements

The electrochemical measurements were performed with a CHI660B electrochemical analyzer and a conventional three electrode electrochemical cell. The prepared Pd-TiO$_2$/C, Pd/C and TiO$_2$/C catalysts were working electrodes. A Pt wire and a saturated Ag/AgCl electrode were used as the counter

electrode and the reference electrode, respectively. The values of all the potentials were quoted with respect to the Ag/AgCl electrode. All the solutions were prepared with triply distilled water. The solution was carefully purged with N$_2$ (99.999%) prior to each experiment in order to remove O$_2$ in the electrolyte.

2.5 Characterization of the catalysts

The composition of catalysts was determined using the Energy Dispersive Spectrometer (EDS, S-4800, Hitachi, Japan). The X-ray diffraction (XRD) measurements of the catalysts were performed on Model D/max-2400 diffractometer using Cu Kα radiation and operating at 40 kV and 150 mA. Transmission Electron Microscopy (TEM) was performed with a Tecnais G2Twins.

3 RESULTS AND DISCUSSION

Figure 1 displays the XRD patterns of the Pd/C and the Pd-TiO$_2$/C catalysts. It was observed from the four curves in the figure that except the characteristic peak of carbon at 24.95°, and the 2θ values of the other three peaks are 40.22°, 46.68° and 68.24°. They correspond to the 2θ values of Pd (111), Pd (200) and Pd (220) crystal faces of the face centered cubic crystalline of the Pd particles in the catalysts, respectively. It was found that the peak of Pd (200), which is used to evaluate the average size of Pd particles in the catalyst. Pd-TiO$_2$/C-21 is wider than that in the Pd/C and other Pd-TiO$_2$/C catalysts, illustrating that the average size of Pd particles in the Pd-TiO$_2$/C-21 catalyst are smaller than that of the Pd particles in all the catalysts and resulting the larger surface area of

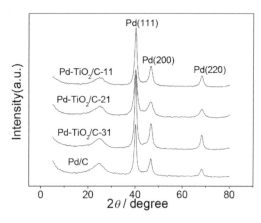

Figure 1. The XRD patterns of the Pd/C and Pd-TiO$_2$/C catalysts.

Pd particles in the Pd-TiO$_2$/C-21. The equation of Scherrer could be used to calculate the mean of Pd particles size [2]. The calculated results of average Pd particles size were showed in Table 1. Compared to other catalysts, Pd-TiO$_2$/C-21 catalyst has the smaller average particles size. In addition, no peaks related to TiO$_2$ were observed. This indicated that TiO$_2$ existed in the state of amorphous.

Figure 2 represents the TEM images of the Pd/C and Pd-TiO$_2$/C catalysts. It can be seen in Figure 2(B) that many Pd particles aggregate together and disperse unevenly in Pd-TiO$_2$/C-11 catalyst. Some agglomeratios of metal particles is also found in Pd-TiO$_2$/C-11 with large particles and the particles size distribution is non-uniform. Compared to Pd-TiO$_2$/C-11, the other catalysts have better dispersion and more uniform metal particles on them and can be seen in Figure 2(A), 2(C) and 2(D). It can be seen in Figure 2(C) that the Pd particles have the smaller size in Pd-TiO$_2$/C-21catalyst. It is in accordance with the calculated results of average particles in XRD. Overall, the above results indicate that adding some ratios of TiO$_2$ may help the dispersion of the Pd particles.

Figure 3 displays the EDS analysis of the Pd-TiO$_2$/C catalysts. The images of A, B and C correspond to Pd-TiO$_2$/C-11, Pd-TiO$_2$/C-21 and Pd-TiO$_2$/C-31, respectively. It is shown in the Pd and Ti that peaks were observed except the carbon peak from the EDS spectrums. From the Figure 3(A), 3(B) and 3(C), we can see the atomic ratios of Pd and Ti in the Pd-TiO$_2$/C catalysts about 1:1, 2:1 and 3:1, respectively. The Figure 3 indicates that PdCl$_2$ added has been completely reduced to Pd and Ti atom in tetrabutyl titanate has been converted to TiO$_2$ thoroughly in the Pd-TiO$_2$/C catalysts.

Figure 4 shows CO stripping linear sweeping voltammograms of Pd/C and Pd-TiO$_2$/C catalysts. CO stripping was often used to evaluate the electrochemically active surface areas of Pd nanoparticles [1,2,18,19]. The CO was adsorbed to the different catalysts in a 0.5 M H$_2$SO$_4$ solution before the measured. The S$_{ESA}$ of catalysts were evaluated were listed in Table 2. Additionally, it can be seen from the Figure 4 that the initial potential of the oxidation of CO almost is at the same potential,

Table 1. The average Pd particles size in Pd/C and Pd-TiO$_2$/C catalysts.

Catalysts	Average particles size (nm)
Pd/C	6.39
Pd-TiO$_2$/C-11	6.47
Pd-TiO$_2$/C-21	4.88
Pd-TiO$_2$/C-31	7.88

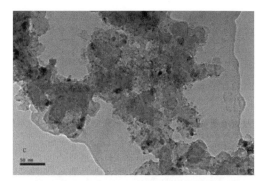

Figure 2. The TEM images of the (A) Pd/C, (B) Pd-TiO$_2$/C-11, (C) Pd-TiO$_2$/C-21 and (D)Pd-TiO$_2$/C-31 catalysts.

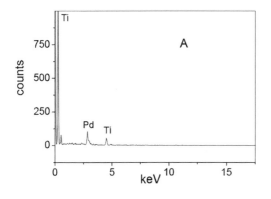

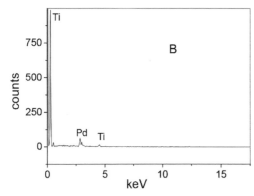

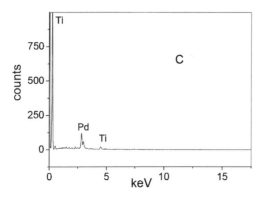

Figure 3. The EDS spectrums of the (A) Pd-TiO$_2$/C-11, (B) Pd-TiO$_2$/C-21 and (C) Pd-TiO$_2$/C-31 catalysts.

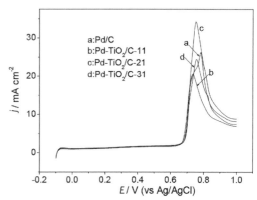

Figure 4. CO stripping linear sweeping voltammograms curves of Pd/C and Pd-TiO$_2$/C catalysts in 0.5 mol·L^{-1} H$_2$SO$_4$ solution at 25°C, scan rate 10 mV·s^{-1}.

Table 2. The S$_{ESA}$ of Pd/C and Pd-TiO$_2$/C catalysts.

Catalysts	S$_{ESA}$ (m^2·g^{-1})
Pd/C	61.19
Pd-TiO$_2$/C-11	48.13
Pd-TiO$_2$/C-21	76.64
Pd-TiO$_2$/C-31	56.87

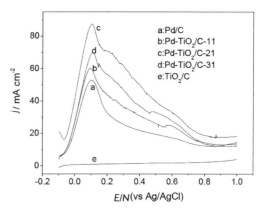

Figure 5. The Linear sweeping voltammograms curves of HCOOH on TiO$_2$/C, Pd/C and Pd-TiO$_2$/C electrodes in 0.5 M HCOOH in 0.5 M H$_2$SO$_4$, at 25°C, scan rate 10 mV·s^{-1}.

however the potential of oxidation peak is at the different. The potential of oxidation peak of the Pd-TiO$_2$/C catalysts were slightly negative than Pd/C catalyst.

Figure 5 presents the linear sweeping voltammograms of 0.5 M HCOOH in 0.5 M H$_2$SO$_4$ solution on the TiO$_2$/C, Pd/C and Pd-TiO$_2$/C catalysts electrodes. It can be observed from Figure 5 that no anodic peak display at the TiO$_2$/C, indicating that

the TiO$_2$/C catalyst has no electrocatalytic activity for the oxidation of HCOOH. The potential of the main peak at the Pd/C and Pd-TiO$_2$/C catalysts are nearly the same 0.1 V, but the current densities of the Pd/C and Pd-TiO$_2$/C catalysts main peak of formic

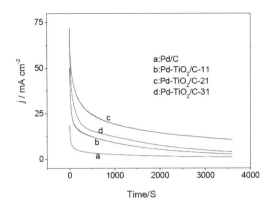

Figure 6. Chronoamperometric curves of 0.5 M HCOOH in 0.5 M H_2SO_4 on the Pd/C and Pd-TiO$_2$/C electrodes at 0.15 V and at 25°C, scan rate 10 mV · s^{-1}.

acid are different. The current densities of the Pd/C, Pd-TiO$_2$/C-11, Pd-TiO$_2$/C-21 and Pd-TiO$_2$/C-31 are 52.86, 60.42, 87.28 and 69.88 mA · cm^{-2}, respectively. The Figure 5 indicates that the electrocatalytic activity of the Pd-TiO$_2$/C-21 catalyst with the atomic ratios of Pd to Ti about 2:1 for the oxidation of formic acid is higher than that of the Pd/C and other Pd-TiO$_2$/C catalysts.

Figure 6 shows the chronoamperometry curves for the Pd/C and Pd-TiO$_2$/C catalysts at 0.15 V in a solution of 0.5 M HCOOH in 0.5 M H_2SO_4 at 25°C. In the curves, an initial rapid decrease in the current density was observed for the Pd-TiO$_2$/C catalysts; however, the all catalysts current subsequently decreased slowly and reached a relative stability state. It was observed that during test procedure the current density of Pd-TiO$_2$/C catalysts are better than that of Pd/C, and the curent density of Pd-TiO$_2$/C-21 catalyst is the best. This clearly illustrated that the electrocatalytic stability of the Pd-TiO$_2$/C-21 catalyst is better than that of Pd/C and other Pd-TiO$_2$/C catalysts.

4 CONCLUSIONS

In the experiment, Pd/C catalyst electrooxidation and stability for formic acid can be improved by adding a certain amount of TiO$_2$. The measurements results displayed Pd-TiO$_2$/C catalysts characteristics is related to the amount of TiO$_2$. When the atomic ratios of Pd to Ti is about 2:1, the Pd-TiO$_2$/C-21 catalyst exhibited excellent electrocatalytic performance and stability which is much better than other catalysts for formic acid oxidation. It is obviously found that TiO$_2$/C has no electrocatalytic for formic acid, and the excellent electrocatalytic performance of the Pd-TiO$_2$/C

catalysts could be attributed to the synergistic effect of TiO$_2$ and Pd.

ACKNOWLEDGMENT

The authors are grateful for the financial support of the Harbin Institute of Petroleum, the Department of Education of Heilongjiang Province (No. 12523027) of China and College of Chemistry and Chemical Engineering, Harbin Normal University.

REFERENCES

[1] L.P. Shen, H.Z. Li, L. Lu, Y.F. Luo, Y.W. Tang, Y. Chen and T.H. Lu. Improvement and mechanism of electrocatalytic performance of Pd–Ni/C anodic catalyst in direct formic acid fuel cell. Journal of Electrochimica Acta, Vol 89 (2013), p497–502.
[2] D.D. Tu, B Wu, B.X. Wang, C. Deng and Y Gao. A highly active carbon-supported PdSn catalyst for formic acid electrooxidation. Applied Catalysis B: Environmental, Vol 103, (2011), p163–168.
[3] Z.L. Liu, X.H. Zhang and L. Hong. Physical and electrochemical characterizations of nanostructured Pd/C and PdNi/C catalysts for methanol oxidation. Electrochemistry Communications, Vol 11 (2009), p925–928.
[4] M. Kosaka, S. Kuroshima, K. Kobayashi, S. Sekino, T. Ichihashi, S. Nakamur, T.Yoshitake and Y. Kubo. Single-Wall Carbon Nanohorns Supporting Pt Catalyst in Direct Methanol Fuel Cells. Journal of Physical Chemistry C, Vol 113 (2009), p8660–8667.
[5] G. Selvarani, S.V. Selvaganesh, S. Krishnamurthy, G.V.M. Kiruthika, P. Sridhar, S. Pitchumani and A.K. Shukla. A Methanol-Tolerant Carbon-Supported Pt-Au Alloy Cathode Catalyst for Direct Methanol Fuel Cells and Its Evaluation by DFT. Journal of Physical Chemistry C, Vol 113 (2009), p7461–7468.
[6] F. Wen and U. Simon. Low Loading Pt Cathode Catalysts for Direct Methanol Fuel Cell Derived from the Particle Size Effect. Chemistry of Materials, Vol 19 (2007), p3370–3372.
[7] V. Mazumder and S.H. Sun. Oleylamine-Mediated Synthesis of Pd Nanoparticles for Catalytic Formic Acid Oxidation. Journal of the American Chemistry Society, Vol 131 (2009), p4588–4589.
[8] S.D. Yang, X.G. Zhang, H.Y. Mi and X.G. Ye. Pd nanoparticles supported on functionalized multi-walled carbon nanotubes (MWCNTs) and electrooxidation for formic acid Journal of Power Sources, Vol 175 (2008), p26–32.
[9] S. Ha, R. Larsen and R.I. Masel. Performance characterization of Pd/C nanocatalyst for direct formic acid fuel cells. Journal of Power Sources, Vol 144 (2005), p28–34.
[10] X.W. Yu and PeterG. Pickup. Deactivation/reactivation of a Pd/C catalyst in a direct formic acid fuel cell (DFAFC): Use of array membrane electrode assemblies. Journal of Power Sources, Vol 187 (2009). p493–499.

[11] W.J. Zhou and J.Y. Lee. Particle Size Effects in Pd-Catalyzed Electrooxidation of Formic Acid. Journal of Physical Chemistry C, Vol 112 (2008), p3789–3793.

[12] Z.L. Liu, L. Hong, M.P. Tham, T.H. Lim and H.X. Jiang, Nanostructured Pt/C and Pd/C catalysts for direct formic acid fuel cells. Journal of Power Sources. Vol 161 (2006), p831–835

[13] W.S. Jung, J.H. Han, and S. Ha. Analysis of palladium-based anode electrode using electrochemical impedance spectra in dire Journal of Power Sources, Vol 173 (2007), p53–59.

[14] G.J. Zhang, Y.E. Wang, X. Wang, Y. Chen, Y.M. Zhou, Y.W. Tang, L.D. Lu, J.C. Bao and T.H. Lu. Preparation of Pd-Au/C catalysts with different alloying degree and their electrocatalytic performance for formic acid oxidation. Applied Catalysis B:Environmenta, Vol 102 (2011), p614–619.

[15] Z.H. Zhang, J.J. Ge, L. Ma, J.H. Liao, T.H. Lu and W. Xing. Highly Active Carbon-supported PdSn Catalysts for Formic Acid Electrooxidation. Fuel Cells, Vol 9 (2009), p114–120.

[16] X. Wang, Y.W. Tang, Y. Gao and T. Lu. Carbon-supported Pd–Ir catalyst as anodic catalyst in direct formic acid fuel cell. Journal of Power Sources, Vol 175 (2008), p784–788.

[17] W.F. Xu, Y. Gao, T.H. Lu, Y.W. Tang and B. Wu. Kinetic Study of Formic Acid Oxidation on Highly DispersedCarbon Supported Pd–TiO$_2$ Electrocatalyst. Catalysis Letters, Vol 130 (2009), p312–317.

[18] J.H. Jiang and A. Kucernak. Electrocatalytic properties of nanoporous PtRu alloy towards the electrooxidation of formic acid. Journal of Electroanalytical Chemistry, Vol 630 (2009), p10–18.

[19] D.C. Chen, Q. Tao, L.W. Liao, S.X. Liu, Y.X. Chen and S. Ye. Determining the Active Surface Area for Various Platinum Electrodes. Electrocatalysis, Vol 2 (2011), p207–219.

Advances in Future Manufacturing Engineering – Yang (Ed.)
© 2015 Taylor & Francis Group, London, ISBN 978-1-138-02817-3

Synthesis of 2,3-dibromo-succinic anhydride and application on polyester/cotton blended fabric

Liyun Lin, Xiang Zhou & Yarong Wu

Key Laboratory of Science and Technology of Eco-Textile, Donghua University, Shanghai, China

ABSTRACT: For a polyester/cotton blended fabric, a reactive-type flame retardant 2,3-dibromo-succinic anhydride (DBrFR) was grafted to cotton and it was also added on polyester as an addative. DBrFR was synthesized via the addition reaction of maleic anhydride and bromine in ethyl acetate solution and extracted by solventing-out crystallization. The application of 2,3-dibromo-succinic anhydride (DBrFR) to cotton fabric and polyester/cotton (65/35) was investigated. The PET/cotton blended treated with DBrFR showed high levels of flame retardant performance. The flame retardancy of fabric treated with DBrFR was studied and the thermal behaviors were investigated.

Keywords: PET/cotton blended fabric; brominated flame retardant; thermogravimetric analysis

1 INTRODUCTION

PET/cotton blended fabrics are widly used for garment materials, work clothes and decor materials due to their excellent physical and chemical properties. It is essential and crucial to impart flame retardance to PET/cotton blends because of their high flammability. However, the flame retardant finishing of these blend fabric especially PET/cotton (65/35) is difficult because of the different burning performance of cotton and polyester fibers, causing the so-called "scaffolding effect" [1–3]. Although great efforts have been made for several decades, no satified practical solution has been developed so far [4–6].

Brominated Flame Retardants (BFRs) are chemicals widely used in consumer products including electronics, plastics and textiles to reduce flammability [7]. But the environmental pollution and toxic effects of BFRs have been attracting great concern for the last two decades and the evaluation of BFRs is undertaken actively. Although the alternatives are available, such as phosphorus and metal based compounds, these are more costly and can pose manufacturing problems, so BFRs are still widely used in many products.

The aim of this study is to find a novel fire retardant which can provide effective bonding with cotton cellulose as a durable flame retardant, and at the same time it can provide flame retardant effect for Polyester. The 2,3-dibromo-succinic anhydride which was a 5-membered cyclic anhydride and has the flame retardant element bromine was selected as the reactive-type flame retardant.

2 EXPERIMENTAL

2.1 Materials

Desized, scoured and bleached plain woven cotton fabric weighing 115 g/m^2 and polyester/cotton (65/35) blended fabric weighing 102 g/m^2 were used in this study. All the chemicals used were AR grade and were purchased from China National Medicines Corporation Ltd.

2.2 Synthesis of DBrFR

Maleic anhydride (19.6 g) and 80 mL ethyl acetate were added into a four-neck flask equipped with a mechanical stirrer, a thermometer, a dropping funnel, and a reflux condenser. The mixture was stirred until maleic anhydride was dissolved. Then 38.4 g bromine was slowly added into the above solution and the reaction was at 48°C for 5 h under stirring. While it turned to colorless transparent solution, the mixture of dimethyl benzene and water was added as an anti-solvent followed by evaporation of ethyl acetate. Through filtrating and drying, the white product 2, 3-dibromo-succinic anhydride was obtained and the yield was above 85%. The product was named as DBrFR in this article.

2.3 Characterization of DBrFR

The FT-IR spectrum of DBrFR was conducted with a FT-IR spectrophotometer (FT-IR 640 Varian) with a potassium bromide pellet technique. The ^{1}H-NMR spectra was obtained on an NMR spectrophotometer (Avance-400, Bruker, Switzerland)

with deuterated acetone as solvent. The elements carbon, oxygen, hydrogen analysis was performed with the automatic elemental analyzer Vario EL III (Elementar Analysensysteme Comp., Hanau, Germany). The elemental bromine was rest by oxygen-flask combustion and Iion Chromatograghy.

2.4 *Flame-retardant treatment of fabrics*

Pad-dry-curing method was applied to finish the fabric. The aqueous bath containing 20% DBrFR with the ratio of DBrFR:SHP = 1:1.5 (mol), pH = 3, The fabric was immersed the bath and padded using two dips and two nips with a wet pickup of about 85%. The padded sample was dried at 80°C for 3 min and subsequently cured at 170°C for 3 min, then rinsed with running water for 3 min and dried at room temperature.

2.5 *Evaluation of the flame retardance of finished fabrics*

The LOI values of the fabrics were measured according to the Chinese standard GB/T 5454-1997 using a LOI-type burning tester (ATS, ATS FAAR Co.). The vertical burning test was conducted according to GB/T5455–1997 using a YG(B)815D-1. The thermal decomposition behavior of the fabrics were

observed at the temperature range of 20–600°C under N_2 at a heating rate of 20°C/min using the TG analyzer (TG 209 F1, Netzsch).

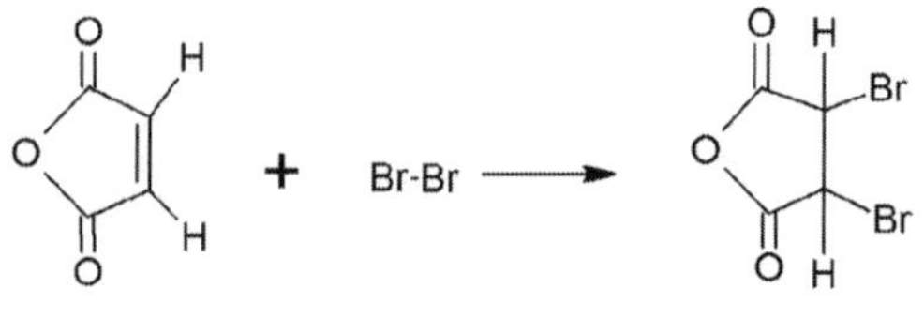

3 RESULTS AND DISCUSSION

3.1 *Characterization of DBrFR*

DBrFR was prepared via the reaction of maleic anhydride and bromine without any catalyst as shown in Scheme 1. The synthesized DBrFR was characterized by FT-IR spectroscopy and the infrared spectrum was shown in Figure 1.

The product was characterized by FT-IR spectroscopy and the FT-IR spectrum was shown in Figure 1. The bands at 1685 cm^{-1} and 1731 cm^{-1} were due to the symmetric and asymmetric stretching of a five-membered cyclic anhydride, respectively. The three explicit bands at 1388 cm^{-1}, 1264 cm^{-1} and 1184 cm^{-1} were attributed to the C-O-C stretching modes of the five-membered cyclic anhydride and the band at 572 cm^{-1} was associated with the stretching mode of C-Br stretching vibration. The band at 3014 cm^{-1} assigned to O-H stretching vibration which would be observed because of hydrolysis of the anhydride.

The ^{1}H-NMR spectrum of the product was shown in Figure 2. There was only one proton resonated peak at $\delta = 4.75$ which indicated that there was only one type of hydrogen.

The synthesized compound was also analyzed with elemental analyzer to study the chemical composition. The elemental analysis results were (%, calculated): C, 18.49 (18.63); H, 0.84 (0.78); O, 18.59 (18.62); Br, 61.97(62.08). The measured results showed good agreement with the theoretical values.

Scheme 1. The synthesis of DBrFR.

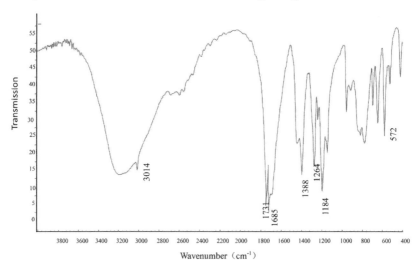

Figure 1. FT-IR spectrum of synthesized DBrFR.

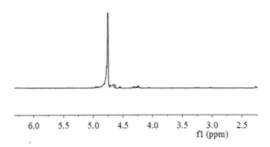

Figure 2. ¹H NMR spectrum of DBrFR.

3.2 Treatment of cotton fabric with DBrFR

DBrFR were applied to the cotton fabric by the method described above. The LOI values and vertical burning test data were compared with untreated fabric and the results were given in Table 1.

From Table 1, it could be seen clearly that the LOI value was increased greatly and the char length was reduced to 4.9 cm. The vertical burning test showed that treated sample was promptly self-extinguished after igniting and without after glow. The excellent flame retardancy maybe attributed to condensed-phase effect of esterification and the gas phase effect of bromine.

3.3 Characterizing ester bond with FT-IR

According to the mechanism that polycarboxylic acids esterify cellulose through the formation of a 5-membered cyclic anhydride under catalysis of SHP [8], the possible esterification of DBrFR with hydroxyl groups of the cellulose would carry out as shown in Scheme 2.

The infrared spectra of the cotton fabric samples before and after finishing were presented in Figure 3. Compare FT-IR spectra of cotton fabrics before and after finishing, the carboxylic acid and the ester overlap at 1732 cm⁻¹ was shown for the finished fabric. Absorption peak at 1400 cm⁻¹ was significantly enhanced (C-O stretching vibration), which indicated that the flame retardant esterified with the hydroxyl groups of cotton cellulose.

3.4. Determination of the amount of ester linkages by titration

Lüdtke [9] suggested that the amount of ester linkages can be determined by titration with calcium acetate solution. Hu et al [10] developed the titration method in our lab. The results of the titration of the free carboxyl groups on cotton were presented in Table 2.

The DBrFR had two carboxyl groups in solution, the amount of DBrFR which available for reaction can be determined by the results of titration after drying, was 704 ÷ 2 = 352 mmol/kg for cotton The amount of DBrFR applied to cotton fabric was also can be calculated by wet pickup and the concentration of the finishing solution, was $20\% \times 85\% \div 257.8 = 618$ mmol/kg, The measured values was lower than the theoretical value. The reason may be due to the addition of the pH adjusting agent NaOH which changed carboxyl groups to carboxylate anion. The difference value of free carboxyl on fabrics after drying and after curing was the amount of flame retardants which esterified with cellulose fiber. By calculating (704 – 405/618 = 48.4%), we can know that there was 48.4% of the flame retardants applied esterified with cellulose fiber of the cotton fabric. If it was true, the treated fabric contained 5.1% bromine by weight.

3.5 Treatment of PET/cotton blended fabric with DBrFR

The flammability and physical properties of the treated PET/cotton blended fabric was compared with untreated fabric and the reaults were presented in Table 3. As shown in Table 3, the LOI value of PET/cotton fabric treated with DBrFR was larger than 27 which were generally known to be self-extinguishing and the char length greatly was shortened to 10.9 cm in strip test which indicated that DBrFR could endow flame retardancy to PET/cotton blended fabric. The breaking strength retentions were about 65%.

3.6 Thermogravimetric Analysis

Thermal analysis technique was used to study the effects of the DBrFR-based finishing system on the thermal properties of cotton and PET/cotton blended fabric. The TG and DTG results were illustrated in Figure 4.

It can be seen from Figure 4 a, the presence of DBrFR on the cotton fabric lowered the decomposition temperature and enhanced the formation of char after pyrolysis of the fabric sample. It is known that phosphorous-containing FRs can reduce cellulose flammability, primarily by dehydration, phosphorylation, and phosphate-ester decomposition mechanisms, further forming a crosslinked network within the cellulose which can inhibit the release of volatile combustible fragments and enhance char formation [11]. The TG and DTG results showed that the esterification of DBrFR with cellulose maybe worked similar to phosphate-ester.

The results of Figure 4b indicated that the amount of char formation of the untreated blend fabric was 9.0% at 600°C which obviously lower

Table 1. Flame retardancy and physical properties of cotton fabric treated with DBrFR.

Samples	Char length (mm)	After flame time/s	After glow time/s	LOI (%)	Whiteness index	Tensile strength (N) Warp	Weft
Untreated	TD[a]	24.7	1.3	17.7	83.6	699	325
Treated	4.9	5.2	0	32.7	80.8	499	227

[a]TD denotes that the fabrics were totally destroyed during the test.

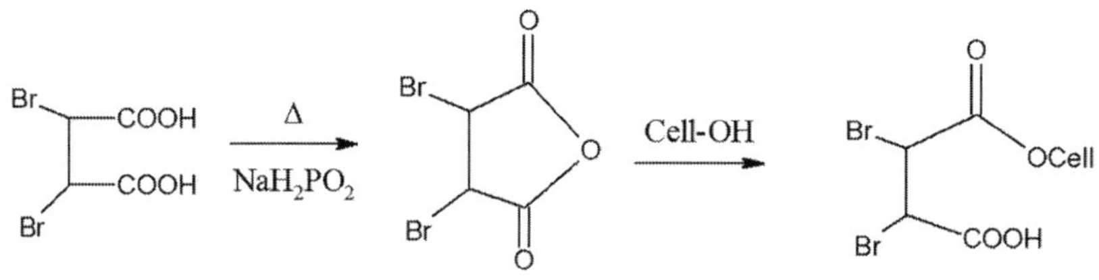

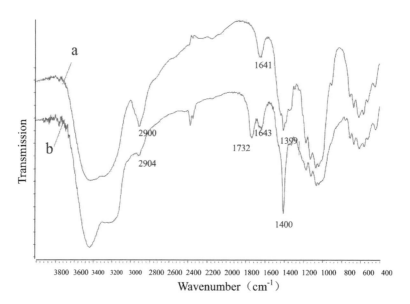

Scheme 2. Reaction of DBrFR and cotton.

Figure 3. FT-IR spectra of untreated (a) and treated (b) cotton.

Table 2. Amount of free carboxyl groups on cotton samples.

Samples	Cotton (mmol/kg)
After drying	704
After curing	305
Curing + rinsing	297

than the theoretical value 19.0% (10.5 × 35% + 23.5 × 65%). The data indicated that the thermal decomposition of cotton and PET were promoted each other which increase the difficulty of flame retardant finishing. After treated with DBrFR, the distance of T_{max} of cotton and PET increased, which indicated that BDrFR had some effect to eliminate "scaffolding effect". After finishing the amount of char formation was 30.5% for PET/cotton which was higher than that of unfinished.

Table 3.   Flammability and physical properties of PET/cotton blended fabric before and after treated.

Samples	Char length (mm)	After flame time/s	After glow time/s	LOI (%)	Whiteness index	Tensile strength (N)	
						Warp	Weft
Untreated	TD	9.2	5.9	17.0	85.1	718.2	301.8
Treated	10.9	3.4	0	27.8	79.8	453.0	198.6

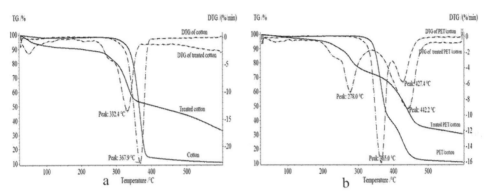

Figure 4.   TG and DTG curves of treated and untreated fabrics (a cotton, b PET/cotton).

4   CONCLUSIONS

The reactive-type DBrFR was synthesized via the addition reaction of maleic anhydride and bromine and was characterized by FT-IR, ^{1}H-HMR and elemental analysis. Flame retardant property was endowed to cotton and PET/cotton blended fabric by pad–dry–curing method. The esterification of DBrFR with hydroxyl of cellulose was determinated by FT-IR and the amount of ester linkages was determined by titration. In TG analysis, the cotton and PET/cotton blended fabrics finished with DBrFR showed condensed phase mechanism. It was speculated that the esterification of cotton with DBrFR worked similar to phosphate-ester decomposition mechanisms. The durability to laundering and the use of DBrFR on the PET were also studied in the flowing experiment.

REFERENCES

[1] M.Lewin: in: Handbook of Fiber Science and Technology, edited by M. Lewin, S. B. Sello, volume II of Chemical Processing of Fiber and Fabrics, chapter, 1, Marcel Dekker: New York, (1983), p.109–117.

[2] A.R. Horrocks: Flame-retardant Finishing of Textiles. Society of Dyers and Colourists. Vol. 1 (1986), p.62–101.

[3] M. Lewin: Unsolved problems and unanswered questions in flame retardance of polymers. Polym. Degrad. Stab. Vol. 88(2005), p. 13–19.

[4] C.T. Giuliana, R. Joseph and R. M. Donald: Flame-Retardant Finishing of Polyester-Cellulose Blends. Ind. Eng. Chem. Prod. Res. Dev. Vol. 2 (1972), p. 164–169.

[5] H. Kubokawa, K. Takahashi, S. Nagatani and T. Hatakeyama: Thermal decomposition behavior of cotton/polyester blended yarn fabrics treated with flame retardants. Sen-i Gakkaishi. Vol. 55 (1999), p. 298–305.

[6] E.J. Blanchard and E.E. Graves: Polycarboxylic Acids for Flame Resistant Cotton/Polyester Carpeting. Text. Res. J. Vol. 72 (2002), p. 39–43.

[7] S. Andreas, G. Donald, Jr. Patterson and B. Ake: A review on human exposure to brominated flame retardants-particularly polybrominated diphenyl ethers. Environ Int. Vol. 29 (2003), p. 829–839.

[8] C.Q. Yang and X.H. Gu: FTIR Spectroscopy Study of the Formation of Cyclic Anhydride Intermediates of Polycarboxylic Acids Catalyzed by Sodium Hypophosphite. Text Res J. Vol. 70 (2000), p. 64–70.

[9] A. Lejaren, Jr. Hiller and P. Eugene: Cellulose Studies: VI. Determination of Carboxyl Groups in Cellulosic Materials. Text. Res. J. Vol. 16 (1946), p. 390–393.

[10] X. Hu, X. Zhou: Determination of Carboxyl Group on Fabric by Acid-base Titration. Journal of China Textile University (Eng. Ed.). Vol. 15 (1998), p. 10–12.

[11] M. Lewin: in: Flame Retardant Polymeric Materials, edited by M. Lewin, S.M. Atlas, E.M. Pearce, volume II, Chaper, 5, Plenum Press: New York (1975), p.134–157.

Advances in Future Manufacturing Engineering – Yang (Ed.)
© 2015 Taylor & Francis Group, London, ISBN 978-1-138-02817-3

Elastic analysis of nanocontact problems with surface effects

L.Y. Wang
Office of Mathematics, Longqiao College of Lanzhou Commercial College, Lanzhou, China

L.J. Wang
College of Chemical Engineering, Gansu Industry Polytechnic College, Tianshui, China

ABSTRACT: Contact problems with surface effects at nano-scale are considered in this paper. The complex variable function method is adopted to derive the fundamental solutions for the contact problem. As examples, the deformations induced by uniformly distributed normal and concentrated force are analyzed in detail. The results reveal some interesting characteristics in contact mechanics, which are distinctly different from those in classical elasticity theory. At nano-scale, the deformation gradient on the deformed surface varies smoothly across the loading boundary as a result of surface effects. In addition, for nano-indentation, the indent depth depends strongly on the surface stress.

Keywords: surface elasticity theory; surface stress; nanocontact; complex variable function method

1 INTRODUCTION

Contact problem has always been one of the most challenging problems in solid mechanics. For solids with large characteristic dimensions, the volume ratios of surface region to the bulk material is small, the effect of surface stress then can be neglected because of its relatively tiny contribution. Owing to the increasing ratio of surface-to-bulk volume, the effect of surface stress has been considered as one of the major factors contributing to the exceptional behaviors at nanoscale.

To account for the effects of surfaces on mechanical deformation, Gurtin et al. presented the surface elasticity theory[1–4]. The surface stress tensor $\sigma^s_{\alpha\beta}$ is about the surface energy density $\Gamma\left(\varepsilon_{\alpha\beta}\right)$ by

$$\sigma^s_{\beta\alpha} = \gamma\delta_{\beta\alpha} + \frac{\partial\Gamma}{\partial\varepsilon_{\beta\alpha}}. \tag{1}$$

where γ is the residual surface tension under unstrained condition.

For the stress tensor σ and the strain tensor ε and the displacement vector u are satisfaction with all the field equations in solid. Then the equilibrium conditions on the surface are expressed as

$$\sigma_{\beta\alpha}n_\beta + \sigma^s_{\beta\alpha,\beta} = 0. \tag{2a}$$

$$\sigma_{ij}n_in_j = \sigma^s_{\alpha\beta}\kappa_{\alpha\beta}. \tag{2b}$$

where n_j denotes the normal to the surface, $\kappa_{\alpha\beta}$ the curvature tensor of the surface.

2 NANOCONTACT MODEL WITH SURFACE EFFECTS

Now we consider a material occupying the upper half-plane $y > 0$ and denote this region by S^+, using the notation S^- for the half-plane $x < 0$ in the Cartesian coordinates system $(O\text{-}xy)$, as shown in Figure 1, where the x axis is along the surface, and the y axis perpendicular to the surface. The material is subjected to pressure $p(x)$ over the region $|x| \le a$.

Due to the second term in Eq. 1 indicates a variation of the surface energy density with respect to elastic strain, which is related to the stretching or compressing the atoms in the surface to accommodate to the bulk phase.

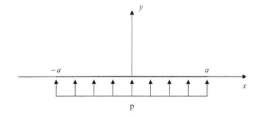

Figure 1. Schematic of contact problem under uniform loading.

If the change of the atomic spacing in deformation is infinitesimal, the contribution from the second term to the surface stresses is negligibly small compared to the residual surface tension[5]. In what follows, for simplicity, we neglect the contribution from the second term in Eq. 1. Then, the surface stresses are given by

$$\sigma_{\alpha\beta}^{s} = \gamma \delta_{\alpha\beta}. \tag{3}$$

In this case, the boundary conditions Eq. 2 on the contact surface ($y = 0$) are simplified to

$$\sigma_{xy}(x) = 0. \tag{4a}$$

$$p(x) + \sigma_{yy}(x) = -\frac{\gamma}{R(x)}. \tag{4b}$$

where $R(x)$ is the radius of curvature of the deformed surface.

For plane problems, the displacement and stress components can be expressed in terms of two analytic functions $\varphi(z)$ and $\psi(z)$ as

$$\sigma_{xx} + \sigma_{yy} = 4\operatorname{Re}\varphi'(z). \tag{5a}$$

$$\sigma_{yy} - \sigma_{xx} + 2i\tau_{xy} = 2[\bar{z}\varphi''(z) + \psi'(z)]. \tag{5b}$$

$$\frac{E}{1+\mu}(u+iv) = \kappa\varphi(z) - z\overline{\varphi'(z)} - \overline{\psi(z)}. \tag{6}$$

where $z = x + iy$, $\kappa = 3 - \mu/1 + \mu$ for plane strain and $\kappa = 3 - 4\mu$ for plane stress.

For convenience, we let $\Phi(z) = \varphi'(z)$ and $\Psi(z) = \psi'(z)$. From Eq. 5 and Eq. 6, once $\Phi(z)$ is determined for a particular problem, a complete solution has been obtained. Define the analytic continuation of $\Phi(z)$ in S^- as

$$\Phi(\bar{z}) = -\overline{\Phi}(\bar{z}) - \bar{z}\overline{\Phi'}(\bar{z}) - \overline{\Psi}(z). \tag{7}$$

By taking the complex conjugate, $\Psi(z)$ in S^+ is given by

$$\Psi(z) = -\Phi(z) - \overline{\Phi}(z) - z\Phi'(z). \tag{8}$$

where $\overline{\Phi}(z)$ is stand for $\overline{\Phi(\bar{z})}$.

On the contact surface ($y = 0$), the stress components and the displacement can be expressed from Eqs. 5 and Eq. 6 and Eq. 7 and Eq. 8, we can get

$$\sigma_{xx} = \Phi(z) + 2\overline{\Phi(z)} - (z - \bar{z})\overline{\Phi'}(\bar{z}) + \Phi(\bar{z}). \tag{9a}$$

$$\sigma_{yy} = (z - \bar{z})\overline{\Phi'}(\bar{z}) + \Phi(z) - \Phi(\bar{z}). \tag{9b}$$

$$\frac{E}{1+\mu}\left(\frac{\partial u}{\partial x} + i\frac{\partial v}{\partial y}\right) = (\bar{z} - z)\overline{\Phi'}(\bar{z}) + \Phi(\bar{z}) + \kappa\Phi(z). \tag{10}$$

As shown in Figure 1, contact will take place over a region L: $|x| \le a$ on the surface ($y = 0$). By letting $y \to 0$ in Eqs. 9 and Eq. 10, it is seen that $\Phi(z)$ must be a function with the property that

$$\Phi_+(x) - \Phi_-(x) = \sigma_{yy}(x). \tag{11}$$

$$\kappa\Phi_+(x) + \Phi_-(x) = \frac{E}{1+\mu}[u'(x) + iv'(x)]. \tag{12}$$

where $\Phi_+(x)$ and $\Phi_-(x)$ denotes the value of $\Phi(z)$ as $y \to 0$ form S^+ and S^-, respectively. That is $\lim_{y\to 0}\Phi(z) = \Phi_+(x)$, $\lim_{y\to 0}\Phi(\bar{z}) = \Phi_-(x)$. Making use of Eqs. 4 and Eqs. 9, we can rewrite Eq. 11 as

$$\Phi_+(x) - \Phi_-(x) = -p(x) - \frac{\gamma}{R(x)}. \tag{13}$$

So the complete expression of half-space contact problem will be described by the appropriate choice of potential function $\Phi(z)$.

On the surface, the radius of curvature due to deformation is given by

$$\frac{1}{R(x)} = \frac{\partial^2 v(x,0)}{\partial x^2}. \tag{14}$$

By differentiating Eq. 10 with respect x, we can obtain

$$\frac{E}{1+\mu}\left(\frac{\partial^2 u}{\partial x^2} + i\frac{\partial^2 v}{\partial x^2}\right) = (\bar{z} - z)\overline{\Phi''}(\bar{z}) + \Phi'(\bar{z}) + \kappa\Phi'(z). \tag{15}$$

By taking the complex conjugate Eq. 15, Then subtract and letting $y \to 0$, we obtain

$$\frac{2E}{1+\mu}i\frac{\partial^2 v(x,0)}{\partial x^2} = \Phi_-'(x) - \overline{\Phi_-'(x)} + \kappa[\Phi_+'(x) - \overline{\Phi_+'(x)}]. \tag{16}$$

Substituting Eqs. 14 and Eq. 16 into Eq. 13, we obtain

$$\Phi_+(x) - bi\Phi_+'(x) - \Phi_-(x) + bi\Phi_-'(x) = -p(x). \tag{17}$$

where $b = \gamma(\kappa - 1)(1 + \mu)/2E$, which is a measure of the residual surface tension under unstrained condition and

$$\Lambda(x) = \Phi(x) - bi\Phi'(x). \tag{18}$$

176

The boundary conditions for the contact problem with surface effects are given by

$$\Lambda_+(x) - \Lambda_-(x) = -p(x). \tag{19}$$

This type of boundary value problem has been considered by Muskhelishvili[6], and the unique solution is

$$\Lambda(z) = -\frac{1}{2\pi i}\int_{-a}^{a}\frac{p(t)}{t-z}dt. \tag{20}$$

2.1 Elastic solution under uniformly distributed pressure

As a particular example, let us consider the effect of a uniformly distributed pressure p over the region $|x| < a$, while the remainder of the boundary $y = 0$ being unstressed as shown in Figure 1. From Eq. 20, with $p(t)$ being constants, then

$$\Lambda(z) = -\frac{p}{2\pi i}\int_{-a}^{a}\frac{1}{t-z}dt = -\frac{p}{2\pi i}\ln\left(\frac{z-a}{z+a}\right). \tag{21}$$

Substituting Eq. 21 in to Eq. 19, we had an ordinary differential equation with respect $\Phi(z)$ and solving the differential equation, we obtain

$$\Phi(z) = -\frac{p}{2\pi i}\left\{\ln\left(\frac{z-a}{z+a}\right) + b\int_{0}^{\infty}\frac{e^{i(z-a)t} - e^{i(z+a)t}}{bt+1}dt\right\}. \tag{22}$$

After differentiating Eq. 22 with respect to augment z, we get stress and displacement gradients component

$$\sigma_{yy} = \frac{yp}{\pi}\left\{\frac{x-a}{(x-a)^2+y^2} - \frac{x+a}{(x-a)^2+x^2}\right.$$
$$\left. -ib\int_{0}^{\infty}\frac{te^{-yt}\left(e^{-i(x-a)t} - e^{-i(x+a)t}\right)}{t+1}dt\right\}$$
$$-\frac{p}{2\pi i}\left\{\ln\left(\frac{x-a}{x+a}\right) + b\int_{0}^{\infty}\frac{te^{-yt}\left(e^{i(x-a)t} - e^{i(x+a)t}\right)}{bt+1}dt\right\}$$
$$+\frac{p}{2\pi i}\left\{\ln\left(\frac{x-a}{x+a}\right) + b\int_{0}^{\infty}\frac{te^{yt}\left(e^{i(x-a)t} - e^{i(x+a)t}\right)}{bt+1}dt\right\}. \tag{23}$$

$$\frac{\partial u(x,0)}{\partial x} = -\frac{bp(\kappa-1)}{4\mu\pi a}\int_{0}^{\infty}\frac{\sin\left(\frac{x}{a}-1\right)\tau - \sin\left(\frac{x}{a}+1\right)\tau}{\frac{b}{a}\tau+1}d\tau. \tag{24a}$$

$$\frac{\partial v(x,0)}{\partial x} = \frac{p(\kappa-1)}{4\mu\pi}$$
$$\times\left\{\ln\left(\frac{1-\frac{x}{a}}{1+\frac{x}{a}}\right) + \frac{b}{a}\int_{0}^{\infty}\frac{\cos\left(\frac{x}{a}-1\right)\tau - \cos\left(\frac{x}{a}+1\right)\tau}{\frac{b}{a}\tau+1}d\tau\right\}. \tag{24b}$$

It is seen, when $b = 0$, that is, the surface influence is ignored in Eq. 23 and Eqs. 24, the stresses of the half-plane are consistent with those in the classical elastic results[7] and the displacement gradients are consistent with the same result[8], respectively.

The distribution of the displacement gradients with respect to x on the contact surface induced by a uniformly distributed pressure is shown in Figure 2 and Figure 3 for various values of b/a, which represent varying degrees of residual surface tension. The displacement gradients are normalized by the dimensionless stress factor M which is defined as $M = -p(k-1)/4\pi\mu$.

Where $U = \partial u(x,0)/\partial x, V = \partial v(x,0)/\partial x$.

Miller and Shenoy tests on some elementary deformation modes of nanosized elements showed that the surface elasticity agrees well with atomic simulations[9]. For the illustration of the effects of surface energy, b/a is taken as 0.01, 0.1, 0.5 and 2 in our calculations. It can be seen from Figure 2 and Figure 3 that the surface shear displacement gradient and the surface indentation gradient transits continuously across the loading boundary $x = \pm a$, as opposed to a singularity predicted by classical elasticity. In addition, the surface shear displacement gradient in the bulk increasing with an increase in surface stress, but the surface indentation gradient in the bulk decreases with an increase in surface stress.

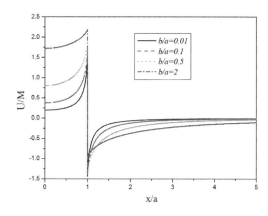

Figure 2. Distribution of surface shear displacement gradient under uniform normal loading.

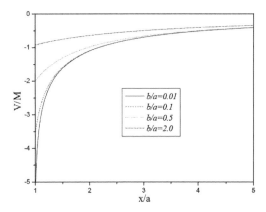

Figure 3. Distribution of surface indentation gradient under uniform normal loading.

2.2 Elastic solution under concentrated tractions

Consider concentrated normal forces acting on the surface at the origin. The corresponding results can be obtained in such a way, as $a \to 0$, $2ap \to P$, where P is real constants.

Letting $\lim_{a \to 0} 2ap = P$, Taking the limit in Eq. 20, we get

$$\Lambda(z) = -\frac{iP}{2\pi z}. \tag{25}$$

Substituting Eq. 25 in to Eq. 19, we obtain

$$\Phi(z) = -\frac{P}{2\pi} \int_0^\infty \frac{e^{zit}}{bt} dt. \tag{26}$$

After differentiating Eq. 26 with respect to augment z, we can get

$$\overline{\Phi'(\bar{z})} = \overline{\Phi'(z)} = \frac{P}{2\pi} \int_0^\infty \frac{ite^{-zit}}{bt+1} dt. \tag{27}$$

Substituting Eq. 27 in to Eq. 9, one has

$$\sigma_{yy} = -\frac{Py}{\pi} \int_0^\infty \frac{te^{-yt}\cos(xt)}{bt+1} dt$$
$$-\frac{P}{2\pi} \int_0^\infty \left(\frac{\left(e^{-yt} - e^{yt} \right)\cos(xt)}{bt+1} \right) dt. \tag{28}$$

It is seen, when $b = 0$, that is, the surface influence is ignored in Eq. 28, the stresses of the half-plane are consistent with those in the classical elastic results[7].

In the same way, Substituting Eq. 28 in to Eq. 15, we should obtain the displacement field of the half-plane as

$$\frac{\partial u(x,0)}{\partial x} = -\int_0^\infty \left(\frac{b}{L}t+1 \right)^{-1} \cos\left(\frac{x}{L}t \right) dt,$$
$$\frac{\partial v(x,0)}{\partial x} = -\int_0^\infty \left(\frac{b}{L}t+1 \right)^{-1} \sin\left(\frac{x}{L}t \right) dt. \tag{29}$$

where $L = P(\kappa-1)(1+\mu)/2\pi E$.

It is seen, when $b = 0$, that is, the surface influence is ignored in Eq. 29, the displacement gradients are consistent with the same result[8].

The variation of the shear displacement gradient on the contact point $x = 0$, which is very obvious form Figure 4. In addition, the surface shear displacement gradient in the bulk decreases with increase in surface stress. Figure 5 shows

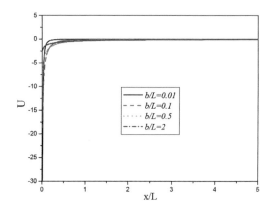

Figure 4. Distribution of surface shear displacement gradient under concentrated loading.

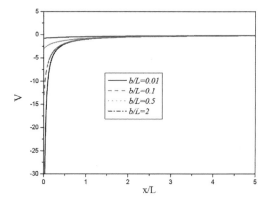

Figure 5. Distribution of surface indentation gradient under concentrated loading.

the indentation gradient for several representative values of surface stresses. The indent gradient decreases as the surface stress increases. It is seen, when the surface stress increases to a maximum, that is, the indent gradient value is zero, which is the properties of the rigid body.

3 SUMMARY

In this paper, we consider the two-dimensional contact problem in the light of surface elasticity theory. The general analytical solution is derived by using the complex variable function method. For two particular loading cases of a uniform distributed pressure and concentrated force, the results are analyzed in detail and compared with the classical linear elastic solutions. It is found that the surface elasticity theory illuminates some interesting characteristics of contact problems at nano-scale, which are distinctly different from the classical solutions of elasticity without surface effects. At nano-scale, the deformation gradient on the deformed surface varies smoothly across the loading boundary as a result of surface effects. In addition, for nano-indentation, the indent gradient depends strongly on the surface stress.

REFERENCES

[1] M.E. Gurtin and A.I. Murdoch: A continuum theory of elastic material surfaces. Archive for Rational Mechanics and Analysis. Vol. 57 (4): 291–323(1975).

[2] M.E. Gurtin and A.I. Murdoch: A continuum theory of elastic material surfaces. Archive for Rational Mechanics and Analysis. Vol. 59 (4): 389–390(1975).

[3] M.E. Gurtin: Effect of surface stress on wave propagation in solids. Journal of Applied Physics. Vol. 47 (10): 4414–4423(1976).

[4] M.E. Gurtin and A.I. Murdoch: Surface stress in solids. International Journal of Solids and Structures. Vol. 14 (5): 431–440(1978).

[5] F.Q. Yang: Size-dependent effective modulus of elastic composite materials: Spherical nanocavities at dilute concentrations. Journal Applied Physics. Vol.95: 3516–3520(2004).

[6] N.I. Muskhelishvili: Some basic problems of mathematical theory of elasticity. Moscow. (1949).

[7] K.L. Johnson: Contact Mechanics. Cambridge University Press. London. (1985).

[8] G.F. Wang and X.Q. Feng: Effects of surface stresses on contact problems at nanoscale. Journal of Applied Physics. Vol. 101: 013510(2007).

[9] R.E. Miller and V.B. Shenoy: Size-dependent elastic properties of nanosized structural elements. Nanotechnology. Vol. 11: 139–147(2000).

Advances in Future Manufacturing Engineering – Yang (Ed.)
© 2015 Taylor & Francis Group, London, ISBN 978-1-138-02817-3

Experimental research on structure stability of expanded-diameter conductor

Long Liu, Jian-cheng Wan, Jia-jun Si & Liang-hu Jiang
China Electric Power Research Institute, Beijing, China

ABSTRACT: In this paper, the sheave test for expanded-diameter conductor has been designed and four types of expanded-diameter conductor with the same cross section (530 mm^2) and the same diameter (33.75 mm) have been trial-manufactured. With the analysis of the experimental results, the critical strand jumping tension of four type expanded-diameter conductors are all greater than 25% RTS. The structure stability of four type expanded-diameter conductors all meets the design requirements.

Keywords: structure stability; expanded-diameter conductor; sheave test

1 INTRODUCTION

Compared with plain conductor of same cross-section area, the expanded-diameter conductor's diameter is bigger and the electrical performance is better [1–2]. On the premise to meet transmission capacity and project requirement, the application of expanded-diameter conductor will decrease the tower load and structure gravity. However, the diameter's expanding will cause the conductor gap increase. And after the construction pay-off, the strand jumping phenomenon becomes a restraining factor in conductor application. In this paper, four types of expanded-diameter conductor [3–4] have been designed and the experimental research on structure stability [5–8] of expanded-diameter conductor after pay-off has been carried out.

2 THE EXPANDED-DIAMETER CONDUCTOR FOR EXPERIMENT

Four types of expanded-diameter conductor with the same cross section (530 mm^2) and the same diameter (33.75 mm) have been designed, as shown in Figure 1.

3 THE SHEAVE TEST FOR EXPANDED-DIAMETER CONDUCTOR

To analyze the conductor structure stability, the sheave test has been designed. The schematic diagram of experimental set-up for sheave test is shown in Figure 2.

The aim of the sheave test is to obtain the critical strand jumping tension of tour type of expanded-diameter conductor and observe the conductor surface state and the aluminum strength of inner layer under different experimental tension.

25% RTS (Rated Tensile Strength), 30% RTS, 35% RTS and 40% RTS are chosen as the experimental tension. The sample length is 18 m. The envelop length is 30°. The bottom diameter of sheave is 700 mm. The recirculation number is 20 times.

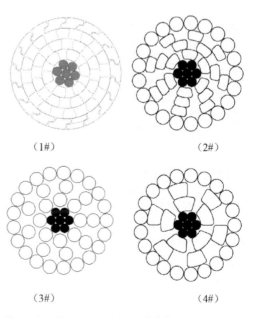

（1#） （2#）

（3#） （4#）

Figure 1. Four types of expanded-diameter conductor.

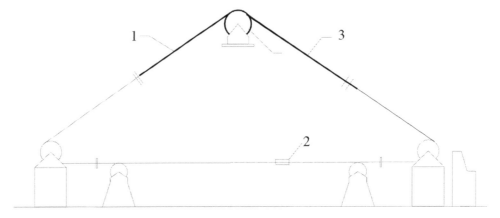

Figure 2. Schematic diagram of experimental set-up.
1—sheave, 2—force transducer, 3—conductor sample.

25%RTS	30%RTS
35%RTS	40%RTS

Figure 3. 1# conductor surface state after sheave test.

25%RTS	30%RTS
35%RTS	40%RTS

Figure 4. 2# conductor surface state after sheave test.

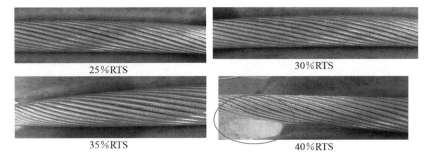

Figure 5.   3# conductor surface state after sheave test.

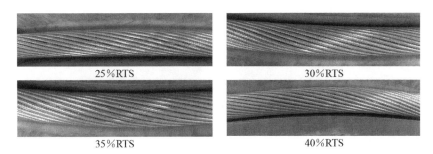

Figure 6.   4# conductor surface state after sheave test.

4   THE EXPERIMENTAL RESULTS

After the experiment, the state of four type of expanded-diameter conductor is shown as Figure 3 to Figure 6. And we can find that the strand jumping appears on 3# conductor surface under the tension of 40% RTS. The strand jumping doesn't appear on 1#, 2# and 4# conductor surface under the tension of 40% RTS. But the surface abrasion is the most serious on 3# conductor.

5   CONCLUSIONS

From the experimental results, the strand jumping appears on 3# conductor surface under the tension of 40% RTS. And the surface abrasion is the most serious on 3# conductor. The critical strand jumping tension of four type expanded-diameter conductors are all greater than 25% RTS. The structure stability of four type expanded-diameter conductor all meets the design requirements.

REFERENCES

[1] Wan Jian-cheng. Classification and Applicability of Diameter-expanded Conductor [J]. Electric Power Construction, 2010, 36(6): 113–118.

[2] Wan Jian-cheng. Design of Expanded Diameter Conductor with Large Section and Stable Structure [J]. Electric Power Construction, 2013, 34(10): 92–96.

[3] Wang Guo-zhong. Constructional Design of Concentric Lay Stranded Formed Wire Overhead Conductors [J]. Electric Wire & Cable, 2010, 4:1–4.

[4] Wan Jian-cheng, Yu Jun. The Feasibility of 900 mm2 Conductor on UHV DC Transmission Wire, Power System Technology, vol. 33 (No. 15), 2009: 64.

[5] Jiang W.G. A concise finite element model for pure bending analysis of simple wire strand [J]. International Journal of Mechanical Science, 2012, 54(1): 69–73.

[6] Jiang W.G., Henshall J.L., Walton J.M. A concise finite element model for 3-layer straight wire rope strand [J]. International Journal of Mechanical Sciences, 2000, 42(1):63–86.

[7] Liu Bin. Constructional Design of Concentric Lay Stranded Formed Wire Overhead Conductors [J]. Electric Wire & Cable, 2010, 4:1–4.

[8] Jiang W.G., Warby M., Henshall J.L. Statically indeterminate contacts in axially loaded wire strand [J]. European Journal of Mechanics A/Solids, 2008, 27(1):69–78.

Advances in Future Manufacturing Engineering – Yang (Ed.)
© 2015 Taylor & Francis Group, London, ISBN 978-1-138-02817-3

Low pH inhibition on High-loaded Anammox Reactor (HAR) and its recovery

Zhixing Li, Lei Zhang, Luo Wang & Hang Li

Department of Marine Technology and Environment, Dalian Ocean University, Dalian, China

ABSTRACT: Anaerobic ammonium oxidation (anammox) reactor, especially the High-loaded Anammox Reactor (HAR), is easily inhibited by some low pH wastewaters. The low pH inhibition on HAR and its recovery were investigated in this paper. HAR has a strong resistibility to low pH shock during a short time, but long-term exposure to low pH still has a significant inhibition on the nitrogen removal performance of HAR. The effect of low pH on HAR was mainly caused by three points, direct impact of low pH, substrate inhibition and toxicity, and lack of effective substrates. Among them, the effect of substrate inhibition and toxicity is the most important factor during the early recovery period. The actual substrate inhibitor is nitrite rather than FNA under non-strong acidic conditions, but FNA has somewhat of impacts under strong acidic conditions. The nitrogen removal performance of low pH inhibited HAR could be restored after 17 days by the adjustment of pH, HRT and influent substrate concentrations.

Keywords: anammox; low pH; high-loaded; ether bond cleavage

1 INTRODUCTION

Anaerobic ammonium oxidation (anammox) is an emerging biotechnology for the treatment of ammonium-rich, low carbon containing wastewater. In this process, ammonium as electron donor and nitrite as electron acceptor are converted anaerobically into mainly nitrogen gas and some nitrate (Eq. 1) by anammox bacteria. Anammox bacteria is chemolithoautotrophic and belongs to the phylum *Planctomycetales*. So far, five anammox "*Candidatus*" genera have been described. "*Candidatus Kuenenia*", "*Candidatus Brocadia*", "*Candidatus Anammoxoglobus*", and "*Candidatus Jettenia*" have all been enriched from activated sludge. The other genus, "*Candidatus Scalindua*", has been enriched from natural habitats, especially from marine sediments and oxygen minimum zones[1]. As the unique pathway of the anammox process entails the shortcut of ammonia removal cycle, anammox has greatly improved the understanding of the nitrogen cycle and further provides a new and promising alternative to the conventional nitrogen removal processes. Application of anammox process as the major nitrogen removal process in wastewater treatment systems is advantageous in terms of energy saving and less secondary pollution. However, how to obtain high rate nitrogen removal in all kinds of wastewaters through anammox process still needs to be studied. pH inhibition on anammox process is one of the main problems need to be resolved.

According to Ji et al.[2], most High-loaded Anammox Reactor (HAR) operators control the influent pH values between 6.8–7.0 in order to make sure that the pH value was kept in the appropriate range of 7.5–8.5. Nevertheless, the performance of anammox reactor, especially HAR, could be easily inhibited by low pH. Thus, it is important to figure out the process, mechanism, and recovery strategies of low pH inhibition in HAR.

$$NH_3 + 1.3NO_2^- + H^+ \rightarrow 1.02N_2 + 0.26NO_3^- + 2H_2O. \quad (1)$$

2 METHODS AND MATERIALS

Reactor operation. The experimental work was carried out in a continuous upflow anaerobic biofilter, which was made of acrylic glass with working volume of 11 L. The Nitrogen Removal Rate (NRR) of the biofilter researched 16.61–21.65 kg-N m^{-3}d^{-1} before low pH inhibition research. The reactor was kept at around 37°C with thermostatic bath and covered with black cloth to avoid light inhibition.

Synthetic wastewater. Synthetic wastewater used in this experiment contained MgSO$_4 \cdot$7H$_2$O 300 mg/L, NaHCO$_3$ 1250 mg/L, KH$_2$PO$_4$ 10 mg/L, CaCl$_2 \cdot$2H$_2$O 5.6 mg/L, trace element solutions I and II each 1.25 ml/L as described by Li et al.[3]. NH$_4$Cl and NaNO$_2$ were provided as demand. pH was controlled by HCl or NaOH solution.

Chemical analysis. The influent and effluent samples were collected on daily basis and analyzed immediately. The ammonium, nitrite and nitrate concentration were carried out following the "Standard Methods for the Examination of Water and Wastewater"[4]. The pH was determined with Mettler Toledo SG68-ELK pH meter.

3 RESULTS AND DISCUSSION

Inappropriate pH levels in the wastewater (too high or too low) were both detrimental to wastewater treatment. They can weaken the reactor performance through creating uncomfortable environment for the bacteria or inducing toxicant production. The inhibition of high pH shock is irreversible while the negative impact of low pH can be artificially recovered by lengthening HRT and reducing the substrate concentration[5]. Therefore, the study of low pH inhibition to HAR process, mechanism, and recovery of HAR is more important for the practical application of anammox process.

Performance of HAR. Before the low pH inhibition research, HAR was continuously operated under relatively short HRT (1.47 h) and high nitrogen loading rate (NLR, 17.32–23.35 kg-N $m^{-3}d^{-1}$) to develop high-loaded anammox reactor which would be used to the following low pH inhibition experiment. Within 30 days, HAR was operated stably with Nitrogen Removal Rate (NRR) as high as 16.61–21.65 kg-N $m^{-3}d^{-1}$, Nitrogen Removal Efficiency (NRE) of 96% ± 3%. The carmine biofilm formed on the polypropylene ring carrier possessed relatively high biomass of 0.03 g VSS/carrier and specific anammox activity of 1.32 mg N/g VSS/h.

Low pH inhibition on HAR. HAR was continuously operated under pH 3.00–4.00 to investigate the effects of low pH shock on HAR during 1–7 d (Fig. 1). The performance of HAR did not altered immediately and its NRE still stayed as high as 97% on the first day. However, reactor performance started to deteriorate on Day 3. The effluent substrate concentration, both ammonium and nitrite, sharply increased. Almost no added ammonium and nitrite were converted on Day 7. NRR and NRE decreased to the lowest point in this period of 1.76 kg-N $m^{-3}d^{-1}$ and 8%, respectively. Both Yu et al.[6] and Xu et al.[5] ever investigatived the effects of pH on anammox reactor at relatively low pH of 4 and 6.40–6.70 respectively and found the same phenomenon as our experiment on the first day of low pH inhibition. The other experimental parameters are shown in Table 1 in detail. These results all indicate that anammox process has a strong resistibility to low pH shock during a short time. However, the liquid of the reactor inner could not be fully replaced by the low pH influent

immediately. What is more, when 1 mol ammonium is converted, 1 mol H^+ is consumed and 0.09 mol OH^- is produced[7]. The consumption of acidity resulted in pH increase in the anammox process. Thus, the acidity brought in through influx could be neutralized by the produced alkalinity during the anammox process if the low pH water was not continuously fed to the reactor. However, if the influent pH persisted at a level lower than the optimum range for a longer time, the low pH inhibition upon anammox process could be observed clearly. Xu et al.[5] also found the same inhibition as our experiment from Day 3 to Day 7 as soon as the experimental pH was lowered to 5.87–5.95.

Mechanisms of low pH inhibition. Low pH inhibition on anammox mainly comes from three aspects: ①Low pH can affect the intracellular proton transfer, thus directly affects cell survival. Low pH may also result in changes in the matrix components, not only ②increases the concentration of the pH-dependent, unionized and presumably more toxic forms of the substrates, (i.e.), HNO_2 (Free nitrous acid, FNA), but also ③decreases the concentration of effective substrates, (i.e.), NH_3 (Free ammonia, FA).

Direct impact of low pH. There is an optimum pH range for microbial growth and metabolism. Once the pH deviates from this range, the growth and metabolic activity were hampered. Anammox bacteria has a physiological pH range of 6.6–8.3 and an optimal pH range around 7.5–8.0[2]. The low pH inhibition on anammox bacteria includes three points as follows: 1. The existence of an intracellular proton gradient over the anammoxosome membrane can force energy (ATP) yielding, while the abundance of H^+ may interfere the proton transfer. Long term alteration of the medium pH could lead to disruption of the Proton Motive Force (PMF), and thus affects the associated energy generation. 2. The PMF in general is constant for a given anammox organism, this generally results in a constant pH difference between the outside and the cytoplasm. Once the proton and electron transfers are hindered, the original pH values between the two sides of the anammoxosome membrane also get changed together. The anammoxosome contains the enzymes involved in anammox catabolism, and so part of them demanding strict pH may be inhibited. 3. Long term exposure to low pH medium may cause denaturation of cellular proteins, original bond cleavage and other biological structures destruction, relatively constant pH values in each cell plan may be gradually changed, and finally leads to enzymes inactivation.

Substrate inhibition and toxicity. The actual inhibition of ammonium and nitrite is generally believed to be caused by the related substances, FA and FNA. The FA and FNA concentration

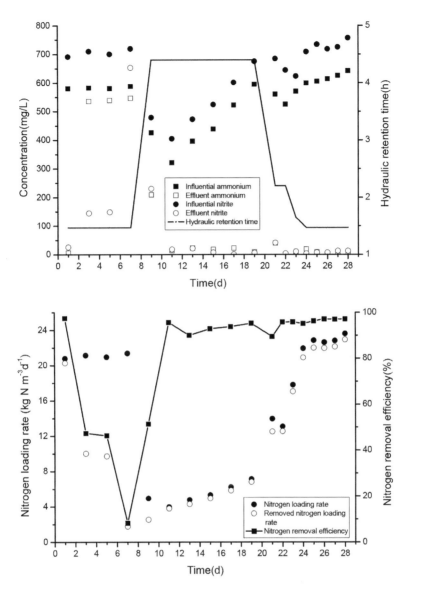

Figure 1. Nitrogen removal performance of high-loaded anammox reactor.

were calculated based on the following Eq. 2 and Eq. 3[8] and the concentrations in this paper were 9.14×10^{-4}–10^{-3} mg/L and 0.235–2.35 mg/L, respectively. However, the inhibition constants of FA and FNA were 75.6 mg/L and 5.3×10^{-3} mg/L respectively[5,8] and the FNA concentration in this paper is nearly 45–444 times of FNA inhibition constant. Thus, FNA strongly inhibited the activity of anammox bacteria. FNA inhibition is widely described by many researchers like Chen et al.[8], Jin et al.[9] and Fernández et al.[10]. As known, ladderane is the major constitute in anammox bacteria cell memebrane lipids, and it is

a strong barrier to material diffusion. Ladderane contains many ether bond. Strong acids (HI > HBr > HCl > HF) could lead to ether bond cleavage at relatively high temperatures, and resulted in FNA penetration into the cell and poisoning the cell.[11] However, Strous et al.[12] investigated the anammox activity with different nitrite concentrations at pH 7, 7.4, and 7.8 separately and found that the anammox activity decreased with the increasing nitrite concentration, which indicated that nitrite itself is the actual inhibitor. Similarities between exposure to nitrite at pH 6.8, 7.5, and 7.8 in the similar experiment carried out

Table 1. The low pH inhibition to high-loaded anammox reactor.

pH	Duration	NLR (kg N m^{-3}d^{-1})	Volume (L)	HRT (h)	Concentration (mg/L)		Recovery time (d)	References
					NH$_4^+$-N	MO$_2$-N		
4	12 h	0.56	4	24	280	280	0	6
6.40–6.70	2 h	9.33	4.5	2.16	370 + 20	470 + 20	0	5
5.87–5.95	3 h	9.33	4.5	2.16	370 + 20	470 + 20	33	
3.00–4.00	7 d	23.35	11	1.47	581	700	17	This paper

by Lotti et al.[13] also indicated that nitrite rather than FNA is the actual inhibiting compound. Furthermore, Lotti et al.[13] observed the recovery after nitrite exposure and stated that the adverse effect of nitrite is reversible and thus the effect should be described as inhibitory rather than toxic in nature. Yet, what should be noted is that both experimental pH of Strous et al. and Lotti et al. were non-strong acids and the ether bond in ladderane membrane lipids is stable. Thence, in author's view, the actual inhibitor is nitrite rather than FNA under non-strong acidic conditions, but FNA has somewhat of impacts under strong acidic conditions.

$$FA(mg/L) = 17/14 \times \text{Total ammonia as N(mg/L)} \times 10^{pH}/[\exp(6344/(273 + T)) + 10^{pH}] \quad (2)$$

$$FNA(mg/L) = 47/14 \times NO_2^--N(mg/L)/ [\exp(-2300/(273 + T)) + 10^{pH}] \quad (3)$$

Lack of effective substrate. According to the previous literatures, the half rate constants of FA and FNA were 3.3 mg/L and 2.7 × 10^{-4} mg/L, respectively[5,8]. But as shown in the last part, "Substrate inhibition and toxicity", the FA concentration in this paper is just 1/3611-1/361 of FA half rate constant. Therefore, it can be determined that lack of effective matrix FA results in the performance decreasing of HAR.

Recovery of low pH inhibited HAR. After one week of low pH inhibition, part of anammox biofilm detachment and decompostion was observed. Anammox activity was inhibited and the NRE was relatively low. So, the regulation of pH, HRT and substrate concentrations was taken.

As shown in Figure 1, in accordance with the recovery strategies mentioned above, operating the inhibited HAR for 17 days could make its NRR restore to the highest level before low pH suppression (22.00 kg-N m^{-3}d^{-1}). With the increase of operating time, a higher NRR (22.94 kg-N m^{-3}d^{-1}) was achieved. This result indicates that as long as the operating parameters are regulated properly, the inhibited HAR could be quickly restored. As evident in Table 2, which listed systematically

from Chen et al.[8], lower influent pH, higher NLR, longer duration, shorter HRT, and higher substrate concentration, all of these could lead to a more serious inhibition. As shown in Table 1, although the low pH shock in this paper is more serious than that of Xu et al.[5], the inhibited HAR has a shorter recovery time. This result indicates that our recovery strategies are reasonably accurate.

To be noted, although the influent pH had been adjusted to 7.0 and the NLR had also been lowered to 4.94 kg-N m^{-3}d^{-1} at the first two days of the recovery period, the nitrogen removal performance was still not so good, the NRR was just 2.54 kg-N m^{-3}d^{-1}, with a 44.3% NRR increase compared to the NRR in Day 7 (1.76 kg-N m^{-3}d^{-1}). However, when the influent substrate concentrations were lowered according to the effluent substrate concentrations, the NRR increased as high as 3.79 kg-N m^{-3}d^{-1} and achieved a 115.3% NRR increase compared to the NRR in Day 7, though the NLR was just 3.96 kg-N m^{-3}d^{-1}. This result indicates that effect of substrate inhibition and toxicity may be more important than direct impact of low pH during the early recovery period. Chen et al.[8] also investigatived the effect of low pH on the performance of HAR and drew an opposite conclusion that the inhibition of acidity is even more serious than FNA when pH value was 5.54. In fact, the study carried out by Chen et al. is similar to the early low pH inhibited period in this paper according to the duration and HRT shown in Table 2. As for the reason, it could well be that nitrite and FA concentration were both relatively low and the high FNA concentration did not have toxicity on anammox bacteria due to the stable existence of ladderane membrane lipids during the early low pH inhibited period; while in the early recovery period, nitrite and FA concentration were both increased due to the higher pH, especially the FA concentration was 10^3–10^4 times higher increased compared to that of low pH inhibited period. What is more, the relatively low FNA concentration may also have toxicity on anammox bacteria with the ether bond breaking in the membrane.

Table 2. The impact of various factors on the low pH shock effect.

pH	Duration (h)	NLR (kg N m^{-3}d^{-1})	HRT (h)	Concentration (mg/L)		Reflux ratio	Nitrogen removal efficiency (%)
				NH$_4^+$-N	MO$_2^-$-N		
7.00							100
6.83							59.5
5.54		9.3		387.2	543.0		27.3
	7.2 h		2.4 h			2:1	46.8
		14.8		–	–		95.7
		19.8		–	–		99.6
6.55				818.0	1092.5	0	0
	4.95 h	27.7	1.65 h	818.0	1092.5	2:1	96
				818.0	1092.5	4:1	99.4

4 CONCLUSIONS

The low pH inhibition on HAR and its recovery were investigated in this paper. Although HAR has a strong resistibility to low pH shock during a short time, the low pH has a significant inhibition on the nitrogen removal performance of HAR as for long-term exposure. The low pH inhibition on HAR is mainly caused by direct impact of low pH, substrate inhibition and toxicity, and lack of effective substrate. Among them, substrate inhibition and toxicity is the most important factor during the early recovery period. Due to the existence of ether bond in ladderane, the actual substrate inhibitor is nitrite rather than FNA under non-strong acidic conditions, but FNA has somewhat of impacts under strong acidic conditions. It has been proved that the low pH inhibition on HAR could be restored within 17 d by taking appropriate recovery strategies, including adjusting pH, reducing NLR and then progressively increasing the influent substrate concentrations and influent flow rate.

ACKNOWLEDGEMENTS

This work was financially supported by the Liaoning Startup Foundation for PhDs (20111073) and College Outstanding Young Scholars Growth Plan of Liaoning Province (LJQ2012065).

CORRESPONDING AUTHOR

Lei Zhang, lisabobo_2009@163.com, 18641109852.

REFERENCES

[1] Laura van Niftrik, Mike S.M. Jetten. Anaerobic ammonium-oxidizing bacteria: unique microorganisms with exceptional properties. Microbiol. Mol. Biol. Rev. Vol. 76 (2012) No. 3, p. 585–596.

[2] Yuxin Ji, Meihong Zhu, Hui Chen, et al. Research progress of high-loaded ANAMMOX reactors. Chem. Ind. Eng. Prog. Vol. 32 (2013) No. 8, p. 1914–1920, 1928.

[3] Hang Li, Lei Zhang, Zhixing Li, et al. Effects of butyrate on autotrophic anaerobic ammonium oxidation (annammox). Appl. Mech. Mater. Vols. 675–677 (2014), p. 410–415.

[4] APHA: Standard Methods for the Examination of Water and Wastewater (American Public Health Association, New York 1995).

[5] Bin Xu, Shaoqi Zhou, Guotao Hang, et al. Effect of pH on performance of anammox UASB reactor at high nitrogen load. Chin. Water Wastewater Vol. 29 (2013) No. 7, p. 10–14.

[6] Jinjin Yu, Rencun Jin. The ANAMMOX reactor under transient-state conditions: Process stability with fluctuations of the nitrogen concentration, inflow rate, pH and sodium chloride addition. Bioresour. Technol. Vol. 119 (2012), p. 166–173.

[7] Chongjian Tang, Ping Zheng, Qaisar Mahmood, et al. Start-up and inhibition analysis of the Anammox process seeded with anaerobic granular sludge. J. Ind. Microbiol. Biotechnol. Vol. 36 (2009), p. 1093–1100.

[8] Jianwei Chen, Ping Zheng, Chongjian Tang, et al. Effect of low pH on the performance of high-loaded ANAMMOX reactor. J. Chem. Eng. Chin. Univ. Vol. 24 (2010) No. 2, p. 320–324.

[9] Rencun Jin, Guangfeng Yang, Jinjin Yu, et al. The inhibition of the Anammox process: A review. Chem. Eng. J. Vol. 197 (2012), p. 67–79.

[10] Fernández, J. Dosta, C. Fajardo, et al. Short- and long-term effects of ammonium and nitrite on the Anammox process. J. Environ. Manage. Vol. 95 (2012), p. 170–174.

[11] Jianzhuang Zhao: Organic Chemistry (Higher Education Press, Bei Jing 2007).

[12] M. Strous, J.G. Kuenen, M.S.M. Jetten. Key physiology of anaerobic ammonium oxidation. Appl. Environ. Microbiol. Vol. 65 (1999) No.7. p. 3248–3250.

[13] T. Lotti, W.R.L. van der Star, R. Kleerebezem, et al. The effect of nitrite inhibition on the anammox process. Water Res. Vol. 46 (2012), p. 2559–2569.

Advances in Future Manufacturing Engineering – Yang (Ed.)
© 2015 Taylor & Francis Group, London, ISBN 978-1-138-02817-3

Effect of post annealing on the structure of Boron Nitride films

Mengyi Zhan, Huayong Fang, Rongrong Cao, Fang Wang, Yulin Feng, Chunjing Li,
Meng Wang, Kai Li, Kailiang Zhang & Baohe Yang
*School of Electronics Information Engineering, Tianjin Key Laboratory of Film Electronic
and Communication Devices, Tianjin University of Technology, Tianjin, China*

ABSTRACT: The phase transformation from hexagonal Boron Nitride (h-BN) to cubic Boron Nitride (c-BN) by post annealing was demonstrated in this paper. From the IR peak intensity and shifting measured by Fourier Transformed Infrared Spectroscopy (FTIR), the cubic phase content was up to 96.2% and the compressive stress had been greatly relaxed after annealing. In addition, according to the morphology and crystal structure measured by Atom Force Microscopy (AFM) and Scanning Electron Microscope (SEM) respectively, 1000°C is concluded as the optimal annealing temperature.

Keywords: phase transformation; cubic phase content; compressive stress; c-BN films

1 INTRODUCTION

Cubic Boron Nitride (c-BN) is of highly desirable mechanical, thermal, and optical properties and has potential application in cutting tools, electronic and photonic devices and so on[1,2]. As known to all, excellent electronics and optics properties of functional material are the key to device performance, and it is that only highly cubic phase content could ensure these properties. Although c-BN film has been received significant attention for their potential use in the field of optoelectronics and microelectronics, only limited studies were placed on electronic and photonic applications other than mechanics because it is difficult to deposite high cubic phase content c-BN film[3–5].

Boron nitride has several structures, c-BN, h-BN, rhombohedral BN (r-BN) and wurtzite boron nitride (w-BN), in general they coexist in growing films. The fabrication of c-BN film requires high temperature and high pressure because the nucleation and growth of the film need high-energy particle[6]. So how to increase the cubic phase concentration of film is essential to c-BN's application. In this paper, the cubic phase concentration of BN film was improved by post annealing, and the effect of different annealing temperature on the properties of BN film was studied.

2 EXPERIMENTAL

BN thin films were deposited on Si (111) substrates by RF magnetron sputtering under same conditions. Samples were annealed at 800°C, 900°C, 1000°C, 1100°C in N_2 atmosphere by rapid annealing equipment respectively.

The composites of samples were characterized by FTIR (JASCO FT/IR-4200). The surface morphologies and crystal structures were investigated by AFM (Agilent-5600LS) and SEM (JSM-6700F).

3 RESULTS AND DISCUSSION

The c-BN contains atomic bonding of sp3 while h-BN is with that of sp2. Relevant researches[7] have confirmed that infrared sensitivity factors of c-BN and h-BN are close, so volume fraction of c-BN in BN films could be expressed as Eq. (1).

$$\varphi_c = I_{1065}/\left(I_{1065} + I_{1380}\right) \tag{1}$$

and represent the peak intensity of infrared absorption spectra at 1065 cm^{-1} and 1380 cm^{-1} respectively. Due to the interface stress between BN film and the under layer, the related peaks position in FTIR often shift, so compressive stresses[8] of the film were calculated by Eq. (2).

$$G = \frac{H_{\text{c-BN}} - 1065}{4.5} \tag{2}$$

Figure 1 is the FTIR spectrum of the samples with different annealing temperatures, Figure 1 (a) is IR spectra of sample unannealing, Figure 1 (b)–(e) are IR spectra of samples annealed at 800°C, 900°C, 1000°C, 1100°C respectively. The

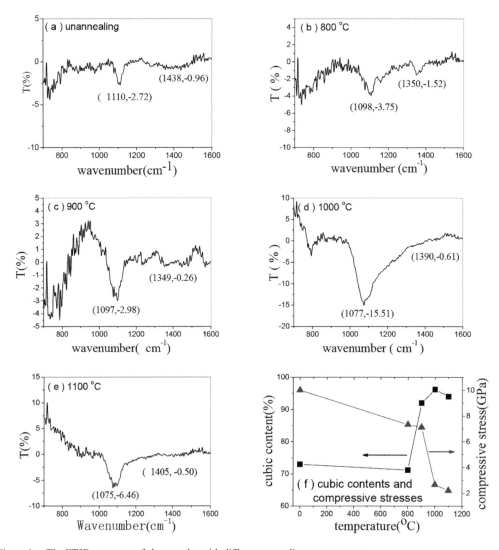

Figure 1. The FTIR spectrum of the samples with different annealing temperatures.

peaks of the FTIR spectrum were auto-picked through Origin software in Figure 1(a)–(e). The closed peak to c-BN and h-BN was used to calculated the c-BN content and compressive stresses. The results were shown in Figure 1(f), in which the c-BN contents are 71%, 71.2%, 92%, 96.2%, 94% and compressive stresses are 10, 7.33, 7.11, 2.67, 2.22 GPa with unannealing, annealed at 800°C, 900°C, 1000°C, 1100°C respectively. Figure 1(f) shows that cubic content increases firstly and then decreases with increasing annealing temperature and up to the maximum value (96.2%) at 1000°C. It also shows that compressive stress decreases with increasing annealing temperature, indicating that more compressive stress could be relaxed

with higher annealing temperature and the highest c-BN content 96.2% appears with 1000°C annealing temperature, the lowest compressive stresses 2.22 GPa with 1100°C. And not only cubic content but also compressive stress, the result are similarity in which 96.2% with 94% of cubic content, and 2.67 GPa with 2.22 GPa of stresses.

Figure 2 are surface topography images by AFM of samples with 2×2 μm^2 scanning size, which Root Mean Square (RMS) are 2.16 nm, 16.4 nm, 26.2 nm, 3.78 nm and 1.96 nm, respectively. The topography of samples has large differences with increasing annealing temperature. The grain size increases from 30–50 nm when unannealing, up to 250 nm at 900°C then decreases at 1000°C and

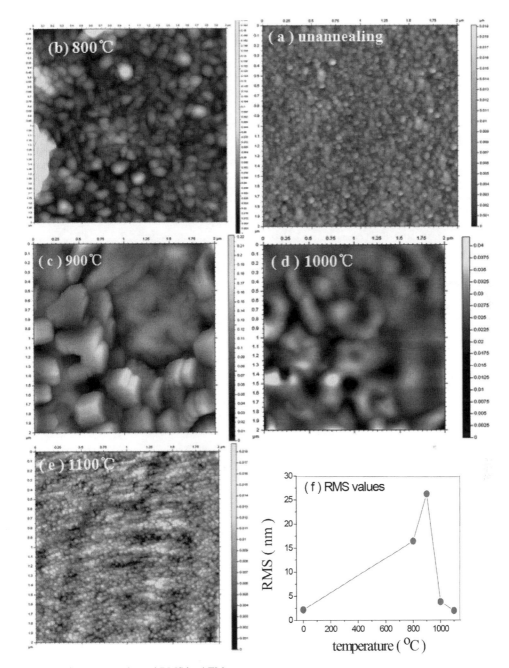

Figure 2.   Surface topography and RMS by AFM.

to 20–30 nm at 1100°C, the particle shape also changed simultaneously.

Figure 3 are images of Scanning Electron Microscope (SEM), and Figures. 3(a)–(e) show the microstructures of different samples, and an image is inserted in Figure 3(c) in order to see the full-scale confetti-like protrusions at 900°C. The surface in Figure 3(a) is smooth with small particle size and there have gaps between blocks, which is composed of several small particles. The uniform particles in Figure 3(b) shaped as rectangle or square, but in Figure 3(c) they are confetti-like

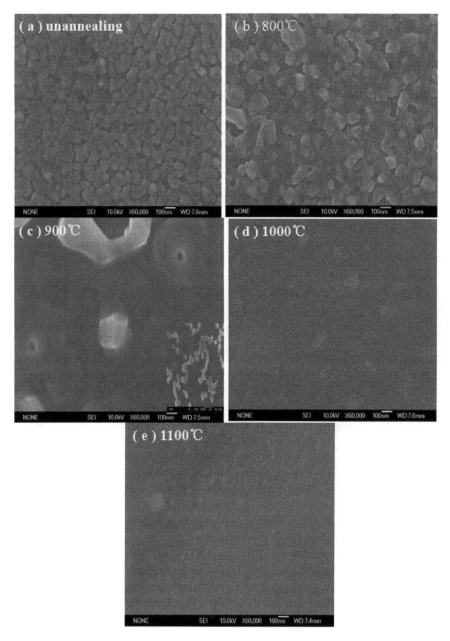

Figure 3. Images of SEM with different temperature.

and have larger size in diameter. Although the surface became smooth and confetti-like protrusions disappeared, there are still large and amorphous protrusions in Figure 3(d). However, in Figure 3(e) the surface is more smooth and particles are uniform and compact with regular circle shape and small size. In short, the microstructures by SEM are consistent with surface topographies by AFM, and surface topography of annealed at 1000°C and 1100°C are smooth and uniform.

From Figure 1, Figure 2 and Figure 3, we could obtain that the change of sample's properties is conspicuous by annealed at 900°C and 1000°C, while the temperature climbing from 1000°C to 1100°C, the degree of change become slowness even decline, the c-BN contents, compressive

194

stresses and surface topography. It may be because that the sample of microscopic particles absorb proper energy, over the energy barrier, complete the transition from h-BN phase and unstable intermediate phase to the stable cubic phase by annealed at 1000°C, at the same time, morphology and stress were also obtained better results.

4 SUMMARY

In summary, the phase transformation from h-BN to c-BN is enhanced by post annealing and 1000°C is the optimal annealing temperature with highest cubic phase content, small compressive stresses and good surface topography. High cubic phase content could guarantee c-BN excellent electric properties and semiconductor properties, small thin film stress is beneficial to device performance stability, and smooth surface is good at device's manufacture. In a word, the post-annealing process at 1000°C would laid the foundations for the use and research of c-BN in the field of optoelectronics and microelectronics.

ACKNOWLEDGMENT

This work is supported by the National Natural Science Foundation of China (Grant Nos 61274113, 11204212 and 61404091), and Program for New Century Excellent Talents in University (Grant No NCET-11-1064), and Tianjin Natural Science Foundation (Grant Nos 13JCYBJC15700 and 13JCZDJC26100, 14JCZDJC31500, 14JCQNJC00800), and Tianjin Science and Technology Developmental Funds of Universities and Colleges (Grant Nos 20100703, 20130701 and 20130702).

REFERENCES

[1] Bornholdt, S. Ye J., Ulrich S., Kersten H., Energy fluxes in a radio-frequency magnetron discharge for the deposition of superhard cubic boron nitride coatings, Journal of Applied Physics, 112(12): 123301, 2012.

[2] Wang Lin C., Yang Wenge, Xiao, Yuming M., Liu Bingbing, Chow Paul, Shen Guoyin Y., Mao Wendy L., Mao Ho-kwang K., Application of a new composite cubic-boron nitride gasket assembly for high pressure inelastic x-ray scattering studies of carbon related materials, Review of Scientific Instruments, 82(7): 073902, 2011.

[3] Ying J., Zhang Xiu-Wen W., Yin Zhigang G., Tan Hairen R., Zhang Shuai G., Fan Y.M., Electrical transport properties of the Si-doped cubic boron nitride thin films prepared by in situ cosputtering, Journal of Applied Physics, 109(2): 023716, 2011.

[4] Yamada, Takatoshi, Nebel, Christoph E., Taniguchi, Takashi, Field electron emission properties of n-type single crystal cubic boron nitride, Vacuum Nanoelectronics Conference (IVNC), 2010 23rd International, 2010, pp: 1–2.

[5] Wang Fang; Yang Baohe; Wei Jun and Zhang Kailiang, Microstructure and Nanometer Scale Piezoelectric Properties of c-BN Thin Films with Cu buffer layer by Piezoresponse Force, IEEE Transactions on Nanotechnology, 2014, 13(3): 442–445.

[6] C.B. Samantaray, R.N. Singh, review of synthesis and properties of cubic boron nitride (c-BN) thin films [J], International Materials Reviews, 2005, 50(6): 313–344.

[7] Friedmann T.A., Mirkarrimi P.B., M edlin D.L, et al. Ion-assisted Pulsed Laser Deposition of Cubic Boron Nitride Films [J]. J. Appl. Phys., 1994, 76(5): 3088–3101.

[8] Fahy S. Calculation of the Strain-Induced Shifts in the Infrared_Absorption Peaks of Cubic Boron Nitride [J]. Phys. Rev. B, 1995, 51: 12873–12875.

Advances in Future Manufacturing Engineering – Yang (Ed.)
© 2015 Taylor & Francis Group, London, ISBN 978-1-138-02817-3

Aluminum silicon alloy rolling method and experimental research

Shuxia Wang
Department of Mechanical Engineering, Dezhou Vocational and Technical College, Shandong, Dezhou, China

ABSTRACT: The grains in 4004 aluminum silicon alloy will take the shape of plate after rolling. Because of this, the internal organization of rolled pieces is bulky and the bulkiness will cause the problems of the defect of the plank crack edge after rolling. This paper aims to give a reasonable rolling method and its process parameters. In order to place the aluminum silicon alloy ingot vertically within the furnace being heated to 340°C, this method requires that the rolled pieces should be heated in advance for every rolling; the roller should be heated to 310°C by the ultraviolet heating apparatus and keeps a constant temperature. The total amount of fabric joint of the rolled pieces should come to 58.3% after rolling. The rolled pieces should be cooled by a rate of 0.5o/s after rolling. Using this method, the internal organization of rolled pieces is uniform and the grains are small without plate-like bulky silicon grains, no crack phenomenon happens. It shows that the performance of the rolled pieces rolled by this method is the best. The reliability and validity of this processing method are verified by a metallographic test and tensile experiments in the paper at last. The research will provide a new idea to analyze the changing rules of the crack edge in the process of hot rolling. Meanwhile, this research will be significant to improve the rate of finished products of 4004 aluminum silicon alloy in hot rolling.

Keywords: aluminum silicon alloy; rolling; the process parameters; hot rolling; edge crack defects

1 INTRODUCTION

4004 Al-Si (aluminum-silicon) alloy, with high content of silicon, has good abrasion resistance and plasticity, thus achieving the requirements of industrial production. Its easy operation and simple procedures enable it to be widely applied in the production of foil, refrigerator fin, stamping sheet metal and extruded profiles[1,2,3].

The rolling process parameters are an important factor in the hot rolling process of Al-Si alloy, for they may impact the rolling effect[4,5,6]. The interaction of different parameters will affect the physical and chemical properties, including the mechanical properties, changes in microstructure and the defects of crack edge. Whether the rolling process is reasonable will exert a direct impact on the energy consumption, thus resulting in the production limitations. Therefore, it is necessary to optimize its process parameters.

Xu Yongqiang[7] conducted in-depth research on the structure and properties of Al-Si alloy which shows the mechanisms to improve its cutting performance. Li Qinglin[8] studied the evolution of microstructure of Al-Si alloy and the results indicate that the refined silicon grains could improve its tensile strength and wear resistance. Zhou Hongwei[9] studied the extrusion of Al-Si alloy, which shows that by controlling the specific pressure of the extrusion, the obtained pieces have less segregation phenomenon and better performance.

The paper established stable conditions for rolling process and conducted hot rolling for the rolled pieces under different temperatures to study the rolling process laws[10] and analyzed the influence of temperature changes on the rolling effect, aiming to improve the properties of products in the subsequent rolling process[11,12].

2 ROLLING CONDITIONS

Engaging conditions. The hot rolling process of 4004 Al-Si alloy is stable. The rolled pieces convey through two rollers at a constant speed and became products after plastic shaping. To complete this rolling process smoothly, the following requirements should be satisfied.

The rolled piece generates a pair of forces in both directions when contacting with the rollers, the normal force N_o and the tangential force T_o, while the rollers generate two counter forces N, T in opposite direction[13,14]. The forward force T_x of T and the backward force N_x of N are as shown in equation (1) and (2), thus the normal engaging of the rolled piece by two rollers is as shown in equation (3).

$$T_x = T\cos\alpha \tag{1}$$

$$N_x = N\sin\alpha \tag{2}$$

$$2T\cos\alpha \geq 2N\sin\alpha \tag{3}$$

Assume that f is the fiction coefficient, β is the angle of fiction, the classic fiction law $T = f \times N$. substitute them in the above equations, and get the equation (4), (5), (6).

$$2fN\cos\alpha \geq 2N\sin\alpha \tag{4}$$

$$\tan\beta \geq \tan\alpha \tag{5}$$

$$\beta \geq \alpha \tag{6}$$

The rolling (engaging) conditions: $\beta \geq \alpha$. To complete the rolling process smoothly, the following engaging conditions should be satisfied, as shown in Figure 1.

The whole rolling process includes three stages. First, the rolled piece is engaged when meeting the rolling conditions. Second, the whole piece is pulled into the roller and the stable rolling begins. Third, the piece is thrown out, as shown in Figure 2.

Rolling process. The 4004 Al-Si alloy ingot refined by the degasser is used in the hot rolling. The dimensions are shown as follows: length of 180 mm, width of 90 mm, and height of 18 mm. The rolling is conducted in the following way: carry out vertical rolling in the dual-roller rolling mill and produce the strip with the thickness of 4–7 mm. The chemical composition of the mass fraction (%) is as shown in Table-1.

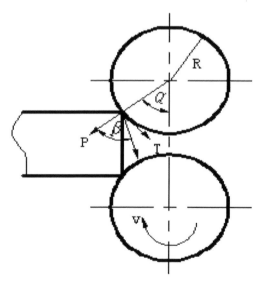

Figure 1. The engaging conditions in the rolling process.

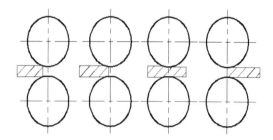

Figure 2. Rolling stages.

Table 1. The element content of 4004 Al-Si alloy.

Si	Fe	Cu	Mn	Mg	Zn	Other
9~13	0.8	0.2	0.2	1.5	0.1	0.1~0.2

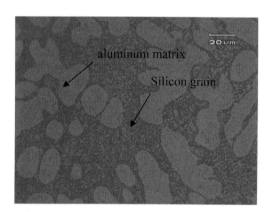

Figure 3. Themetallograph of 4004 Al-Si alloy.

The silicon content of 4004 Al-Si alloy is high. The grains are shaped like plate and bulky, with its microstructure shown in Figure 3.

The CB151. D300-100. DZ2600 dual roller rolling mill is adopted in this paper. It is equipped with the heater unit which could provide continuous heating. The roller has the length of 230 mm and the diameter of 210 mm. Heat the roller to 310°C by the heating device at the speed of 0.14 rad/s. The uniform furnace of aluminum silicon and the dual-roller rolling mill are as shown in Figure 4.

Put the 4004 Al-Si alloy ingot into the furnace to heat to 340°C and keep it warm for 30 minutes. Then place it to the engaging position with tangs and make it contact with rollers. The total press amount of the rolled pieces reaches 58.3%. The rolled pieces are heated to 340°C under passes, and then cooled by a rate of 0.5°C/s. The press

Figure 4.   The furnace and rolling mill.

Table 2.   The press amount scheme under different passes.

Pass	1	2	3	4	5
Press amount (%)	10	20	20	20	10
Width of rolled pieces (mm)	16.2	12.9	10.4	8.3	7.5

amount scheme under different passes is as shown in Table 2.

Conduct experiments to further verify the physical and chemical properties of rolled pieces.

3   ROLLING TEST

Tensile test. Conduct a tensile test on plates after rolling. First, process the rolled pieces by linear cutting and the resulting tensile samples and related parameters are shown in Figure 5.

The tensile strength is expressed in equation (7).

$$\sigma_b = F_b / S_0 \tag{7}$$

The tensile strength σ_b is then obtained, and in this equation, S_0 represents the cross-sectional area of the tensile sample, which measures 25 mm^2 in this test.

4004 Al-Si alloy has relatively poor plasticity, and its maximum load when it fractures is not the yield stress in a standard sense[15,16]. Normally, with 0.2 percent of the residual strain of materials as a reference, a line parallel with the straight line of the deformation of the rolled piece will intersect with the curve during strip rolling at point A. The stress sustained by point A is the yield stress of 4004 aluminum-silicon alloy, with the tensile test curve shown in Figure 6.

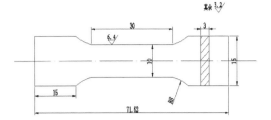

Figure 5.   Tensile samples.

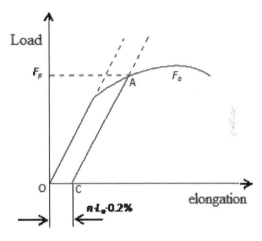

Figure 6.   Tensile test curve.

It can be seen that AC is parallel to a straight line of the deformation. A bears a stress of $F_{0.2}$, and based on equation (7), we can obtain the expression of the tensile strength of 4004 aluminum-silicon alloy, as is shown in equation (8) below.

$$\sigma_{0.2} = F_{0.2} / S_0 \tag{8}$$

Hot rolling process parameters of Al-Si alloy and comparison of the rolling results are shown in Table 3.

The analysis of the edge crack distance and mechanical properties helps to obtain a new way of rolling as well as the comparison with the original processing results. It can be seen that as all of process 3, 5, and 6 have produced edge crack defects, it is necessary to trim the strip after rolling in the industrial production, which will add to processing steps and increase production costs. After they are improved to process 2, 4, and 7, the strip after rolling does not have edge crack defects. The improvement from process 4 to process 7 has enhanced the rolling efficiency by 100%, and process 2, compared with process 7, has increased

Table 3. Rolling process and results.

Process parameters	Roll speed (rad/s)	Temperature of rolled piece (°C)	Press distance (mm)	Tensile strength (Mpa)	Elongation (%)	Edge crack amount (mm)
1	0.14	340	58	200	5.4	0
2	0.14	300	58	190	5.0	0
3	0.14	400	58	200	5.3	3.5
4	0.07	340	58	187	4.9	0
5	0.21	340	58	195	5.4	2.5
6	0.14	340	65	198	5.4	6.8
7	0.14	340	50	180	5.2	0

Figure 7. Metallograph of rolled pieces.

the tensile strength by 5.3% while reducing the energy consumption by 13.3%. We can see a 5% increase in tensile strength and a 7.4% increase in elongation rate through comparing Process 1 with Process 2. The strip after process 1 does not show edge cracks, and its tensile strength and elongation have significantly improved.

Metallographic test. To further study the reason causing changes in microstructure of the rolled materials and the defects of crack edges, conduct metallographic test for the rolled 4004 Al-Si alloy and observe tis microstructure. The device is PM-C4X4-5 electron microscope. The microstructure of the pieces rolled by process 1 is as shown in Figure 7.

The metallograph indicates that the microstructure takes the shape of small spherical particles, with the dynamic recrystallization percentage of them up to more than 50%, the average size of the silicon grains is less than 10 μm, and the aluminum matrix is elongated dispersively. The improved mechanical properties and refined microstructure of the rolled strip could meet the requirements of 4004 Al-Si alloy. In the rolling process, the temperature variation is slight and the rolling state is stable.

4 CONCLUSIONS

To achieve stable rolling, this paper explored the normal engaging conditions and hot rolling process. The established geometric model of engaging conditions indicates that the friction angle β must be greater than or equal to the engaging angle α. This could optimize the rolling process parameters, reduce energy consumption and improve productivity.

The optimized rolling process is shown as follows: Put the Al-Si alloy ingot into the furnace to heat to 340°C, and the total press amount of the rolled pieces reaches 58.3%. The rolled pieces under different passes are all heated to 340°C at the speed of 0.14 rad/s and then cooled by a rate of 0.5°C/s. In this way, the rolled pieces have the tensile strength of 200 Mpa, the elongation of 5.4%, and the amount of fission of 0 mm. The dynamic recrystallization percentage of the microstructure is up to more than 50%, the average size of silicon grains is less than 10 μm and the aluminum matrix is elongated dispersively.

This paper presented an efficient rolling method to solve the defects of crack edges of the rolled strip, reduce the manufacturing processes and improve the productivity. The superior tensile strength and elongation of the material could improve the mechanical properties of the rolled strip, thus satisfying the requirements of the products. The recrystallization could reduce the grain size and improve the performance of the microstructure, thus solving the problem of coarse grains of silicon.

The paper studied the rolling effects of 4004 Al-Si alloy and explored the influences of different parameters on the microstructure and mechanical properties of rolled strip, including rolling speed, rolling temperature, press amount, etc. By the test and analysis of the microstructure and mechanical properties, coupled with the variance analysis, an efficient rolling method of 4004 Al-Si alloy has been obtained which could better control the microstructure and mechanical properties of the rolled materials. This method is superior in improving the accuracy and optimizing the rolling parameters.

REFERENCES

[1] Fan Xiang, Yang Yitao. Microstructure and properties of semi-solid and die-cast hypereutectic Al-Si alloys [J]. Special Casting and Nonferrous Alloys, 2014, 2: 2–11.

[2] Sun Ju. Microstructure and properties of squeeze casting hypereutectic Al-Si alloy [D]. Shenyang University of Technology, 2014: 3–16.

[3] Sunhu. The impact of melt mixed processing on the morphologies of the primary silicon of Al-20% Si alloy [J]. Thermal Processing, 2014, 3: 5–9.

[4] Jiang Jun, Liu Yu, Wu Peng, Zeng Bing, He Hangmin, Jian Qiping, Zhou Jie. Research and development of Al-Si alloy modifier [J]. Light Technology, 2014, 3: 1–12.

[5] Fu Xiaotao, Li Xiaojian, Zhang Xin, Zhao Yong. Theoretical research of electric production of aluminum-silicon alloy [J]. Light Metals, 2014: 3–8.

[6] Dang Bo. Research on the solidification characteristics and microstructure of Al-25% Si alloy [D]. Hangzhou: Xi'an University of Technology, 2014: 9–22.

[7] Xu Yongqiang. The impact of composite modification on the microstructure and properties of hypereutectic Al-Si alloy [D]. Jilin University, 2014: 3–20.

[8] Li Qinglin. Si phase morphology evolution and properties of hypereutectic AL-20% Si alloy [D], 2014: 5–19.

[9] Zhou Hongwei. Research on the controlled diffusion and solidification behavior of hypereutectic Al-Si alloy and extrusion molding [D]. Lanzhou University of Technology, 2014: 8–17.

[10] Tae-HaengLee, Han-JinKo, Tae-SikJeong, Soon-Jik Hong. The Effect of Sc on the Microstructure and Mechanical Properties of Hypereutectic Al-Si Alloy Fabric [J]. Current Nanoscience, 2014, Vol. 10(1), pp. 146–150.

[11] Matthew M. Schneider, JamesM. Howe. Study of Dynamic Solid-Liquid Interfacial Phenomena in Hypereutectic Al-Si Alloy Using in situ HRTEM [J]. University of Newcastle. Microscopy and Microanalysis, 2014, Vol. 20(S3), pp. 1638–1639.

[12] Martin Halmann. VacuumCarbothermic Production of Aluminum and Al-Si Alloys From Kaolin Clay: A Thermodynamic Study [J]. Mineral Processing and Extractive Metallurgy Review, 2014, Vol. 35(2), pp. 106–116.

[13] Yan Yuxi. Research on the cracking behavior of strip edge defects in the rolling process [D]. East China University of Technology, 2014: 12–34.

[14] Zhou Dejing, Chen Zhi, Zhang Xinming, Tang Jianguo. The impact of the content of Si on the microstructure and morphology of the compound of the rolled composite aluminum steel laminated composite interface [J]. Metal heat treatment, 2014: 38–41.

[15] Qian Xiaolei. Experiment-based magnesium alloy constitutive model and its applications in the numerical simulation of rolling [D]. Jilin University, 2014: 8–22.

[16] Wu Liang, JinQinglin, Jiang Yehua. The impact of iron phase morphology on Al-Si alloy tensile fracture mechanisms at high temperature [J]. Materials and Heat Treatment, 2013, 4: 18–22.

Advances in Future Manufacturing Engineering – Yang (Ed.)
© 2015 Taylor & Francis Group, London, ISBN 978-1-138-02817-3

Research on preparation and properties of manganese oxide Nano materials

Jinyu Liu & Lingju Meng
Chemistry Department, Hebei Normal University for Nationalities, Chengde, Hebei, China

Mingliang Lu
Hebei Iron & Steel Co. Ltd., China

ABSTRACT: Taking $MnSO_4 \cdot H_2O$ and $K_2S_2O_8$ as raw materials, the use of liquid phase co-precipitation method at a temperature of 60°C and PH = 1 environment to prepared barbed spherical manganese dioxide particles with radius 1–2 μm, analyze the crystal type and basic structure using SEM, XRD and other detection methods, the results show that method for preparing is tetragonal alpha-MnO_2 manganese oxide; The alpha-MnO_2 electrode materials is used in super capacitors, taking 6 mol/L KOH as the electrolyte solution, we study its constant current charge discharge, cyclic voltammetry and AC impedance electrochemical properties, and the specific capacitance is 66F/g at the scan rate of 100 mV/s, power density 3.5 kW/kg. At the same time, and we study the effects of discharge properties of the prepared manganese dioxide Nano rods in alkaline zinc manganese battery. Electrochemical properties were compared with reported results of commercial electrolytic manganese dioxide.

Keywords: manganese oxide; nano rods; hydrothermal synthesis; electrochemical property

1   INTRODUCTION

Manganese dioxide (MnO_2) is a kind of important transition metal oxides, which has unique structure and non-toxic, cheap, friendly to environment, can be used as a redox catalyst, molecular sieve (ion sieve) as well as the original battery, two batteries and super capacitor electrode material. Bach[1] noted in particular the electrochemical performance of MnO_2 is mainly influenced by its structure parameters, such as the influence of morphology, particle size, and phase structure and body density. The one-dimensional Nano structure is by far the smallest structure that can be used as efficient electron transport. Therefore, in recent years the study on Synthesis and properties of one-dimensional MnO_2 nanostructures has attracted extensive attention of materials scientists and chemists.

Wang[2] first reported the selective MnO_2 Nano rods control synthesis, and further put forward the "rolling mechanism" to explain the mechanism of its formation. After that, scientists in different countries, the synthesis of one-dimensional MnO_2 Nano-material has done a lot of research and development of a variety of methods, it can be categorized into the below methods[3–5]: (1) direct hydrothermal treatment of commercial C-MnO_2 or Mn_2O_3 in neutral or alkaline system; (2) coordination polymer precursor method; (3) hydrothermal reduction of $KMnO_4$ to prepare C-$MnO \cdot OH$ nanowire/rods after annealing

treatment to obtain the MnO_2 nanowire/corresponding rod; (4) the $KMnO_4$ or Mn^{2+} oxidation of neutral or acidic water thermal reduction; (5) the sol-gel template method; (6) surfactant assisted hydrothermal method; (7) low temperature solid phase method; (8) synthesis of MnO_2 Nano needles and Nano rods in ionic liquid.

Compared with the sol-gel method, hydrothermal method, and low temperature solid phase synthesis method and so on, liquid phase co precipitation method for the preparation of Manganese Dioxide (MnO_2) has the advantages of simple technique, high yield, simple device, so it is more suitable for industrial production. Liquid phase co precipitation method refers to the aqueous solution of metal salts (generally use the Potassium Permanganate), under certain condition, the reducing agent and metal cations (generally using manganese sulfate, manganese chloride, manganese acetate) reaction, to obtain precipitate powder. The basic steps are: two kinds of metal salt solution with a certain concentration, adjustment to the appropriate pH value, filtering under intensive stirring reaction after a few hours, and the resulting dry, reoccupy agate mortar full grinding, required sample can be obtained. Research shows that, with this kind of manganese dioxide powder as an active material is made into electrode of super capacitor, which has excellent electrochemical performance.

Considering the preparation of MnO_2-CNTs composite material, this experiment adopts a longer reaction time for the preparation of manganese dioxide liquid phase precipitation method, in order to extend the two mixed time, and enable the CNTs to maximize the embedded in MnO_2 materials. Selection of potassium persulfate ($K_2S_2O_8$) as the oxidant, $MnSO_4$ as a reducing agent, controlling the mixed solution of pH is 1, the reaction temperature is 60°C, and this method can prepare the alpha type manganese dioxide.

2 THE EXPERIMENTAL PART

2.1 The experimental materials

Table 1. The main raw material for preparing manganese dioxide particles.

The raw material name	Remarks	Chemical formula	Molecular weight	Manufacturers
Potassium per sulfate	AR	K2S2O8	270.32	Tianjin Shentai Chemical Reagent Co., Ltd.
Manganese sulfate monohydrate	AR	MnSO4·H2O	169.02	Taishan City Chemical Plant Co., Ltd.
Concentrated sulfuric acid	95-98%	H2SO4	98.08	Shanghai Zhenxing Chemical Factory Co., Ltd. two
Anhydrous ethanol	AR	CH3CH2OH	46.07	Anhui ante Biological Chemical Co., Ltd.

2.2 The experimental instrument

Table 2. The main instrument for preparing manganese dioxide particles.

Name	Model	Manufacturers
Vacuum drying box	DZF-6050	Shanghai Jinghong Experimental Equipment Co., Ltd.
The constant temperature heating magnetic agitator	DF-101s	Gongyi Yuhua Instrument Co., Ltd.
Ultrasonic cleaning apparatus	BUG25-06	Solvent, dispersing powder
Electronic balance	FA1104	Weighing drug
Microporous membrane	0.45μ	Shanghai Xingya purifying material factory

2.3 Technological process

A. Moll ratio is 1:1 weighing of manganese sulfate monohydrate ($MnSO_4 \cdot H_2O$) and potassium per sulfate ($K_2S_2O_8$).

B. Respectively to weigh $MnSO_4 \cdot H_2O$ and $K_2S_2O_8$ dissolved in an appropriate amount of deionized water. Use a glass rod after mixing into the ultrasonic cleaning instrument to accelerate the dissolution rate. Discovery of manganese sulfate solution is in light pink, potassium per sulfate solution is colorless.

C. It will be dissolved $MnSO_4 \cdot H_2O$ and $K_2S_2O_8$ solution and $K_2S_2O_8$ solution removed slowly poured into the manganese sulfate solution, the measured pH was 2.5, after dropping concentrated sulfuric acid slowly adjusting pH to 1.

D. The mixed solution is into the thermostat magnetic stirrer, the temperature is 60°C. Adjust the appropriate speed.

E. We down into the reaction time, and took out after 22 h.

F. And then removed after cooling to room temperature, it was observed that a small amount of black powder stick in the wall of the beaker, but can be easily scraped off with a glass rod.

Table 3. Reaction conditions of MnO_2.

Sample no.	Volume of $MnSO_4 \cdot H_2O$	Volume of $K_2S_2O_8$	pH	Reaction temperature
Sample 1	100 mL	100 mL	1	40°C
Sample 2	100 mL	100 mL	1	60°C
Sample 3	150 mL	150 mL	1	80°C

After filtration, and deionized water and ethanol wash 3 times.

G. The sample is into 110 constant temperature drying box drying 5h after washing.

2.4 Sample group

In order to discuss the effects of solution concentration and reaction temperature on the particle morphology of the manganese dioxide, select the following 3 samples randomly, divide into research, in which the quality of manganese sulfate and potassium per sulfate respectively 5.634 g and 9.01 g.

3 THE SURFACE MORPHOLOGY AND XRD ANALYSIS OF MANGANESE DIOXIDE

Figure 1 is the sample 1, the reaction temperature is 40°C, reaction solution of total volume is 200 ml, scan the manganese dioxide particles. From the Figure 1, the manganese dioxide particles as sphere with other non-spherical shape, and the larger particle size, appear without thorns.

Figure 2 is the sample 2, the reaction temperature is 60°C, reaction solution of total volume is 200 ml, scan the manganese dioxide particles. It can be seen from the figure, the reaction temperature is 60°C, manganese dioxide shows the aristate spherical morphology, is the alpha-MnO_2, consistent with literature reports[3]. Thorn ball diameter is about 5 m.

Figure 3 is the sample 3, the reaction temperature is 80°C, reaction solution of total volume is 300 ml, scan the manganese dioxide particles. Figures 2 and 3 with the magnification of 3000 times known that the concentration of reaction solution is small, thorn ball diameter relative to smaller, longer thorns.

Figure 2 is XRD spectrum of manganese dioxide in the reaction temperature of 60°C, the control standard card (JCPDS No. 44-0141) can see the main characteristic peaks at 28.2 degrees, 37 degrees, 50 degrees and 60.2 degrees etc.,

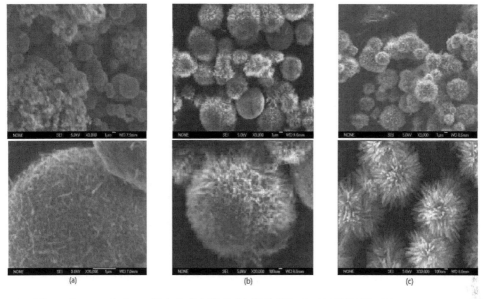

Figure 1. The surface morphology of MnO_2 ((a) 40°C; (b) 60°C, 200 mL; (c) 60°C, 300 mL).

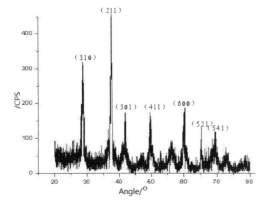

Figure 2. The XRD spectra of MnO_2 in 60°C.

characteristic peak corresponds to the alpha type manganese dioxide crystal, show that prepared samples is Quartet alpha-MnO_2, the peak of alpha-MnO_2 characteristic peak in the experiment obtained more sharp, the sample shows good crystallization.

4 THE DISCHARGE PERFORMANCE OF PRODUCTS IN THE ZINC MANGANESE BATTERY

This paper studies the discharge behavior of addition 1 mL thick H_2SO_4 to get alpha-MnO_2 Nano rods in 150°C and α-MnO_2 nanorods are prepared by KMnO4 in the condition of 140°C and $MnSO_4$ (mass ratio 5:2) water and the simulation in alkaline zinc manganese battery at different rates, and with the commercial electrolytic MnO_2 (EMD) compared. Figure 3 is the discharge curves of samples at different rates.

From Figure 3 (a) visible, it has good discharge performance through water-thermal decomposition of $KMnO_4$ to synthesize α-MnO_2 Nano rods in 6 mol/LKOH solution at acidic conditions. The open circuit voltage of MnO_2 electrode is around 1.5 V and has a long discharge platform. In the discharge rate of 50 mA/g, the discharge platform in nearly 1.1 V, when the discharge cut-off voltage is 0.6 V, corresponding to the discharge specific capacity is about 278 mAh/g and discharge specific capacity also has 274 mAh/g in the 0.8 V, higher than the literature value [3]. As the discharge rate increase, exacerbating the electrode polarization phenomenon, the discharge platform of MnO_2 electrode is reduced, corresponding to the discharge specific capacity also decreased, but not obviously. In 300 mA/g rate discharge to 0.6 V, the discharge specific capacity is still 235 mAh/g, shows that the synthesized MnO_2 Nano rods also has a large good current discharge ability.

From Figure 3 (b) can be seen, the synthesis of MnO_2 Nano rods in hydrothermal reaction of $KMnO_4$ and $MnSO_4$ has better discharge capacity in 6 mol/LKOH solution. The open circuit voltage is around 1.5 V, the discharge platform is in 1.0~1.1 V. However, the discharge capacity relative to Figure 4 (a) is slightly lower. In 50 mA/g

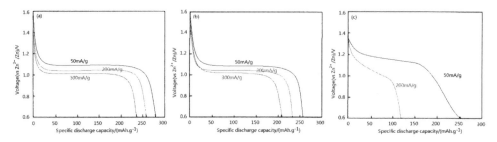

Figure 3. The discharge performance of different sample in 6 mol/L KOH.

discharge rate discharge to discharge cut-off voltage of 0.6 V when the MnO_2 electrode is about 256 mAh/g than the corresponding discharge capacity, discharge to 0.8 V when the ratio of the capacity of about 250 mAh/g, and the reported value equivalent to [3]. When 300 mA/g discharge to 0.6 V, corresponding to the discharge specific capacity of about 208 mAh/g, shows that the synthesized MnO_2 Nano rods also have large current discharge characteristics better.

From Figure 3 (c) shows that the open circuit voltage, the EMD samples is 1.3~1 −1.4 V, discharge voltage platform of sample is in 1.1~1.2 V with 50 mA/g rate, slightly higher than the previous value. the discharge specific capacity is 254 mAh/g when discharge termination voltage is 0.6 V; however, with the increasing rate of discharge, discharge voltage and specific capacity of EMD samples decreased, the discharge specific capacity is only 121 mAh/g under the 200 mA/g rate discharge to the termination voltage of 0.6 V, which shown that large current discharge capacity has poor performance of commercial EMD.

5 CONCLUSIONS

By hydrothermal decomposition $KMnO_4$ to synthesized single crystal α-MnO_2 Nano rods in acidic conditions, we study the effects of various experimental conditions (hot water temperature, the dosage of acid, acid and mixing) on the production of single crystal, and the results findα-MnO_2

Nano rods can be in a wide range of temperature (100~150°C) synthesis. The amount of acid has certain influence on crystal surface and morphology of the resultant MnO_2. When the concentration of H_2SO_4 is less than 0.1 mL, we obtained layered MnO_2 products; and we can get a mixture of α-MnO_2 and $MnPO_4 \cdot H_2O$ when the use of concentrated H_3PO_4. at the same time we study discharge performance of synthetic product in zinc manganese battery, find the synthesis of α-MnO_2 Nano rods have excellent discharge performance that obtained according to the literature method. The MnO_2 Nano rods have good performance to commercial electrolytic MnO_2, especially in the condition of high current discharge.

REFERENCES

[1] Bach S., Henry M., Baffier N., et al. Sol-gel synthesis of manganese oxides [J]. J Solid State Chem, 1990, 88 (2): 325–333.
[2] Wang X., LI Y.D. Rational synthesis of α-MnO_2 single-crystal Nano-rods [J]. Chem Commun, 2002: 764–765.
[3] Zhang Zhian, Deng Meigen, Hu Yongda, et al. Characteristics and applications of electrochemical capacitors [J]. Electronic components and materials, 2003, 22 (11): 2–5.
[4] Xia Xi. The physical and chemical properties and electrochemical activity for manganese dioxide related [J]. Cell, 2005, 35 (6): 432–436.
[5] Wu Zhongshuai, Zhang Xiangdong, Zang Jian, et al. Research progress and synthesis of Birnessite type manganese oxide [J]. Chem, 2006, 69 (6): 2–6.

Materials engineering and applications

Advances in Future Manufacturing Engineering – Yang (Ed.)
© *2015 Taylor & Francis Group, London, ISBN 978-1-138-02817-3*

Establishment of frost resistance model for autoclaved fly ash bricks

Shuang Lu & Zheng Wang
Key Laboratory of Structures Dynamic Behavior and Control, Ministry of Education,
Harbin Institute of Technology, Harbin, China

ABSTRACT: In cold area, the freeze-thaw cycling durability of building materials is one of the primary influence factors. This paper studied on the loss ratio of the relative dynamic elastic modulus of the autoclaved fly ash bricks, which is according to the ultrasonic speed after suffering different freeze-thaw cycles. After that, the corresponding numerical model has been established accordingly. The results indicate that the calculated results by the model can fit with the experimental data very well. Furthermore, the model can be applied to predict the service life by using environment parameters as the factors in the model.

Keywords: fly ash brick; freeze-thaw cycling; model; dynamic elastic modulus; durability

1 INTRODUCTION

Autoclaved Fly Ash Brick (AFAB) is a very attractive environment-friendly building material for its advantages in saving energy and using industrial end products [1]. Much attention has been paid to autoclaved brick in its resistance to dry shrinkage and desirable strength performance. Normally, it is used to produce brick with high strength by optimizing the curving condition, changing the molding craft, adding the activator or burnt gypsum in the mixture [2–5]. The effect of water absorption on the frost-resistance performance of the AFAB has been studied by Liu [6], the results showed that the weight loss and strength loss decreased with the water absorption increasing, and pointed out the upper limit of water absorption should be less than twenty percentages.

In particular, the abnormal phenomenon has been found by Li [7], that the compressive strength of the AFAB may increase under the freeze-thaw cycling. This phenomenon means that the traditional method to evaluate the frost-resistance of this novel building material is not suitable. Zhao [8] points out that the compressive strength of AFAB shouldn't be over twenty percentages after 50 times freeze-thaw cycling, the weight loss shouldn't be over two percentages comparing to its regular state. However, the traditional method using weight loss criteria or strength loss criteria has not taken into account for the time-dependent cracks growth and water absorption of AFAB, which will underestimate the side-effect of the freeze-thaw cycling to the AFAB. The technology of ultrasonic inspection is the mature application technology. It is widely used in the inner defect of building materials [9]. This paper has mainly measured the speed variation of ultrasonic propagation in the AFAB during different freeze-thaw cycling, thus the loss rate of the relative dynamic elastic modulus of the AFAB could be determined. Another purpose of the paper is to establish a damage model for predicting the service life of AFAB to endure the freeze-thaw cycling damage.

2 MATERIALS AND EXPERIMENTAL

2.1 Materials

Composite cement used was from Harbin Cement Factory. The river sand with a fineness modulus of 2.4 and Calcium Hydroxide $(Ca(OH)_2)$ were used in this study. Fly ash was obtained from Harbin San-Fa Building Materials Ltd., China. SJ-2 were prepared and used as entraining agent in AFAB. The compositions of materials used in this research are given in Table 1.

2.2 Molding techniques

Firstly, optimum water demand, brick forming pressure, steam pressure and autoclaving time were determined. Ten different types of brick specimens were produced as shown in Table 2. Brick specimens were prepared with 75% water demand for normal consistency of the resultant cement and formed with the aid of a hydraulic press of 26 MPa into mold with $120 \times 240 \times 53$ mm. The specimens were steam autoclaved in an autoclave with constant operating pressure of 1 MPa and curving temperature of 160~170 °C for 12 hours. Autoclaved specimens were curved under the circumstance of 20 ± 2 °C, RH 60%. And then the specimens were

Table 1. Chemical composition of materials used in this research.

Oxide	Composite cement (%)	Fly ash (%)	Calcium hydroxide (%)
SiO_2	21.08	66.57	1.32
Al_2O_3	5.47	18.95	1.01
Fe_2O_3	3.96	4.44	0.79
CaO	62.28	3.05	67.12
MgO	1.73	1.22	3.22
SO_3	2.63	0.31	4.21
Loss on Ignition (LOI)	1.61	3.10	21.12

Table 2. Mix proportion of AFAB.

Specimens	Fly ash (%)	Calcium hydroxide (%)	Cement (%)	Gypsum (%)	Sand (%)
MU 1	45	8	10	2	35
MU 2	43	8	12	2	35
MU 3	41	8	14	2	35
MU 4	45	12	6	2	35
MU 5	40	15	8	2	35

rectangular bars of size $100 \times 100 \times 53$ mm for testing their properties. Specimens were stored in water until they reached constant weights.

2.3 Measurement

All the ultrasonic wave velocity tcs were tested by Koncrete NM-4B nonmetal supersonic test meter at an ambient temperature of 20 ± 1 °C and RH of 50% after the specimens cured in water for 4 days. The freeze thaw cycling regulation is Quick-frost method, that one cycle include freezing under −15~20 °C for 5 hours and then thawing under 10~15 °C for 3 hours. For different cycles, the ultrasonic wave velocity $t_{t,cs}$ (μs) in specimens and ultrasonic wave velocity t_c (μs) in coupling medium could be measured and calculated as follows:

$$\Delta E_{dyn,n} = \left[1 - \left(\frac{t_{t,cs} - t_c}{t_{t,nflc} - t_c} \right)^2 \right] \times 100\% \qquad (1)$$

where $\Delta E_{dyn,n}$ is the loss of relative dynamic modulus of elasticity (%), n is cycling numbers.

3 RESULTS AND DISCUSSION

3.1 Effect of cement content

Figure 1 depicts the loss of relative dynamic modulus of elasticity varied with the different cement

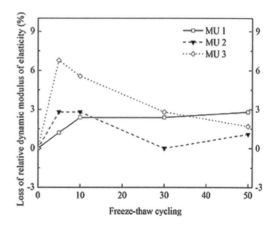

Figure 1. Effect of cement content on the loss of relative dynamic modulus of elasticity.

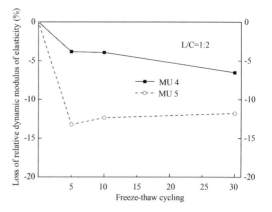

Figure 2. Effect of fly ash content on the loss of relative dynamic modulus of elasticity.

content of the AFAB during freeze-thaw cycling. Test results show that, for the cement content lower than 6%, the loss of relative dynamic modulus of AFAB after 50 times freeze-thaw cycling is almost more than 50%. This means that for the lower cement content couldn't confirm the frost-resistance ability of AFAB. While the cement content exceeds 6%, the freeze-thaw cycling damage will not influence the inner structure visibly. For these specimens, the loss of relative dynamic modulus of AFAB is less than 5% compared with the original value. That is to say, the cement content should be more than 6% for the frost-resistance AFAB.

3.2 Effect of fly ash content

Figure 2 shows the effect of fly ash content on the loss of relative dynamic modulus of elasticity. The results show that, while the lime to cement ratio is fixed at 1:2 in the mixture, the loss of relative dynamic modulus of elasticity deceases with the freeze-thaw cycling increasing. Especially at the early stage of the freeze-thaw cycling, the relative dynamic modulus of elasticity increases about 13% compared with its original value. For the compacts with 40% fly ash, the loss of relative dynamic modulus of elasticity is always less than that of compacts with 45% fly ash. That is because of the higher compressive strength for the 40% fly ash. The higher compressive strength could reduce the inner damage caused by the crystal compress and stop destroying the cement stone.

4 SERVICE LIFE PREDICTION MODEL

4.1 Establishment of prediction model

Based on the data related to the loss of relative dynamic modulus of elasticity for AFAB varied with different cement and fly ash content, a conclusion could be summarized that, the ultrasonic wave velocity deceases with the addition of cement, but increases with the addition of fly ash along with the freeze-thaw cycling. These two kinds of phenomena together confirmed that the cement will improve the hydration degree of AFAB definitely; however, the addition of fly ash will make it lower. Thus through discussing the relationship between the loss of relative dynamic modulus of elasticity and the cement content in the mixture, and then establish a service life prediction model related to the relative dynamic modulus and frost-resistance ability is a desirable manner for AFAB health condition evaluation.

$$y = \gamma_1 7.604 e^{-0.05N} + \gamma_2 \Delta R_d \qquad (2)$$

where ΔR_d is the reduction ratio due to the freeze-thaw cycling, y is the Loss of relative dynamic modulus of elasticity. γ could be varied with the AFAB composition. Both γ_1 and γ_2 are relative factors related to the cement and lime content. With more cement addition ratio, a higher value of γ_1 is available. Figure 3 depicts the model established according to the tested values related to MU 3. In this equation, γ_1 and γ_2 have the same value 0.5, which indicates the similar contribution level made for the cement and lime to the relative dynamic modulus loss.

4.2 Exportation of prediction model

In view of visible fluctuation of the relative dynamic modulus during the first five freeze-thaw cycling, we deduce these data from the equation and model. After that, by combining with regression analysis and error analysis according to the tested results, Figure 4 related to the service life

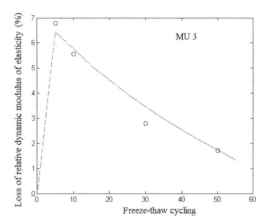

Figure 3. Service life prediction model of MU3.

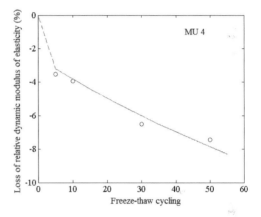

Figure 4. Service life prediction model of MU 4.

prediction model of MU 4 could be obtained. Accordingly, the line in the figure is the calculated results based on the equation 2, and the value of γ_1 and γ_2 is 0.3 and 0.7 for MU 4 respectively.

The fitting error R2 is equal to 0.9034, which is acceptable for service life prediction of AFAB. Till now, no reference related to service life prediction based on loss of relative dynamic modulus of elasticity for AFAB. When t5 is adopted as the critical point, the loss of relative dynamic modulus of elasticity is taken as 5% according to equation (2).

$$t_5 = \frac{390}{12} = 32.5 \qquad (3)$$

The equation 3 is established according to the loss of relative dynamic modulus caused by freeze-thaw cycling. However, the application of this model should take the temperature difference caused by the seasonal connections into account.

$$t = \zeta \cdot t_5 \qquad (4)$$

where ζ is the impact factor of temperature difference.

4.3 Determination of ζ

The temperature difference impact factor ζ is a function of crack temperature T_c, crack width C_w and crack length C_l.

$$\zeta = f(T_c, C_w, C_l) < 1 \qquad (5)$$

In the cold area, the surface temperature of AFAB is relative low than its freeze temperature in the late spring; however, the inner of AFAB is in its freeze condition. Thus, a higher temperature stress is observed in AFAB. For service life prediction, the amplification coefficient Am should be induced.

$$t = A_m \cdot \zeta \cdot t_5 \qquad (6)$$

During the application period of AFAB, the observation time is usually after 3 years. So ζ is determined as 0.1. And the service life prediction model was obtained.

$$t = 0.1 F(C) \cdot t_5 \qquad (7)$$

5 CONCLUSIONS

The loss of relative dynamic modulus of elasticity of AFAB decreases with the cement content increases, but increases with the addition of fly ash. It infers that the addition of cement can improve the hydration degree of AFAB. After decreasing the influence of the early stage data, the prediction model has been established. In the model, the value of γ_1 and γ_2 is related to the cement content, which reflects damage behavior of AFAB during the freeze-thaw cycling. Though γ_1 usually relates to the crack damage, but γ_2 relates to the surface damage. And the final model for life prediction of AFAB is expressed as $t = 0.1 F(C) \cdot t_5$.

ACKNOWLEDGEMENT

This work is funded by the National Natural Science Foundation of China (No. 51108133) and China Postdoctoral Science Foundation (No. 20110491077).

REFERENCES

[1] P.Z. Yu, Effect of technological parameter on AFAB, Master' thesis of Harbin institute of technology, (2010), p.49.
[2] Y.M. Song, J.S. Qian, Z. Wang, The self-cementing mechanism of CFBC coal ashes at early ages, J Wuhan Univ Techol-Mater Sci, (2008), 24, pp.338–41.
[3] J.Y. Cao, D.D.L. Chung. Damage evolution during freeze-thaw cycling of cement mortar, studied by electrical resistivity measurement, Cement and Concrete Research, (2002), 32(10): pp.1657–61.
[4] S. Goñi, A. Guerrero, M.P. Luxán et al. Activation of the fly ash pozzolanic reaction by hydrothermal conditions. Cement and Concrete Research, (2003), 33(9): pp.1399–1405.
[5] Y. Min, Qian Jueshi, et al. Activation of fly ash–lime systems using calcined phosphogypsum. Construction and Building Materials, (2008), 22(5): pp.1004–1008.
[6] X. Liu, J.G. Liang, S.H. Cheng, L. Hong, Frost-resistance ability of AFAB, Building blocks and block construction, (2008), 6: pp.30–2.
[7] Q.F. Li, The discussion related to the reason of compressive strength increase of the AFAB after freeze-thaw cycling. Bricks, (2009), 3: pp.53–6.
[8] C.W. Zhao, H. Ying, Study on the freeze-thaw cycling test of AFAB. Bricks, (2010), 11: pp.15–6.
[9] M. Molero, I. Segura, M.A.G. Izquierdo, J.V. Fuente, J.J. Anaya, Sand/cement ratio evaluation on mortar using neural networks and ultrasonic transmission inspection. Ultrasonics, (2009), 49: pp.231–37.

Advances in Future Manufacturing Engineering – Yang (Ed.)

Innovative design and application of woody materials in silver jewelry

Xiaojing Yang & Kunqian Wang
Faculty of Art and Communication, Kunming University of Science and Technology, Yunnan, China

ABSTRACT: Design inspiration is obtained from traditional Chinese culture and folk craft in order to break the limitations of material selection and design of traditional jewelry. The data based on the material analysis and consumer survey indicates that the combination of precious metals and woody materials can bring harmony to product modeling, which could contribute to the creation of classic jewelry with Chinese charm and provide a basis for the modern jewelry design with Chinese characteristics.

Keywords: jewelry design; mahogany; silver jewelry; combination of different materials

1 INTRODUCTION

As a major country of traditional jewelry design and manufacturing, jewelry in ancient China was characterized by exquisite craftsmanship and rich artistic expression [1]. Jewelry materials were not limited to traditional ones such as precious metals and gems, but also incorporated superior wood and metals with specific crafts, as were seen in girl's hairpin and man's bracelets.

2 TRADITIONAL SILVER JEWELRY IN CHINA

Silver is irreplaceable in the Chinese people's heart. Significant in carrying forward traditional culture and preserving cultural heritage of ancient China, it not only serves as the currency for thousands of years, but also is commonly used as ornaments and ritual supplies. It is even regarded as a symbol of status and good taste in some particular stages or for a certain group.

2.1 *National silver jewelry*

Silver has simple color and it is easy to be inlaid and processed to perform a variety of texture effects. Therefore, designers often use symbolic patterns and auspicious symbols of ancient China to create shape patterns for jewelry. Then it could be processed as headdress, necklace, bracelet, costume decoration and other silver products [2].

The silver decorations have various forms, generally including abstract, realistic patterns, as well as the three-dimensional and plane performance techniques. The combination of these decorative patterns could make jewelry exquisite. The auspicious decorations are mainly based on natural objects or myths, legends, folk proverbs. The key point is to seize the main features of the image and combine plants and animals through exaggeration, distortion, symbolism and other techniques, thus designing images with auspicious meaning [3]. In the traditional concept of Chinese people, the silver could be used to avoid evil, so wearing silver decorations could bring us good luck. For example, children usually wear the longevity lock made of silver (as shown in Fig. 1), and it represents the blessing of the elders, hoping them to remove the disease and live a long life [4].

2.2 *The combination of silver and other materials*

In this contemporary society which advocates personalization, the jewelry with exaggerated and personalized characteristics has won the appreciation of the public. Compared to the traditional precious metals and jewelry materials, consumers prefer the combination of integrated materials [5].

Due to its good flow ability, silver could be used to create unpredictable texture effect by creative designers, and then to decorate the wood, embroidery, porcelain and so on. It makes a difference in the design of personalized jewelry. The combination of different materials has become a popular trend in modern jewelry design. This kind of simple and frank style enables consumers to better understand the design. Match the wood in warm colors with the precious silver. Although these two materials may seem irrelevant, they have common characteristics, that is, solemn and quiet, coupled with the Chinese style.

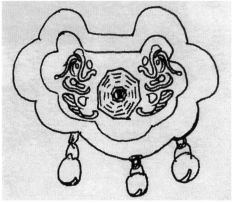

Figure 1.   Longevity lock made of silver.

3   QUESTIONNAIRE SURVEY

This survey regarded the dense crowd on Nanping Street of Kunming as the research objects. It issued 50 questionnaires and recovered 50 valid questionnaires with the pass rate of 100%. The questionnaire mainly includes consumers' expected price, style preference, purchasing and consumption habits of mahogany, and other aspects. The structure samples of consumers are as shown in Table 1.

3.1   *Consumer demand and preference for jewelry*

A comprehensive data analysis indicates that residents in Kunming are very interested in the silver jewelry decorated with woody material, and they usually wear rings, bracelets and necklaces, as shown in Figure 2.

3.2   *Purchasing characteristics of consumers*

Classify the modern jewelry styles. The funny cartoon design is classified as lovely type, the exaggerated design is classified as personalized style, the jewelry inlaid with jade or other gemstones is classified as mature style, the jewelry decorated with animal patterns is classified as retro style, and the abstract style with point, line, surface or other geometric patterns is classified as the simple style. The survey shows that most of the consumers prefer the simple style and personalized style, as shown in Figure 3.

3.3   *Acceptable price of consumers*

Mahogany jewelry inlaid with silver is created by mounting exquisite symbolic pattern made from handcrafted silver on mahogany jewelry. Mahogany follows a long growth cycle, which is

Table 1.   The structure samples of consumers.

Sample partitioning	Frequency	Effective proportion
Gender		
Female	17	43.33%
Male	33	56.67%
Age		
Under 18 years old	2	6.67%
18–25 years old	30	47.70%
25–35 years old	12	29.00%
35–50 years old	6	13.30%
Above 50 years old	2	3.33%

a very precious resource and has extraordinary values in terms of the quality and collection. Research data from various sources show that the pain threshold of consumers for mahogany jewelry inlaid with silver is about 300 Yuan. Figure 4 illustrates the proportion of the acceptable price of consumers.

3.4   *Summary of the survey*

The survey also looked at whether consumers like mahogany jewelry inlaid with silver, habits of wearing jewelry, preferred buying manners, etc., providing quantified basis for market prospects of mahogany jewelry inlaid with silver. The main subject of the surveys is college students and white-collars. They share in the pursuit of fashion, unique personality, and a certain amount of purchasing power for fashion commodities. Due to limitations of time and space, it is impossible to carry out a large-scale investigation. Therefore, the data inevitably has some elements of instability, but generally, it is accurate and credible.

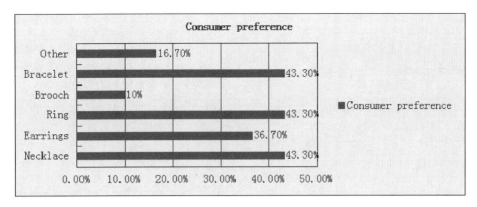

Figure 2. Consumer demand and preference for jewelry.

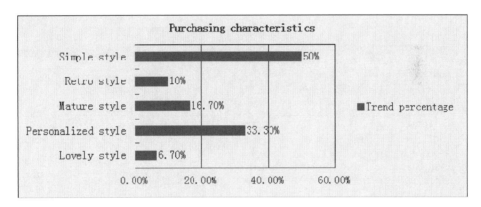

Figure 3. Purchasing characteristics of consumers.

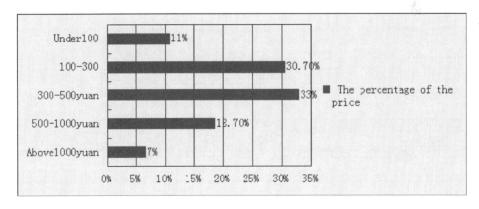

Figure 4. The proportion of the acceptable price of consumers.

4 DESIGN OF MAHOGANY JEWELRY AND SILVER JEWELRY

Wood is one of the natural raw materials, commonly used in furniture making. In jewelry design, mahogany, a precious material, is generally used [6]. As wood is easily processed and polished, smooth, and comfortable to the skin, it is suitable for wearing. Jewelry design is inseparable from the right material, color selection and matching.

215

Rosewood, sandalwood, and siam rosewood are generally used in jewelry making. They are characterized by density, glossiness, clear texture and durability. Sometimes they become more attractive because of years of wearing [7]. Mahogany jewelry inlaid with silver has cumbersome process. First, choose materials, which are processed by cutting, rough grinding, fine grinding, polishing, drilling, trimming, carving, silver inlaying, and then remove the dust [8]. For jewelry made of mahogany, each piece of wood has unique texture and does not need coloring and painting, which does not produce adverse effects on the skin [9]. The warm color echoes with the soft gloss of silver, but also gives consumers a sense of grace and being close to nature.

5 CONCLUSION

Social development is inseparable from the traditional culture as well as the continuity and development of craftsmanship. Jewelry design requires the designer's insight into consumer trends, clever use of product design, and combination of a variety of materials to lead a new design trend. The design of mahogany jewelry inlaid with silver requires bold redesign and shaping of traditional elements, the incorporation of modern jewelry design concepts into the traditional process in an effort to fit modern public aesthetic and consumer demand [10]. Therefore, this traditional craft and the incorporation and application of the two precious materials to modern jewelry design reflect the classical beauty of China. Mahogany jewelry inlaid with silver is bound to move in a richer and more diverse form of direction, with abundant jewelry market.

CORRESPONDING AUTHOR

RefName: Kunqian Wang, Email: 1601292016@qq.com, Mobile phone: 13170622004.

REFERENCES

[1] Wei Jiang: Art and Literature for the Masses. Vol. 1 (2009), p.100–101.
[2] Feng Qin, Ronghong Zhang, Yijuan Zhao: Journal of Gems and Gemmology. Vol. 9 No. 1 (2007), p.31–33.
[3] Tie Yang: The Appllcation of Comprehensive Materials in Modern Jewelry, Sheji, BeiJing. Vol. 2 (2012), p.66–67.
[4] Yanwen He: China Gold Economy. Vol. 1 (2002), p.87–89.
[5] Xiaojun Wu, Ronghong Zhang, Liu Liu: Journal of Gems and Gemmology. Vol. 15 No. 1 (2007), p.52–56.
[6] Yi Shi, Ruohui LI, Shulan Yu: Art and Literature for the Masses. Vol. 15 (2013), p.75–76.
[7] Xiaomei Jiang: China Wood Industry. Vol. 24 (2010), p.45–47.
[8] Kai Hua: Public Communication of Science Technology. Vol. 10 (2010), p.100–101.
[9] Kun Xue, Boming Xu: Furniture Interior Design. Vol. 10 (2011), p.102–103.
[10] Yong Zhou, Xuan Wang, Qiujuan Bian: Journal of Gems and Gemmology. Vol. 6 (2004), p.37–39.

Advances in Future Manufacturing Engineering – Yang (Ed.)
© *2015 Taylor & Francis Group, London, ISBN 978-1-138-02817-3*

Building a reasonable path of wall materials industry development under the goals of circular economy

Xu Jiang & Jianxin Huang
Center for Assessment and Development of Real Estate, Harbin Institute of Technology, Harbin, Shenzhen, China

ABSTRACT: Modern architecture is an important part of the consumption of energy and materials. With the deteriorating global environment and reduced resources, maintaining the sustainable development of the building materials and improving the comprehensive utilization rate of construction resources have become a widespread concern. At present, China's research on the recycling of the building material resources has made breakthrough achievements, but there are still technical and social identity issues. The development of wall materials industry should follow: the recycling of the materials, the recycling of the general waste in the construction and the application of the new recycling building materials should be led by the government, and form the protection of building resources in the "top-down" manner.

Wall materials industry is an important basic raw material industry, which plays an important role in the development of the national economy. The traditional wall materials industry is the typical resources and energy consumption oriented industries. During its rapid development at the same time, it causes excessive consumption of resources and serious pollution of the environment. Wall materials industry accounts for the bulk of China's energy use, which accounts for 9% of the national energy consumption, 13% of the national industrial energy consumption, and the energy consumption of ten thousand Yuan is nearly 5 tons of standard coal, and the energy consumption is basically non-renewable fossil fuels—coal and oil.

Taking the cement industry as an example, the recoverable deposits of the limestone resources of the country which is the main raw material to produce cement is only 54.2 billion tons. Calculated from the amount of annual consumption of the limestone cement production in our country at present, it can only last for decades and the resource depletion problems are close at hand. From the environmental point of view, the annual emissions of CO_2, SO_2 of the wall materials industry accounted for 1/4 and 1/10 of the national industry's total emissions. The dust emissions also accounts for a large proportion in the national industrial emissions, which has greatly affected the emission of more and more serious greenhouse gas, the formation of acid rain and the air quality. Therefore, to proceed along a new road to industrialization of a high technological content, good economic returns, low resource consumption, less environmental pollution, human resource advantages full played, in the sustainable development of wall materials industry at the same time, the building materials market should realize the coordinated development of economy, society and the environment, and make the ecological gradually into the benign cycle.

In view of the above, the following will discuss the topic.

Keywords: wall material industry; circular economy; rational path; construct

1 THE CONCEPT OF CIRCULAR ECONOMY

With the goals of circular economy, to develop the wall materials industry first of all need to grasp the inherent requirements of circular economy, and then construct a reasonable path in the development model in line with its internal requirements. The following defines it from the broad sense and narrow sense.

1.1 *The generalized concept of circular economy*

Wu Zongjie and Li Jianmin thinks that the circular economy covers three aspects of economic development, social progress, ecological environment, and the pursuit of the ideal combination state between the three systems. Ma Shijun thinks that the essence of sustainable development is the coordinated development of subjectivity in life and its habitat work environment, production environment and the social cultural environment.

Wu Shaozhong thinks that the circular economy is to control the waste in the production of human activities, establish the repeated and natural recycling mechanism, count human production activities into the natural cycle, and maintain the natural ecological balance. Feng Zhijun thinks that the development of circular economy is a profound revolution of paradigms. This new paradigm and the production process end control mode have essential distinctions: changing the emphasis from human productivity to natural capital, emphasizing to improve resource productivity, with the realization of "wealth doubled, resource use reduced less than half", namely the so-called "four times leap". Wu Jisong thinks that circular economy is to continuously improve the efficiency of resource use in the whole process of resources, production, product consumption and waste in the system of human, natural resources and science and technology, to change the traditional development increasing relying on resource consumption to economic development relying on Ecotype resource recycling. Zhang Luqiang and Zhang Lianguo put the circular economy as a complex system research composed of economic system, social system, natural system of society—the economy—nature, and point out that this system is not a purely spontaneous evolution, but a man-made ecosystem after the grasp of self-organization of the natural ecological system, economic cycle system and social system. The generalized circular economics is to study the self-organization of the artificial ecosystem and integrated knowledge system of material, energy, information circulation.

1.2 The concept of circular economy in a narrow sense

Zhu Dajian points out that circular economy is spoken for the linear economy of the high consumption and high emission since the campaign of industrialization and is a kind of economic development mode friendly to the earth. It requires the organization of economic activity to be the closed loop process of "natural resources—products and supplies—renewable resources". All the raw materials and energy can be reasonably utilized in the ongoing cycle of the economy, so as to limit the impact of economic activities on the natural environment in a degree as small as possible. Mao Rubai thinks that circular economy is the closed material flow mode of "resource consumption—products—renewable resources" compared to open (or single) material flow pattern of "resource consumption—products—waste" in the traditional economic activities. Xie Zhenhua thinks that the circular economy is a new economic paradigm at the stage that the ecological environment becomes the restrictive elements of the economic growth and the good ecological environment becomes the public wealth. It is a new economic form on the basis of the equality of the human condition and welfare and the maximum of the welfare of all members of the society as the goal, and its essence is to adjust the human relations of production. Ma Kai thinks that circular economy is a key to the efficient use of resources and recycling. Duan Ning thinks that circular economy is short for a closed-loop flow of material economy. Ren Yong thinks that circular economy is a new economic development model to implement "reduction, reuse, recycle and harmless" management regulation of resource flow mode in the social production and reproduction activities, which has a high ecological efficiency.

2 ANALYSIS OF THE CURRENT DEVELOPMENT SITUATION OF WALL MATERIALS INDUSTRY

Through the comparison of the inherent requirements of circular economy, the current wall materials industry in China is still facing many problems. Wall materials industry is the main industry in the development of circular economy "three wastes", especially the use of industrial waste residue. Most of the waste residue is eliminated by the wall materials industry. In the utilization of waste residues, China's cement industry achievement is very significant. Since the 50–60's in the twentieth century, we already began to use industrial waste residue, and in the 70–80's, the variety and quantity of industrial waste residue is in growing. Since the 90's we dispose the city garbage, sewage sludge and some toxic and harmful substances, which has made considerable economic benefits and social benefits of enterprises.

The production of new building materials using various industrial waste residues is an important way for the wall materials industry to disposal waste. For example, we can use coal gangue, fly ash, carbide slag, slag and other industrial waste slag to produce a variety of new wall materials, roofing and insulation materials. We can also use a variety of industrial gypsum (including flue gas desulphurization gypsum) to produce plaster board gypsum; use the bagasse and other plant fibers to produce fiber boards. According to incomplete statistics, in 2004 the national comprehensive utilization of industrial waste is 0.36 billion tons, making the new wall proportion reach 35%. The annual energy which can be saved reaches about 72 million tons of standard coal, saving 1.1 million acres of land. Among them, the largest waste residue utilization is the volume of fly ash and coal gangue.

3 BUILDING A REASONABLE PATH UNDER THE GUIDANCE OF THE ANALYSIS

Circular economy is the product of close combination of the environment and the economy, and is essentially a kind of ecological economics. It is based on the closed material flow, using the ecology rule to organize economic activity into the recycling model of "resources—products—renewable resources" and the feedback process of "low mining—high utilization—low emissions", to bring the economic system harmoniously into the circulation of substances of the natural ecological economic system, and to achieve the ecologicalization of the economic activities.

The guiding ideology of circular economy development of wall materials industry is: take the scientific outlook on development as the guidance. To save energy and resources and protect the environment as the center, be with the cement, wall and other industries as the focus, the clean production as the foundation. To improve the utilization rate of resources and reduce emissions, be with new technology and system innovation as the power, relying on the national laws and regulations and policy measures, relying on the market operation, relying on all the participation and unremitting efforts. Construct the wall materials industry into a new green industry with stronger capacity for sustainable development, and economic, social and environmental coordinated development.

Specifically, the reasonable path of wall material industry development is as follows.

3.1 Improve the quality of products and prolong the service life

Improving the quality of products and prolonging the service life is the most fundamental method, because only prolonging service life is the most environment-friendly and economical method. Simply speaking, while the service life is long, the cost of the material is relatively reduced, the material consumption is reduced, the resource consumption is reduced, and the damage to the environment is reduced. But to improve the quality of products needs to increase scientific research strength and raise the level of management.

Taking the concrete for example, we know that prolonging the service life is to increase the durability of the concrete. And the durability of the concrete in a narrow sense mainly includes 6 aspects of impermeability, frost resistance, resistance to corrosion, carbonization, alkali-aggregate reaction and the corrosion of the reinforcing bar. So we should start in the 6 factors and increase scientific research strength, so that the service life of the concrete will be increased from 50 years to 100 years or even longer time.

Constantly improving the quality and function of wall material products not only is the eternal theme of both enterprises and the management, but also the superior choice of promoting the development of circular economy and improving the sustainable development ability of the building materials enterprises.

3.2 Save energy and improve the energy efficiency

As far as the non-renewable energy, although we cannot ultimately avoid the depletion of this kind of resources, we can prolong the service life for the development of new energy, especially renewable energy to create time conditions, and also reduce the environmental loads. Therefore, energy conservation is the preferred strategy for the development of circular economy.

China wall materials industry is the significant consumer of energy. The average consumption of unit product is much higher than the world advanced level, and the potential of the energy saving is huge. In the field of production, relying on scientific and technological progress and strengthening management will reduce half of the existing energy consumption level in a relatively short period of time. Taking the modern precalcining cement as an example, it can make the heat consumption of the cement clinker close to the theoretical heat consumption, only half of the cement wet kiln heat consumption. Energy saving is not only the strategy of the enterprise long-term development, but also has become a pressing matter of the moment of wall material enterprises.

3.3 Speed up and improve the legislation, establish and promote the development of circular economy laws and regulations and policy system

Wall materials industry is an important industry in the national development of circular economy. The accelerating of the development needs the support of the national environment, including the laws and regulations of the state, economy, technology and environmental policy, industrial policy and standards and norms. In the process of practicing the development of circular economy, we must accelerate the formulation and improvement of relevant laws and regulations and policy systems, to make sure that there are laws and rules to abide by. I believe that in the near future, the state will continue to introduce a series of laws and regulations and policies promoting the development of circular economy.

In summary, the above composes of the theme the author discusses.

4 DRAWING LESSONS FROM THE DEVELOPMENT MODEL OF WALL MATERIAL INDUSTRY IN THE DEVELOPED COUNTRIES

The western developed countries went through the industrial development for 200 years, and have reached a very high level in the use of resources and energy rate. In the pattern of circular economic development in many countries, building materials industry not only has achieved a steady and rapid development, gradually created a new green industry, but also contributes more to the development of circular economy in the related industries, to speed up construction of a conservation-oriented society. All the countries in the world are committed to promoting the development of circular economy of the building materials industry, which basically has covered all aspects of the building materials, such as waste concrete, waste building plastic, waste waterproof roll, waste road asphalt concrete, waste plant fiber, waste bricks, industrial waste and other waste etc.

In the ASTMC-33-82 "concrete aggregate standard", the United States included the coarse aggregate in the broken hydraulic cement concrete. The CYCLEAN company of the United States of America uses microwave technology, and the recovery can be 100% of the old concrete pavement materials with the same quality and fresh concrete pavement materials, and the cost can be reduced by 1/3. And at the same time, it saves costs of the trash pickup and the processing, greatly reducing the environmental pollution in the city.

Holland is the earliest one of the countries carrying out the research and application of recycled concrete. In the nineteen eighties, Holland made clear requirements on the production technology of recycled aggregate concrete using the specification, and points out that if the content of recycled aggregate in the weight of the aggregate is not more than 20%, then, the concrete production can be done completely according to the design and production of natural aggregate concrete preparation method.

In addition, the developed countries as Germany, Japan, and South Korea have to formulate relevant laws and regulations to supervise and manage the concrete.

5 CONCLUSION

At present, although China's resources recycling research on the building materials has made breakthrough achievements, there are still technical and social identity issues. The development of wall materials industry should follow: the recycling of the material, the recycling of the general waste in the construction and the application of new building materials. And what's more, it should be led by the government, to form the protection of building resources with the "top-down" manner.

REFERENCES

[1] Zhang Yongjuan, Zhang Xiong. Civil engineering materials [M], Shanghai: Tongji University press, 2008.
[2] Dong Mengchen. Civil engineering materials [M], Beijing: China electric power press, 2008.
[3] Yang Changming, Zhang Juan. Recycling of the building material resources [J], Journal of Harbin Industry University, 2007, (6), 27–6.
[4] Feng Zhijun. China's Circular Economy Forum [M], Beijing: People's publishing house, 2005.

Advances in Future Manufacturing Engineering – Yang (Ed.)
© 2015 Taylor & Francis Group, London, ISBN 978-1-138-02817-3

Research and analysis on the laboratory test of frozen rocks from Jurassic strata

Haibin Ma, Sheng Du & Junhao Chen
School of Engineering and Architecture, Anhui University of Science and Technology, Huainan, Anhui, China

ABSTRACT: Jurassic strata are mainly located in the northwest, northern and northeast China, coupled with the red layer widespread in southwest area. Based on the frozen works of ventilation shaft in Bojianghaizi mine, this paper tested the mechanical properties of the Jurassic rocks at low temperature, which could lay a foundation for the promotion and application of frozen sinking in the Inner Mongolia region.

Keywords: Jurassic strata; frozen; laboratory test

1 PREPARATION OF THE LABORATORY TEST OF FROZEN ROCKS FROM JURASSIC STRATA

The test materials are taken from the inspection hole, which means to select the representative rock stratum from the Jurassic strata and test the rock mechanical properties. The rock stratum is divided into eight groups according to the lithological characters as well as the depth variation, as shown in Table 1.

Since the samples taken from the inspection hole are limited, it is necessary to select just some of these tests, including the thermal conductivity, frost heaving ratio, uniaxial compressive strength and other parameters of the above 8 samples. Before the texts, determine the basic physical performance

of the rocks, such as the moisture index, density, as shown in Table 2.

2 THE LABORATORY TEST AND ANALYSIS OF FROZEN ROCKS FROM JURASSIC STRATA

Determination of the thermal conductivity of rocks. The thermal conductivity refers to the heat transfer through $1m^2$ area of materials with the thickness of 1m in 1s when the temperature difference between both surfaces is 1°C in a stable heating transferring condition. It is related to the structure of rocks, water content, density, temperature and other factors. It is regarded as an evaluation index of the heat capacity of

Table 1. The sample grouping.

Serial number	D1	D2	D3	D4	D5	D6	D7	D8
Lithology	Coarse-grained sandstone	Coarse (medium) sandstone	Fine sandstone	Coarse sandstone	Mudstone	Fine sandstone	Mudstone	Fine sandstone
Depth (m)	402.38~419.55	473.00~479.00	486.00~524.60	525.00~526.93	528.20~530.57	531.50~532.20	533.00~533.40	533.80~536.10

Table 2. The test index of basic physical performance.

Serial number	D1	D2	D3	D4	D5	D6	D7	D8
Water content w (%)	6.33	8.51	7.13	2.56	3.8	4.72	5.76	5.06
Wet density ρ (g/cm³)	2.38	2.33	2.41	2.49	2.39	2.37	2.34	2.36
Dry density ρ (g/cm³)	2.24	2.21	2.25	2.43	2.3	2.26	2.21	2.25

a material. Therefore, it can be used to determine the thermal conductivity properties of soil and grasp the development rate of permafrost, providing basic parameters for the determination of the frozen time and the calculation of frozen wall thickness. The QTM-PD2-type conductivity meter from Japan is selected for the test, with the measurement of the sample shown as follows: length × width × height = $120 \times 60 \times 30$ mm^3.

It can be obtained from the test that the thermal conductivity of rocks is between 2.3~2.7 $kcal \cdot m^{-1} \cdot h^{-1} \cdot {}^{\circ}C^{-1}$ at room temperature, while that at low temperature ranges from 2.7 to 3.1 $kcal \cdot m^{-1} \cdot h^{-1} \cdot {}^{\circ}C^{-1}$. For details, please refer to Table 3.

Determination and analysis of the frost heaving ratio of frozen rocks. The frost heaving ratio test, also known as unconstrained axial frost heaving test, expresses the frost heaving capacity of measuring point samples when it carries on free expansion in the axial direction. During the test, the relationship between the sample's axial displacement and time is measured according to the required time, and the maximum frost heaving capacity of the sample δmax is also obtained. The ratio between δmax and the original length of the sample is the frozen heaving ratio. The sample size is 50×25 mm, and the test is carried out at −5°C, −10°C, and −20°C. The test is as shown in Figure 1.

The formula of the frost heaving ratio is:

$\varepsilon_{th} = (\Delta h / h_0) \times 100$

In the formula:

ε_{th}—the frost heaving ratio of the sample at t, %;

Δh—the axial deformation of the sample between 0-t, mm;

h_0—the height of the sample before the test, mm;

The frost heaving ratios of different strata are obtained through the test as shown in Table 4.

Test load and test data are all controlled and acquired by the computer program. The test is carried out according to the Ministry of Coal Industry Standard *Physical and Mechanical Performance Test of Artificial Frozen Soil* (MT/T593.4-2011), and the specific procedures are shown as follows:

1. Before the test, maintain the sample for 24 hours under the test temperature to achieve the consistent temperature inside and outside the sample;
2. Do the test in the manner of strain control loading and set the strain rate to 1%. The load time at this moment is about (30 ± 5) seconds;
3. 3–5 samples are generally tested in the experiment;
4. The determination of uniaxial compressive strength;

Strain calculation:

$\varepsilon_1 = \Delta h / h_0$

Calibration calculation of cross-sectional area:

$A_a = A_0 / 1 - \varepsilon_1$

Stress calculation:

$\sigma = F / A_a$

In the formula,

ε_1—Axial strain;

Δh—Axial deformation; mm;

Table 3. The test index of thermal conductivity.

Serial number	Sampling depth/m	Sample names	At room temperature		At low temperature	
			Sample surface temperature/°C	Thermal conductivity kCal/mh°C	Sample surface temperature/°C	Thermal conductivity kCal/mh°C
D1	402.38–419.55	Coarse-grained sandstone	21	2.58	−10	3.05
D2	473.00–479.00	Coarse (medium) sandstone	22	2.33	−10	2.84
D3	486.00–524.6	Fine sandstone	22	2.32	−10	2.76
D4	525.00–526.93	Coarse sandstone	21	2.30	−10	2.91
D5	528.20–530.57	Mudstone	22	2.37	−10	2.88
D6	531.50–532.20	Fine sandstone	22	2.35	−10	3.01
D7	533.00–533.40	Mudstone	21	2.61	−10	2.89
D8	533.80–536.10	Fine sandstone	22	2.58	−10	3.10

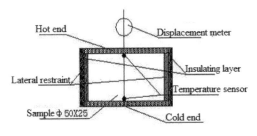

Figure 1. Schematic diagram of the frost heaving ratio test.

h_0—the height of the sample before the test, mm;
A_a—Sectional area of the sample after the calibration, mm^2;
A_0—Sectional area of the sample before the test, mm %;
σ—Axial stress, MPa;
F—Axial load, N.

5. The determination of the elastic modulus of frozen rocks

The method of determining the elastic modulus of frozen rocks is as follows: Take the ratio of half

Table 4. The test results of the frost heaving ratio.

Serial number	D1	D2	D3	D4	D5	D6	D7	D8
Water content W (%)	6.33	8.51	7.13	2.56	3.8	4.72	5.76	5.06
Frost heaving ratio (%)								
−5°C	0.11	–	0.2	−0.28	−1	−1.1	−0.21	0.13
−10°C	0.11	0.76	0.2	−0.28	−1.7	−1.2	−0.21	0.36
−20°C	0.12	–	0.19	−0.29	−1.8	−1.3	−0.22	0.43

Table 5. Test results of uniaxial compressive strength of rocks.

Serial number	Lithology	m-sampling depth/m	Test temperature/°C	Average compressive strength/MPa	Average elastic modulus/MPa
D1	Coarse-grained sandstone	402.38~419.55	At room temperature	14.89	492.151
			−5°C	21.89	534.16
			−10°C	22.50	561.341
			−15°C	24.11	595.804
D2	Coarse (medium) sandstone	473.00~479.00	−5°C	6.56	119.03
			−10°C	9.19	316.87
			−15°C	11.89	328.75
D3	Fine sandstone	486.00~524.60	At room temperature	13.37	510.31
			−5°C	15.68	541.64
			−10°C	20.07	447.41
			−15°C	19.86	485.6
D4	Coarse sandstone	525.00~526.93	At room temperature	13.75	519.36
			−10°C	12.55	502.89
			−15°C	15.23	692.27
D5	mudstone	528.20~530.57	At room temperature	14.68	674.7
			−5°C	15.53	295.7
			−10°C	17.08	691.53
			−15°C	17.14	583.59
D6	Fine sandstone	531.50~532.20	At room temperature	21.84	825.71
			−10°C	26	765.22
D7	mudstone	533.00~533.40	At room temperature	13.19	293.11
			−10°C	16.1	367.89
D8	Fine sandstone	533.80~536.10	At room temperature	22.43	772.3
			−5°C	23.55	385.62
			−10°C	28.29	485.11
			−15°C	29.96	916.29

compressive strength to its corresponding strain, namely:

$$E = \frac{\sigma_s / 2}{\varepsilon_{1/2}}$$

In the formula,

E—the elastic modulus of the sample, MPa;

σ_s—the ultimate compressive strength of the sample, MPa;

$\varepsilon_{1/2}$—the corresponding strain to half the ultimate compressive strength of the sample.

Tests are carried out according to the above procedures, and the corresponding test data are recorded, with detailed test results shown in Table 5. The curve of uniaxial compressive strength versus time of each group of rock layers is charted, as shown in Figure 2.

The following trends can be drawn through the comparisons of the uniaxial compressive strength of rock samples of Jurassic strata under different temperatures:

1. The strength of the sample shows some small increase with the falling of temperature, and even the strength of D4 group (coarse sandstone) is slightly lower than normal at −10°C. The reason is that the test sample is between the alternation in coarse sandstone and mudstone at −10°C, which affects its compressive strength;

2. It can be seen from the test results that the changes in ambient temperature have varied impact on the change rate of the strength of rock samples.

3. Results of the test data alone do not reflect obvious regularity, but the overall trend is that the elastic modulus is larger in negative temperature than at room temperature, and the elastic modulus is increasing with the decrease of temperature.

The stress-strain curves of frozen rocks. The stress-strain curves of each representative rock stratum are obtained by the experiment. Select stress-strain curves of group D3, D5, and D8 at different temperatures, as shown in Figures 3, 4, and 5.

Figures 3, 4, and 5 indicate that the stress-strain curves show similar trends in general at different temperatures and with different lithologies. This regularity can be used to draw the uniaxial loading stress-strain curves of frozen rock samples, including four stages: porous and fractured compaction stage (phase OA), stable development stage from elastic deformation to slightly elastic fracture (phase AC), non-steady rupture stage (phase CD), the post-rupture stage (phase DE), as shown in Figure 6.

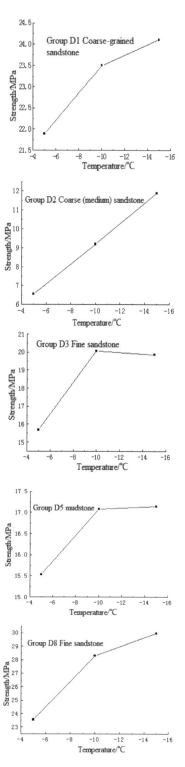

Figure 2. The curve of uniaxial compressive strength vs. time.

224

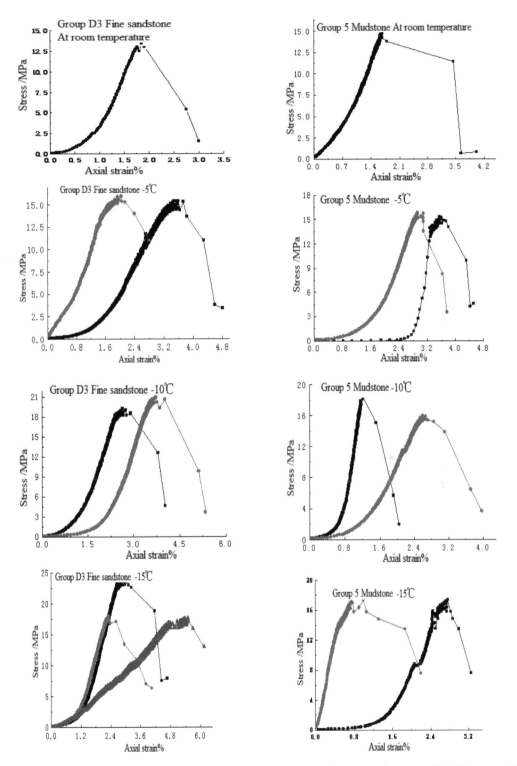

Figure 3. The stress-strain curves of the D3 group of fine sandstone under different temperature.

Figure 4. The stress-strain curves of the D5 group of fine sandstone under different temperature.

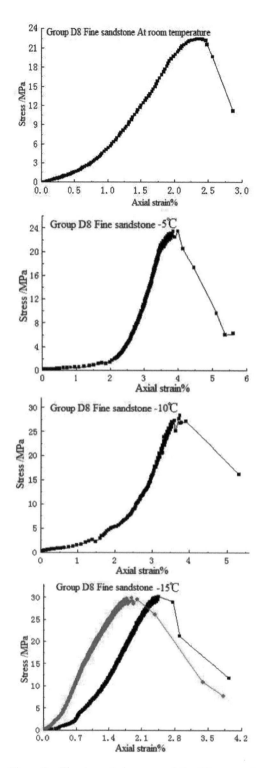

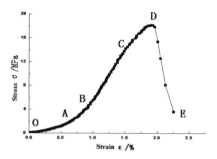

Figure 6. The stress-strain curves of frozen soil under uniaxial loading conditions.

3 CONCLUSION

The laboratory test of rock's physical properties and low temperature mechanical performance indicates the following points:

1. The Jurassic rocks have high thermal conductivity, showing that the temperature field of rocks spreads fast under frozen conditions, which is conducive to frozen on site;
2. There is no frost heaving for the Jurassic strata and even the frozen shrinkage phenomenon occurs;
3. Rock itself has high strength at room temperature, but the strength is no more than 15 MPa, still belonging to soft rock;
4. The stress-strain curves at different temperatures and with different lithologies all experience four stages: pore and fracture compaction phase, stable development stage from elastic deformation to slightly elastic fracture, non-steady rupture stage, the post-rupture stage;
5. As the Jurassic strata has certain strength and stability, coupled with the low frozen frost heave rate, the displacement of surrounding rock after the wellbore digging is small, thus generating less frozen stress. Therefore, the frozen design should focus on the waterproof capacity, while the outer wall design should concentrate on stability control.

REFERENCES

[1] Yao Zhishu. *Design and Optimization of Frozen Shaft Lining of Deep Bedrock in the Western Region* [J]. Coal Society, 2010 (5): 760–764.
[2] Ma Qiankun. *Analysis of the Interaction of Frozen Wall and Well Wall of Vertical Shaft in Cretaceous Strata* [D]. Huainan: Anhui University of Science and Technology, 2011.
[3] Wang Tao. *Field Measurement and Numerical Simulation of Frozen Temperature in Cretaceous Strata* [J]. Coal technology, 2009 (3): 121~123.
[4] Deng Weiguo. *On the Problems of Coal building in Western Regions* [J]. Well Construction Technology, 2010 (4): 29~33.

Figure 5. The stress-strain curves of the D8 group of fine sandstone under different temperature.

Advances in Future Manufacturing Engineering – Yang (Ed.)
© *2015 Taylor & Francis Group, London, ISBN 978-1-138-02817-3*

Research on structure, properties and advantages of prefabricated environmental protection rubber running track

Ming Yang
Guangxi International Business Vocational College, Nanning, Guangxi, China

Bikai Dong
Department of Physical Education, Guangxi Normal University for Nationalities, Chongzuo, Guangxi, China

Weidong He
Sports and Health Science Academy, Guangxi University for Nationalities, Nanning, Guangxi, China

ABSTRACT: Prefabricated environmental protection rubber running track keeps up with the green concept currently advocated by the society and meets the developmental needs for construction of modern athletic fields. In the paper, research methods are used, such as literature information method, mathematical statistics method, and field investigation method, for analysis of popularization, structure, performance, advantages and other factors of prefabricated environmental protection rubber running track. After studying, the conclusions are drawn as follow: 1) Prefabricated rubber running track has become the product designated by the main venues of the modern Olympic Games, and leads the developing direction of high-level track and field venue construction. Based on the green concept, series products of prefabricated environmental protection rubber running track have been publicized and applied successfully with its advanced performance and excellent quality, and rises successfully in China. 2) Prefabricated rubber running track is a kind of environmental protection rubber coiled material with a double-layer structure. The upper layer is a high toughness wear-resisting surface layer, and the lower layer is a high-elastic bottom layer. It features scientific structure and high resilience. 3) From the aspect of performance, prefabricated rubber running track highlights its variety, uniqueness and hierarchy in the aspects of professional property, safety, environmental protection, excellent weather resistance, economical efficiency and durability, multi-color effect, better reflecting fair competition, convenient installation, free maintenance, wide range of application, etc. 4) Compared with traditional Polyurethane (PU) plastic running track, series products of prefabricated environmental protection rubber running track have five significant advantages. Due to its stable performance and uniqueness, it has won more market shares, and achieved good social benefit and economic benefit. With R&D of its products and recognition from the market, it still has huge market space and advantages in the future.

Keywords: prefabricated; environmental protection rubber running track; products; structure; performance; advantages; benefits

1 INTRODUCTION

Environmental protection refers to the general term that describes a variety of actions, which human beings take to solve realistic or potential environmental problems, to coordinate the relationship between human and environment, and to guarantee the sustainable development of economic society. It is involved in many fields such as natural science and social science, and its unique research object. Environmental protection is to prevent environmental pollution and destruction through reasonably using natural resources, to achieve the balance between the natural environment, cultural environment and economic environment, as well

as sustainable development, to expand reproduction of useful resources, to guarantee the social development. People do physical exercises and participate in reasonable fitness activities for the pursuit of happiness and health, so a number of athletic fields are required. Furthermore, athletic fields require rubber running track balancing specification with environment.

At the end of 1999, Beijing Green World Sports Industry Development Center collaborated with colleges and universities in Beijing, and international professional organizations, and successfully developed R (ZEPHYR) series products of prefabricated environmental protection rubber running track. All technical indicators meet or exceed the International

Amateur Athletic Federation (IAAF) standards, and pass the tests by the Athletic Field Synthetic Materials Testing Center of China Athletic Association and National Sports Goods Quality Supervision and Inspection Center. It has become dedicated products of China Athletic Association with advanced performance and high quality. On July 1, 2004, R (ZEPHYR) prefabricated environmental protection rubber running track became the only dedicated product of running track of China having the qualification for bidding for 2008 Beijing Olympic Games, marking China has the technology of running track up to international level [1] (2004 (9), R (ZEPHYR) series products of prefabricated environmental protection rubber running track has been certified by the IAAF—The running track products of China first won the qualification for bidding for the Olympic Games). In recent years, with the rapid development and needs of economy and culture in various regions of China, series products of prefabricated environmental protection rubber running track are brought to market comprehensively.

The research team consulted relevant literature of "rubber running track", and found that the research center and scholars, such as Beijing Green World Sports Industry Development Center [1], Ma Xiao [2], He Weidong [3], Haiming Cui [4], Gao Minglei [5], Li Changlu [6], Tian Wenbao [7], Guan Ying [8], Chen Zhanhua [9], and Jiang Yuxia [10], have carried out the research on physical and chemical properties, structural design, athletics, fitness and other applications of rubber running track, and have made academic achievements. However, the research on the environmental protection rubber running track is seldom conducted. Therefore, aimed at the green concept and the current developmental needs for construction of modern athletic fields, the research team will focus on structure and characteristics of prefabricated environmental protection rubber running track, further understand the connotation, types and characteristics of corresponding products, and search its market value and advantages, so as to promote successful popularization and application of prefabricated environmental protection rubber running track and its developed products in the market.

2 SUCCESSFUL POPULARIZATION AND APPLICATION OF PREFABRICATED ENVIRONMENTAL PROTECTION RUBBER RUNNING TRACK IN CHINA

In the 1970s, prefabricated rubber running track became the product for main venues designated by all previous Olympic Games, and led the trend and mainstream of construction of high-level athletic field. At that time, in the international market, only an Italian company named MONDO can produce such high-quality dedicated products of running track representing the trend and mainstream. In the late 1990s, developed countries and regions such as Europe, the United States and Taiwan prohibited the use of athletic fields containing PU material in various schools. Poisonous PU plastic running track easy to be threshing and fading, however, was still paved in many athletic fields of China. With continuous growth and deepening of environmental awareness, the demands for environmental performance of athletic field material are growing in the international society, so it is inevitable to develop upgraded and updated products advanced with environmental performance.

Mr. Wang Shen, General Manager of Beijing Green World Sports Industry Development Center, set his sights on the development of high-quality prefabricated rubber running track with his keen market insight and profound strategic decision-making. In order to follow up the international advanced design philosophy, Mr. Wang Shen has cooperated with universities in Beijing and foreign professional organizations, to jointly develop prefabricated rubber products. After nearly two years of development and testing, R (ZEPHYR) series products of prefabricated environmental protection rubber running track were successfully launched at the end of 1999. Compounded rubber is its raw material, after press polish, vulcanization and stabilizing prefabrication, it is produced. All technical indicators have passed the tests by the IAAF, Athletic Field Synthetic Materials Testing Center of China Athletic Association and National Sports Goods Quality Supervision and Inspection Center. The product has been certified by the IAAF with advanced performance and excellent quality, and has received the product certification accredited by the China Athletic Association. It becomes the only dedicated running track surface certified by the IAAF in China, the only dedicated product of running track which can be used in all kinds of international athletic fields, and also the only running track surface qualified for bidding for 2008 Olympic Games in China. Practice indicates that it is of epoch-making significance to successfully popularize and apply prefabricated environmental protection rubber running track and series products in the track and field industry of China, and prefabricated environmental protection rubber running track is rising successfully in China.

3 STRUCTURE OF PREFABRICATED ENVIRONMENTAL PROTECTION RUBBER RUNNING TRACK

Prefabricated rubber running track is a kind of environmental protection rubber coiled material with a double-layer structure.

The upper layer is a high toughness wear-resisting surface layer, with a depth of 1–2 mm, and irregular pattern is produced through continuous rolling by the special professional equipment. The surface layer adopts the process formula with high rubber content, high density and high elasticity. The main component is composed of natural rubber, synthetic rubber, aging-proof agent, ultraviolet absorption stabilizer, wear-resistant additives, etc. The upper layer has the characteristics of ageing resistance, non-discoloration, sting-resistance, wear-resistance and stretch-resistance, to ensure that its service life is longer than 10 years.

The lower layer is a high-elastic bottom layer, a microcellular foam layer with a reasonable lean and a teardrop-shaped honeycomb geometric construction. Such a structure can provide excellent ability to absorb impact energy, smooth energy transition and continuous energy feedback. After inert gas closed-cell foaming, there are many tiny air bags of inert gas in the elastic bottom layer, so as to ensure high resilience performance of the bottom layer. Owing to specific underlying structure of prefabricated environmental protection rubber running track, it can be guaranteed that rubber coiled products have good resilience and energy feedback properties, as well as the aging of products cannot have a strong impact on its high resilience after outdoor use for years.

4 CHARACTERISTICS OF PREFABRICATED ENVIRONMENTAL PROTECTION RUBBER RUNNING TRACK

1. Professional property: In combination of sports science and material science, the product adopts the prefabrication process, so as to fully satisfy and reflect professional requirements of sports participants for running track, and has comprehensively passed the test in the physical and chemical properties of running tracks by the IAAF.
2. Safety: The design of unique embossed patterns on the surface and grooves at the bottom provides the appropriate system of shock absorption and energy feedback, so that sports performance can be improved and sports safety can be guaranteed.
3. Environmental property: Poisonous and harmful substance in the material is abandoned, really for physical and psychological health of sports participants. Long-term quality assurance and secondary renovation design prolongs its service life, and also saves earth resources. As manufacturers, we guarantee that we will recycle resources after the expiration of service life.
4. Excellent weather resistance: Under normal climatic conditions, the product has the properties of good acid resistance, alkali resistance, UV resistance, fungal resistance. Under the wet condition, it has anti-sliding performance required in the professional games. Even under the severe weather (such as low temperature, plateau climate, high ultraviolet region, etc.), it also has excellent sports performance.
5. Economic durability: Due to its high cost performance, the returns to the investment of athletic fields can be maximized. In addition, the product's high wear resistance, sting-resistance, aging resistance, aging resistance and other excellent physical properties guarantee that athletic fields in the colleges, middle and primary schools can be frequently used.
6. Multicolor effect: The combination of non-dazzling multicolor increases the beauty of athletic fields, and brings pleasant physical and psychological experience to sports participants.
7. Better reflecting fair competition: The productive technology of prefabricated rubber running track can ensure the consistency of product thickness and uniformity of ground system in the same game, and also reflect the Olympic spirit of fair play.
8. Convenient installation: Under the conditions of a few manpower and machines, only special adhesives provided by us are used to pave coiled material of running track on the dense foundation. Running track can be put into normal use 24 hours later after installation.
9. Free maintenance: There is no threshing flaky phenomenon. The product has a certain self-cleaning performance, so just daily cleaning can keep it clean, and there is no need for professional maintenance.
10. Wide range of application: The product is applicable to athletic fields of colleges and middle schools, as well as sports venues of domestic and international events.

5 ADVANTAGES OF PREFABRICATED ENVIRONMENTAL PROTECTION RUBBER RUNNING TRACK

Series products of prefabricated environmental protection rubber running track have the remarkable quality. Compared with traditional PU plastic running track, it has five significant advantages as follows: The first is outstanding sports safety performance; the second is excellent environmental performance; the third is stable performance and providing fair space of technique use; the fourth is convenient construction and maintenance, as well

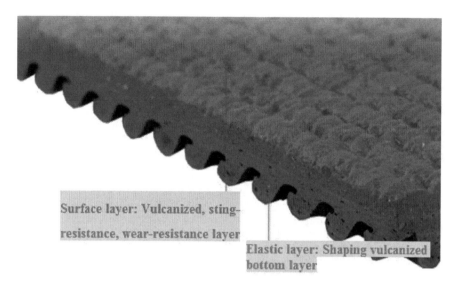

Figure 1.   Structural diagram for coiled material of prefabricated environmental protection rubber running track.

as low maintenance cost; the fifth is frequent use, long service life.

6   ADVANTAGES OF PREFABRICATED ENVIRONMENTAL PROTECTION RUBBER RUNNING TRACK

Series products of prefabricated environmental protection rubber running track have the remarkable quality. Compared with traditional PU plastic running track, it has five significant advantages as follows: The first is outstanding sports safety performance; the second is excellent environmental performance; the third is stable performance and providing fair space of technique use; the fourth is convenient construction and maintenance, as well as low maintenance cost; the fifth is frequent use, long service life.

With the aging and obsolescence of traditional PU plastic running track, all-around innovation and giant leap of series products of prefabricated environmental protection rubber running track in the aspects of technology and performance, as well as more attention to environmental protection in the society, people have a deeper understanding of environmental protection. Advanced products made from superior quality materials in athletic field can fully satisfy the construction and development of sports venues. Due to its stable performance and uniqueness, series products of prefabricated type environmental protection rubber running track have won customers' trust, gradually gain more market share in the construction of athletic fields of China, and are developing the market in South Asia. In nearly decades, series products of prefabricated environmental protection rubber running track have obtained better social benefit and economic benefit, and have broad market prospects. With the products upgrading, popularization and application in the future, the market prospects will produce more advantages.

7   CONCLUSIONS

1. Prefabricated rubber running track has become the product for main venues designated by the modern Olympic Games, and leads the developmental direction of construction of high-level sports venues. From an Italian company named MONDO to Beijing Green World Sports Industry Development Center, they have developed the materials of athletic fields with environmental performance. At present, China has successfully launched R (ZEPHYR) series products of prefabricated environmental protection rubber running track. Due to its advanced performance and excellent quality, the products has been certified by the IAAF and affirmed by the Chinese Athletic Association, and is the only dedicated surface layer of running track of China having the qualification for bidding for 2008 Beijing Olympic Games. In recent period,

the R&D centers and relevant enterprises focus on the green concept, successively launch new products, and successfully popularize and apply series products of prefabricated environmental protection rubber running track. Prefabricated environmental protection rubber running track rises successfully in China.

2. Prefabricated rubber running track is a kind of environmental protection rubber coiled material with a double-layer structure. The upper layer is a high toughness wear-resisting surface layer, and the lower layer is a high-elastic bottom layer. It features scientific structure and high resilience. From the aspect of performance, prefabricated rubber running track highlights its variety, uniqueness and hierarchy in the aspects of professional property, safety, environmental protection, excellent weather resistance, economical efficiency and durability, multi-color effect, better reflecting fair competition, convenient installation, free maintenance, wide range of application, etc.

3. Compared with traditional PU plastic running track, series products of prefabricated environmental protection rubber running track have five significant advantages. Due to its stable performance and uniqueness, it has won more market shares, and has brought good social and economic benefit. Prefabricated environmental protection rubber running track is continuously upgraded, recognized, popularized and applied in the market. Its market prospects will produce more market space and advantages.

ACKNOWLEDGEMENTS

Project Fund: 2014 Science and Technology Studies of Guangxi Universities (Project No.: YB2014423).

ABOUT THE AUTHOR

Brief Introduction of First Author: Yang Ming (1981-), male, Rongxian County of Guangxi, Teacher. Research orientation: sport teaching and sport industry.

About corresponding author: Dong Bikai (1973-), male, from Tiandeng County, Chongzuo, Guangxi, associate professor, master degree. Research orientation: Science of Ethnic Traditional Sports.

REFERENCES

[1] Athletics Editorial Department, (ZEPHYR) Prefabricated type environment-friendly rubber runway series obtained IAAF certification—Chinese Track Products Bid for the Olympic Games qualification for the First Time [J]. Track and Field. 2006, (9), pp:7.

[2] Ma Xiao. Prefabricated Rubber Track and Its Manufacturing Process [J]. Rubber Industry. 2012, (6), pp.378.

[3] He Weidong. Introduction and Application Status of Track and Field Sports Instrument and Equipment in the Colleges and Universities in Guangxi [A]; Symposium Proceedings of 2006 National Sports Instrument and Sports System Simulation [C]. 2006, (7), pp:34–39.

[4] Cui Haiming, He weidong. Superiority of Prefabricated Green Rubber Track and Analysis of Tongxin Products [A]; Advanced Materials Research [C]. 2014, (5), pp:24–28.

[5] Gao Minglei, Yan Bo. Italian Mondo of Prefabricated Rubber Track Used in Ji'nan Olympic Sports Center Stadium [J]. 21st Century Building Materials. 2009, (6), pp:88–90.

[6] Li Changlu, Li Binyung. Application of Tasteless Green Plasticized Rubber Plastic Cover in Conveyor Belt [J]. Chinese High-tech Enterprises. 2011, (03), pp:86.

[7] Tian Wenbao, Zhang Qi, Huo Yongsheng. Research on of Paving Technology Based on Concrete Cement Prefabricated Rubber [J]. Construction Technology. 2012, (6), pp:144–147.

[8] Guan Ying. Plastic Sports Ground and the Corresponding Plastic Materials [J]. New Chemical Materials. 2011, (1), pp:36–41.

[9] Chen Zhanhua. Plastic Ground Structure Design of Sports Venues [J]. Construction Technology. 2001, (9), pp:611–612.

[10] Jiang Yuxia, Zhang Jiliang. Application of Runway Ddge in the Amateur Sports Training [J]. Sports, 2011, (5), pp:57–58.

[11] Rubber Technology and Equipment Editorial Department. Tracks Structure with Coalescence Lines Rubber Environmental Prefabricated Coil [J]. Plastics Technology and Equipment. 2010, (06), pp:64–65.

[12] Editorial Department of Journal of Modern Rubber Technology. US Anti-tank Complex Prefabricated Synthetic Rubber Track [J]. Modern Rubber Technology. 2005, (10), pp:39.

Advances in Future Manufacturing Engineering – Yang (Ed.)
© *2015 Taylor & Francis Group, London, ISBN 978-1-138-02817-3*

Monitoring and analysis of expansive soil road cutting slope's pile-plank wall

Xiao Yang
Department of Civil Engineering, Monash University, Melbourne, Australia
School of Civil Engineering, Central South University, Changsha, China

ABSTRACT: Take expansive soil road cutting slope's pile-plank wall in Yun-Gui railway as engineering background. Take a long-term field monitoring for the testing components which set in pile-plank wall. By analyzing the measured data, acquiring the mechanics characteristics of the expansive soil slope pile-plank wall, it shows that the pile-plank back's soil pressure distributions with depth are K type and "C" type. The rules compare with Coulomb's soil pressure. The result shows that there were greater differences between them. Due to the soil arching effect, the bottom plank between two piles back soil pressure horizontal direction distribution is "both sides is big and the centre is small". The pile's bending moment distributions in depth in line with the mechanical characteristics of a cantilever beam. In addition, because the crack is in the top pile's platform, and the invagination in wall toe's side ditch, they lead to infiltration of surface water.

Keywords: expansive soil road cutting slope; pile-plank wall; filed monitoring

1 INTRODUCTION

At this time, the main research methods of expansive soils slope retaining structure are model test and numerical simulation. It is hard to reflect the actual work situation of supporting structure. The main reason is the laboratory model test always uses remolded soil and its shape is limited. That is different compared with the actual situation. Meanwhile the soil's stress condition is hard to simulate. So the base of laboratory model test, it is necessary to do the prototype measurement at construction site. For example, Richards[1] in Australia monitored a rigid foundation supporting wall which length is 25 metre, height is 7.5 metre, the results show that earth pressure cell in bottom it's test value is enhanced with the growth of time; it is 4 times of the depth theoretical calculation of pressure. The cause for this situation is groundwater which in the 3 metre depth. It leads to the expansion at the bottom of earth-pressure cell and large lateral pressure, Li Ming[2] aim at the cutting slopes retaining wall of expansive soil at Nanning–Kunming Railway Line's DK146 + 350 to take the monitor of soil pressure which behind the wall's different depth. The result shows that, in the depth range of 0–1.5 metre, lateral swelling force increases gradually beneath 1.5 metre, with increased of depth the soil's lateral swelling force reduced gradually. Researcher thinks soil behind

wall only generates expansive force in depth range which atmospheric influence.

When using cantilever anti-slide pile reinforce slope in the middle conterminous hanging panels, it forms pile-plank wall, which is widely used in embankment and cutting slope reinforce of coastal engineering. And earthquake area, pile-plank wall compare with gravity retaining wall, they have definite advantages, convenient construction, beautiful construction and low charge of maintenance[3]. Now, the research report of anti-slide pile's in-situ monitoring is more[4–9] but pile-plank wall's monitor is little. In this paper, through stress of piles and board and soil between piles moisture field monitoring, get field's first-hand data. By analyzing the relationship between the relevant test data, can provide reference data and suggestion for design of pile-plank wall and design of cutting pile-plank wall distribution and value of expansion force.

2 PROJECT OVERVIEW

The test project is at the left of Yunnan-Guiling Railway Line. The geomorphic type is basin; through the form of excavation, skin layer is cover with expansive soil, the argillaceous siltstone and mudstone clip and lignite which with a modest and more expansive are under it. Main soil's property is, 1. soil (2) swelling soil, $\gamma = 19.5 \text{ kN/m}^3$,

$c = 25$ kPa, $\varphi = 13°$, $[\sigma] = 150$ kPa, is belong to moderate to strong expansive soil. 2. soil (4-w4). The argillaceous siltstone and mudstone clip and lignite (E2-3n) $\gamma = 20$ kN/m³, $c = 20$ kPa, $\varphi = 16°$, $[\sigma] = 200$ kPa, is belong to moderate to strong expansive soil. 3. soil (4-w3) the argillaceous siltstone and mudstone clip and lignite (E2-3n) $\gamma = 22$ kN/m³, $\varphi = 18°$, $[\sigma] = 300$ kPa soil (4-w2) the argillaceous siltstone and mudstone clip and lignite (E2-3n) $\gamma = 23$ kN/m³, $\varphi = 20°$, $[\sigma] = 350$ kPa.

3 ARRANGEMENT OF MONITORING DEVICES

3.1 *Earth-pressure cell*

At the interface of anti-slide pile, pile-plank and soil, arrange a number of earth-pressure cells in accordance with one-meter distance along with depth direction. Can test pile-back, plank-back's soil pressure in different depth; arrange three equidistant earth-pressure cells behind the bottom of pile-plank; monitor the change of soil pressure along with horizontal which behind the bottom of plank.

3.2 *Soil moisture meter*

Arrange a number of soil moisture meters with depth in soil which behind pile, measured parameter take central probe as center, test volumetric water content of soil which in a cylinder scale. This cylinder scale's diameter is seven centimeter, and height is seven centimeter. It can test the change of soil moisture in different depth which behind pile.

3.3 *Concrete strain meter*

Bind the concrete strain gauges, plates and pile reinforcing bar in tension together. By testing the strain gauge value, then according to the internal forces of bending formula[10], get the pile and plank internal force indirectly.

4 THE RESULTS OF MONITORING AND ANALYSIS

4.1 *Soil moisture monitoring results and analysis*

Figure 1 is pile-back's soil moisture value fluctuating law over time. It shows that after 2011 year, pile-back's soil moisture value mostly agrees with different test points, and soil moisture meter value, which is in slope, appears downturn slowly. The main reason is there is top of pile-plank wall exit complete and blocked mortar bed, and at this time, the soil behind pile is not influenced essentially by rainfall. But in the year 2012, the soil moisture value in different slope's depth has changed significantly compare with that in 2011 year. It is because at top of the pile platform, the mortar bed, which used for closed and waterproof, is crack due to lack of strength, thus constituting a channel Surface water infiltration. The circulation function makes the mortar bed's water-resisting affection decrease.

4.2 *Pile-back's soil pressure monitoring results and analysis*

Figures 2 and 3, the pile-back's measure soil pressure changes with depth. It shows that the pile-plank's measure soil pressure with depth is

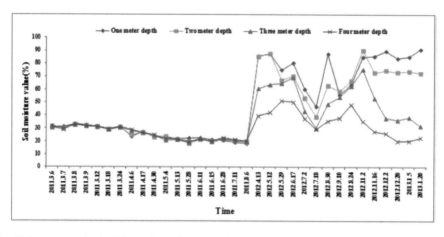

Figure 1. Soil moisture value in different depth fluctuating law over time.

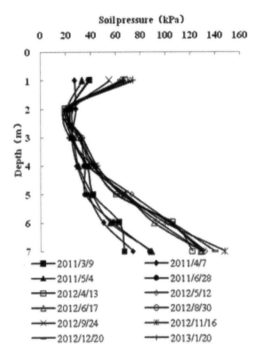

Figure 2. 5# pile-back's measurement of soil pressure distribution with depth.

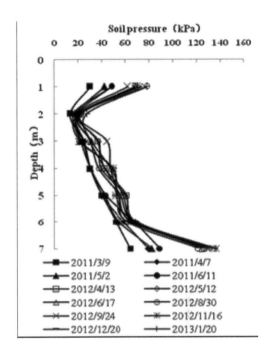

Figure 3. 6# pile-back's measurement of soil pressure distribution with depth.

non-linear relationship, it render K type distribution which the top and bottom is large, the centre is small, according to the comparison with Coulomb's soil pressure. The depth in one meter to three meter's measure soil pressure is large than Coulomb's soil pressure. It shows that in this depth is the extent of the atmosphere, the depth in three meter to five meter measure soil pressure is equal to Coulomb's soil pressure. It shows that in this depth, the effects of atmospheric action of soil are unapparent, and the depth in one meter to three meter's soil pressure is appearing mutate. This area is wall toe, according to the field observation, this area's roadside ditch platform appears some concave, leads to surface water flow backward, so appears the situation of soil pressure mutate.

4.3 Pile's bending moment monitoring results and analysis

Figures 4 and 5 are 7# and 8#'s bending moment distribution rule with depth. It shows that, pile's bending moment is positive value, 5# and 6# pile's bending moment is approaching a maximum value near the depth of seven meters. It accords with a characteristic of cantilever anti-slide pile, 2011 year's observation show measure bending moment

is not much different compared with the bending moment cause by Coulomb's soil pressure, 2012 year's observation appear larger change, due to the change of soil moisture lead to swelling pressure of soil, the calculative bending moment only cause by Coulomb's soil pressure is smaller than measured value.

4.4 Plank-back's soil pressure monitoring results and analysis

Figures 6 and 7 are plank-back's soil pressure distribution rules with depth. It shows that, plank-back's measure soil pressure's distribution with depth appears C type which in both sides is large, the centre is small. Its maximum value appears the bottom plank between the piles. Plank-back's maximum soil pressures are 97 kPa and 96 kPa. Compared with Coulomb's soil pressure, the depth in 0.75 meter to 1.75 meter measure soil pressure is large than Coulomb's soil pressure. It is the result of the dramatic atmosphere effect. The depth in 1.75 meters to 3.75 meters measure soil pressure is close to Coulomb's soil pressure. The 2012 year's monitoring situation shows that, the soil pressure in 0.75 m~2.75 m, 4.75 m is increasing significantly; main reason is soil's expansion caused by the surface water's infiltration at the top and the side-channel.

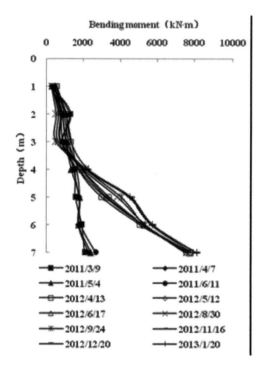

Figure 4. 5# pile's bending moment change with depth.

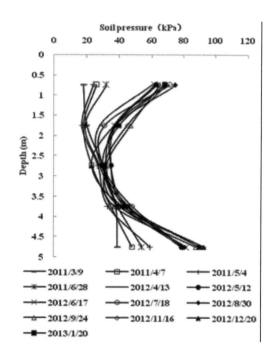

Figure 6. 5#-6# pile-plank back measurement of soil pressure distribution with depth.

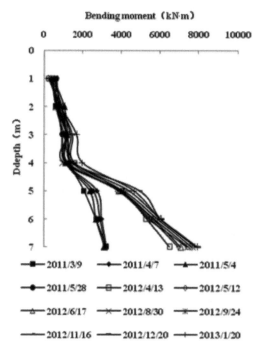

Figure 5. 6# pile's bending moment change with depth.

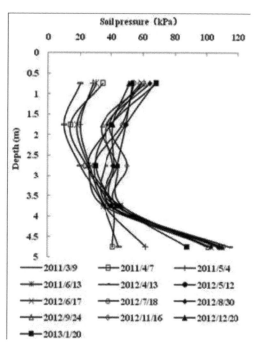

Figure 7. 7#-8# pile-plank back measurement of soil pressure distribution with depth.

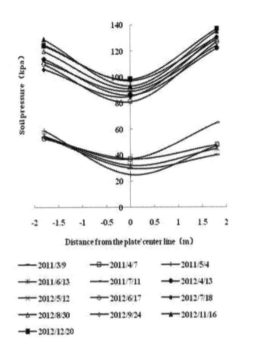

Figure 8. 5#-6# pile-bottom plank back soil pressure horizontal distribution.

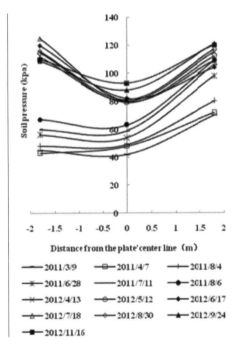

Figure 9. 6#-7# pile bottom plank back soil pressure horizontal distribution.

4.5 *Bottom plate-back soil pressure monitoring results and analysis*

Figures 8 and 9 is the bottom plank between pile's-back soil pressure horizontal distribution rule with depth. It shows that the bottom plank between pile's back measure soil pressures along the horizontal direction appears lenticular distributions which in both sides are large, the centre is small. The main reason is existing soil arching between the pile-plank; midspan's soil pressure is smaller than both sides. The observed value of 2012 compared is increasing by a fairly big margin; main reason is surface water's infiltration leads to the bottom plank-back's soil expansive force increase.

5 CONCLUSIONS

1. Pile-back's expansive force distribution with depth is K type, it's rule compare to Coulomb's soil pressure exist greater differences, at meantime, according to the soil moisture meter value and the pile-back's soil pressure distribution rule with depth, can get the expansive force's atmosphere influenced depth is 3 m.
2. Plank-back's measure soil pressure's distribution with depth appears C type. At meantime, the environmental factor and the difference of different pile's displacement lead to different plank-back's soil pressure values and the distribution rules with depth are not completely integrated. Because of the soil arching effect, the bottom plank-back's soil pressure appears a phenomenon which in both sides is large, the middle is small.
3. 7# 8# pile cantilever section's bending moment distribution with depth is in line with the mechanics characteristic of cantilever beam, the maximum bending moment is appear at the depth of 7 meters, the both of two are not exceed the allow bending moment which according to reinforcement design value.
4. Due to the stress concentration, which leads to pile-plank wall toe appears a degree of invagination, appears a degree of surface water flow backward, leads to this area's wall-back soil pressure increase. When make a design calculation, this area should be taken as a weak point to consider.

ACKNOWLEDGEMENTS

Thanks for the support of National Natural Science Foundation of China (No. 51278499; No. 51478484).

REFERENCES

[1] Richards B.G. Pressures on a Retaining Wall by Expansive Clay [A]. Proc of 9th Int Conf on Soil Mechanics and Foundation Engineering. Tokyo, 1977, 705–710.

[2] Li Ming, Jiang Zhongxin Qing Xiaolin. An explore on design principle of cut slope in expansive rock at Nanning-Kunming Railway [J]. The Chinese Journal of Geological Hazardand Control, 1995, 6(1): 60–69.

[3] Li Haiguang. Design of new retaining structures and project example [M]. Beijing China Communications Press, 2004.

[4] Ruan Bo, Li Liang, Liu Bao-chen, Nie Zhi-hong. Analysis on deformation monitor and tendency of XuJiaDong landslide [J]. Chinese Journal of Rock Mechanics and Engineering. 2005, 24(8), 1445–1449.

[5] Lu Jianhong, Yuan Baoyuan. Study on slope monitoring and quick feedback [J]. Journal of Hohai University, 1999, 27(6): 98–102.

[6] Chen Qiang, HanJun, Ai Kai. Monitoring and analysis of slope deformation along a speedway [J]. Chinese Journal of Rock Mechanics and Engineering, 2004, 23(2): 299–302.

[7] SHEN Qiang, Chen Cong-xin, Wang Ren, Liu Xiao-wei. Monitoring and analysis of reinforcement effect on slope anti-slide pile [J]. Chinese Journal of Rock Mechanics and Engineering, 2005, 24(6), 934–938.

[8] Xiong Zhaohui. New technology of landslide treatment-Yuan' an anti-slide piles and cut-and-cover tunnel [J]. Chinese Journal of Rock Mechanics and Engineering, 2001, 20(4): 532–537.

[9] Zhang Youliang, Feng Xiating, Fan Jianhai, et al. Study on the interaction between landslide and passive piles [J]. Chinese Journal of Rock Mechanics and Engineering, 2002, 21(6): 839–842.

Advances in Future Manufacturing Engineering – Yang (Ed.)
© 2015 Taylor & Francis Group, London, ISBN 978-1-138-02817-3

Study on dynamic character of high-speed railway subgrade in expansive soil

Xiao Yang

Department of Civil Engineering, Monash University, Melbourne, Australia
School of Civil Engineering, Central South University, Changsha, China

ABSTRACT: On the basis of the case study of the construction project of Yun-Gui high-speed railway in expansive soil area, adopting the Finite Difference software (FDM) FLAC3D to establish three-dimensional road subgrade mechanical model, the dynamic response and sensitivity factors of water-proof structure layer were studied. The analysis results show that the layer of water-proof could accelerate the attenuation of dynamic stress in subgrade, and reduced the displacement of subgrade surface. Increase the water-proof structure layer thickness and elastic modulus could reduce the dynamic displacement of subgrade and the dynamic stress of beneath the water-proof layer, but this will increase dynamic stress level of water-proof layer surface. This will enforce dynamic stress of subgrade surface when the position of water-proof layer laying down, but it would not have impacted the dynamic displacement of subgrade surface. It was recommended that the thickness of water-proof layer was not less than 15 cm for requirements of the specification. The conclusion could be used and referenced for the design, construction and study on dynamic response of high-speed railway cutting subgrade in expansive soil area.

Keywords: expansive soil; new cutting subgrade; dynamic response; numerical simulation; sensitivity analysis; high-speed railway

1 INTRODUCTION

There are many results about the dynamic responses laws and influence factors of traditional subgrade structure,[1–4] this need to further explore and improve to new subgrade structure impact on the dynamic response. Yang Guolin,[5–7] etc, based on the model test and field test method, this paper discusses the new water-proof structure layer influence on the dynamic characteristics of subgrade, but the parameters of water-proof structure layer influence on the dynamic response wasn't analyzed. The cutting subgrade project of Yun-Gui high-speed railway in expansive soil area as the research background, and the three dimensional dynamic calculation model of subgrade was established by using finite difference software FLAC3D[7] to study the dynamic response of new subgrade structure under train load compare with traditional subgrade structure, which it has some advantages such as salvation dynamic, large deformation and nonlinear problems, and the sensitivity factors of water-proof structure layer was analysis. The conclusion could be used and referenced for the design, construction and study on dynamic response of high-speed railway cutting subgrade in expansive soil area.

2 PROJECT DESCRIPTION AND NUMERICAL MODEL

Engineering geology conditions. Yun-Gui high-speed railway is the national grade I double-track railways, the design speed of 200 km/h. It has expansion road about 129.7 km, accounts for 18.3% of all, expansive soil subgrade about 23.7 km. There are weak, medium and strong expansion soils along with line, and the rainfall and evaporation are larger for in a subtropical humid monsoon climate, it has obvious seasonal characteristics of dry-wet circulation throughout the year.

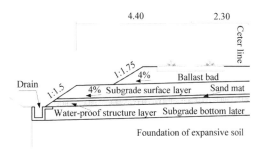

Figure 1. Sketch of new type cutting subgrade.

Cutting subgrade structure. Combining Yun-Gui high-speed railway engineering, the new type structure layer of water-proof was designed in expansive soil areas.[7] The cutting subgrade structure diagram is shown in Figure 1. The foundation is medium-weak expansive soil, and the subgrade structure from top to bottom as follows: 0.7 m subgrade surface layer (0.65 m grading macadam + 0.05 m coarse sand) + 0.2 m new water-proof structure layer (modified cement-based composite water-proof material) + 0.5 m replacing layer of A and B group + foundation of expansive.

Three-dimensional numerical model. The numerical calculation models of new and traditional subgrade structure were established for more effective judgment the new type water-proof structure layer effect on dynamic response of subgrade. At the same time, A half of the subgrade numerical calculation model is set up, considering symmetry of subgrade structure and convenience of numerical calculation.

Model one: new subgrade structure, laying new water-proof structure layer, and its structural layer in order: 0.7 m subgrade surface layer + 0.2 m water-proof structure layer + 0.5 m replacing layer, as shown in Figure 2.

Model two: traditional subgrade structure, not laying new water-proof structure layer, and its structural layer in order: 0.7 m subgrade surface layer + 0.7 m replacing layer, as shown in Figure 3.

In the dynamic calculation, it caused great influence to the calculation results by wave reflects. Through set up the bigger model to reduce the influence of reflection wave, this will greatly increase the computation time. FLAC3D software provides two boundaries of static boundary and free field boundary, in order to effectively solve the contradiction, the two boundaries can effectively prevent the reflection of wave. The static boundary was selected; consider the convenience of numerical calculation. The damp forms are adopted in the numerical calculation by the low frequency of train load. All of calculation factors of model are shown in Table 1.

Calculation and Simulation of the vibratory load of train. Train load is a very complicated problem, involving train car body, track and road subgrade coupling effect. With the developing of computer

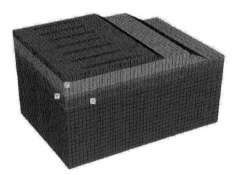

Figure 2. Finite difference model of new type cutting subgrade.

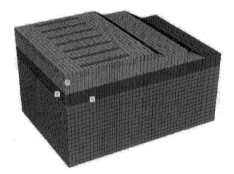

Figure 3. Finite difference model of traditional cutting subgrade.

Table 1. Material factors of model.

Name	Constitutive model	Thickness (m)	Elastic modulus (MPa)	Poisson	Density (kg/m³)	Cohesion (kPa)	Friction (°)	Damping
III type sleeper	Elastic model	0.15	30000	0.17	2300	–	–	0.063
Ballast	M-C model	0.35	200	0.25	2200	0	38	0.094
Subgrade surface of graded broken stone	M-C model	0.70	190	0.27	2140	15.0	19	0.088
Replacing layer	M-C model	0.50	110	0.32	1950	20.0	17	0.110
Foundation of weak expansive soil	M-C model	3.50	67	0.33	1860	41.4	11.2	0.088
Water-proof structure layer	Elastic model	0.20	1000	0.25	1900	–	–	0.157

performance, numerical analysis method was become a key means for research on dynamic of train-track-roadsubgrade. The train load equation was proposed to the corresponding high, medium and low frequency in reference, and the train vibration load is loaded through the fish language programming in the FLAC3D.

3 THE DYNAMIC CHARACTERS OF NEW SUBGRADE STRUCTURE

Dynamic stress change laws of subgrade. The curve of subgrade dynamic stress of new cutting and traditional subgrade is show in Figure 4. The figure shows: the attenuation curve of subgrade dynamic stress along with depth is exponential distribution, and these changing laws of the traditional subgrade structure and new cutting subgrade structure are alike. At the top and bottom of water-proof structure layer, the dynamic stress values are 32.10 kPa, 23.23 kPa, while the dynamic stress values of traditional subgrade structure are 31.61 kPa, 28.16 kPa. Therefore, the attenuation rate of dynamic stress in the new type of water-proof structure layer are increased, it can to speed up attenuation of dynamic stress in subgrade by set up water-proof structure layer.

Dynamic displacement change laws of subgrade. The curve of subgrade dynamic displacement of new cutting and traditional subgrade is show in Figure 5. The figure shows: the attenuation curve of subgrade dynamic displacement is linear in above 0.7 m (water-proof structure layer location), and the attenuation curve of subgrade dynamic

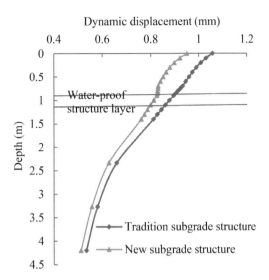

Figure 5. Curve of dynamic displacement along with depth.

displacement is exponential distribution in below 0.7 m. The dynamic displacement of the top and bottom of water-proof structure are 0.83 mm. The water-proof structure layer was occurred whole deformation on rail vibration load and the layer has distribution function and resistance deformation ability. Contrast new subgrade structure and traditional subgrade structure, the roadbed surface dynamic displacement of traditional subgrade and new subgrade are 1.06 mm and 0.95 mm. The dynamic displacement of roadbed surface is no more than 1 mm by laying new water-proof structure, and the values to meet 《TB 10621-2009 code for design of high speed railway》 (hereinafter referred to as "specifications") requirement, so that the water-proof structure layer can effectively reduce the dynamic displacement of roadbed surface.

4 ANALYSIS OF INFLUENCING FACTORS

4.1 *The influence of water-proof layer thickness*

Dynamic stress changes. The structure layer thickness are selected as follow 0.1 m, 0.15 m, 0.2 m and 0.25 m, for analysis the water-proof structure layer thickness effect on the subgrade dynamic response.

The curve of subgrade dynamic stress along with depth is show in Figure 6, when the thickness of new type water-proof structure layer was changed. The figure shows that the vertical dynamic stress along with depth are almost consistent attenuation in different thickness case. When the thickness of

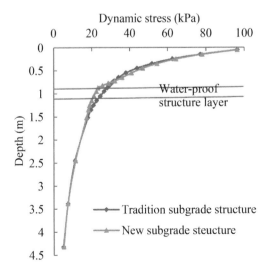

Figure 4. Curve of dynamic stress along with depth.

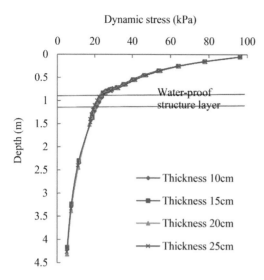

Figure 6. Curve of dynamic stress with the thickness of water-proof structure.

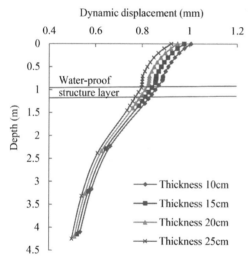

Figure 7. Curve of dynamic displacement with the thickness of water-proof structure.

water-proof layer are increased, the dynamic stress of the top of the structure layer is increased from 31.41 kPa to 32.43 kPa, but the dynamic stress of the bottom of the structure layer is decreased from 25.68 kPa to 25.68 kPa. Therefore, increase the water-proof structure layer thickness could reduce the subgrade dynamic stress of beneath the water-proof layer.

Dynamic displacement changes. The curve of subgrade dynamic displacement along with depth is show in Figure 7, when the thickness of new

type water-proof structure layer was changed. The figure shows that new type of water-proof structure layer thickness change does not affect the attenuation law of dynamic displacement along the depth, but as the water-proof structure layer thickness increases, the dynamic displacement of the attenuation curves shift to the left, dynamic displacement value is reduced, and the new type of water-proof structure layer displacement value within the scope of basic unchanged; With water-proof structure layer thickness increase, the top of the subgrade surface layer and water-proof and drainage structure dynamic displacement of the top and bottom are linear decreases, and shows that laying semi-rigid structure water-proof layer can reduce to a certain extent the dynamic displacement of foundation. When the water-proof structure layer thickness is equal to 10 cm, on the surface of the roadsubgrade dynamic displacement is 1.01 mm, not meet the specification, to the requirement of the subgrade surface displacement is less than 1 mm, so the laid of water-proof structure layer thickness of not less than 15 cm.

4.2 The influence of water-proof layer elastic modulus

Dynamic stress changes. Take water-proof structure layer modulus of 0.1 GPa and 1 GPa, respectively under the condition of 5 GPa and 10 GPa, the vertical dynamic stress analysis of subgrade depth attenuation curve is shown in Figure 8. The vertical dynamic stress attenuation mainly occurs within the water-proof structure layer (that is

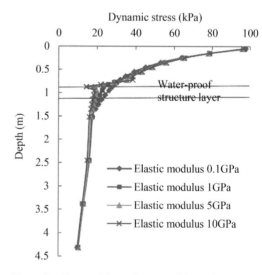

Figure 8. Curve of dynamic stress with elastic modulus of water-proof structure.

under the roadsubgrade surface 0.7 m to 0.9 m) with elastic modulus of water-proof structure layer. The modulus of elasticity, the greater the attenuation of dynamic stress in the structure of water-proof structure layer more quickly, but on the surface of the roof water-proof and drainage structure dynamic stress slightly increased; When elastic modulus at more than 5 GPa, the vertical dynamic stress curve appeared inflection point on the underside structure water-proof structure layer, bottom layer and the water-proof and drainage structure under dynamic stress change difference is smaller and smaller. Indicates that elastic modulus can improve the water-proof structure layer in a certain extent, reduce the underlying subgrade dynamic stress, but increase the structure water-proof structure layer on the surface of the top level of dynamic stress.

Dynamic displacement changes. The curve of sub-grade dynamic displacement is shown in Figure 9, when the elastic modulus of new type water-proof structure layer was changed. The figure shows that with the increase elastic modulus of water-proof structure layer, dynamic curves of displacement and water-proof structure layer thickness changes caused by the dynamic displacement change rule is consistent; When the modulus of structure layer of water-proof and drainage is greater than or equal to 1 GPa, the subgrade surface displacement is less than 1 mm.

The influence of water-proof layer laying position. In the process of the actual subgrade construction, new type of water-proof and drainage structures can be laid at any location in the foundation soil layer. But from the point of view of safety and economy, there are two major aspects as follow.

First, the structure of water-proof structure layer laying position control of side ditch drainage design elevation and its position from the deeper the subgrade surface, earthwork excavated volume and the greater the slope support of quantities will be; Second, the structure of water-proof structure layer position from the nearer the subgrade surface, may cause the failure of structure layer of water-proof and drainage, the greater the probability will affect its durability. So it need to the determination of reasonable water-proof structure layer laying position. Here will choose water-proof structure layer laying in the middle or bottom of the subgrade surface layer, and in middle layer and bottom of fill layer, have four typical location, the water-proof structure layer laying position change on the influence of dynamic response.

Dynamic stress changes. The curve of subgrade dynamic stress is show in Figure 10, when the position of new type water-proof structure layer was changed. The figure shows that as the water-proof structure layer laying down position, subgrade attenuation curves of dynamic stress in 0.35 m to 1.4 m depth within the scope of change is more obvious, at the same depth of dynamic stress increases gradually. Four kinds of laying position corresponding to the top of the subgrade surface of dynamic stress were 95.17 kPa, 96.09 kPa, 98.13 kPa, 99.7 kPa. When the water-proof and drainage laying in the middle of the replacement layer and structure layer bottom, subgrade almost consistent attenuation curves of dynamic stress, to show that the structure of water-proof structure layer location in filling layer change don't effect subgrade dynamic stress.

Dynamic displacement changes. The curve of subgrade dynamic displacement is show in Figure 11,

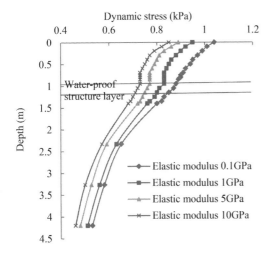

Figure 9.   Curve of dynamic stress with elastic modulus of water-proof structure.

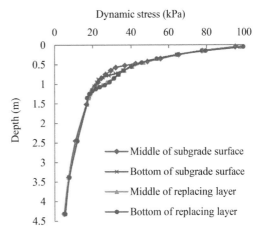

Figure 10.   Curve of dynamic stress with the position of water-proof structure.

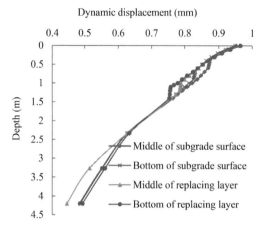

Figure 11. Curve of dynamic displacement with the position of water-proof structure.

when the position of new type water-proof structure layer was changed. The figure shows that as the water-proof structure layer laying down position, subgrade dynamic displacement attenuation curve in 0.35 m to 1.4 m depth within the scope of change is more obvious, at the same depth of dynamic displacement decreases; Change the position of water-proof structure layer laying dynamic displacement of the subgrade surface, its value is about 0.96 mm.

5 CONCLUSIONS AND SUGGESTIONS

1. The new type of subgrade structure compared with the traditional subgrade structure, the water-proof structure layer could accelerate the attenuation of dynamic stress, and reduced the dynamic displacement of subgrade surface.
2. Increase the water-proof structure layer thickness and elastic modulus could reduce dynamic displacement and dynamic stress of subgrade of beneath the water-proof layer, but this will increase dynamic stress level of water-proof layer surface. It will enforced dynamic stress of subgrade surface when the position of water-proof layer laying down, but it would not have impacted the dynamic displacement of subgrade surface. It was recommended that the thickness of water-proof layer was not less than 15 cm for requirement of the specification what the dynamic displacement of subgrade surface is less than 1 mm.

3. Based on analysis of sensitivity factors of the water-proof structure layer show that the laying position of water-proof structure layer that affects it greatly is the dynamic stress of subgrade surface, the thickness of water-proof structure layer that affects it greatly is the dynamic stress of foundation surface, the elastic modulus of water-proof structure layer that affects it greatly is the dynamic displacement of subgrade surface. Consider the effects of factors of water-proof layer to dynamic responses of subgrade, it was the optimal solution, as follow the laying thickness of 20 cm, the elastic modulus of 1 GPa, the water-proof structure the position selection at the bottom of subgrade surface.

ACKNOWLEDGMENT

Foundation item: National Natural Science Foundation of China (51478484, 51278499).

REFERENCES

[1] Liang Bo, Luo Hong, Sun Chang-xin. Simulated Study on Vibration Load of High Speed Railway [J]. Journal of the China Railway Society, 2006, 28(4): 89–94.
[2] Hu Y.F., Garamg E., Prühs H., et al. Evaluation on dynamic stability of railway subgrade under traffic loading [J]. Geotechnical Engineering, 2003, 26(1): 42–56.
[3] Pedro A.C., Rui C., Antonio S.C., et al. Influence of soil non-linearity on the dynamic response of high-speed railway track [J]. Soil Dynamics and Earthquake Engineering, 2010, 30(4): 221–235.
[4] Kong Xianghui, Jiang Guanlu, Li Anhong, et al. Analysis of Dynamic Characteristics of Railway Subgrade Based on Three dimensional Numerical Simulation [J]. Journal of Southwest Jiaotong University, 2014, 49(3): 406–411.
[5] Wang Liangliang, Yang Guolin Model test on effects of subgrade dynamic characteristics by using semi-rigid water-proof structure layer [J]. Journal of Central South University (Science and Technology). 2013, 44(10): 4244–4250.
[6] Yang Guolin, Wang Liangliang, Fang Yihe, etc. In-situ test on dynamic characteristics of cutting subgrade with different water-proof layers along YUN-GUI high-speed railway [J]. Chinese Journal of Rock Mechanics and Engineering, 2014, 33(8): 1672–1678.
[7] Wang Liang-liang, Yang Guo-lin, Fang Yi-he, et al. In-situ tests on dynamic character of fully-enclosed cutting subgrade of high-speed railways in expansive soileareas [J]. Chinese Journal of Geotechnical Engineering, 2014, 36(4): 640–645.

CNC machining and machinery

Advances in Future Manufacturing Engineering – Yang (Ed.)
© 2015 Taylor & Francis Group, London, ISBN 978-1-138-02817-3

Longitudinal control law design and flight test for a small UAV based on Eigenstructure Assignment

Xiaobo Qu, Weiguo Zhang & Fuyu Cai
School of Automation, Northwestern Polytechnical University, Xi'an, China

ABSTRACT: Focusing on the longitudinal control system design, a miniature unmanned aerial vehicle is used for research in this work. The basic approach of Eigenstructure Assignment (EA) is studied. The feedforward gain and error integral control schemes, which can eliminate tracking errors, are proposed and discussed. Using the linear model trimmed at typical cruise conditions, the longitudinal control law is designed based on the two control schemes. Six-degree-of-freedom nonlinear simulation and flight testing are presented for comparison. The results show that the controller designed with EA demonstrates good control performance, and longitudinal flying qualities are improved.

Keywords: eigenstructure assignment; feedforward gain; error integral; nonlinear simulation; flight testing

1 INTRODUCTION

Eigenstructure Assignment (EA) is developed from 1960s and has been proved as an excellent method for incorporating classical specifications on damping and mode decoupling into a modern multivariable control framework [1,2]. The assignment of eigenvalues and the assignment of eigenvectors are consisted in the main idea of EA, in which the stability of response is decided by eigenvalues and eigenvectors decide the component of different modes in the output of system [3,4]. Decoupling control and expectations dynamic performance of the closed-loop system can be realized with the appropriate closed-loop eigenvalues and eigenvectors. Comparing with other modern control approach, the eigenstructure assignment method has been widely used in the aerospace sector, especially in the design of aircraft flight control systems for assigning the dynamics of multivariable systems [5,6]. Andry et al. applied eigenstructure assignment to the design of a constant gain output feedback flight control system of the L-1011 aircraft, and proposed a choice for the desired eigenvectors for the lateral dynamics based upon mode decoupling [7]. EA approach is also adopted in the inner loop of the airbus A320 lateral control system in order to improve the dutch roll and spiral modes and to put restrictions on the lateral perturbation for the plane in an event of engine failure [8,9]. EA is very suitable for the low order systems design. However, as the order of the system increased, the zero and pole placement of system becomes increasingly difficult. Therefore, in actual flight control system design, the low-order model is often adopted to design controller based on EA method. When the satisfactory results are obtained, classic or modern control algorithms can be used for the outer loop controller design.

In current work, EA in multivariable linear systems via output feedback is investigated. A small unmanned aerial vehicle is developed for the study. The longitudinal control system is design based on *Feedforward Gain (FG)* and *Error Integration (EI)* control scheme of EA. The full Six-Degree-of-Freedom (6DOF) *nonlinear simulation* and actual *flight testing* are also presented in this paper. Simulation and flight testing results show that the satisfactory control performance is obtained with the two control schemes, while achieving all of the design requirements.

2 UAV SPECIFICATION

A small unmanned aircraft named Cessna-140 is developed for the current study, as shown in Figure 1. The length of wingspan is 2.77 m, and the maximum take-off gross weight reaches to 10 kg with a 3 kg payload. The cruise velocity is 28 m/s at altitude of 1000 m. A miniature four-stock methanol engine is installed on the front of the body, which generates approximately 5 kg thrust with a 2-blades fixed-pitch propeller. The maximum thrust-to-weight-ratio of the whole vehicle reaches to 0.5. A miniature flight control system (STA3x) and a digital data link are installed in the center of fuselage. Flight testing data are

Figure 1.   Cessna-140 UAV.

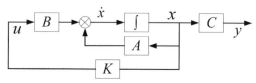

Figure 2.   State feedback control block diagram.

transmitted through the antenna mount on the back of the aircraft.

The basic longitudinal and lateral aerodynamic characteristics of the UAV and control derivative of the control surfaces are studied via Computational Fluid Dynamics (CFD) simulation, which is playing an increasingly important role in the aerodynamic design of aircraft and reduce the wind tunnel test outlay and design cycle time [10]. The moments of inertia are output by CATIA software based on the physical aspects of every component contained in the aircraft. With these data, the 6DOF nonlinear model was built in MATLAB software. After trimming and linearization, the longitudinal linear model of the UAV is obtained in the form of state-space equation [11].

During the actual flight testing, the feedback signals that EA based longitudinal control system required are received from the sensors equipped on the airplane. The angle of attack signal is measured directly by the weather vane sensors mounted on the nose of the airframe. The pitch rate and pitch attitude angle signals are obtained from the MEMS integrated in STA3x. The airspeed and altitude are obtained from the pressure sensor by measuring static pressure and dynamic pressure during flight.

3   EIGENSTRUCTURE ASSIGNMENT

In the theory of eigenstructure assignment, the eigenvalue focus on the improvement of the system dynamic response, while the eigenvector is for the decoupling of dynamic response. The desired control effect can be achieved via the eigenvalues and the corresponding eigenvectors assigned together. Consider the linear multivariable continuous system with the state-space description:

$$\begin{cases} \dot{x} = Ax + Bu \\ y = Cx \end{cases} \tag{1}$$

where $A \in R^{n \times n}, B \in R^{n \times m}, C \in R^{p \times n}$, n is state variable number, p is output number which can be observed, m is output number, $rank(B) = m$ and $rank(C) = p$.

The mathematical description of the eigenstructure assignment can be summarized as follows: For the given self-conjugate scalar sets $\{\lambda_i^d\}$ and corresponding self-conjugate-dimensional vector set $\{v_i^d\}$, to find a $m \times n$ dimensional real matrix K, so that the eigenvalues of $A + BK$ is the self-conjugate vector $\{\lambda_i^d\}$, and the corresponding eigenvectors are $\{v_i^d\}$ $(i = 1, \cdots n)$.

The structure of state feedback control schematics is shown in Figure 2. The polarity of the feedback signal is determined by the symbol of each element in the feedback gain matrix, which can be calculated via $u = Kx$.

In Moore [2], it was indicated that if the system is controllable, and each eigenvectors satisfied the following qualifications:

1. $\{v_i\}_{i=1}^n$ is a linearly independent vectors set in complex domain C^n
2. if $\lambda_i = \lambda_j^*$, then $v_i = v_j^*$
3. $v_i \in span\{N_{\lambda_i}\}$

And then the feedback gain matrix $K = [-M_{\lambda_i}z_i \quad -M_{\lambda_i}z_i \quad \cdots \quad -M_{\lambda_i}z_i][v_1 \quad v_2 \quad \dots \quad v_n]^{-1}$, if $rank B = m$, the feedback gain matrix exists and is unique.

In general, a desired eigenvector v_i^d will not reside in the prescribed subspace and hence cannot be achieved. Instead, a "best possible" choice for an achievable eigenvector is made [5]. This best possible eigenvector is the projection of v_i^d onto the subspace spanned by the columns of $(\lambda_i I - A)^{-1}B$.

4   EA BASED CONTROL SCHEME

EA can improve the stability and the dynamic response of the system. A certain tracking error usually exists in the system when the eigenstructure is assigned. Common approach is to bring in feedforward gain or error integral in order to eliminate the error. Considering the actual flight test requirement, two typical EA control schemes are described in this article.

4.1   Feedforward gain scheme

By introducing a gain in front of the control loop, the Feedforward Gain (FG) scheme does not change the roots distribution of the system if the

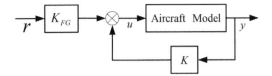

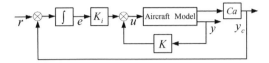

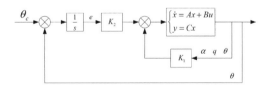

Figure 3. Feedforward gain control scheme.

Figure 4. Error integral control scheme.

closed-loop gain from input to output equals to 1. And the purpose of eliminating the tracking error can be successfully achieved. The block diagram of FG control scheme is shown in Figure 3.

4.2 Error integral scheme

In the Error Integrator (EI) scheme, an integral part of the tracking error is introduced, which is expanded to the state-space equation of the original system as a new state. The impact to the other states are reduce as much as possible using decoupling, thus guaranteeing the same root distribution and eliminating the steady state error. The block diagram of EI control scheme is shown in Figure 4.

5 LONGITUDINAL CONTROL LAW DESIGN

For the selected eigenvalues, by just specifying the eigenvectors required for modal decoupling elements corresponding to zero, the decoupling can be obtained to meet the design requirement [5,7]. Consider the aircraft modeled by the linear time invariant system with the state-space description:

$$\begin{bmatrix} \dot{V} \\ \dot{\alpha} \\ \dot{q} \\ \dot{\theta} \end{bmatrix} = \begin{bmatrix} -0.07 & 5.79 & 0 & -9.81 \\ -0.25 & -6.29 & 0.96 & 0 \\ 0 & -70.07 & -6.63 & 0 \\ 0 & 0 & 1 & 0 \end{bmatrix} \begin{bmatrix} V \\ \alpha \\ q \\ \theta \end{bmatrix}$$

$$+ \begin{bmatrix} 0.26 \\ -0.46 \\ -117.11 \\ 0 \end{bmatrix} [\delta_e] \qquad (2)$$

where V is airspeed, α is angle of attack, and q is pitch rate and θ is pitch angle. The control variable δ_e represents the deflection of elevator.

The longitudinal model consists of two modes, the short-period mode whose eigenvalues is $-6.45 \pm 8.27i$; long-period mode whose eigenvalues are $-0.03 \pm 0.4i$. Obviously, the short-period and long-period of the aircraft are stable.

5.1 Using FG scheme

The eigenvalues of the angle of attack pitch rate and pitch angle are assigned together with the corresponding eigenvectors, and then feedforward gains are computed in order to assure the closed-loop gain of pitch angle equals to 1. The controller structure is shown in Figure 5.

Where, k_q, k_α, k_θ are the gain parameters of the pitch rate, angle of attack and the pitch angle respectively. For the pitch angle control, the feedforward gain $K_{FG} = -k_\theta$. Considering the flying qualities as described in the MIL-STD-1797A [12],

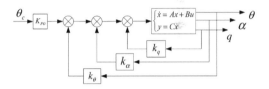

Figure 5. Pitch angle control with FG scheme.

Table 1. Feedforward gain eigenvectors.

	α	q	θ
α	1	×	0
q	×	1	0
θ	0	0	1

Figure 6. Pitch angle control with EI scheme.

Table 2. Error integral eigenvectors.

	α	q	θ	e
α	1	×	0	0
q	×	1	0	0
θ	0	0	1	×
e	0	0	×	1

the desired poles of short-period are placed at $-3 \pm 2i$. In this case, the selected eigenvectors are listed in Table 1. Where, × represents an arbitrary value.

5.2 Using EI scheme

In error integral scheme the pitch angle error is integrated as a new state of the system and expanded to the system matrix. The controller structure is shown in Figure 6.
where $\begin{bmatrix} K_1 & K_2 \end{bmatrix}$ are the gain parameters of $\begin{bmatrix} (\alpha & q & \theta) & e \end{bmatrix}$ respectively. The corresponding eigenvectors are listed in Table 2.

For the two-control scheme, the eigenvalues of angle of attack, pitch rate and pitch angle can be selected with the same value, while for the eigenvalues of the error term is based on the actual selection requirements. Generally, farther the value from the imaginary axis, faster the response will be.

6 SIMULATION AND FLIGHT TESTING

In nonlinear simulation and flight testing, the lateral control loop and airspeed hold loop are designed using classical control algorithm. Figure 7 is the flight testing photo with pitch angle hold at trimming value ($\theta_{trim} = 1.4°$). The snapshot in Figure 8 represents the pitch angle control mode ($\theta_{ref} = 15°$) at the altitude of 1000 m and velocity of 100 km/h.

For the comparison of simulation and flight testing with FG & EI scheme, the same initial conditions are set at altitude of 1000 m (airport is 850 m above sea level, relative flight altitude approximate 150 m). Given a step input command $\theta_{ref} = 15°$, 6DOF nonlinear simulation and flight testing results are presented in the same graph for comparison, as shown in Figure 9(a) and Figure 9(b), respectively. In the two figures, the flight test data were recorded real time with 50 Hz sampling rate by the onboard flight control computer.

Figure 8. Pitch angle control ($\theta_{ref} = 15°$).

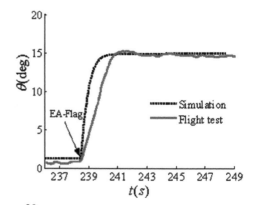

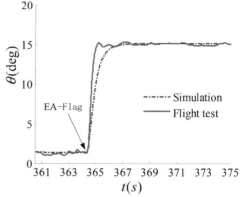

Figure 9. Nonlinear simulation and flight testing test results comparison. (a) FG control scheme; (b) EI control scheme.

Nonlinear simulation illustrates that by using the FG and EI based on EA approach, the response of pitch angle, θ, tracks the desired value very well within 2 seconds. Due to modeling error, in actual flight testing, obvious overshoot and tracking error are existed for the FG control scheme. While for the error integral control scheme, the pitch angle response rapidly tracks the command.

Figure 7. Level flight testing ($\theta_{ref} = \theta_{trim}$).

7 SUMMARY

The basic eigenstructure assignment control method is introduced, and the feedforward gain and error integral schemes based on eigenstructure assignment are studied. The longitudinal control law is designed using the linear model of the Cessna-140 UAV with the two control schemes. Results of nonlinear simulation and flight testing are presented for the comparison. Results show that the longitudinal control system designed with feedforward gain and error integral control scheme shows good controlling effects and dynamic performance. The lateral control law design and flight testing work based on EA approach will be carried out in recent time.

ACKNOWLEDGMENT

The current work was supported by the National Natural Science Foundation of China (No. 61374032) and the Natural Science Foundation of Shaanxi Province (No. 2013JQ8026).

CORRESPONDING AUTHOR

The corresponding author named Xiaobo Qu, a Ph.D. student at school of automation, Northwestern Polytechnical University. His area of research includes new concept unmanned air vehicle design, flight control system design and application. E-mail address: quxiaobo4665838@163.com. Mobile phone: (+86)18729298043.

REFERENCES

[1] Lockheed Martin Skunk works. Application of Multivariable Control Theory to Aircraft Control Laws [Z]. 1996.

[2] Moore B.C. On the flexibility offered by state feedback in multivariable systems beyond closed-loop eigenvalue assignment[J]. IEEE transactions on Automatic Control, 1976, 21: 689–692.

[3] Sobel K.M., Yu W., Lallman F.J. Eigenstructure assignment with gain suppression using eigenvalue and eigenvector derivatives[J]. Journal of Guidance, Control and Dynamics, 1990, 13(6).

[4] Maximilian Merkel, Marcus Heinrich Gojny and Udo B. Carl. Enhanced eigenstructure assignment for aeroelastic control application[J]. Aerospace Science and Technology, Vol.8, 2004,8(9): 533–543.

[5] Lester F. Faleiro. The Application of Eigenstructure Assignment to the Design of Flight Control Systems[D]. PhD thesis, Department Aeronautical and Automotive Engineering and Transpotr Studies. Loughborough University, November 1998.

[6] D. Hee Seob Kim, Youdan Kim. Partial Eigenstructure Assignment Algorithm in Flight Control System Design [J]. IEEE Transactions on Aero space and Electronic System, 1999, 35 (4): 1 403–1409.

[7] Andry A.N., Shapiro Jr E.Y., Chung J.C. Eigenstructure Assignment for Linear Systems[J]. IEEE TransAerosp and Elec-tron, 1983, 19 (5): 711–729.

[8] Erik Karlsson, Stephan Myschik, and Florian Holzapfel. Eigenstructure Assignment and Robustness Inprovement Using a Gradient-Based Method[J]. Advances in Aerospace Guidance, Navigation and Control, 2011: I: 67–77.

[9] Duvala, G. Clercb0, Y. Le Gorrec. Induction machine control using robust eigenstructure assignment[J]. Control Engineering Practice, 2006,14:29–43.

[10] QU Xiao-bo, LI Zhong-jian, Zhang Wei-guo. Application of Model Predictive Control in the Flight Control System of Flying-Wing UAV. Measurement & Control Technology, 2013, 32(3): 57–61. (in Chinese).

[11] Brian L. Stevens, Frank L Lewis. Aircraft Control and Simulation [Z]. 1992.

[12] Standard M. Flying Qualities of Piloted Aircraft[R]. MIL-STD-1797A, Department of Defense, 1990.

Advances in Future Manufacturing Engineering – Yang (Ed.)
© *2015 Taylor & Francis Group, London, ISBN 978-1-138-02817-3*

An optimal design on the oil conservation switch for the "6 × 4" train for traction

Lin Deng

Vehicle Engineering, Chongqing College of Electronic Engineering, Chongqing, China

ABSTRACT: This research refers to the loading situation of the train for traction and conducts the simulation calculation through the AVL-CRUISE software to produce the data regarding the total oil consumption of the 6-mode test approach which would be put under the further examination. Furthermore, this research selects the relevant power range of the engine in pursuit of the lower oil consumption of the train for traction.

Keywords: train for traction; economic efficiency; simulation calculation; cruise

1 INTRODUCTION

As a result of the heavy investment into the highways in China, the transportation and logistics performance have been significantly improved. In the year of 2010, China's production and sales volumes of the heavy trucks reached 1 million in which the train for traction occupied over half of the sales volume. Meanwhile, Chinese government confirmed the "Energy Consumption and Emission Reduction" as its national policy. In 2010, China's dependency on the import of crude oil still exceeded 50%, which suggested that the energy security should be considered as one of the significant concerns of the country. Currently, the diesel oil consumption for automobiles accounts for 35% to 45% of the total figure. Prediction reveals that the shortage of the oil for automobile will mount to 130 million tons by 2020.

This research employs the example of "6 × 4" train for traction and uses the simulation calculation through the AVL-CRUISE over the train. Also, the optimal design will be conducted over the multi-power range of certain engine. In different loading situations, the best power range will be selected for matching in pursuit of the lower oil consumption per centum mile.

2 ESTABLISHMENT OF THE SIMULATION MODEL AND THE PARAMETERS OF THE TRAIN

The software of AVL-CRUISE is generally used to simulate the driving force, oil consumption, emission performance and braking system of the automobiles. AVL-CRUISE will simulate assembling and establishing the train model through putting the automobile body, engine, clutch, gearbox, differential mechanism, tyres and driving control system together. The train, quality and axle load parameters, as well as the sliding experiment data, will be employed to calculate the driving force and oil efficiency through the full-load accelerating performance and the steady-state driving performance.

2.1 Establishing the simulation model

This research will simulate the total oil consumption for different engine powers of 6 × 4 trains for traction through the cycle computation.

According to the detailed layout of the 6 × 4 train for traction, the AVL-CRUISE will establish the simulation analysis model (Fig. 1) and input the major data regarding the driving force and economic efficiency into each module of the model.

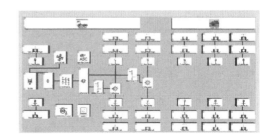

Figure 1. Simulation model for 6x4 train for traction.

Table 1. Parameters of the train and the principal assembly.

Engine power/ PS	Transmission	Principal reduction ratio	Tyre	Wheel base	Total quality	Total transmission efficiency
380	12 JS160TA	4.11	12.00R20	3225+1350	49000	0.862

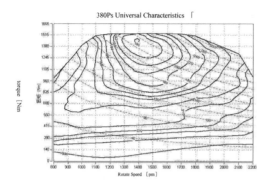

Figure 2. Universal characteristics of the 380Ps engine.

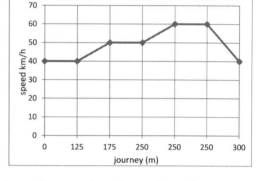

Figure 3. Cyclic six-mode operation mode.

2.2 Parameters of the train

The major focus of this research is to optimize the engine power range and therefore produce the simulated driving force and economic efficiency of the train for traction in empty load. Meanwhile, there are various and complicated factors affecting the driving force and oil consumption performance. This research tends to simplify the simulation calculation and therefore employs the parameters of the train listed in Table 1 and universal parameters of the engine without the oil conservation switch in Figure 2.

3 CYCLIC OPERATION MODE FOR SIMULATION EXPERIMENT

Currently, there are various methods to test the oil consumption of the heavy automobiles. This research employs the six-mode method to do the simulation calculation for the train for traction, illustrated by Figure 3. The detailed information is depicted in Table 2.

4 OPTIMAL DESIGN FOR THE OIL CONSERVATION SWITCH

4.1 Design for the oil conservation switch

According to the loading situation of the drivers, there are three categories of operation modes for the 6 × 4 train for traction, including the empty load, half load and full load. As for 380Ps engine, the change of the parameters of the engine can match the power, maximal torque and maximal rotate speed with the related load, illustrated in Table 3.

Thus, the universal features of the engine with three kinds of power range are depicted in Figure 4.

4.2 Simulation analysis on the oil conservation switch

The shortage of measures towards the oil consumption of heavy automobiles leads this research to employ the six-mode method for calculation. AVL-CRUISE is able to calculate the total oil consumption through CYCLE calculation task. When the train is in empty or half load condition, the total oil consumptions based on the power range of the engine are shown in Table 4.

Table 3 suggested that when the train is in empty load, the total oil consumption of the 250Ps engine was 1.89 L/100 km smaller than that of the 380Ps one without the oil conservation switch. The oil conservation rate reaches 7.6%. In half load condition, the oil consumption of the 310Ps engine was 1.56 L/100 km smaller than that of the 380Ps one, approximately 4.2% as the oil conservation rate.

5 SUMMARY

This research configured the oil conservation switch, increased the power ranges of the engine,

Table 2. Data for cyclic six-mode operation mode.

No	Operation status (km/h)	Journey (m)	Accumulative journey (m)	Time (s)	Accelerated speed (m/s^2)
1	40	125	125	11.3	–
2	40~50	175	300	14	0.2
3	50	250	550	18	
4	50~60	250	800	16.3	0.17
5	60	250	1050	15	–
6	60~40	300	1350	21.6	0.26

Table 3. Load power for the oil conservation switch.

Load condition	Total quality for the train (Kg)	Power of engine	Maxima torque (Nm)	Maximal rotate speed (rpm)
Empty	18000	250	1060	2200
Half	34000	310	1280	2200
Full	49000	380	1540	2200

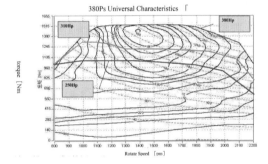

Figure 4. Universal features of multiple power range.

Table 4. Total oil consumption of each power range under different loads.

Load condition	Total quality (Kg)	Power of the engine (Ps)	Total oil consumption per centum mile (L/100 km)
Empty	18000	250	25.03
		380	23.14
Half	34000	310	36.78
		380	35.22

matched different loads, and utilized the AVL-CRUISE for simulation analysis in order to confirm the powers of the train related to the empty, half and full load conditions. Typically, when the train is in empty load, the oil conservation rate was up to 7.6%, which will further reduce users' transportation cost.

ACKNOWLEDGMENT

Fund is from science and technology research project of Chongqing Education Commission. (No. KJ1402908).

REFERENCES

[1] Qin D.T., Wei H.B., Duan Z.H., Chen S.J. Multiple-objective Real-time Optimum Control Strategy for Fuel Consumption and Emission of Full Hybrid Electric Vehicle. Jouranl of mechanical engineering. Vol. 45(2012), p.78.
[2] Song L.F., Zheng G.S., Wang J.X. Performance Optimization and Cycle Simulation Calculation for Engine B231. Chinese internal combustion engine engineering. Vol. 32(2006), p.80.
[3] Zhang J.H., Ma Z.Y., Zhang G.C., He Z.P., Ma L. Optimization Matching for the Dynamical Characteristic and the Fuel Economic Characteristic of the Truck's Engine. Transactions of the chinese society for agricultural machinery. Vol. 41(2010), p. 40.
[4] Fu Y.Z., Cui Y. Study Fuel-Consumption of Some Engine. Equipment Manufacturing Technology. Vol.1(2014), p. 38.
[5] Zu L., Wang J.S., Jiang Y.C., Guo D. Detection of Air-fuel Ratio with Gasoline Engine Fuel Consumption Study. Journal of changchun university of science and technology. Vol. 31(2008), p. 43.

Advances in Future Manufacturing Engineering – Yang (Ed.)
© 2015 Taylor & Francis Group, London, ISBN 978-1-138-02817-3

An analysis of different types of standard basketball stands

Chuansheng He
College of Physical Education, Qinzhou University, Qinzhou, China

Weidong He
Sports and Health Science Academy, Guangxi University for Nationalities, Nanning, Guangxi, China

ABSTRACT: This paper studied six different types of standard basketball stands in China and their appropriate products with the literature consultation, mathematical statistics, expert investigation method and logic analysis and other research methods. Take the "Jinling" basketball from the China Jiangsu Jinling Sports Equipment Co., Ltd as an example. The results indicate that 1) Six different types of standard basketball stands mainly include the fixed single-armed basketball stand, the movable single-armed basketball stand, the petrel-type buried basketball stand, the movable box-type basketball stand, the manual hydraulic basketball stand and the electro-hydraulic basketball stand. All of them have their own dimensions, material and technical parameters. 2) These standard basketball stands have the main bar, so the explosion-proof tempered glass backboard can be installed on them. 3) They are safe and secure and could be applied in various fields; among them, the electro-hydraulic basketball is the most representative. 4) The single-armed, manual hydraulic and electro-hydraulic basketball stands have their own structural principles. The 24-second timer of the electro-hydraulic basketball stand could improve its performance and integration. Therefore, the role played by different types of basketball stand as well as the value of the corresponding products could be embodied.

Keywords: basketball; standard; basketball stand; 24-second timer; type; value

1 INTRODUCTION

Basketball stand, as the important equipment in basketball court, develops along with basketball sports, from the initial fixed single-armed basketball stand to the single-armed basketball stand, movable single-armed basketball stand, petrel-type buried basketball stand, movable box-type basketball stand and electro-hydraulic basketball stand. Currently, the movable box-type basketball stands are frequently used in the outdoors, while for the indoor gymnasium, the electro-hydraulic basketball stands are adopted, but in the high-end sports venues, the advanced electro-hydraulic basketball stands and related configurations are necessary to organize the high-level basketball tournament.

The research group reviewed the relevant literature of basketball stands and found that the Jiangsu Jinling Sports Equipment Co., Ltd [1], (US) • Bob Murray, [2], Zhang Xiuhua [3], Tang Kairong [4], Wu Jibin [5], Bai Lan [6], Li Yuanwei [7], Huang Guoqiang [8], Yu Daifeng and Gao Jian [9], Dong Jie [10] and Geoff Manaugh and other institutions and scholars studied the training, teaching, athletics, fitness and facilities of the basketball sports and made a number of academic achievement. However, few of them focus on the basketball stand. Therefore, our research group

intends to sort the development process of basketball stands, especially the electro-hydraulic stands and its related configuration. This could provide a theoretical basis for the application of basketball sports in the universities as well as the introduction of basketball stand and the corresponding facilities in the sports venues and stadiums.

2 THE DIMENSIONS, MATERIALS AND TECHNICAL PARAMETERS OF DIFFERENT TYPES OF STANDARD BASKETBALL STANDS

2.1 *The fixed single-armed basketball stand*

1. Dimensions: the national standard on the length of the arm is 2.25 m and the height is 3.05 m.
2. The materials and technical parameters: the fixed arm, buried type, the diameter of 18×18 cm, the thickness of arm of 4 mm, the shape of square tube, the outrigger of 1.8 m/ 2.25 m. For the surface treatment, the electrostatic spraying is adopted and the basic configuration includes tri-point support, smooth glass backboard and elastic ring.
3. For the fixed single-armed basketball stand, the backboard is made of the high-strength tempered glass enclosed by the aluminum alloy

perimeter frame and the dimension is 180×105 cm; the basketball stand is seamlessly welded with high-hardness steel and has the arm length of 2.25 m, which could avoid the collision caused by the inertia in limited distance. The embedded parts are the concrete sets with the dimensions of $60 \times 60 \times 100$ cm, which have better stability; electrostatic spray process is adopted to decorate the surface, coupled with the basket with high elasticity used for the higher-ranking games as shown in Figure 1.

2.2 The movable single-armed basketball stand

1. The main bar of basketball stand: the high-quality square steel tube with the diameter of 150 mm.
2. The movable bottom box of the basketball stand is made of the high-quality steel plate with the dimensions of 30 cm (height) × 100 cm (width) × 180 cm (length); heavy loads are placed in the box to ensure the stability.
3. The pull rods between the main bar and backboard, two round steel bars with high quality could form three triangles with the main bar, which could ensure the stability of the backboard.
4. The pull rods between the main bar and the base, two round steel bars with high quality, could form three triangles with the main bar, which could ensure the stability of the whole stand.
5. The ring is made of the high-quality steel with an inner diameter of 450 mm, which meets the international standards.
6. The standard height of the ring to the floor is 3.05 m. The default color is blue, but customization based on user requirements is feasible as shown in Figure 2.

2.3 The petrel-type buried basketball stand

1. Dimensions: backboard has the length of 1800 mm and width of 1050 mm; for the col-

umn, the tube size is 150×150 mm and the wall thickness is 3 mm, which all meet the industry standard.
2. Material: the backboard is made of fiber-reinforced polymer, generally including the fiber reinforced polymer and high-quality tempered glass as shown in Figure 3.

2.4 The movable box-type basketball stand

The box of this type of stand is forged and jointed by several steel plates with the thickness of 2–3 mml. The column is made of the high-quality 130×130 mm square tubes and the outriggers with the length of 1.8 m; the distance between the ring and the ground is 3.05 m, the lower edge of the backboard is 2.9 m off the ground; for the back stay, use the 4×4 cm thick round tubes to form a trapezoid; there are four fixed pull rods with stable tension; CNC bending is conducted for the rods to enhance the stability; the dimension of the box base is 180×100 cm; the spray process is adopted for the appearance.

The standard diameter of the basketball is 24.6 cm, so the ring, made of the solid iron materials, has the diameter of inner edge ranging from 45 cm to 45.7 cm. The net should be white

Figure 2.   The movable single-armed basketball stand.

Figure 1.   The fixed single-armed basketball stand.

Figure 3.   The petrel-type buried basketball stand.

Figure 4.   The movable box-type basketball stand.

Figure 6.   The electro-hydraulic basketball stand.

Figure 5.   The manual hydraulic basketball stand.

Figure 7.   The high-level hydraulic basketball stand.

and hanging on the ring, which aims to generate slight resistance for the ball into the ring. The net should have 12 meshes with the length varying from 40 cm to 45 cm as shown in Figure 4.

2.5   *The electro-hydraulic basketball stand*

The hydraulic basketball stand could ascend and descend according to the set height through the hydraulic lifting system. It is the latest product produced under the FIBA Standard.

1. Specifications: the length of the base is 2.2 m, the width is 1.2 m and the wingspan is 3.25 m.
2. Material: the backboard is made of high-strength tempered glass.
3. The hydraulic basketball stand includes the manual type, electric type and high-level type, FIBA certified. The length of their bases and outrigger is respectively 2.5 m 3.35 m. The most appropriate venue should have an area of 37.4 m at least. They are equipped with the high-

strength safety glass and the 24-second timer. In addition, the electric type and high-level type of basketball stand have the microcomputer control and light ban as show in Figures 5–7.

3   CHARACTERISTICS, APPLICATION RANGE AND ADVANTAGES OF DIFFERENT TYPES OF STANDARD BASKETBALL STANDS

3.1   *Fixed single-armed stand*

It maintains good stability in the main bar, with a safe and secure backstop, suitable for a variety of professional basketball games, and mainly used in the outdoor venue.

3.2   *Movable single-armed stand*

It has solid main bar and delivers good performance, which is movable, adjustable and flexible with a safe and secure backstop. With a wide range of applications, it is capable of applying to a variety of professional basketball games, convenient and flexible to move to outdoor or indoor venues.

3.3 *Petrel-type buried stand*

It possesses stable main bar, featuring two-way shooting practice area, with a safe and secure backstop. Saving space for setting up two poles, it is more suitable for use in schools, mostly installed in outdoor venues.

3.4 *Movable stand with counterweight box*

It is equipped with a safe and secure backstop and four fixed bars, which have stable pull and CNC bending. The movable bottom case is made from high quality steel plate, in which weight materials are loaded to ensure great stability. With wide applications, it is perfect for a variety of professional basketball games and can be installed in outdoor and indoor venues.

All above four kinds of stands have explosion-proof tempered glass backboards installed, which are highly transparent, impact resistant, and with elegant appearance and good safety performance. All the surfaces are processed with electrostatic spray, characterized by the advantages of anti-corrosion and anti-oxidation, free from losing varnish and fading, etc. In addition, they are all equipped with high-elasticity baskets.

3.5 *Electro-hydraulic stand*

With a safe and secure backstop, it combines manual, electric, and remote rise of wheels, which is convenient, flexible, and durable. The high-strength tempered glass backboard is marked by great impact resistance, high transparency, durability, flat surface and security, etc., which is supplied with a high-elasticity basket. It mainly applies to teaching, training and competition for basketball professionals, which is installed in the stadium and especially for higher professional basketball games.

4 STRUCTURAL PRINCIPLES OF SINGLE-ARMED STAND, MANUAL HYDRAULIC STAND AND ELECTRO-HYDRAULIC STAND

4.1 *Single-armed stand*

Customized embedded iron parts are provided, and they are buried before the installation of the rack, which will not be ready for overall installation until the maintenance period is over.

4.2 *Manual hydraulic stand*

Taking YLJ-SB manual hydraulic basketball stand as an example, it is equipped with a lifting system,

the telescopic mechanism of traveling wheels, and a hydraulic system. The stand has a four-bar linkage, whose movement is controlled by the hydraulic control cylinder. It thus controls the vertical movement of the post and the expanding and contracting of traveling wheels.

Both the YLJ-SB-type manual hydraulic basketball stand and telescopic oil passage of traveling wheels have the hydraulic self-locking oil passage, so that the stand can be stopped in any position. Setting up a single-way throttle valve in the descending oil passage of the stand, we can adjust check the size of the throttle to control the speed of the basket lowering.

4.3 *The electro-hydraulic stand*

YLJ-5 electric hydraulic basketball stand, used as an example, has the lifting system, the telescopic mechanism of traveling wheels, electrical equipment and hydraulic systems.

The vertical movement of the stand adopts the four-bar linkage. Hooked up to 220 V, 50 HZ single-phase power, it will start working to drive the oil pump, which then conducts the process of feature conversion by the electrical control system. After that, telescopic movement occurs to the cylinder, thus vertically moving the post and traveling wheels. Please refer to Figure 5: YLJ-5 electro-hydraulic basketball, Figure 6: YLJ-5 electro-hydraulic basketball, and Figure 7: YLJ-5 electro-hydraulic basketball.

5 THE 24-SECOND TIMER OF THE ELECTRO-HYDRAULIC BASKETBALL STAND

Take the Jinling ZJS-3A 24-second timer for an example and it is the only designated product of CBA (Chinese Basketball Association).

5.1 *Features*

This timer is designed according to the latest international basketball competition rules. The imported light-emitting diode with high-brightness is used as the display source, for it has long service life and could save electricity. Two displays share with one clock source, so it could synchronize with the large screen and TV relay, coupled with the easy-maintenance, which make it the ideal equipment for basketball games as shown in Figure 8.

5.2 *The main functions*

1. Display the game time and the count clock could achieve the timing range of 0–99"59'.
2. Preset the countdown in seconds and suspend it arbitrarily. The time clock could be accurate to 0.1 second and the timer expires along with the red light signal and beep sounds.

Figure 8. ZJS-3A 24-second timer

24-secondController+ ZJS mind controller.

Figure 9. 24-second controller; ZJS mind controller.

3. The 24-second controller can be used to preset time with the countdown mode. It could be paused and reset arbitrarily. The timer expires along with the red light signal and beep sounds, and the game is suspended automatically.
4. When the match time is suspended, the 24-second timer also stops automatically. When the game continues, the 24-second controller would be started manually. For the 24-second violation, the counting of the game time will continue.
5. This timer provides two serial ports, so the game time and 24-second controller could synchronize with the computer and TV relay.

5.3 Technical specifications

1. The size of display: $74 \times 56 \times 10$ cm
2. The light tube: $\Phi 5$ mm highlighted white touched with red, white touched with green light tube
3. Voltage: 220 V $\pm$ 10%
4. Power: 100 W

5.4 Application methods

1. Connect the connecting lines of the seven-core aerial linker at both ends of communication cable between the display and controller, and the two controllers. Then plug the power supply into a 220 V electrical outlet and turn on the power switch.
2. To preset the game time and 24-second timer, use the button "Time +", "Time −" and "24 seconds +", "24 seconds −" and then press the time saving button and the "reset" button. In this way, the preset times and the 24-second controller could be shown on the display.
3. When the game starts, press the "time/pause" button on the controller to count down the game time; press the "time/pause" button of the 24-second controller to start the countdown of 24 seconds.
4. To pause the game time, press the "time/pause" button on the controller; to pause the 24 seconds, press the "time/pause" button on the 24-second controller.

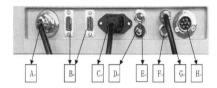

Figure 10. A: Signal line interface of the 24-second timer; B: 232 serial ports; C: Power interface; D: Fuse; E: Power switch; F: External signal interface; G: 24-second console interface; H: Signal line interface of the 24-second timer.

Figure 11. The display of 24-second timer.

5. If only the 24 seconds are needed, the game time should not be shown on the display; if the 24 seconds are not needed during the game, just press the button of "clean 24 seconds". Please do not press the key of "clear screen" during the game, otherwise, all the data on the screen will disappear as shown in Figures 9–11.

6 CONCLUSION

6.1 Different types of standard basketball stands have their own corresponding dimensions, materials and technical parameters

These six types of standard basketball stands— the single-armed basketball stand, the movable

261

single-armed basketball stand, and the petrel-type buried basketball stand, the movable box-type basketball stand, the manual hydraulic basketball stand and electro-hydraulic basketball stand—have their own dimensions, materials and technical parameters. Besides, they are easy to install, and could meet user needs according to different objects and sites.

6.2 Different stands with a variety of features and benefits and a wide application range

These standard basketball stands have the stable main bars, safe and secure stands and explosion-proof tempered glass backboard. They have high transparency, strong impact resistance, elegant appearance, good security performance and other characteristics, coupled with anti-rust property, resistance to oxidation, fastness, etc., and high elasticity used for the higher-ranking games. With a safe and secure backstop, the electro-hydraulic stand combines manual, electric, and remote rise of wheels, which is convenient, flexible, and durable. The high-strength tempered glass backboard is marked by great impact resistance, high transparency, durability, flat surface and security, etc., which is supplied with a high-elasticity basket. It mainly applies to teaching, training and competition for basketball professionals, which is installed in the stadium and especially for higher professional basketball games.

6.3 Main structural principles of single-armed, manual hydraulic and electro-hydraulic stands

Single-armed basketball stand is characterized by pre-burying embedded main lever parts, and then conducting the overall installation of the basketball board and basket; manual hydraulic stand adopts a four-bar linkage, whose movement is controlled by the hydraulic control cylinder. It thus controls the vertical movement of the post and the expanding and contracting of traveling wheels; the vertical movement of the electro-hydraulic stand adopts the four-bar linkage. Hooked up to 220 V, 50 HZ single-phase power, it will start working to drive the oil pump, which then conducts the process of feature conversion by the electrical control system. After that, telescopic movement occurs to the cylinder, thus vertically moving the post and traveling wheels.

6.4 Twenty-four-second timer provided for the electro-hydraulic basketball stand improves its performance, unifies with it, and reflects its value

It features decent appearance, vivid color, clear screen, good synchronization performance and easy maintenance with targeted primary function. The light signal and beep sound components cooperate perfectly, reaching the design requirements of latest international basketball competition rules. As it is convenient and easy to use, it improves the performance of the basketball stand and unifies with it, thus reflecting the value of the corresponding products.

PROJECTS

Project of science and technology research projects in Guangxi Universities in 2014: teamwork research on the participation in traditional sports events of ASEAN international students-A Case Study of Guangxi University for Nationalities (Project Grant No: YB2014090), initial results

About the first author: He Chuansheng (1977-), male, zhuang, guangxi qinzhou, lecturer, director of basketball teaching and research section. Research direction: physical education and national traditional sports culture.

About corresponding author: He Weidong (1968-), male, Han nationality, from Chong zuo, Guangxi, professor, graduate instructors. Research interests: sports industry, traditional sports and physical education.

Postcode: 530006; Tel: (086)07713260115; Mobile: (086)13978857035; E-mail: yy3384@163.com.

REFERENCES

[1] Sports Equipment of China Jiangsu Jinling Sports Equipment Co., Ltd [EB/OL]. [Nov. 10th, 2014]. China Jiangsu Jinling Sports Equipment Co., Ltd
[2] Edited by (US) • Bob Murray, translated by Tan Zhenbin, *Selection of Basketball Training Program.*[M]. People's Sports Publishing House. 2004: 1–12.
[3] Zhang Xiuhua, Liu Yulin. *Basketball Tactical system* [M]. People's Sports Publishing House, 2005: 1–15.
[4] Tang Kairong, Peng Juan. Research on the Application of Electronic and Movable Devices in the Basketball Passing Training [J]. Science and Technology Information. 2011, (30): 294.
[5] Wu Jibin. Inspiration of Basketball Stand Effect on the Education and Teaching in Vocational Schools [J]. Scientific Public, 2008, (10): 52.
[6] Bai Lan. *Hydraulic Basketball* [J]. Modern Marketing, 2006, (01): 22.
[7] Li Yuanwei. Science and Sports—on the Development of Sports Science and Technology in the New Century [J]. China Sport Science and Technology 2002, (06): 3–8.
[8] Huang Guoqiang. *Pros and Cons of High-tech Sports Equipment* [J]. China Modern Educational Equipment. 2011, (10): 43–44.
[9] Yu Daifeng, Gao Jianhe. *20-year Review and Prospect of Sports Instrument* [J]. China Sports Science. 2003, (03): 61–63.
[10] Dong Jie. The Development of Technological Progress and Contemporary Sports [J]. Sports and Science. 2002, (03): 3–5.

Advances in Future Manufacturing Engineering – Yang (Ed.)
© 2015 Taylor & Francis Group, London, ISBN 978-1-138-02817-3

Microstructural evolution of AlSi9Mg alloy prepared by the electromagnetic stirring

Chenyang Zhang, Shengdun Zhao, Yongfei Wang & Peng Dong
Xi'an Jiaotong University, Xi'an, Shaanxi Province, China

ABSTRACT: Different stirring frequencies were introduced into the electromagnetic stirring process as well as the different stirring currents to prepare semisolid AlSi9Mg alloy. The microstructural evolutions of semisolid AlSi9Mg alloy under the as-stirred condition and the partial remelting condition were investigated. The results show that the microstructures of the as-stirred alloy change from the predominantly non-dendritic solid particles to fine equiaxed grains as both the stirring current and frequency increases, though there are some aggregations at either high stirring current or high stirring frequency due to their relatively strong EMF. Furthermore, the higher stirring current and frequency lead to finer and more spherical solid particles during the partial remelting of the as-stirred AlSi9Mg alloy.

Keywords: aluminum alloy; semisolid; electromagnetic stirring; microstructure

1 INTRODUCTION

The key factor to Semisolid Metal (SSM) processing is to obtain non-dendritic microstructure with fine and spheroidal particles uniformly dispersed in the liquid matrix [1]. Among different kinds of production technologies, the Electromagnetic Stirring (EMS) is still an ideal technology with commercial potentiality, due to its high local shear rate, high controllability and non-contamination [2]. Desired raw material prepared via this route for thixoforming process includes two important steps, i.e. EMS step and partial remelting step [3, 4], which have been paid a lot of attention. C.G. Kang et al. [5] applied horizontal EMS to control grain size of A356 alloys for various rheological forming processes and ideal semisolid microstructure was produced by optimized stirring current, stirring time and pouring temperature. The high-quality semisolid microstructure of A357 alloy was obtained by introducing Annular Electromagnetic Stirring (A-EMS) into the stirring process [6]. Wei-min Mao et al. [7] put forward that weak electromagnetic stirring and low superheat pouring was applied into the EMS process to produce the uniform and fine microstructure of semisolid A365 alloy. The two-stages of Electro-Magnetic (EM) stirring process were successfully employed to manufacture semi-solid slurries of a near eutectic Al–Si based piston alloy [8]. Il-Gab Chung et al. [9] employed vacuum pump punch to EMS system in the EMS process. The results indicated that at the optimal vacuum pressure the presence of the

porosities greatly reduced to improve the quality of the slurry. Vanluu Dao et al. [10] proposed the EMS Combined with Mechanical Vibration (EMSCMV) process, in which the material was stirred electromagnetically and vibrated mechanically to avoid the heterogeneity of the microstructure due to skin effect. Although many methods have been applied to the EMS step, the application of stirring frequency is less presented, which also affects the spheroidization and distribution of globules during EMS [11].

The present paper aims to employ different stirring current and frequency to the EMS process for the production of semisolid AlSi9Mg alloy. The microstructural evolution of semisolid alloy in two different conditions: as-stirred and partial remelting, namely the as-stirred material is partial remelted for different soaking time, is investigated.

2 EXPERIMENTAL PROCEDURE

Figure 1 shows the schematic diagram of the EMS system. A frequency converter is used to supply the desired current and frequency. The temperature of crucible is held at 350°C to prevent the formation of initial solidification shell [8]. Damage to the stator from heat radiating from the melt and resistive heater is prevented by insulation layer and cooling system. Commercial AlSi9Mg alloy was used as experimental material, with its chemical composition and thermal characters shown in Table 1. Ingots of AlSi9Mg alloy was

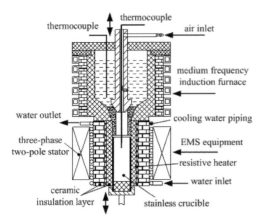

Figure 1. The schematic diagram of the EMS system.

Table 1. Chemical compositions (wt. %) and thermal characters of AlSi9Mg.

Si	Mg	Mn	Cu	Fe	Zn	Al	T_L [°C]	T_S [°C]
9.23	0.25	0.3	0.05	0.2	0.2	Bal.	595	555

melted in the medium frequency induction furnace and held at $625 \pm 5°C$ for 30 min. After being degassed for 20 min, the melt was poured into the crucible (60 mm in diameter and 170 mm in length). Then the EMS treatment was conducted. After EMS process, these billets were quenched in water. This is designated as the as-stirred condition. During partial remelting, the samples were isothermal heat treated at 580 °C for 5–30 min under a protective atmosphere. Once removed from the furnace, the samples were immediately quenched in water. This is designated as the partial remelting condition. For the microstructural observation, etching operation was carried out with an aqueous solution of 0.5% HF after all samples were grounded and polished. The average Globule Size (GLS) and the Shape Factor (SF) were determined based on the measurements of at least 50 randomly selected dendrites or cells in each case.

3 RESULTS AND DISCUSSION

3.1 Microstructural evolution of as-stirred AlSi9Mg alloy after different EMS treatments

Figure 2 shows the microstructure of as-stirred AlSi9Mg alloy under different stirring current, while the stirring frequency is 10 Hz. The microstructure

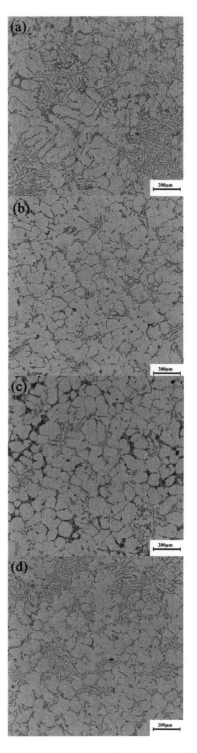

Figure 2. Microstructures of as-stirred AlSi9Mg alloy with different stirring currents: (a) 20 A, (b) 30 A, (c) 40 A, (d) 50 A.

264

of as-stirred AlSi9Mg alloy with different stirring frequency is shown in Figure 3, while the stirring current is 20 A. As the stirring current and frequency increasing, the solid particles present different degrees of fragmentation and tended to disintegrate into smaller and more spherical particles.

As shown in Figure 2(a), there are still some coarse dendritic structures maintained owing to the weaker stirring intensity as the current was 20 A. When the current is 30 A and 40 A, the dendritic are mostly crushed and presented predominantly quasi-globular solid particles evenly distributed among the eutectic (Fig. 2(b) and (c)). The solid particles at 40 A are more spherical and self-contained as the grain boundaries are much smoother and wetted well. While the stirring current is increased to 50 A, the primary α-Al phase particles collides and recombines leading to many aggregations (Fig. 2(d)). Too strong EMF generated by higher stirring current contributes to the recombination of the cracked particles. Moreover, the solid particles are distributed unevenly. When the stirring frequency is 10 Hz and 20 Hz, the microstructure has not been obviously crushed and coarse dendritic characteristic is almost maintained, as shown in Figure 3(a) and (b). With increasing the stirring frequency 30 Hz, the dendrites are obviously broken up and be detached into fine equiaxed grains as shown in Figure 2 (c). The particles have obviously undergone spheroidization when the stirring frequency increases to 40 Hz as shown in Figure 3(d). But they are coarser than the low frequency (20 Hz and 30 Hz). This may attribute to the increased heterogeneity of EMF which creates strong turbulent flow to provide enough energy for the big particles to merge the adjacent small ones and form coarser ones. If higher current and frequency applied in the EMS process, the stirring efficiency decreases. The melt can overflow outside of stirrer owing to the violent turbulent fluid. Even worse, some small pores would form by the absorption of atmospheric air through the deep vortex formed during the strong stirring and then a defect of shrinkage cavity forms which can lead to inferior slurry.

From the Figure 2 and Figure 3, it can be seen that both the stirring current and frequency have strong effect on the microstructural evolution of the as-stirred AlSi9Mg alloy. Stirring current is mainly representative of stirring intensity, while local shear rate is mainly dependent on stirring frequency. However, the impact of stirring frequency on the stirring process is nonlinear [12]. Higher frequency means the larger shear rate, which strengthens the stirring velocity of molten/semi-solid metal and heat transmission. But high

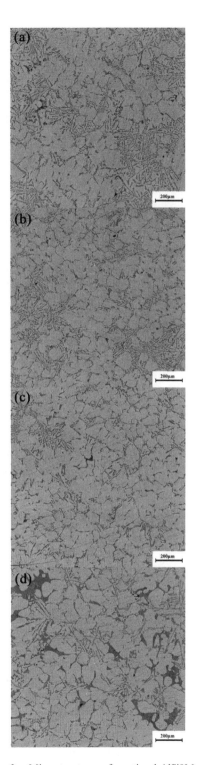

Figure 3. Microstructures of as-stirred AlSi9Mg alloy with different frequencies: (a) 10 Hz (b) 20 Hz (c) 30 Hz (d) 40 Hz.

frequency reduces the depth of penetration zone where the heterogeneity of EMF is increased, which can create highly turbulent flow of the melt. Inversely, low frequency permits more homogeneous distribution of EMF, but considerably high current will be required to have similar effects on microstructure due to its low local shear rate.

3.2 Microstructural evolution of as-stirred AlSi9 Mg alloy during partial remelting

Figure 4 shows the microstructures of as-stirred alloys with different stirring current after isothermal holding at 580 °C for 10 min. Figure 5 shows the microstructures of as-stirred alloys with different stirring frequency after isothermal holding at 580 °C for 10 min. Figure 6 shows the variations of the average particle size and shape factor for as-stirred alloys with different stirring current and frequency respectively after partial remelting and subsequent isothermal holding at 580 °C for 10 min.

As shown in Figure 4(a), when the current is 20 A, the microstructure consists of many coarse solid particles with irregular shape. Figure 4(b) shows the coexistence of rosette-shaped particles and spheroidal particles. With further increasing the stirring current, solid particles became more spheroidal gradually and uniformly distributed in a liquid matrix as shown in Figure 4(c) and (d). Such microstructures are good enough for semisolid processing. As shown in Figure 5(a), when the stirring frequency is 10 Hz, the microstructure is analogous to that when the stirring current was 20 A. As the stirring frequency increases, the microstructure attains many uniformly concomitant rosettes or globules (Fig. 5(b) and (c)). But the globules suspended in the liquid phase in Figure 5(c) are finer and much more spheroidal than that in Figure 5(b). However, when the stirring frequency reached 40 Hz, the major globule size is relatively larger than lower frequency. This phenomenon can be explained that during the partial remleting process, coalescence and coarsening mechanism may play a major role. It can be seen from Figure 6 that average particle size gradually decreases with increasing the different stirring current and frequency, especially in the initial growth stage of current and frequency. However, its particle size is slight larger when the frequency is 40 Hz, as well as when the current is 50 A, though its variation is not obvious. In addition, both increasing stirring current and frequency promote the degree of spheroidization as shown in Figure 6.

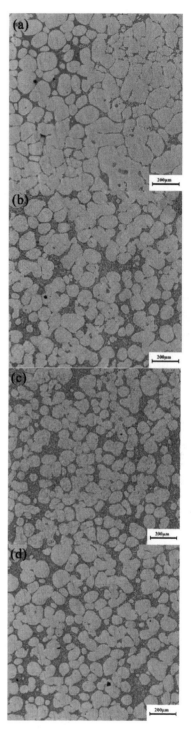

Figure 4. Microstructures of as-stirred alloys with different stirring currents after isothermal holding at 580 °C for 10 min: (a) 20 A (b) 30 A (c) 40 A (d) 50 A.

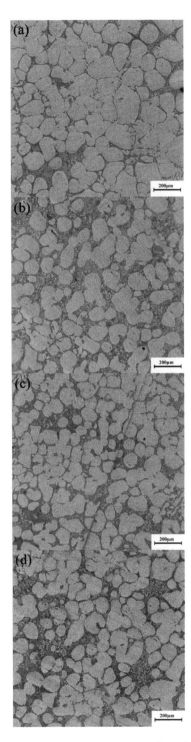

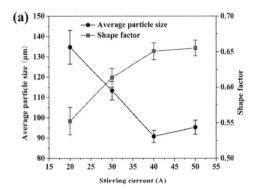

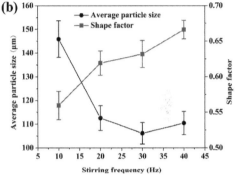

Figure 6. Variations of average particle size and shape factor for as-stirred alloys after isothermal holding at 580 °C for 10 min with stirring current (a) and stirring frequency (b).

4 CONCLUSIONS

As both the stirring current and frequency increase, the microstructures of as-stirred alloy evolved from the coarse dendritic structures to fine equiaxed grains. However, either too high stirring current or stirring frequency leads to many aggregations of solid particles and makes no obviously improvements of the microstructure. During the partial remelting process, higher stirring currents and frequencies result in the finer and more solid particles in the partial remelting microstructures of the as-stirred alloy, though the average particle size is slight larger when the frequency is 40 Hz or the current is 50 A. Meanwhile, increasing both the stirring current and frequency promote the degree of spheroidization.

ACKNOWLEDGEMENTS

The authors gratefully acknowledge the support of the National Natural Science Foundation of China

Figure 5. Microstructures of as-stirred alloys with different stirring frequencies after isothermal holding at 580 °C for 10 min: (a) 10 Hz (b) 20 Hz (c) 30 Hz (d) 40 Hz.

for key Program (Grant No. 51335009), Basic Research Priorities Program of Science andTechnology of shanxi (Grant No. 2012 JQ7032) and Science and Technology Project of Xi'an (CX1249(1)) for funding this project.

CORRESPONDING AUTHOR

Name: Chenyang Zhang; Email: zhangchenyang-1333@126.com; Mobile phone: 18792746298.

REFERENCES

[1] H.V. Atkinson: Semisolid processing of metallic materials, Mater. Sci. Tech-lond Vol. 26 (2010), p. 1401–13.

[2] S. Nafisi, D. Emadi, M.T. Shehata and R. Ghomashchi: Effects of electromagnetic stirring and superheat on the microstructural characteristics of Al–Si–Fe alloy, Mater. Sci. Eng. A Vol. 432 (2006), p. 71–83.

[3] E.J. Zoqui, M. Paes and E, J. Es-Sadiqi: Macro-and microstructure analysis of SSM A356 produced by electromagnetic stirring, Mater. Process. Technol Vol. 120 (2002), p. 365–373.

[4] E.J. Zoqui, M.T. Shehata and M. Paes: Morphological evolution of SSM A356 during partial remelting, Mater. Sci. Eng. A Vol. 325 (2002), p. 38–53.

[5] S.W. Oh, J.W. Bae and C.G. Kang: Effect of electromagnetic stirring conditions on grain size characteristic of wrought aluminum for rheo-forging, J. Mater. Eng. Perform Vol. 17 (2008), p. 57–63.

[6] G. Zhu, J. Xu, Z.F. Zhang: Annular electromagnetic stirring—a new method for the production of semisolid A357 aluminum alloy slurry, Acta, Metall, Sin Vol. 22 (2009), p. 408–414.

[7] S. Li, W.M. Mao: Preparation of semi-solid AlSi7 Mg alloy slurry with big capability, Rare Metals Vol. 29 (2010), p. 642–645.

[8] Y.S. Jang, B.H. Choi, C.P. Hong: Effects of Two Stages of Electro-Magnetic Stirring on Microstructural Evolution in Rheo-Diecasting of a Near Eutectic Al—Si Alloy, ISIJ Int Vol. 53 (2013), p. 468–475.

[9] I.G. Chung, A. Bolouri, C.G. Kang: A study on semisolid processing of A356 aluminum alloy through vacuum-assisted electromagnetic stirring, Int. J. Adv. Manuf. Tech Vol. 58 (1–4) (2012), p. 237–245.

[10] V.L. Dao, S.D. Zhao and W.J. Lin: Preparation of AlSi9Mg Alloy Semi-Solid Slurry by Electromagnetic Stirring Combined with Mechanical Vibration, Solid. State. Phenom Vol. 192 (2013), p. 398–403.

[11] C. Stelian, D. Vizman: Numerical modeling of frequency influence on the electromagnetic stirring of semiconductor melts, Cryst: Res. Technol Vol. 41 (2006), p. 645–652.

[12] F.C. Robles Hernandez and J.H. Sokolowski: Comparison among chemical and electromagnetic stirring and vibration melt treatments for Al–Si hypereutectic alloys, J Alloy Compd Vol. 426 (2006), p. 205–212.

Advances in Future Manufacturing Engineering – Yang (Ed.)
© 2015 Taylor & Francis Group, London, ISBN 978-1-138-02817-3

Study on the improvement of surface quality through ultrasonic elliptical vibration cutting

Wen Li & Fangfei Lin
College of Mechanical and Electrical Engineering, North China University of Technology, Beijing, China

Deyuan Zhang
School of Mechanical Engineering and Automation, Beijing University of Aeronautics and Astronautics, Beijing, China

ABSTRACT: The periodic impact of ultrasonic Elliptical Vibration Cutting (EVC) can form plastic deformation layer on the surface of the material, having the function of improving the surface quality of engine parts. From the angle of mechanics, this article analyzed the mechanism of ultrasonic elliptical vibration cutting ameliorates the surface quality. Then the thesis studied the superiority of the ultrasonic elliptical vibration cutting from the point of impact loads. From the perspective of kinematics and energetics ultrasonic, this paper presented the work-hardening effect of components produced by the percussive action of ultrasonic elliptical vibration cutting, and established physical model of the characteristics. Through the experiment, it was verified ultrasonic elliptical vibration cutting can improve the surface quality.

Keywords: ultrasonic elliptical vibration cutting; surface quality; surge pressure

1 INTRODUCTION

The aircraft engine has much complicated shaft parts, which are the ultra-weak stiffness currently. The machining effect makes the finished surface of the part produce residual tensile stress and compressive stress. The residual tension will accelerate the fatigue invalidation of parts. On the other side, the residual compressive stress can improve the surface properties and extend the fatigue life of parts. It not only makes the elements produces surface hardening, but also can remove the surface defect caused by machining process and offset the harmful tensile stress.

Ultrasonic elliptical vibration cutting, as a new type of metal machining technology, has a series of advantages of reducing the cutting force[1], improving the machining accuracy[2] and quality of the machined surface. In recent years, the domestic and foreign scholars focused on the research including the establishment of the ultrasonic elliptical vibration cutting system[3], the effect of ultrasonic elliptical vibration cutting on the workpiece's surface roughness and tool life[4,5], the enhancement of machining precision by ultrasonic elliptical vibration cutting[6]. Besides, the cutting process and cutting force of ultrasonic elliptical vibration cutting have been analyzed through large number of experiments[7,8]. However, studies about the effect of ultrasonic elliptical vibration cutting on the surface quality of components are few. On the basis of the theory of fracture mechanics, this paper researched the mechanism of impact load can reduce the cutting force and develop the surface quality of the machining surface, and got the conclusion that the impact load can improve hardness of processed surface. The objective of this study is to improve the quality of the machined surface by applying the ultrasonic machining technology.

2 ANALYSIS OF ULTRASONIC ELLIPTICAL VIBRATION CUTTING

Ultrasonic elliptical vibration cutting can be divided into separation type (K < 1) and non-separation type (K > 1)[9]. When the critical cutting speed of the cutting tools is greater than the linear velocity of the workpiece, which is called the separating ultrasonic elliptical vibration cutting[10]. When the critical cutting speed of the cutting tools is less than the linear velocity of the workpiece, which is defined the unseparated ultrasonic elliptical vibration cutting[11]. In the process of ultrasonic elliptical vibration cutting, cutting force has a great influence on the machining precision of parts, processing ability and tool life. Therefore, mechanical analysis of ultrasonic elliptical vibration cutting is very necessary. Figure 1 showed the stress condition of cutting an area when entering the cutting state. In the process of machining, the surface of the workpiece metal material will produce plastic deformation and elastic deformation. Part of the cutting force comes from the resisting force caused by plastic deformation and elastic deformation of the metal of cutting lay and workpiece surface layer. The other part comes from the frictional resistance between chips and rake face of the cutting tool, the machined surface and the flank surface of the cutting tool. Where a_p is depth of cut (DOC), γ_0 is rake angle.

When the cutting tool and the workpiece in a state of separation, the separation of the machined surface and flank surface of cutting tool occurs[12], as shown in Figure 2. The separation can prevent the recurrence of friction between the cutting tool and the machine surface. Therefore, the machined surface quality is under guaranteed.

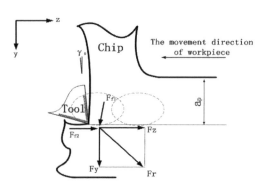

Figure 1. Schematic diagram of ultrasonic elliptical vibration cutting motion.

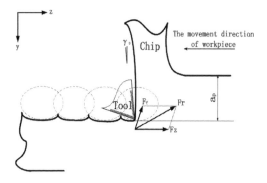

Figure 2. Schematic diagram of ultrasonic elliptical vibration cutting separation state.

With the separation of the tool and the workpiece, the friction between chips and rake face of the cutting tool turned the detrimental resistance into useful cutting force[13]. The force made the turning resistance and cutting thickness decreased, and suppressed the production of built-up edge and flake's burr at the tool nose. It is obvious that the separation reduced the surface residual stress and the processing surface roughness.

Ultrasonic elliptical vibration cutting enhanced separation characteristics. This would enable the cutting zone received sufficient lubrication cooling by the cutting fluid, thus greatly improving the tool life and cutting performance.

3 STUDY ON THE IMPROVEMENT OF SURFACE QUALITY THROUGH ULTRASONIC ELLIPTICAL VIBRATION CUTTING

3.1 Mechanism study of the effect of impact loading is useful for processing

Impact load is the load imposed on the components in a very short period or suddenly. During the ultrasonic elliptical vibration cutting, the tool tip moves in accordance with the elliptic trajectory, and does reciprocate variable speed intermittent cutting[14]. The workpiece is subjected to the effect of impact loading when the tool into the cutting condition. At this point, the strain rate of cutting materials is:

$$\dot{\varepsilon} = \frac{de}{dt} \tag{1}$$

It is observed that strain rate expresses the dependent variable per unit time. When the workpiece under impact load, strain rate is greater than $10^{-2}\,\mathrm{s}^{-1}$, the mechanical properties of metal changes significantly.

The cutting form of ultrasonic elliptical vibration cutting is a type of discontinuity; this makes the crack in the processed material or periodically appears unstable propagation phenomena or growth steadily when the disk is formed. When the workpiece moves closer to the cutting tool, the disk moves along the rake face. It can produce much compression and shear strain. If the chip formation turns into an unstable state at the same time, it is easy to develop cracks and separate chips. Usually the cutting speed is lower than 10 m/s, while the rate of crack propagation is up to 5000 ms^{-1}.

Due to the loading speed of the workpiece in cutting is very large, it is need to consider the inertial effect of material, so the traditional theory of mechanics and fracture mechanics are inappropriate. According to the dynamic fracture mechanics theory, the stress wave velocity caused by impact load is much greater than the moving speed of tool[15]. When the cutting tool does not reach the place to do the machining, the internal pressure at this point has reached the ultimate strength of the material that is enough to make the workpiece fracture damage because of the dynamic stress intensity factor. So when the cutter movement to this position, only need a small force can realize cutting.

3.2 The examination and simulation of ultrasonic elliptical vibration cutting

The tool does simple harmonic vibration in the cutting depth y direction. When the cutting tool and workpiece contact instantly, the tool impacted the workpiece with maximum speed, and then the speed rapidly reduced to zero after a quarter cycle. Due to the tool's vibration period is very short, it can use the momentum theorem at this moment, and calculate the instantaneous impact force. In the ideal cutting conditions, the trajectory equation of the cutting tool is:

$$y = A\sin 2\pi ft \qquad (2)$$

where f is the vibrational frequency, A is the amplitude.

The rate expression can be written as follows:

$$v = y' = 2\pi fA\cos 2\pi ft$$

List the momentum equation as follows:

$$mv = Ft \qquad (3)$$

where m is the total quality of cutting tool and amplitude transformer, t is the contact time of cutter and workpiece.

Cutting parameters are selected as f = 20 kHZ, A = 2 μm, m = 0.5 kg and t = 1.25 × 10^{-5} s. Finally calculated the instantaneous impact force F = 10048 N.

Selecting cylinder model (diameter of 8 cm, height of 2 cm) of high temperature nickel base alloy GH4169 as shown in Figure 3. The numerical simulation analysis of the finished surface residual stress of high temperature alloy GH4169 is mainly divided into two stages: one is the impact loading stage, with a constant force loaded on the workpiece; the other is the unloading stage.

Exerted the impact force F on the workpiece, and simulated by ANSYS. Finally got the results as shown in Figure 4.

It can be seen from the Figure 4, the stress distribution layer is uniform, and the shape

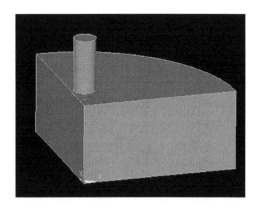

Figure 3. The schematic model of high temperature nickel base alloy GH4169.

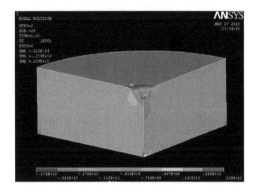

Figure 4. Residual stress diagram.

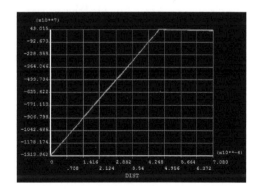

Figure 5. Residual stress curves of A = 2 μm.

is symmetrical. Figure 5 shows the distribution of residual stress of the workpiece.

The stress value is divided into positive and negative as shown in Figure 5. A negative value indicates that the residual compressive stress and

Table 1. Experimental conditions.

Experimental installation	HARDINGE precision meter lathe Micro hardness meter			
Experimental material	LY12 duralumin, A3 steel and GH4169 alloy			
Tool	Hard alloy cutter			
Cutting and vibration conditions	Vibrational frequency (f) [Hz]	Rotate speed (n) [r/min]	Depth of cut (a_p) [mm]	Feed rate (f) [mm]
	20086	150	0.2	0.5

a positive value indicates that the tensile residual stress. In the process of cutting, the surface layer has been machined produces plastic deformation in the cutting direction and the inner metal is in the condition of elastic deformation. At this moment, the inner metal is subjected to the restriction of surface layer that has already produced plastic deformation, and cannot restore to its original state, so the residual compressive stress is produced in surface layer, which is balanced by the residual tensile stress in the inner layer. Residual stress has great influence on the fatigue strength of metal surface. It can prevent the expansion of the fatigue crack, and delay the generation of fatigue damage. If residual compressive stress is produced, fatigue strength can be improved by about 50%. From the two pictures, the maximum residual stress is 278 MPa, and the depth of residual compressive stress layer is about 0.4 mm. The simulation results showed that the workpiece surface of high temperature alloy GH4169 after ultrasonic elliptical vibration cutting can produce surface hardening. However, if the thickness of residual compressive stress layer is too big to bring the work hardening, this can have a negative impact on the tool wear and cutting force.

4 EXPERIMENTAL VERIFICATION

4.1 The experimental system and conditions

Three kinds of different materials for cutting operation by using ultrasonic elliptical vibration system are selected. The experimental conditions are summarized in Table 1. The feed direction is from left to right.

4.2 The test of the influence of ultrasonic elliptical vibration cutting on the machining surface quality

Building the experiment platform, and respectively using conventional cutting and ultrasonic elliptical vibration turning to cut three different kinds of experimental materials under the same

(a) GH4169   (b) A3 Steel   (c) Duralumin LY12

Figure 6. Comparison chart of the quality of machined surface.

experimental conditions. Then measuring changes in surface hardness under different cutting conditions through microscopic hardness meter.

The surface quality of the three materials after cutting is shown in Figure 6. Smooth surface is the surface quality after ordinary cutting, and the other is the surface quality after ultrasonic elliptical vibration cutting.

The surface hardness and the residual stress of the machined workpiece have an inverse relationship. When the sample surface had tensile stress, its hardness value decreased. On the other hand, hardness value increased when the sample surface had compressive stress. The test used the indentation method with microscopic hardness meter, and applied a precise impact or static pressure on the polished surface of the sample with conical diamond indenter hardness tester to produce an indentation, and then measured its hardness.

It can be seen in the figures that hardness of the surface after cutting is different, and the hardness is bigger after ultrasonic elliptical vibration cutting, the same results with previous simulation. So it can be concluded ultrasonic elliptical vibration cutting can improve the surface hardness has been processed.

5 CONCLUSIONS

1. When the impact load produced by ultrasonic elliptical vibration cutting effects on the workpiece, it can bring a shock wave. The wave velocity is much larger than the cutting speed, so the shock wave can arrive on unprocessed surface in

advance, and change the unprocessed surface by force. Then it can be easier to cut when the knife arrives, and the cutting force is reduced.

2. The simulation and experimental verification concluded that the effect of impact load can make the processing surface hardening.

3. Under the same preparing conditions, the hardening effect is different when processing different hardness materials by ultrasonic elliptical vibration cutting.

CORRESPONDING AUTHOR

Name: Fangfei Lin, Email: linfangfei9036@163.com, Mobile phone:13683381051.

REFERENCES

[1] Ma C.X., Shamoto E. and Moriwaki, T. Suppression of burrs in turning with ultrasonic elliptical vibration cutting[J]. Int. J. Mach. Tool. Manu., 2005, 45, (11), 1295–1300.

[2] Zhou M., Ngoi B.K.A. and Yusoff M.N., Wang X.J. Tool wear and surface finish in diamond cutting of optical glass[J]. J. Mater. Process. Tech., 2006, 174, 29–33.

[3] Ma C.X. and Hu D.J. Ultrasonic elliptical vibration cutting[J]. Chin. J. Mech. Eng., 2003, 39, (12), 67–70.

[4] Wang G.L. and Duan M.L. Study on the characteristics of ultrasonic elliptical vibration turning[J]. Chin. J. Mech. Eng., 2010, 21, (4), 415–419.

[5] Li X. and Zhang D.Y. Research on micro-surface formation mechanism of ultrasonic elliptical vibration cutting[J]. Chin. J. Mech. Eng., 2009, 20, (7), 807–811.

[6] Kim G.D. and Byoung G.L. An ultrasonic elliptical vibration cutting device for micro V-groove machining: Kinematical analysis and micro V-groove machining characteristics[J]. J. Mater. Process. Tech., 2007, 190, 181–188.

[7] Li X. and Zhang D.Y. Research on experiments of single actuator driven ultrasonic elliptical vibration cutting ultra-thin wall parts[J]. Acta Aeronautica et Astronautica Sinica, 2006, 27, (4), 720–723.

[8] Ma C.X., Shamoto E. and Moriwaki, T. Study on the improvement of machining system stability by applying ultrasonic elliptical vibration cutting[J]. Acta Armamentaii, 2004, 25, (6), 752–756.

[9] Li W. and Xu M.G. Study on the unseparated ultrasonic elliptical vibration cutting force[J]. Acta Aeronautica et Astronautica Sinica, 2013, 34, 1–8.

[10] Ma C.X., Wang Y., Shamoto E. and Moriwaki. T. Influence of ultrasonic vibrated diamond tool on the critical depth of cut of brittle materials[J]. Chin. J. Mech. Eng., 2005, 41, (6), 198–202.

[11] Li X. and Zhang D.Y. Study on the unseparated ultrasonic elliptical vibration cutting[J]. J. Mech. Eng., 2010, 46, (19), 177–181.

[12] Ji Y., Li X. and Zhang D.Y. Precision ultrasonic elliptical vibration cutting[J]. Aeronautical Manu. Tech., 2005, 4, 92–95.

[13] Xiao M., Sato K. and Karube S. The effect of tool nose radius in ultrasonic vibration cutting of hard metal[J]. J. Mach. Tool. Manu., 2003, 43, 1375–1382.

[14] Lu Z.S. and Yang L. Analysis and simulation of mechanism in ultrasonic-vibration cutting based on dynamic fracture mechanics[J]. J. Harbin Inst. Tech., 2008, 40, (9), 1400–1403.

[15] Xiao H., Li X. and Zhang D.Y. Single driven ultrasonic elliptical vibration turning with natural diamond tools[J]. Aeronautical Manu. Tech., 2007, 9, 81–84.

Advances in Future Manufacturing Engineering – Yang (Ed.)
© *2015 Taylor & Francis Group, London, ISBN 978-1-138-02817-3*

FPED, a fine-grained parallel edge detection mechanism

Qi Shao
College of Computer Science and Information Engineering, Zhejiang Gongshang University, Zhejiang, China

ChunHua Ju
Gongshang University, Hangzhou, China

ABSTRACT: Edge detection is a fundamental task in image processing. The core process of edge detection is typically a serial of matrix operations that are very time-consuming. Some studies suggest accelerating the process using multi-core architectures. However, their parallel schemes are all simple and coarse-grained. The cache behavior, which is very important to the performance of multi-core systems, is not taken into account in previous schemes. This paper introduces FPED, a fine-grained parallel edge detection mechanism that can achieve better performance. By introducing fine partition of images and global balance, FPED can achieve higher rate of data reuse and reduce cache miss. Compared with previous work, FPED can achieve an average speedup of 1.4x.

Keywords: edge detection; parallel edge detection; fine-grained; cache miss; data reuse

1 INTRODUCTION

Edge detection [1] is a fundamental task in image processing and has received many research efforts in the last decades. Edge, as one of the most obvious features in images, is crucial for many image analyzing tasks such as object detection, recognition, image enhancement, image compression and so on [2]. Edge detection is always adopted as the fore-end of many image-processing tools.

Current edge detection algorithms include Sobel [3], Prewitt [4], Kirsch [5], and canny operators [6]. The core process of these edge detection algorithms is matrix operation in which some special masks (small matrixes typically sized in 2×2 or 3×3) are introduced to operate on the whole image (a large matrix contains pixel information of color or gray level). A major problem of this process is that it is typically computation intensive and will be very time-consuming when processing large images. The relatively low processing speed may constrain its usage in some real-time situations. Some work is needed to be done to speedup it.

The fast development of multi-core systems has enlightened us a new way to accelerate edge detection. Some work has proposed [7] that, the matrix operation in edge detection is very fit for parallel processing. The whole image can be divided into several parts and processed in several cores separately. During the processing, there would be almost no communication between the cores and thus the parallelism of the process will be very good. Work [7] shows that this parallel scheme can achieve near-linear speedup.

However, the parallel scheme in previous work [7] is very naive and simple. It just divides the images and distributes them to different cores. Some other important facts (such as cache contention, load-balance and so on) which are crucial for the performance of multi-core systems are not taken into account. We call this parallel scheme Coarse-grained Parallel Edge Detection (CPED).

This paper introduces FPED, a fine-grained parallel edge detection mechanism that could achieve further improvement on performance. We introduce two mechanisms to achieve fine-grained parallelism and thus achieve better performance. (1) When dividing images, we introduce fine-grained partition to reduce the cache contention among cores. (2) When processing image parts, we introduce global synchronization to balance the speeds of different cores. Experiments show that our FPED can effectively reduce cache miss and improve performance compared with previous coarse-grained parallel mechanism.

The rest of this paper is organized as follows: we give a brief introduction of a typical edge detection algorithm (Sobel) and the coarse-grained parallel version of it in section 2. In section 3, we introduce our design of the fine-grained parallel edge detection mechanism. Experimental results are shown in section 4 and we conclude in section 5.

2 SOBEL EDGE DETECTION AND COARSE-GRAINED PARALLEL SOBEL

Edge detection aims to find the points at which the image information changes sharply. Typically, the whole image information is stored in a matrix. The Sobel algorithm uses two masks (shown in Fig. 1) to convolve with the image to get the derivatives in two directions (horizontal and vertical).

The traditional computational process of Sobel algorithm is shown in Figure 2. We use the Sobel masks to convolve with every 3×3 matrix in the large matrix, which represents the whole image to get the filtered result. The filtered result matrix stores the edge information of the image.

Previous work [7] shows that the Sobel algorithm can be easily parallelized. The parallel version is shown in Figure 3. It just divides the whole image into two parts and processes the two parts in two different cores and then merges the final results.

This coarse-gained parallel scheme can make full use of the computational resources of multi-core. During the processing, different cores are operating on different data and will communicate no information with each other, making the process undisturbed and fast. However, this parallel scheme may lead to severe contention of the Last Level Cache (LLC) among cores. As all know, the last level cache is shared by all cores in multi-core architecture [8]. If the cores are all operating on totally different data,

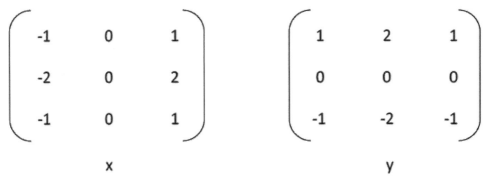

Figure 1.   The Sobel operator in two directions.

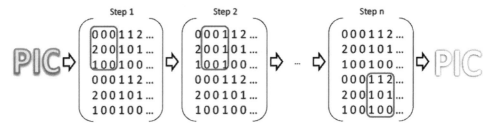

Figure 2.   The computational process of Sobel algorithm.

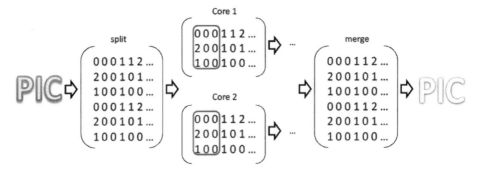

Figure 3.   Coarse-grained parallel Sobel algorithm.

they will be competing to use the last level cache to store their own data. This contention situation will lead to performance degradation and thus the coarse-grained parallel scheme is not our best choice, especially when we are processing large images.

3 FINE-GRAINED PARALLEL SOBEL ALGORITHM

As described in section 2, the main performance hazard of coarse-grained parallel Sobel is the contention for the last level cache. Thus, the main purpose of this paper is to reduce the contention. We firstly introduce fine-grained partition mechanism to increase data reuse among different cores to reduce cache contention. Additionally, we introduce global synchronization to balance the speeds of different cores to achieve further improvement.

3.1 Fine-grained partition

Unlike the coarse-grained method, that just cuts the image into two or several different parts. Our mechanism makes the partition in a fine-grained way.

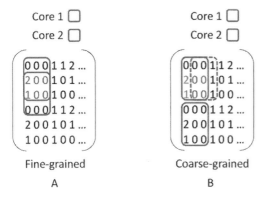

Fine-grained   Coarse-grained

A   B

Figure 4.   Comparison between fine-grained and coarse-grained partition.

The main purpose of fine-grained partition mechanism is that, when a core loads a data into the cache, the data would probably be reused by it or another core later. Thus, the last level cache is better made use of. Holding this in mind, our fine-grained partition mechanism is designed as shown in Figure 4A.

As shown in Figure 4A, in our fine-grained partition mechanism, we do not cut the image into separated parts. In order to make full use of the last level cache, we want the data used in different cores to overlap as much as possible. In Figure 4A, after *core 1* loads the data in the blue box for calculation, *core 2* can reuse the 6 data (in red) when *core 2* loads its data in the red box.

As for the coarse-grained parallel mechanism shown in Figure 4B, there is no data overlap between the two cores and thus the data reuse is relatively poor. For *core 1,* after it loads the data in the blue box, it can reuse the 4 data (in red) in the next step (shown in the dotted blue box). However, the 2 data (in green) will have no chance to be reused by any core and will probably be evicted from the cache before *core 1* gets back to the second line. Thus compared with our fine-grained parallel mechanism, the coarse-gained mechanism will theoretically lost 1/3 of data reuse. The lost 1/3 data will have to be re-cached from main memory when they are accessed again.

3.2 Global balance

The cores in multi-core systems may fluctuate on performance due to the undeterministic scheduling. Thus given equal amount of tasks, different cores may finish in different time. As shown in Figure 4A, if *core 1* runs much faster than *core 2,* then some data loaded and cached by *core 1* may not have the change to be reused by *core 2* before they are evicted. In order to avoid this problem, we introduce global synchronizations to control the steps and make the cores in the same pace.

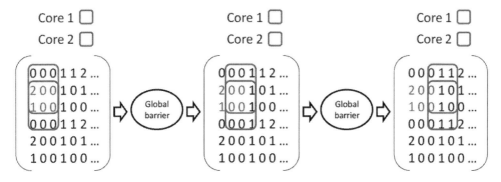

Figue 5.   Global balance.

277

Our global balance mechanism is simple and shown in Figure 5. In order to keep all the cores in the same pace, we introduce a global barrier after every processing step. This mechanism can ensure the largest data reuse and make all the cores finish their task at the same time.

4 EXPERIMENTAL RESULTS

Previous section introduces that our fine-grained parallel mechanism has higher rate of data reuse compared with traditional coarse-grained mechanism and thus would achieve better performance. In this section, we mainly test the performance of the proposed mechanism in different workloads.

The tested mechanisms are the *base,* which is the sequential version of Sobel algorithm; the *CPED,* which is the coarse-grained parallel version; and the FPED, which is the fine-grained parallel version proposed in this paper. Our benchmarks are three pictures of different sizes (1k × 1k, 2k × 2k, 3k × 3k). All the pictures just have the information of grey level.

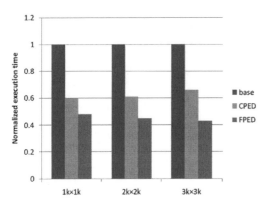

Figure 6.   Results on 2 cores.

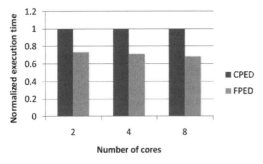

Figure 7.   Results on different number of cores.

Figure 6 shows the experimental results when running on two cores. CPED and FPED both achieve a near-linear speedup compared with the sequential version. Moreover, as the increasing scale of the workloads (from 1k × 1k to 3k × 3k), our FPED performs increasingly better than CPED. The average speedup of FPED over CPED is 1.38x.

Figure 7 shows the experimental results when running on different number of cores on workload 2k × 2k. We can see as the number of cores increases, the speedup of FPED is not obvious when compared with CPED. This is mainly because the last level cache is shared by all cores. The rate of data reuse will not be better when it come to 4 or more cores. Overall, our FPED can achieve an average speed-up of 1.4x.

5 CONCLUSIONS

This paper introduces FPED, a fine-grained parallel edge detection mechanism that can achieve fast edge detection. By introducing fine partition of the image matrix and global balance that can keep all cores in the same pace, FPED achieves better data reuse and thus achieves better performance. Compared with traditional coarse-grained parallel edge detection, FPED can achieve an average speed-up of 1.4x.

ACKNOWLEDGMENTS

This project supported by The National Key Technology R&D Program of China (Grant 2012BAI34B01-5), Major Project for Science and Technology Plan of Hangzhou (No. 20122511A26), Natural Science Foundation of Zhejiang Province (Grant LY14F020002), Key Research Institutes of Social Sciences and Humanities Ministry of Education (Grant 14 JJD630011).

CORRESPONDING AUTHOR

Name: Qi Shao, Email: shaoqitony@gmail.com, Tel: +8618768122385.

REFERENCES

[1] Ziou D., Tabbone S. Edge detection techniques-an overview[J]. Pattern Recognition And Image Analysis C/C Of Raspoznavaniye Obrazov I Analiz Izobrazhenii, 1998, 8: 537–559.

[2] Kumar T., Sahoo G. A novel method of edge detection using cellular automata[J]. International Journal of Computer Applications, 2010, 9(4): 0975–8887.

[3] Perra C., Massidda F., Giusto D.D. Image blockiness evaluation based on Sobel operator[C]//ICIP (1). 2005: 389–392.

[4] Dong W., Shisheng Z. Color image recognition method based on the prewitt operator[C]//Computer Science and Software Engineering, 2008 International Conference on. IEEE, 2008, 6: 170–173.

[5] Wang L., Wong T.T, Heng P.A., et al. Template-matching approach to edge detection of volume data[C]//Medical Imaging and Augmented Reality, 2001. Proceedings. International Workshop on. IEEE, 2001: 286–291.

[6] Vianna C.J., De Souza E.A. Combining Marr and Canny operators to detect edges[C]//Microwave and Optoelectronics Conference, 2003. IMOC 2003. Proceedings of the 2003 SBMO/IEEE MTT-S International. IEEE, 2003, 2: 695–699.

[7] Khalid N.E.A., Ahmad S.A., Noor N.M., et al. Parallel approach of Sobel edge detector on multi-core platform[J]. International Journal of Computers and Communications, 2011, 5(4): 236–244.

[8] Jaleel A., Mattina M., Jacob B. Last Level Cache (LLC) performance of data mining workloads on a CMP-a case study of parallel bioinformatics workloads[C]//High-Performance Computer Architecture, 2006. The Twelfth International Symposium on. IEEE, 2006: 88–98.

Advances in Future Manufacturing Engineering – Yang (Ed.)
© *2015 Taylor & Francis Group, London, ISBN 978-1-138-02817-3*

Mechanical analysis of RuT300 steel by using modified 3D RVE model

Linfeng Sun
Department of Mechanical Engineering, University of Michigan, Ann Arbor, USA

Ridong Liao
School of Mechanical Engineering, Beijing Institute of Technology, Beijing, China

ABSTRACT: The complex behavior and failure of mechanism of casting material which contains varieties of defect in the range from macroscopic to microscopic, has made multi-scale modeling a necessity when determining the response of the specific structure under different loading conditions. In this work, a more practical RVE, which can be considered as the fundamental of semi-concurrent multi-scale method, has been established by using Python code. Related elastic and plastic analysis of the RVE has also been operated. The results have proved that mean stress value of RVE will decrease with the enlarging of the volume fraction of voids when loading the same strain value. Meanwhile, the failure extent of the RVE has also been revealed.

Keywords: FEM RVE PBC multi-scale

1 INTRODUCTION

Due to the limitation of casting technique, conditions and material property, micro voids and other defects are inevitably and randomly existed among the whole structure. As a typical foundry structure which has complicated shape, cylinder head always undertake high thermo-mechanical loads. There is no doubt that the condensed and large-area appeared casting defects will apparently influence the material's mechanical performance. However, microscopic voids which are sometimes being neglected could also become a disturbing part of the structure's life period, even though these micro-defects has the characters of uncontrollable, hard-detectable and disperse distribution. How to incorporate the effects of micro-defects on the overall macroscopic behavior has become more and more serious to the engineering designers.

Over the last decades of years, mainly three categories of multi-scale methods (shown in Fig. 1) have been developed: concurrent, semi-concurrent and sequential modeling.

Sequential modeling methods contain standard homogenization strategies which can calculate the macroscopic material properties through the microstructure and constituents properties[1][2][3][4]. Unfortunately, they can only provide initial elastic properties instead of saving the microstructure information after the process of homogenization. Concurrent modeling methods utilize domain-decomposing in order to directly couple the micro-structure information at a homogenized macroscopic domain or limited-size regions of interest[5][6][7]. Apparently, when the interest region becoming larger or more complicated (i.e. crack growth), more computation time and parallel computing sources will be needed, even the adaptive re-meshing strategy.

On the other hand, semi-concurrent modeling methods can avoid the difficulties mentioned above by bonding each integration point at macroscopic model with a unique unit microscopic cell Boundary Value Problem (BVP). The process of information exchange between macro model and micro structure could be operated by

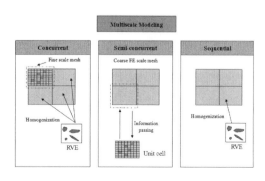

Figure 1. Categories of multi-scale modelling.

User Material Defined Subroutine (UMAT), and the related model files are established by Python codes. The proposed work aims to develop a methodology which can reveal the interacted relationship between the micro voids and macro material strength performance.

2 MULTI-SCALE MATHEMATICAL FRAMEWORK

In semi-concurrent modeling methods, two main components—homogenization and localization—should be considered. Homogenization means the determination of macroscopic data from the Representative Volume Element (RVE) to a related global integration point; on the contrary, localization means the information transformation from the macro scale of homogenized area to the micro scale of RVE. As one of the most impressive achievements in the 20th century, FECM[8][9] (Finite Element Computational Micromechanics) has the advantage of acquiring the stress and strain field of inclusions under different scales in order to reflect the material's macro response characteristics. For instance, quantitative description of the micro structure parameters' (i.e. voids shape, size, distribution and volume fraction) impact on the macro mechanical performance[10].

2.1 Representative volume element

As is shown in Figure 2 (a), the casting void defects could still exist even after the optimized scheme of casting process[11]. So, during the computation, the casting material which contains void defects is assumed to be macroscopically homogeneous but microscopically inhomogeneous, which means the structure is consist of matrix and randomly distributed voids. The limited defect area where contains varieties of micro voids can be treated as being composed periodically by RVEs (Fig. 2 (b)) that perform similar mechanical characteristic.

Instead of calculating the macroscopic strain at every global integration point of macro FE meshed structure, $\mathbf{F}_M$, the macroscopic deformation gradient will be used in the following work to specify the RVE's BVP. The volume average relationship between the deformation at the scale of global and RVE can be formed as:

$$\mathbf{F}_M = \frac{1}{V_0}\int_{V_0}\mathbf{F}_m dV_0 \qquad (1)$$

The superscripts M and m represent macro and micro quantities respectively, V_0 is the initial (or reference) volume of RVE. Similarly, microscopic stress which is written by the first Piola-Kirchhoff stress tensor $\mathbf{P}$ can be formulated to the macroscopic stress as:

$$\mathbf{P}_M = \frac{1}{V_0}\int_{V_0}\mathbf{P}_m dV_0 \qquad (2)$$

2.2 Boundary value problem

Considering the deformation of adjacent elements, periodical boundary conditions should be applied on RVE. Xia[12] has summarized the generalized method of PBC, and developed the explicit formulation which can be used in FEM. Meanwhile, it has been proved that the PBC can satisfy the continuity of both strain and stress. Figure 3 shows the deformation of 2D RVE under constrain of PBC. The difference value of the corresponding nodes along the RVE's boundary is a constant, and that can assure the same shaping deformation between the corresponding boundaries.

In the proposed work, PBC will be applied in the FE calculation with Python codes by the following method: coupling the nodal number of corresponding faces and lines together and make the equivalence of each other. This method can reduce the constrain strength and enhance the accuracy of the calculation result, although it will bring some

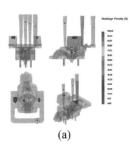

(a)

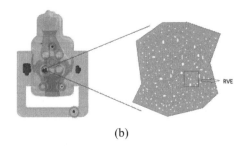

(b)

Figure 2. (a) Void defects distribution after the optimized casting process; (b) RVE selection.

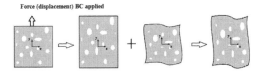

Figure 3. PBC in 2D RVE deformation process.

difficulties on numbering the nodes. The complete PBC formulation is as follows:

$$u_i = \overline{\varepsilon_{lk}} x_k + u_i^* \qquad (3)$$

$\overline{\varepsilon_{lk}}$ is the averaged strain, u_i^* is the periodical displacement on the boundary faces, i means the three coordinate directions in RVE as shown in Figure 4. Equation (3) can be formulated as follows on the three sets of faces (top—bottom, right—left, and front—back surfaces) whose normal directions are along the vectors of $x_k (k = 1, 2, 3)$:

$$u_i^j = \overline{\varepsilon_{lk}} x_k^j + u_i^* \qquad (4)$$

$$u_i^{j-} = \overline{\varepsilon_{lk}} x_k^{j-} + u_i^* \qquad (5)$$

The superscript "$j+$" "$j-$" respectively means the positive and negative direction. Subtracting (4) and (5), we can get:

$$u_i^{j+} - u_i^{j-} = \overline{\varepsilon_{lk}}(x_k^{j+} - x_k^{j-}) = \overline{\varepsilon_{lk}} \Delta x_k^j \qquad (6)$$

Especially, x_k^j is a constant when the two corresponding surfaces are parallel. So (6) can be expressed as:

$$u_i^{j+}(x_1, x_2, x_3) - u_i^{j-}(x_1, x_2, x_3) = c_j^i \qquad (7)$$

In ABAQUS standard, this formulation can be used to operating FE analysis for RVE model by use of the multipoint constraint function accompanied with relevant codes.

In this work, the very three edges that contain the coordinates are assumed to be the master edges, the related surfaces are master faces and the rest of corresponding edges and surfaces are slave edges and slave faces. Point A, which is bounded by complete constrain, is the origin of coordinate. Point B, C and D are the datum points which are bounded by constrain of two DOFs, respectively shown in Figure 5. The PBC equations used upon RVE are as follows:

$$u_i^{b'} = u_i^b + u_i^B \qquad (8)$$

$$u_i^{c'} = u_i^c + u_i^C \qquad (9)$$

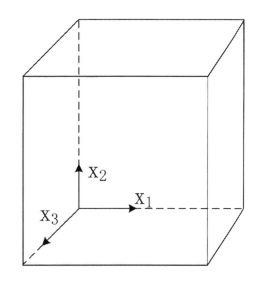

Figure 4. Simplified 3D RVE.

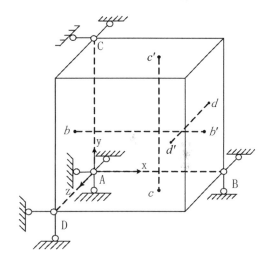

Figure 5. Geometric description of PBC on RVE.

$$u_i^{d'} = u_i^d + u_i^D \qquad (10)$$

In these equations, $i = x, y, z$. b c d are the nodes on the master faces, b', c', d' are the nodes on the corresponding slave faces. When operating the uniaxial tension or compression calculation, we only need to apply loads or displacement boundaries on the related datum points, the rest of the nodes on other edges and faces can be calculated by the corresponding connections and constrains.

283

3 NUMERICAL IMPLEMENTATION AND SIMULATION

3.1 Establishing RVE model

In most of the past references, the shape of the inclusions is always fixed regularly (i.e. sphere, ellipsoid or cylinder). In order to describe the micro defects structure more veritably, we have built up three-ellipsoid orthogonal form of voids which are controlled by multiple parameters by use of Python coding. These mentioned parameters are Spatial coordinates, Rotation angle along the axis of X Y Z, Major axis $b1$ of the ellipsoids which consist of the voids' form and ratio of major and minor axis $f = b1/b2$. Figure 6 has shown different types of voids under different f, Figure 7 shows the detailed procedure of establishing the RVE in this work.

As is mentioned above, volume distribution of the voids is an important characteristic parameter to researching elastic behavior. In fact, diversities of casting voids' morphology and volume could be present due to different casting techniques and

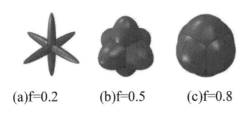

(a)f=0.2   (b)f=0.5   (c)f=0.8

Figure 6.   Different types of voids.

materials. Basing on T. Taxer's theory, logarithmic normal distribution form of the voids' volume control has been used in this work.

$$f_p(v_p; \mu, \sigma) = \frac{1}{v_p \sigma \sqrt{2\pi}} \cdot \exp\left(-\frac{(\ln v_p - \mu)^2}{2\sigma^2}\right) \quad (11)$$

$$\mu = 5.968 + 0.052\,pp$$

$$\sigma = 1.484 + 0.028\,pp \quad (12)$$

$f_p(v_p; \mu, \sigma)$ is the relative arising frequency of voids' volume, pp is the volume fraction of voids, v_p is the number of pixel occupied by void, each volume of cubic pixel is 79.507 μm^3. Considering the size scale of the voids, the maximum value of void volume is set to 4×105 μm^3, minimum value is 3200 μm^3, nearly be equal to a sphere void whose diameter is 18.4 μm. Figure 8 shows the RVE model used in this work, the void volume fraction is 2%, 4%, 6% and 10% respectively.

3.2 Elastic behavior analysis

In this work, we use RuT300 steel as the experimental material, the macro elastic behavior has been performed by calculating the equivalent elastic modulus of RVE, and also had the comparison with the classic ceiling limit value of Voigt and Hanshin Shitrikman (shown in Fig. 9).

From the curves in Figure 9, we can see that the elastic modulus of RuT300 appear to be linear relationship with the void volume fractions. The reason why the FEM result has obvious difference with Voigt, meanwhile, basically identical with Hanshin

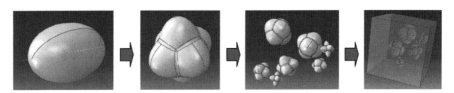

Figure 7.   Specific process of establish the RVE model.

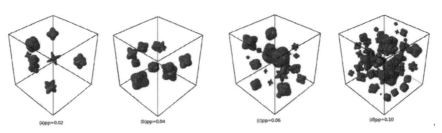

(a)pp=0.02   (b)pp=0.04   (c)pp=0.06   (d)pp=0.10

Figure 8.   RVE models whose void volume fraction is 2%, 4%, 6% and 10%.

Shitrikman, is that the Voigt solution is an unbalanced approximated solution which has neglected the anisotropy inhomogeneous characters. However, the equivalent elastic modulus calculation method claimed by Hanshin Shitrikman can obtain the effective analytical solution basing on principle of minimum potential energy. Also, this result shows that the shape of the voids has limited influence on material's elastic performance.

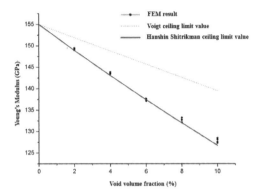

Figure 9. Different types of RuT300 (f = 2%, 4%, 6% and 10%)'s elastic modulus.

3.3 Plastic behavior analysis

Varieties of experimental data show that the location of flame face, oil jet holes and exhaust port of cylinder head which is under the condition of high-temperature environment are the dangerous part of plastic deformation occurring. Besides, they are the very parts that the casting defects have the maximum probability of emerging. After optimized scheme of casting process, the percentage of void defects could still be up to 5%. So, we have selected the RVE which contains 6% fraction of voids to operate the plastic performance simulation. Table 1 shows the mean stress, strain and plastic deformation volume percent under different tension loading conditions.

Figure 10 (a) and (b) shows the equivalent stress distribution contour plotting and plastic deformation field under the condition of tensile deformation being 0.25, respectively. Combining Table 1, it can be concluded that noticeable high stress and even plastic deformation has emerged around the voids, although the mean stress of RVE has not reach the yield-stress value. Strips of high-stress area have been formulated between some of the voids, and this phenomenon could provide the inducing condition for the voids and micro cracks' growth.

In order to synthetically analyze the relationship between the volume fraction of voids and plastic

Table 1. Plastic performance of RVE which contains 6% volume fraction of void.

Uniaxial tensile deformation (%)	Mean strain (%)	Mean stress (MPa)	Plastic deformation volume percent (%)
0.2	0.1998	270.8180	7.2260
0.25	0.2497	312.7100	51.7430
0.3	0.2996	330.3211	65.4155

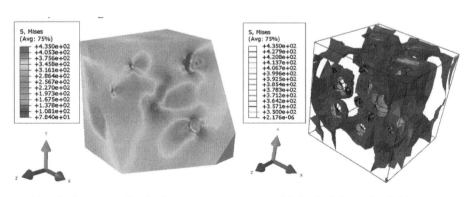

(a)equivalent stress distribution contour   (b)plastic deformation field

Figure 10. Plastic simulation results.

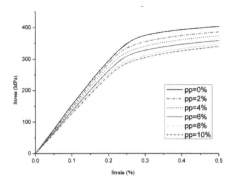

Figure 11. Mean stress-strain curves of RVE whose volume fractions of voids changing from 0% to 10%.

performance of RuT300, ten different uniaxial tensile deformation value has been applied on different kinds of RVE whose volume fraction of voids is 2%, 4%, 6%, 8% and 10% respectively. Each group of the test has acquired the mean stress and strain value, as shown in Figure 11.

From this set of curves, we can conclude that the mean stress value of RVE will decrease with the enlarging of the volume fraction of voids when loading the same strain value. Comparing with the curve of the 0% fraction, the maximum mean stress difference of the 10% fraction has reached to 19.06%, which shows the fact that the material's mechanical property has been seriously reduced.

4 CONCLUSION

By using the PBC applied on the RVE, deformation compatibility problem could be solved, and then guarantee the accuracy of the computational results from RVE. Basing on Python programming and Random Sequential Adsorption modeling method, algorithm of parameterized RVE auto-modeling has been established to simulate different types of void defects, which has the stochastic characteristics of volumes, positions and shapes, existing in the casting structure. According to the results of elastic and plastic simulation, deformation fields and stress or strain distribution turn out to be more close to the real situation. The proposed method can also provide the necessary fundamental of multiscale mechanical calculation more effectively for the casting materials, or even the composite materials.

CORRESPONDING AUTHOR

Linfeng Sun, sunlinfengbit@gmail.com, +17347096086.

REFERENCES

[1] Eshelby J.D. The Determination of the Elastic Field of an Ellipsoidal Inclusion, and Related Problems: Proceeding of the Royal Society of London, Series A. Mathematical and Physical Science. 241(1226). 376–396(1957).

[2] Mori T. and Tanaka K. Average Stress in Matrix and Average Elastic Energy of Materials with Misfitting Inclusions [J]. Acta Metallurgica. 21(5). 571–574(1973).

[3] Hill R. Elastic Properties of Reinforced Solids: Some Theoretical Principles [J]. Journal of the Mechanics and Physics of Solids. 11(5). 357–372(1963).

[4] Whitney J. and McCullough R. Micromechanical Materials Modeling. [M]. Technomic Publ Co Inc, Lancaster, 1993.

[5] Ghosh S., Lee K. and Ragharan P. A Multi-level Computational Model for Multi-scale Damage Analysis in Composite and Porous Materials[J]. International Journal of Solids and Structures. 38(14). 2335–2385(2001).

[6] Raghavan P., Moorthy S., Ghosh S. and Pagano N.J. Revisiting the Composite Laminate Problem with an Adaptive Multi-level Computational Model [J]. Composites Science and Technology. 61(8). 1017–1040(2001).

[7] Rao M.P., Nilakantan G., Keefe M., Powers B.M. and Bogetti T.A. Global/Local Modeling of Ballistic Impact onto Woven Fabrics [J]. Journal of Composite Materials. 43(5). 445–467(2009).

[8] P.R. Dawson and A. Needleman. Issues in the finite element modeling of polyphases plasticity[J]. Materials Science & Engineering. A. 175. 43–48(1994).

[9] M.I. Baskes, R.G. Hoagland, A. Needleman, et al. Summary report: computational issues in the mechanical behavior of metals and intermetallics [J]. Science and Engineering. A. 159. 1–34(1992).

[10] D.N. Fang and C.W. Zhou. Numerical Analysis of the Mechanics Behavior of Composites by Finite Element Micromechanic Method [J]. Progress of Mechanics. 28(2). 173–188(1998).

[11] M. Ries, C. Krempaszky, B. Hadler, et al. The influence of porosity on the elastoplastic behavior of high performance cast alloys. Proceedings in applied mathematics and mechanics. 7. 2150005–2150006(2007).

[12] Z.H. Xia, Y.F. Zhang and E. Fernand. A unified periodical boundary conditions for representative volume elements of composites and applications [J]. International Journal of Solid & Structure. 40(8). 1907–1921(2003).

Advances in Future Manufacturing Engineering – Yang (Ed.)
© 2015 Taylor & Francis Group, London, ISBN 978-1-138-02817-3

Accurate positioning foil stamping technology

Qinghua Chen, Yingjun Chen & Yanmei Li
Institute of Electronic Information and Electrical Engineering, Zhaoqing University, Guangdong, China

Miaogang Su & Yibin He
Leibao Optoelectronics Technology Co. Ltd., of Guangdong, Zhaoqing, China

Wengang Wu
National Key Laboratory of Micro/Nano Fabrication Technology, Institute of Microelectronics, Peking University, Beijing, China

ABSTRACT: In this paper, we propose an environment friendly accurate positioning foil stamping technology. The whole idea is to meet the country's high-end anti-counterfeiting technology, consumer products packaging appearance, high-end gifts demand for the target material. The State Quality Supervision and Testing Center (Guangzhou) detection results shows that it includes many merits such as high accuracy, functionality, security, reliability, ease of use, stability and other indicators, the performance meets or exceeds the requirements of enterprise technical conditions.

Keywords: positioning; foil stamping; accuracy; mold; film

1 INTRODUCTION

In recent years, with the improvement of people's living standards and national economic development, people's appearance and variety of products are becoming more emphasis on beautifully wrapped by hot foil stamping pattern showing strong metallic luster, colorful and anti-fade characteristics. Among these technologies, the laser pattern transfer in the packaging printing industry more widely, it makes use of natural light into the colorful packaging effect, which will play an important role in the packaging applications.

During the production process of the Laser holographic hot stamping film, the membrane needs to be wound up. Currently, the constant tension take-up control is generally divided into open-loop control and closed-loop control of two categories.

Open-loop control is the use of soft and winding characteristics similar to some of the mechanical properties of the motor itself has direct use of such motors to drive the winding mechanism to obtain approximate constant tension control. On the other hand, closed-loop control is divided into direct tension control, tension control and composite indirect tension control in three ways. Direct tension control is one of the most direct

and most effective ways to control. This tension control advantages are high precision tension control, can theoretically achieve zero error control. The disadvantage is that the control accuracy depends on the accuracy of tension detection device. Indirect tension control system does not have the tension sensing elements, and the tension control is static and dynamic analysis through the winding mechanism of physical equations. Then, it can find out the impact of physical tension of all electrical, control of these quantities, so as to achieve a constant tension control. The advantage of this constant tension control method is to reduce the tension roller and the corresponding detection devices with reducing system cost. The disadvantage is that the control is more complicated than direct control of accuracy, and is completely dependent on the accuracy of the controller.

In this paper, we originally propose an intelligent tension control hologram positioning hot stamping technology principle, structure and manufacturing method for producing laser holographic hot stamping film. The high degree of automation technology, detection speed, and high detection accuracy can avoid downtime losses caused by manual measurement; positioning information can ensure holographic film size consistency.

2 DESIGN AND PRINCIPLES OF THE NOVEL ACCURATE POSITIONING FOIL STAMPING TECHNOLOGY

The on line detection control device prototype is composed by the encoder, photoelectric detectors and computing unit together, and its working principle is as follows (shown in Fig. 1).

We assume the perimeter of the installation of the encoder active role is C, the number of pulses per revolution encoder is N. Since the encoder is mounted coaxially on the active roll, the encoder sends each pulse width, drive roller conveyor or the amplitude pattern of the printed film on the length of radium (C/N), the edge line within the signal lines and the rear edge of the first interval of the signal, calculate the number of pulses received from the encoder unit is M. Based on the calculation unit after calculation, it is obtained a printing pattern of a film or web of the upper radium actual length $L = (C/N) \times M$, then the calculation unit calculates the actual length L obtained with a predetermined target length L_0, which can be compared to results amplitude real-time error and the actual length L of the target pattern set in advance the length L_0 between the quantity e. Next, according to the size of the real-time correction error e of the process parameters, a closed loop control is achieved. Finally, the mentioned real time controls the amount of error in the size of the e in the allowable range, so as to ensure the length of the pattern mold is formed to reach the actual amplitude requirements.

The on-line detecting means are provided with a second-stage photo-detector and a second-stage encoder. The second encoder is mounted coaxially on the root of the driving roller of the second conveyor which is formed of molded rigid printed film or a holographic film installed at the second photo-detector shelves. The light irradiated from the second photo-detector in the root of the driving roller transfers through the second film or a holographic film printed pattern of amplitude. The signal input terminals of the third encoder and a second computing unit connected to a second the fourth input terminals of the photo-detector signal and the calculation unit is connected. According to the received second photo-detector for the calculating unit, the second detection signal of the encoder pulse signal to calculate the actual length of the newly formed molded amplitude pattern. To calculate the actual length of the pre-set target, the obtained length can be obtained by being compared to the actual length of the amount of error in real time amplitude pattern and set in advance between the target lengths. Next, the error correction process parameters according to the real size of the volume to achieve the purpose of review, which can also further ensure the formation of mold amplitude pattern length to reach the actual requirements.

3 EXPERIMENTS

In this paper, a prototype accurate positioning foil stamping film has been successfully achieved and tested. Based on this technique, several laser stamping film have been carried out (as shown in Fig. 2).

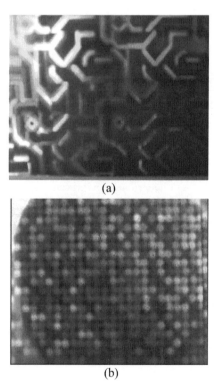

(a)

(b)

Figure 2. Two kinds of accurate positioning laser stamping film fabricated by the novel technique.

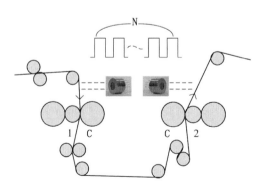

Figure 1. The proposed process of the accurate positioning foil stamping technology.

Table 1. Specifications of the proposed accurate positioning foil stamping film.

Parameters	Values
Positioning accuracy	2 μm
Flatness	>12000 meters without folds
Reflectivity	95%
Rally indicator	2.0 KG*9.8 N/KG
5% CINA salt spray test	>24 hours

Table 1 shows the specifications of the proposed accurate positioning foil stamping film. Measurements show the merits of the film such as excellent flatness, high positioning accuracy, high reflectivity, high Rally indicator and impressive durability. Based on the results of research and intelligent tension control hologram positioning hot stamping technology to promote the project will enhance the technological level and competitiveness of our products stamping products to market high-end printing industry development has brought vast opportunities for the province to create huge economic wealth and social benefits, the formation of a new high-tech industry growth.

4 SUMMARY

This paper proposes a novel accurate positioning foil stamping film based on the novel technique. This technique can meet a wide range of applications. The State Quality Supervision and Testing Center (Guangzhou) detection results shows that it includes many merits such as accuracy, functionality, security, reliability, ease of use, flexibility, stability and other indicators, the performance meets or exceeds the requirements of enterprise technical conditions.

ACKNOWLEDGEMENTS

This work was supported in part by the Science and Technology Project Foundation of Zhaoqing and the Innovation Fund for SMEs of Guangdong Province.

REFERENCES

[1] Kim, J., Kim, N., Lee, D., Park, S., Lee, S. Watermarking two dimensional data object identifier for authenticated distribution of digital multimedia contents. Signal Processing: Image Communication, 2010, 25: pp.559–576.

[2] Mori K., Okuda Y. Tailor die quenching in hot stamping for producing ultra-high strength steel formed parts having strength distribution [J]. CIRP Annals-Manufacturing Technology, 2010, 59(1): pp.291–294.

[3] Li, D., Kim, J. Secure Image Forensic Marking Algorithm using 2D Barcode and Off-axis Hologram in DWT-DFRNT Domain. Applied Mathematics and Information Sciences, 2012, 6(2S): pp.513–520.

[4] Tian X., Zhang Y., Li J. Investigation on tribological behavior of advanced high strength steels: influence of hot stamping process parameters [J]. Tribology Letters, 2012, 45(3): pp.489–495.

[5] Schieck F., Hochmuth C., Polster S., et al. Modern tool design for component grading incorporating simulation models, efficient tool cooling concepts and tool coating systems [J]. CIRP Journal of Manufacturing Science and Technology, 2011, 4(2): pp.189–199.

[6] Cui J., Lei C., Xing Z., et al. Analysis of Phase Transformation Simulation and Prediction of Phase Distribution in High Strength Steel in Hot Stamping [J]. Journal of Computational and Theoretical Nanoscience, 2012, 9(9): pp.1309–1314.

[7] Liu W.L., Chiang T.H., Tseng C.H. Analysis of Hot Stamping for Automotive Structure [J]. Applied Mechanics and Materials, 2014, 446: pp.279–283.

[8] Hu P., Ma N., Liu L., et al. Hot Forming Simulation Algorithms of High-Strength Steels [M]//Theories, Methods and Numerical Technology of Sheet Metal Cold and Hot Forming. Springer London, 2013: pp.113–151.

Advances in Future Manufacturing Engineering – Yang (Ed.)
© 2015 Taylor & Francis Group, London, ISBN 978-1-138-02817-3

Research of an LED lamp with quasi-resonant mode driving circuit

Qinghua Chen & Lunqiang Chen
Institute of Electronic Information and Electrical Engineering, Zhaoqing University, Guangdong, China

Fuyi Ou
Zhaoqing New Sun Lighting Co. Ltd., Zhaoqing, China

Yingjun Chen
Institute of Electronic Information and Electrical Engineering, Zhaoqing University, Guangdong, China

ABSTRACT: An LED lamp with a novel quasi-resonant mode driver for interior-lighting applications is reported in this paper. High power factor is achieved by using the transformer with back winding coil and the transformer with "sandwich" winding coil. Furthermore, the proposed driver demonstrates cost-effectiveness, high circuit efficiency, low input current ripples and a reduced components count. A fabricated driver is developed to supply a 20W LED interior-lighting module with a 0.95 power factor. Experimental results demonstrate the functionalities of the proposed circuit.

Keywords: quasi-resonant mode; driver; inductance; circuit; led

1 INTRODUCTION

LEDs (Light Emitting Diodes) are being used as backlights for notebook PCs, medium and large size LCD TVs, colorful outdoor signs and lighting engineering because of their high-energy efficiency and very long operation time. LED lamps presented are mainly comprised of the electronic driver circuit and the group of LED chips [1]. Most LED driver is consists of three parts, which is the Power Factor Correction (PFC) circuit, the DC/DC voltage regulate circuit and the constant current circuit [2]. The boost converter working in the DCM state is usually adopted for the PFC circuit, while a flyback converter or a forward converter is usually adopted for the voltage regulator circuit. However, the energy efficient factor of the LED driver is not enough high for some interior-lighting applications [2].

In these days, the SY5800 had been brought in the utility of the LED current control module. It is an accurate primary-side-controlled LED driving IC without applying any secondary feedback circuit for low cost. The SY5800 can increase both the equivalent output resistance and sampling accuracy to obtain better output current characteristics.

In paper [3], the SY5800A driver is designed at LED lighting applications as shown in Figure 1.

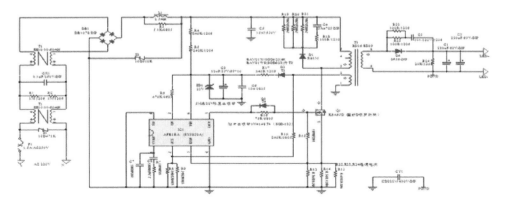

Figure 1. The proposed LED driver based on SY5800A for high power factor applications.

It consists of a SY5800A chip, short circuit protection, line regulation modification, Over Voltage Protection (OVP), protection, constant current control, inductor, output capacitor and MOSFET. However, due to the output inductor and its power conversion, the circuit efficiency and output power is limited and more power components are required in the conventional driver.

In the case of PFCs based on the flyback family of converters, a simple solution to obtain a high-energy efficient factor is presented in this paper. This solution is based on the use of the transformer with back winding coil and the transformer with "sandwich" winding coil instead of a normal one.

2 PRINCIPLES OF THE IMPROVED DRIVER CIRCUIT

Figure 2 shows the proposed driver, it combines with an improved SY5800A LED driver for supplying a rated LED power of larger than 15 W for interior-lighting applications. Compared to conventional SY5800A LED driver, the improved driver used the transformer with back winding coil and the transformer with "sandwich" winding coil to enhance the output power and the efficiency.

3 DESIGN EQUATIONS OF KEY COMPONENTS

The power is transferred from AC input to output only when the input voltage is larger than output voltage in Buck converter. Primary side control is applied to eliminate secondary feedback circuit, which reduces the circuit cost. The switching waveforms I_{OUT} can be induced finally by

$$I_{OUT} = \frac{K_1 \times K_2 \times V_{REF} \times N_{PS}}{R_S} \quad (1)$$

where K_1 is the output current weight coefficient; K_2 is the output modification coefficient; V_{REF} is the internal reference voltage; R_S is the current sense resistor. K_1, K_2 and V_{REF} are all internal constant parameters, I_{OUT} can be programmed by N_{PS} and R_S.

4 EXPERIMENTS

An experimental prototype of the proposed circuit with a rated output power of 20 W has been successfully fabricated (Fig. 2) and tested in order to power the LED interior-lighting module (BSX-IQHT-F(D1)-Z1), which includes 20 LEDs in three parallel paths and four LEDs in series connection for each parallel path. Steady output performance is obtained on the key parameters. The dimension of the driver is 140 mm (in length), 34 mm (in width) and 23 mm (in height), respectively.

Table 1 shows the specifications of the proposed LED driver. Measured relationships of input voltage and the key output performances of the driver are shown in Figure 3. Figure 3(a) presents the measured efficiency and the harmonic factor of the driver. Figure 3(b) shows the measured output power factor. The measured averaged power factor is 0.95, the measured averaged harmonic factor is 0.1 and the measured averaged circuit efficiency is 0.85, respectively. Measured output voltage and current along with the input voltage are shown in Figure 3(c) and Figure 3(d). The average values of output voltage and current are approximately 40 V and 497 mA, respectively. In addition, steady output performance is obtained on both items. Excellent EMC measurements of the proposed driver based on SY5800A LED are shown in Figure 4.

5 SUMMARY

An LED lamp with a novel quasi-resonant mode driver for interior-lighting applications is reported in this paper. The high power factor is achieved by using the transformer with back winding coil and the transformer with "sandwich" winding coil. A 20 W prototype driver has been developed and tested with a 40 V output voltage. The experimental results of the presented LED driver for interior-lighting applications have demonstrated

Figure 2. The proposed fabricated LED.

Table 1. Specifications of the proposed LED driver.

Parameters	Values
Input voltage	AC 100–264 V
Output voltage	DC 32–50 V
Start-up delay time	0.5 S
Output power	16–25 W

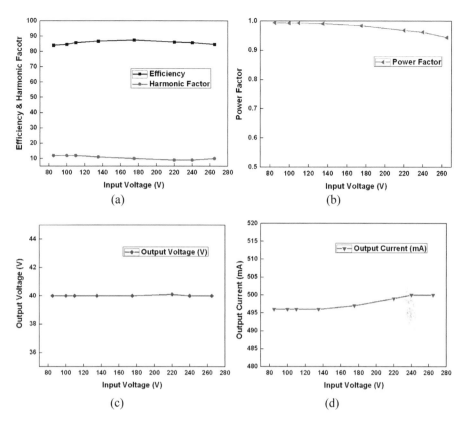

Figure 3.   Measurements of the proposed SY5800A LED driver for interior-lighting applications: (a) the input voltage vs the efficiency and harmonic factor; (b) the input voltage vs the power factor; (c) the input voltage vs the output voltage; (d) the input voltage vs the output current.

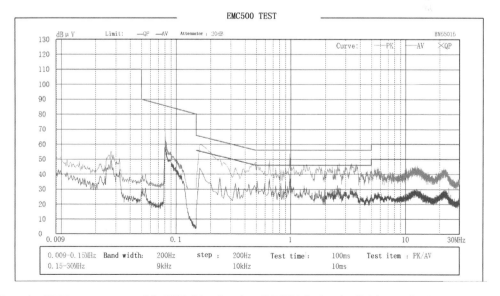

Figure 4.   EMC measurements of the LED driver based on SY5800A for interior-lighting applications.

low harmonic factor (<0.1), high circuit efficiency (>0.85), high power factor (>0.95) on output rectifier diodes.

ACKNOWLEDGEMENTS

This work was supported in part by the Science and Technology Project Foundation of Zhaoqing City and Duanzhou District.

REFERENCES

[1] Almeida, P.S.; Soares, G.M.; Pinto, D.P.; Braga, H.A.C. "Integrated SEPIC buck-boost converter as an off-line LED driver without electrolytic capacitors" in Proceedings of the 2012 IEEE Annual Conference on IEEE Industrial Electronics Society (IECON 2012), pp. 4551–4556.

[2] J. Sebastian, D.G. Lamar, M. Arias, M. Rodriguez, and M.M. Hernando, "A very simple control strategy for power factor correctors driving high-brigtness LEDs," in Proc. of IEEE APEC'08, 2008, pp. 537–543.

[3] Fanghua Zhang; Jianjun Ni; Yijie Yu. "High Power Factor AC–DC LED Driver With Film Capacitors", IEEE Trans. on Power Electronics, vol. 28, no. 10, 2013, pp. 4831–4840.

Advances in Future Manufacturing Engineering – Yang (Ed.)
© 2015 Taylor & Francis Group, London, ISBN 978-1-138-02817-3

Measurements of the LED driver with steady output characteristics

Yu Fan & Lunqiang Chen
Institute of Electronic Information and Electrical Engineering, Zhaoqing University, Guangdong, China

Fuyi Ou
Zhaoqing New Sun Lighting Co. Ltd., Zhaoqing, China

Yingjun Chen
Institute of Electronic Information and Electrical Engineering, Zhaoqing University, Guangdong, China

ABSTRACT: A high-efficiency, high-power factor LED driver, capable of driving two standard white LEDs in series with steady output characteristics, is presented in this paper. The LED current is sensed using voltage-mode feedback to maintain a constant load current and drove by PWM controller. Experimental results demonstrate the functionalities of the proposed circuit. The measured averaged power factor is larger than 0.95; and the measured averaged circuit efficiency is 0.92, respectively. In addition, the average fluctuation values of output voltage and current are approximately within 0.3 V and 3 mA, respectively. Moreover, excellent EMC radiation performance is achieved by the proposed driver.

Keywords: steady output; driver; interior-lighting; circuit; LED

1 INTRODUCTION

Recently, the requirements of industry product efficiency are getting more and more important. For lighting applications to be considered, the Light Emitting Diode (LED) possesses many advantages, such as no Hg pollution, small size, long life, high luminous efficacy, fast time response, lows power consumption [1], [2], etc. LEDs are becoming more common in energy-saving lighting applications, such as Liquid Crystal Display (LCD) background lighting, automotives, motorcycle lighting, and traffic lighting, decorative lighting, and so on, and may even be considered for general lighting applications in the near future.

Up to now, digital dimming is preferred because operating at low load currents with analog dimming may result in low efficiency and may produce unpredictable results [3]. However, even though some trends in commercial interior-lighting LEDs are toward high-power, high-current devices, the power factor and efficiency of the LED driver is not enough for some interior-lighting applications [4]. As the digital-dimming-type IC, the SY5814A had been brought in the utility of the LED current control module. The SY5814A is a single stage Buck PFC controller targeting at LED lighting applications. The chip has a high PF value, high efficiency, current harmonics less than 20% of the performance, non-isolated buck constant current

driver chip to design a high PF value and high efficiency, current harmonics and low cost of the LED lamp. Nevertheless, due to the output capacitor and its power conversion used in the driver based on SY5814A, the circuit efficiency and output power is limited and unsteady and more power components are required in the conventional driver.

The important challenging parameters in LED driver design are efficiency and power factor. In response to these challenges, this paper presents a novel single-stage LED driver based on SY5814A for interior-lighting applications with high levels of power factor and efficiency for interior-lighting applications.

2 DESIGN OF THE PROPOSED DRIVER CIRCUIT

Figure 1 shows the proposed driver which combines an improved SY5814A LED driver. In this design, the improved driver used the "Pen" output capacitor to enhance the output power and the filter component to improve the efficiency.

3 EXPERIMENTS

In this experiment, a sample driver with a rated output power of 10–15 W has been successfully

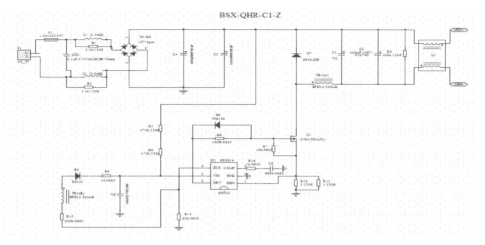

Figure 1. The proposed LED driver for interior-lighting applications.

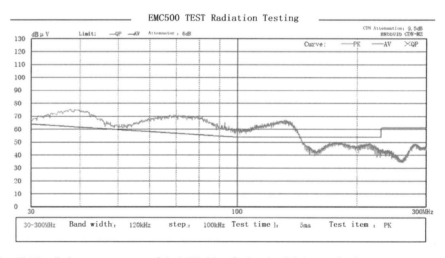

Figure 2. The proposed fabricated LED.

Figure 3. EMC radiation measurements of the LED driver for interior-lighting applications.

built and tested in order to power the LED interior-lighting module (BSX-QHR-C1-Z), which includes 18 LEDs in two parallel paths and nine LEDs in series connection for each parallel path. Steady output performance is obtained on the key parameters. The dimension of the driver is 156 mm (in length), 18 mm (in width) and 10.5 mm (in height), respectively. Excellent EMC radiation measurements of the proposed driver are shown in Figure 3.

Table 1 shows the specifications of the proposed LED driver. Measured relationships of input voltage and the key output performances of the driver at different input power states are shown in Figure 4. Figure 4(a) presents the measured harmonic factor of the driver. Figure 4(b) presents the measured efficiency of the driver. Figure 4(c) shows the measured output power factor. The measured averaged power factor is larger than 0.95; and the measured averaged circuit efficiency

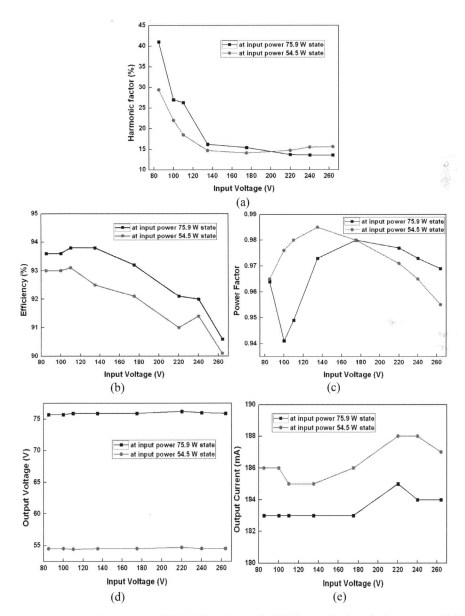

Figure 4. Measurements of the proposed LED driver for interior-lighting applications (at input power 75.9 W and 54.5 W, respectively): (a) the input voltage vs the harmonic factor; (b) the input voltage vs the efficiency; (c) the input voltage vs the power factor; (d) the input voltage vs the output voltage; (e) the input voltage vs the output current.

is 0.92, respectively. Measured output voltage and current along with the input voltage are as shown in Figures 3(d) and 4(e). The average fluctuation values of output voltage and current are approximately within 0.3 V and 3 mA, respectively. Therefore, steady output performance is obtained on both items.

4 SUMMARY

This paper has proposed a high-efficiency, high-power factor LED driver, capable of driving two standard white LEDs in series with steady output characteristics. From measured waveforms, it can be seen that the proposed LED driver has a good performance on dimming. The measured averaged power factor is larger than 0.95; and the measured averaged circuit efficiency is 0.92, respectively. In addition, the average fluctuation values of output voltage and current are approximately within 0.3 V and 3 mA, respectively. Moreover, excellent EMC radiation performance is achieved by the proposed driver.

ACKNOWLEDGEMENTS

This work was supported in part by the Science and Technology Project Foundation of Zhaoqing City and Duanzhou District.

REFERENCES

[1] H.P. Zhao, G.Y. Liu, J. Zhang, J.D. Poplawsky, V. Dierolf and N. Tansu, "Approaches for high internal quantum efficiency green InGaN light emitting diodes with large overlap quantum wells," *Optics Express*, vol.19 (S4), pp. A991–A1007, 2011.

[2] Y.-K. Lo, K.-H. Wu, K.-J. Pai, and H.-J. Chiu, "Design and implementation of rgb led drivers for lcd backlight modules," IEEE Transactions on Industrial Electronics, vol. 56, no. 12, pp. 4862–4871, 2009.

[3] J. Sebastian, D.G. Lamar, M. Arias, M. Rodriguez, and M.M. Hernando, "A very simple control strategy for power factor correctors driving high-brigtness LEDs," in *Proc. of IEEE APEC'08*, 2008, pp. 537–543.

[4] Fanghua Zhang; Jianjun Ni; Yijie Yu, "High Power Factor AC–DC LED Driver With Film Capacitors", IEEE Trans. on Power Electronics, vol. 28, no. 10, 2013, pp. 4831–4840.

Advances in Future Manufacturing Engineering – Yang (Ed.)
© *2015 Taylor & Francis Group, London, ISBN 978-1-138-02817-3*

Microemulsions of Coumarins from Cortex Daphnes for enhanced transdermal permeability: Preparation and *in vitro* evaluation

Li Ma
Department of Pharmacy, Tianjin Institutes of Pharmaceutical Research, Tianjin, China

Ya fei Wang
Department of Chemistry, School of Science, Tianjin University, Tianjin, China

Xing cui Qu
Bo Teng Biological Pharmaceutical Technology Co. Ltd., Guangzhou, China

Feng Han
Department of Pharmacy, Tianjin Institutes of Pharmaceutical Research, Tianjin, China

Jin li Deng & Xia Feng
Department of Chemistry, School of Science, Tianjin University, Tianjin, China

ABSTRACT: Microemulsions (MEs) with good skin bioavailability for Coumarins from Cortex Daphnes (CCD) were prepared using a titration method. The systems were optimized using pseudo-ternary phase diagrams. The physical and chemical characteristics of the MEs including morphology, particle size, zeta potential and stability were studied. The transdermal permeation ability of the CCD microemulsions was evaluated by mounting *in vitro* rabbit skin on Franz diffusion cells. The optimized oil-in-water ME formulation was composed of 2.86% CCD, 17.14% ethyl oleate (oil), 40% cremophor RH-40 (emulsifier) and 40% glycerol (co-emulsifier). The MEs were quite stable and the average particle size was 14.81 nm with a zeta potential of −3.03 mV. Compared to the control, the permeation rates of daphnin, daphnetin-8-O-β-D-glucoside and daphnetin in the ME increased by 4.34, 4.04 and 2.08 times respectively. The use of a ME significantly enhanced the transdermal permeability of CCD and may be useful as a new drug delivery system for CCD.

Keywords: coumarins from cortex daphnes; microemulsion; water titration method; pseudo-ternary phase diagram; transdermal permeability

1 INTRODUCTION

Cortex daphnes is a traditional Chinese medicine which comes from the dried root and stem bark of *Daphne giraldii Nitsche, Daphne retusa Hemsl.,* or *Daphne tangutica Maxim.* It has been used clinically to treat rheumatism and other similar diseases and it is an effective analgesics. [1, 2] Cortex daphnes extracts contain several classes of natural products including coumarins, diterpenoids, lignans, flavonoids, anthraquinones and sterols and particular attention has been paid to the coumarins because of their extensive biological activity. Daphnin, daphnetin-8-O-β-D-glucoside and daphnetin [3] are the three index components of coumarins.

Currently, coumarins from cortex daphnes (CCD) are used clinically in the form of tablets, pastes, and injections. [4, 5] However, applications are still limited because of the low absorption rate and bioavailability of the drug in normal administration route, which is mainly due to its lipophilic nature. Many attempts have been made to improve the transportation of CCD through the skin. [6–8] Among the various drug delivery systems, microemulsions (MEs) are considered to be an ideal vehicle for the transdermal delivery of poorly water-soluble drugs. [9–11] Therefore MEs may be suitable to conveniently and effectively deliver CCD.

MEs are monophasic, optically isotropic and thermodynamically stable liquid systems. They can be considered as ideal liquid vehicles for drug delivery as they have most of the requirements for this including the thermodynamic stability, ease of formulation, high drug-loading capacity and small droplet size. The very small droplet size and very low interfacial tension of ME can provide excellent surface contact with the skin surface, thereby transporting drugs in a controlled manner. [12, 13]

Until now, ME formulations of CCDs have not been investigated. The aim of this work was to develop a ME system which could serve as a vehicle for the topical delivery of CCD with improved skin bioavailability. A method to formulate CCD MEs was developed and the physical-chemical properties of the MEs such as surface morphology, globule size, zeta potential and stability were measured. In addition, the in vitro permeation abilities of the active components in the CCD MEs were evaluated using rabbit skin.

2 MATERIALS AND METHODS

Materials. Ethyl Oleate (EO) and Tween 80 were from Tianjin Guangfu Fine Chemical Research Institute (China). Triolein (LCT) and soybean oil were from Sinopharm Chemical Reagent Co, Ltd and Shangzhi Earth Oil Co (China), respectively. Cremophor RH-40 came from BASF (Germany), Cremophor EL-40 from Nanjing Duly Biotech Co, Ltd (China) and polyethylene glycol 400 from Hunan Kang Pharmaceutical Co, Ltd (China). Daphnetin was purchased from China Pharmaceutical and Biological Products. The coumarins, daphnin and daphnetin-8-O-β-D-glucoside were extracted from Cortex daphnes plants in our lab using 60% aqueous ethanol as the solvent. The water used in this study was double distilled. All other chemicals and reagents were also of analytical grade and were used without further purification.

Construction of pseudo-ternary phase diagram. Oil-in-water type microemulsions were prepared to improve the solubility of the CCD and to reduce the usage of organic solvents. The type of oil phase, emulsifier, co-emulsifier and the weight ratio of emulsifier to co-emulsifier (K_m) are all key factors that influence the properties of drug-loaded MEs. In this study, these factors were determined using pseudo-ternary phase diagrams. Based on previous studies, [14] soybean oil, EO and LCT were selected as candidates for the oil phase and Tween 80 (HLB = 15), cremophor RH-40 (HLB = 14–16) or cremophor EL-40 (HLB = 13.5) were used as the emulsifier. Four kinds of short-chain alcohols including ethanol, propylene glycol, butanol and glycerol were tested as co-emulsifiers.

The pseudo-ternary phase diagrams were constructed using a titration method at 37 °C. First, a mixture of emulsifier and co-emulsifier were prepared with the desired K_m value. The oil phase

was then mixed with the above mentioned mixture to give weight ratios of 1:9, 2:8, 3:7, 4:6, 5:5, 6:4, 7:3, 8:2, or 9:1. [15] Next, distilled water was added dropwise to the oil and surfactant mixture with magnetic stirring (78HW-1, Hangzhou Meter Motor Factory, China) until the clear ME region appeared and then disappeared again. The quantities of water required to make the mixture clear and turbid were then recorded and the percentages of the different components were calculated. Pseudo-ternary phase diagrams were then constructed from these values. The phase diagrams were used to identify the ME regions and based on this, the optimum oil phase, emulsifier, co-emulsifier and K_m value were determined.

Preparation of CCD-loaded MEs. Based on the results of the pseudo ternary phase diagrams, three CCD-loaded MEs with CCD to oil ratios of 1:4, 1:5 and 1:6 were prepared. The CCD, oil and surfactants were mixed at 37 °C and a measured amount of distilled water was added dropwise to the mixture until a clear solution was obtained. The particle size and zeta potential of the MEs were measured using a potential and nano-particle size analyzer (Zetasizer Nano ZS, Malvern Instruments Ltd, England) and these results were used as the criteria to optimize the drug-loaded ME formulation. [16]

HPLC method. HPLC was used to determine the drug content of the CCD-loaded MEs. The analyses were performed with an Agilent 1100 HPLC (Aglient Technologies, America) equipped with a G1310 pump and a G1314A UV detector. A Prodigy ODS (3) (4.6 × 250 mm, 5 μm) column was used to separate and test the drug ingredients. The mobile phase was methanol and 0.05% phosphoric acid (18:82, v/v) and pumped at a flow rate of 1.0 mL/min. The column temperature was 35 °C. The effluent was monitored at 327 nm. [17]

Preparation of the calibration standards. A standard stock solution was prepared by dissolving 24.14 mg of daphnin, 22.45 mg of daphnetin-8-O-β-D-glucoside and 21.16 mg of daphnetin in 5 mL of 50% aqueous methanol to give final concentrations of 4.828, 4.49 and 4.232 mg/mL, respectively. The stock solution was stored at 4 °C before use.

Specificity. The specificity of the method was checked by performing chromatographic analysis of a blank ME, a drug-loaded ME, the standard solution and the CCD by itself.

Stability. The stability of the drug-loaded ME was investigated by monitoring the concentration change and Relative Standard Deviation (RSD) of the analyses of daphnin, daphnetin-8-O-β-D-glucoside and daphnetin every 2 h for a total of 10 h.

Linearity. Working solutions were prepared by diluting 5, 10, 20, 30, 40 and 50 μL of the standard solution with 5 mL of blank ME solution. Calibration curves were prepared by plotting the HPLC peak-area (y) versus the concentration (x) of daphnin, daphnetin-8-O-β-D-glucoside or daphnetin. The linearities of the calibration curves were determined by linear regression analysis using Origin 85 Software.

Recovery. The absolute recoveries of the three CCD components were determined by adding known amounts of daphnin, daphnetin-8-O-β-D-glucoside and daphnetin standards to low, medium, and high concentrations of drug-loaded MEs with known CCD contents. The peak areas of the three ingredients in each sample were recorded. The recoveries at the three concentration levels were measured at least five times.

Precision. The precision of the method was assessed by analysis of five replicates of the drug-loaded ME samples under the same operating conditions over a short period of time. The precision was determined by calculating the RSD of the peak areas of daphnin, daphnetin-8-O-β-D-glucoside and daphnetin. The method was accepted when the values were lower than 2.0%.

Morphology studies. The morphology of the ME systems was observed using Scanning Electron Microscopy (SEM) (S-4800, Hitachi, Japan). Several drops of a 100-fold diluted CCD-loaded ME were air dried on a copper grid and then gold was deposited under high vacuum on the sample. The microstructure of the sample was examined at an accelerating voltage of 3 kV.

In vitro skin permeation study of CCD-loaded MEs. The Microemulsion Formulation (MF) of CCD was prepared and studied its transdermal delivery ability for the drug. The MF was prepared via the drug-loaded ME preparation method described in Section 2.3. Normal Formulation (NF) was prepared with the same amount of CCD and spiked with proportional amounts of glycerol and polyethylene glycol (PEG) 400, and used as the control.

New Zealand rabbits (2.5 ± 0.5 kg, Tianjin Chunle experimental animal farm) were used to study the transdermal penetration of the three components of CCD from the MEs. Skin permeation studies were conducted with Franz diffusion cells (TK-12B Transdermal Diffusion Instrument, Shanghai Kai Kai Technology Trade Ltd, China) with an active area for diffusion of 2.6 cm² and a receptor volume of about 6.1 mL. The whole rabbit dorsal skin was removed and carefully separated from the subcutaneous fat and connective tissues and then examined for integrity. After being washed with Normal Saline (NS), the skin was cut to the appropriate sized pieces and clamped between the donor and receptor cells with the stratum corneum facing upward toward the

donor cell. The receptor cells were then filled with NS and the temperature was maintained at 37 ± 1 °C using a circulating water bath, which was continuously stirred at 270 rpm. At time zero, 1 mL of MF or NF was added to the donor cell, which was occluded with parafilm. At pre-determined time intervals (2, 4, 6, 8, 10, 20 and 24 h), 1 mL samples were withdrawn from the receptor cell and replenished immediately with an equal volume of fresh NS. During the whole manipulation process, the formation of bubbles was avoided. The samples were filtered through a 0.45 mm membrane filter and analyzed for drug content by HPLC as described in Section 2.4. For all animal studies, the experimental procedures conformed to the ethical principles of Tianjin Institute of Pharmaceutical Research. All animals were treated according to the principles of laboratory animal care.

Data analysis. All experiments were performed at least three times (n = 3–5). All data is presented as the mean ± Standard Deviation (SD). The amount of cumulative active components that penetrated through the rabbit skins (Q_n, μg/cm²) was calculated from the following equation:

$$Q_n = \left(C_n V + \sum\nolimits_{i=1}^{n-1} C_i V_i \right) \Big/ A$$

where, C_n is the drug concentration in the receptor solution (μg/mL) at each sampling time, C_i is the drug concentration of the sample (μg/mL), V and V_i are the volumes of the receptor solution (mL) and the sample (mL), respectively, and A is the active diffusion area. Q_n and time (t_h) were calculated from the linear regression. The linear fitting equation is the penetration kinetic equation [18] and the slope of the equation is the transdermal penetration rate (J_s, μg·h⁻¹·cm⁻²).

3 RESULTS AND DISCUSSION

Screening formulation ingredients of MEs. The construction of phase diagrams makes it easy to determine the concentration range of ingredients where MEs exist. The suitable ingredients to formulate MEs were selected from the pseudo-ternary phase diagrams shown in Figures 1–3.

Oil phase selection. Using cremophor RH-40 as the emulsifier and glycerol as the co-emulsifier, soybean oil, EO and LCT were each tested as the oil phase to form MEs. The pseudo-ternary phase diagrams for each are shown in Figure 2. When soybean oil was used as the oil phase, a ME was only formed when the weight ratio of oil to surfactant was 1:9, which suggests that soybean oil is not very suitable to prepare MEs. A comparison of the ME regions for the other two oils shows that

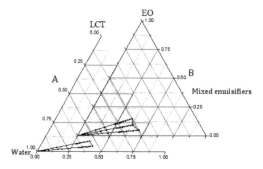

Figure 1. Pseudo-ternary phase diagrams for oil phase selection (A. LCT, B. EO). The K_m value was set to be 1.

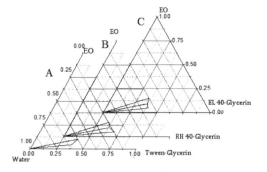

Figure 2. Pseudo-ternary phase diagrams for emulsifier selection (A. Tween 80, B. Cremophor RH-40, C. Cremophor EL-40). The K_m value was set to be 1.

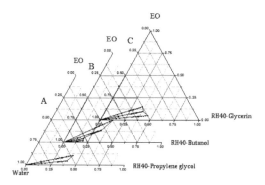

Figure 3. Pseudo-ternary phase diagrams for co-emulsifier selection (A. Propylene glycol, B. Butanol, C. Glycerin). The K_m value was set to be 1.

the one with EO has a larger ME area than that with LCT. So EO was selected as the oil phase.

Emulsifier selection. Next three surfactants including Tween 80, cremophor RH-40 and cremophor EL-40 were screened as emulsifiers using EO and glycerin as the oil phase and co-emulsifier,

respectively. The pseudo-ternary phase diagrams in Figure 3 show that the size of the ME region for cremophor RH-40 and cremophor EL-40 were about the same and that the region for Tween 80 was much smaller. The ME formed with cremophor RH-40 was the most transparent and stable so cremophor RH-40 was chosen as the emulsifier.

Co-emulsifier selection. Finally ethanol, propylene glycol, butanol and glycerin were tested as co-emulsifiers using EO as the oil phase and cremophor RH-40 as the emulsifier. The pseudo-ternary phase diagrams shown in Figure 4 show that ethanol is not a suitable co-emulsifier because no ME region was formed. For the other co-emulsifier candidates, the size of the ME regions with glycerin and butanol are about the same and both are larger than that with propylene glycol. Glycerin was chosen as the co-emulsifier in view of its convenience, stability and nontoxicity.

After selecting the three above ingredients which form stable MEs over a broad area, five more phase diagrams were constructed using different weight ratios of the emulsifier to the co-emulsifier (K_m). The ratios were set to be 1:2, 1:1.5, 1:1, 1.5:1 and 2:1. The area of the ME decreased as the amount of co-surfactant increased (data not shown), which is consistent with a previous report. [19] Taking into account both the overall drug content and the amount of surfactant used in the formulation, the optimal value of K_m was determined to be 1.

Based on the above results, EO, cremophor RH-40 and glycerin were selected as the oil phase, emulsifier and co-emulsifier, respectively, to prepare MEs for the topical delivery of CCD. The weight ratio of surfactant to co-surfactant (K_m) was selected to 1.

Optimization of CCD-loaded ME formulation. Three ME formulations were prepared with CCD/EO ratios of 1:6, 1:5 and 1:4. The particle sizes and zeta potentials of the MEs were measured and

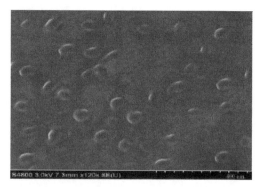

Figure 4. Scanning Electron Microscopy (SEM) image of TCCD-loaded ME.

used as the criteria to optimize the CCD-loaded MEs. The results are listed in Table 1.

When the drug content was increased from 1:6 to 1:4, the absolute zeta potential of the MEs decreased from 3.03 to 1.53, the particle size increased from about 15 to 62 nm and the PDI doubled. PDI is a measurement of particle homogeneity and smaller numbers (i.e. close to zero) indicate more homogeneity. A decrease in the zeta potential value and an increase in the particle size can lead to unstable MEs and so a drug content of 1:6 (CCD to EO ratio) was selected.

In summary, the optimal drug-loaded ME composition is as follows: 2.86% CCD, 17.14% EO, 40% cremophor RH-40 and 40% glycerol.

Specificity. The chromatograms of daphnin, daphnetin-8-O-β-D-glucoside and daphnetin showed that the three major components of CCD are well separated and there was no interference from any endogenous substance that may exist in the MF. Therefore, the HPLC method is specific for analyzing the drug content of the MEs.

Linearity and calibration curve. The statistical data demonstrated a good linear relationship between the areas of the peaks and the corresponding concentrations of the daphnin, daphnetin-8-O-β-D-glucoside and daphnetin standards. The plots (n = 6) were linear in the concentration ranges of 4.828–48.28, 4.49–44.9 and 4.232–42.32 μg/mL for the three components, respectively. The linear regression equations for the three components

Table 1. Particle sizes and zeta potentials of CCD-loaded MEs.

CCD/ EO ratio	Average particle size [nm]	Polydispersity index (PDI)	Average zeta potential [mV]
1:6	14.81	0.141	−3.03
1:5	35.38	0.178	−2.18
1:4	61.97	0.294	−1.53

Table 2. Percutaneous permeation parameters (mean ± SD; n = 5).

Components	Type	Q_n [μg·cm^{-2}]	J_s [μg·h^{-1}·cm^{-2}]
Daphnin	MF	94.242 ± 22.397*	4.0875 ± 1.031*
	NF	18.351 ± 8.882	0.94201 ± 0.425
Daphnetin-8-O-β-D-glucoside	MF	179.122 ± 56.2*	7.58827 ± 2.395*
	NF	35.302 ± 16.825	1.88761 ± 0.876
Daphnetin	MF	162.51 ± 4.805*	8.1354 ± 0.398*
	NF	76.44 ± 25.659	3.909 ± 1.47

*Significantly different from the NF groups (P < 0.05).

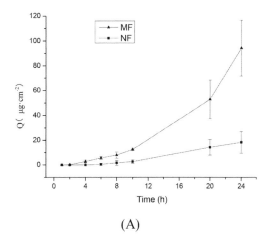

(A)

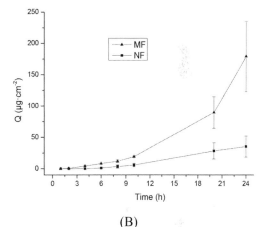

(B)

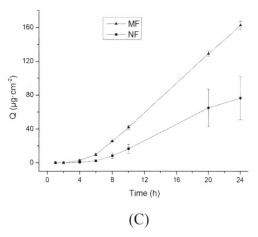

(C)

Figure 5. Permeation profiles of three active ingredients of TCCD in different formulation types (A) Daphnin; (B) Daphnetin-8-O-β-D-glucoside; (C) Daphnetin. Each point and vertical bar represents mean and standard deviation, respectively.

303

were y = 16.978x + 7.8071 (r = 0.9996), y = 22.081x + 9.1153 (r = 0.9996) and y = 36.81x + 12.32 (r = 0.9995), respectively.

The RSD for the recovery of daphnin, daphnetin-8-O-β-D-glucoside and daphnetin were all less than 2% which indicates that the HPLC analytical method is a good way to determine the concentrations of CCD in drug-loaded MEs.

Characterization of the CCD-loaded MEs. The CCD-loaded MEs were clear and transparent. When the MEs were diluted with substantial amounts of water, they turned sky-blue opalescent. The morphology of the ME was observed by SEM and the results are shown in Figure 4. Homogeneous and spherical shaped ME droplets with mean particle sizes of 40–60 nm are clearly seen.

In vitro skin permeation studies. The skin permeabilities of the three active components in the CCD MF were evaluated and compared with those of the NF. The percutaneous permeation parameters of the tested formulations are presented in Table 2.

According to Table 2, all the MFs have significantly higher Q_n values for all three active components ($P < 0.05$). The Q_n values for daphnin and daphnetin-8-O-β-D-glucoside were about five times greater than those in the NFs ($P < 0.05$) and the Q_n values of daphnetin were about twice as large (162.51 ± 4.805 for MF compared to 76.44 ± 25.659 $\mu g \cdot cm^{-2}$ for NF). Generally, drug-containing MEs interact with the lipid layers in the stratum corneum and change the structural integrity of the layers [20] which leads to enhanced permeation of the drug. This makes the drug more suitable for transdermal delivery.

The skin permeation profiles of the three active components in the CCD in MF and NF formulations are presented in Figure 5.

The skin permeation profile of daphnin was quite similar to that of daphnetin-8-O-β-D-glucoside which is due to their similar structures. When formulated in MEs, the permeation rates (J_s) of daphnin and daphnetin-8-O-β-D-glucoside were both about four times higher than those in NF. For daphnetin, the J_s in the MF increased from 3.909 ± 1.47 (in NF) to 8.1354 ± 0.398 $\mu g \cdot h^{-1} \cdot cm^{-2}$. These results demonstrate that the permeation of the three active ingredients of CCD through rabbit skin were significantly accelerated with the MF compared to those for the NF ($P < 0.05$).

4 CONCLUSIONS

CCD-loaded MEs were prepared using a water titration method and pseudo-ternary phase diagrams. The optimal drug-loaded ME composition was determined to be: 2.86% CCD, 17.14% ethyl oleate, 40% cremophor RH-40 and 40% glycerol. The drug-loaded ME had long-term stability, uniform particle sizes and a narrow particle size distribution. The MF can be used for the transdermal administration of poorly soluble CCD and the permeation of the drug through rabbit skin was greatly improved over that obtained for a NF. Because of the complexity of CCD components, micro-emulsifying the drug not only could reduce the damage to skin, but could also improve the efficacy of the active ingredients. This study provides the experimental foundation for the successful development of new transdermal delivery vehicle for CCD.

ACKNOWLEDGEMENTS

This study was supported by the Key Science-technology Plan Project of Tianjin "Research on Daphnes Giraldii Cortex and its gel" (12ZCDZSY12000).The authors are also grateful to the Traditional Chinese Medicine Department of Tianjin Institute of Pharmaceutical Research for providing the experimental apparatus.

CORRESPONDING AUTHORS

Tel.: +86 22 27890272; fax: +86 22 23006851. E-mail address: zhihua504@163.com(L. Ma). f_x@tju.edu.cn (X. Feng).

REFERENCES

[1] Pu D.C. Cortex Daphnes formulation for clinical application. Gansu J Tradit Chin Med. 1992; 5(3): 37–37. (in Chinese).

[2] Xia W.Q. Clinical Observation of the treatment for Arthropathy with Cortex Daphnes tablet. Gansu J Tradit Chin Med. 1998; 5(10): 57–58. (in Chinese).

[3] Wang M.S., Yu M., Zhang Y.J. Study on chemical constituents of Cortex Daphnes, Chin Tradit Herb Drug. 1976; 7 (10): 13–15. (in Chinese).

[4] Zhang X.Z., Di L.Q., Shan J.J., et al. Study on the preparation of Zushima total coumarins effective components sustained release pellets and its release behavior. Chin J Hosp Pharm. 2010; (11): 891–894. (in Chinese).

[5] Li G., Li J.Z., Xie J.W. Study on the preparation of the compound Cortex Daphnes injection and the quality control method. Chin J Chin Mater Med. 1992; 17(11): 664–665. (in Chinese).

[6] Fan B., et al. Study the Effects of penetration enhancers on the permeability of Cortex Daphnes through skin. Chin Tradit Patent Med. 2014; 36(3): 631–634. (in Chinese).

[7] Wang N.N., Zhang C., Lin G.T. Investigation of in vitro Transdermal Permeability of Total Coumarins in Daphne giraldii. Chin J Exp Med Formul. 2013; 19(20): 18–20. (in Chinese).

[8] Yu Y.Y. Pharmaceutical Studies on ZSM Cataplasm [D]. Masters Dissertation of Nanjing University of Traditional Chinese Medicine. Nanjing, 2007. (in Chinese).

[9] Araya H., Tomita M., Hayashi M. The novel formulation design of O/W microemulsion for improving the gastrointestinal absorption of poorly water soluble compounds. Int J pharm. 2005; 305(1): 61–74.

[10] Kawakami K., Yoshikawa T., Moroto Y., et al. Microemulsion formulation for enhanced absorption of poorly soluble drugs: I. Prescription design. J Control Release. 2002; 81(1): 65–74.

[11] Kawakami K., Yoshikawa T., Hayashi T., et al. Microemulsion formulation for enhanced absorption of poorly soluble drugs: II. In vivo study. J Control Release. 2002; 81(1): 75–82.

[12] Kreilgaard M. Influence of microemulsions on cutaneous drug delivery. Adv Drug Deliv Rev. 2002; 54: 77–98.

[13] Kogan A., Garti N. Microemulsions as transdermal drug delivery vehicles. Adv Colloid Interfac. 2006; 123–126: 369–385.

[14] Warisnoicharoen W., Lansley A.B., Lawrence M.J. Nonionic oil-in-water microemulsions: the effect of oil type on phase behaviour. Int J Pha. 2000; 198(1): 7–27.

[15] Wilk K.A., Zielińska K., Hamerska-Dudra A., et al. Biocompatible microemulsions of dicephalic aldonamide-type surfactants: Formulation, structure and temperature influence. J Colloid Interf Sci. 2009; 334(1): 87–95.

[16] Neslihan Gursoy R., Benita S. Self-Emulsifying Drug Delivery Systems (SEDDS) for improved oral delivery of lipophilic drugs. Biomed Pharmacother. 2004; 58(3): 173–182.

[17] Yin D.D., Wang X.L., Zhao G.R., et al. Determination of Etoposide Microemulsion by HPLC. Chin J Pharms. 2007; 38(2): 121–122.

[18] Wang L., Zheng N., Fang L., et al. Effects of penetration enhancers on the in vitro percutaneous absorption of piperine in Extractum Piperis Longi. J Shenyang Pharm Univ. 2010; 27(1): 20–23.

[19] Zhang Q.Z., Jiang X.G., Jiang W.M., Lu W., Su L.N., Shi Z.Q. Preparation of nimodipine-loaded microemulsion for intranasal delivery and evaluation on the targeting efficiency to the brain. Int J Pharm. 2004; 275: 85–96.

[20] Azeem A., Khan Z.I., Aqil M., et al. Microemulsions as a surrogate carrier for dermal drug delivery. Drug Dev Ind Pharm 2009; 35(5): 525–547.

Advances in Future Manufacturing Engineering – Yang (Ed.)
© 2015 Taylor & Francis Group, London, ISBN 978-1-138-02817-3

The feasibility research of steam flooding assisted by air for heavy oil reservoir

Jie Fan & Xiangfang Li
School of Petroleum Engineering, China University of Petroleum (Beijing), Changping, Beijing, China

ABSTRACT: With the intent of solving problems that appear at the later period of steam flooding, different displacement methods at the later period of steam flooding are studied. Steam flooding assisted by air is one of these important methods. In this paper, laboratory experiments basing on an actual heavy oil reservoir were used to study the steam flooding assisted by air. Some parameters, such as, the weight fraction of the catalytic agent, temperature of reaction, injection pressure and injection approach were optimized using laboratory experiment. What's more, steam flooding assisted by air was tested in the target block, the results show that technology can improve the development efficiency of heavy oil reservoir.

Keywords: steam flooding assisted by air; heavy oil reservoir; laboratory experiment; simulation

1 INTRODUCTION

During the later period of steam flooding in heavy oil reservoir, a collection of serious situations such as the lower oil rate, lower OGR and higher water cut appear. In this case, it is important to study some conversion displacement to improve the development efficiency of heavy oil reservoir. At the beginning of the 20th century injection air has be studied, the technology has some advantages, such as, rich gas source and low cost [1–2]. The first project of air injection is in the former Soviet Union. Air injection has been used as the main EOR method since 1931 [3]. Later, air injection was initiated as the third oil recovery in light-oil reservoirs in the USA by Amoco, Gulf, or Chevron, and increased oil production has been obtained in these projects. There are a number of ongoing successful air injection field projects, such as, the West Hackberry field, Buffalo field, Medicine Pole Hills field, and Horse Creek reservoir in the USA [4–7]. Finally, some high-pressure air injection projects are also planned or have carried out the pilot test [8–11].

After air is injected into the reservoir, heat energy, methane, and flue gas will be produced because of the Oxidation reaction. CO_2 and CO generated in Oxidation reaction dissolve into heavy oil and water to reduce the oil-water interfacial tension and crude oil viscosity; the main part of the flue gas is N_2 which is an ideal gas to displace and lifting crude oil, surfactants generated in the reaction, such as aldehyde, ketone, alcohol and acid can improve the displacement efficiency.

What's more, this process will make full use of flue gas drive and the heating effect [12–13]. The steam flooding assisted by air is a high efficiency and low cost exploiting technology for heavy oil which combines the recovery mechanism of thermal recovery, flue gas flooding and surfactant flooding.

In this paper, we studied the flues gas composition of catalytic oxidation reaction, optimized the impact of weight fraction of catalytic agent, temperature, injection pressure and injection approach on the catalytic oxidation reaction by laboratory experiment, what's more, analyzed the change of temperature and pressure for an actual heavy oil reservoir by numerical simulation.

2 EXPERIMENT RESEARCH

2.1 Experiment equipment and material

In the experiment, experimental equipment mainly included beakers, graduated cylinders, high-pressure and-high temperature reaction kettle, homogenizer, steam generator, pumps, calorstat, sand tubes, back-pressure valve, digital display viscometer, electrooptic analytical balance and gas chromatograph, etc.

Materials included crude oil from the Qi block of Liaohe Oilfield, configured water according to the actual components of the connate water from the Qi block (salinity range is 2000 mg/L ~ 3000 mg/L), distilled water, quartz sands, high-temperature catalytic agent and high-pressure air.

2.2 Experimental method

Put 50 g oil, 50 ml water and a measured amount of catalytic agent into the experimental instrument for heavy oil catalysis oxidation. Heat up the mixture with the injected air to a specific temperature (220°C) to allow a period of isothermal stir and reaction. After that, gauge the oil viscosity, acid number and production gas composition.

2.2.1 The evaluation of catalytic agent

Firstly, a set of catalytic oxidation experiment was carried out. Air was continuously injected into high-pressure and high-temperature reaction kettle which had equipped with crude oil, water and catalytic agent, and keep constant 220°Cof reaction temperature and 600 r/min rotating speed. The acid number was 19.1 mg KOH/g after reaction, meanwhile, the results show that flue gas consisted of carbon dioxide, nitrogen and light hydrocarbon, and a lot of surfactant were generated from the catalytic oxidation reaction, as shown in Table 1, all of the compositions had a comprehensive displacement effect for heavy oil. There was litter CO_2 in the tail gas that implied most of oxygen had reacted with heavy oil, however, the reaction was oxygen absorption not combustion.

2.3 Experimental results and discussion

2.3.1 The impact of weight fraction of catalytic agent on the catalytic oxidation reaction

To optimize weight fraction of catalytic agent, putted 50 g crude oil, 50 ml configured water and different weight fraction of catalytic agent (0.1%, 0.2%, 0.3%, 0.4%, 0.5%, 0.6%) in the reaction kettle, meanwhile, keep a constant temperature at 220°C. The consequences were demonstrated in Figure 1. As can be seen from the figure, the acid number increases firstly, and decreases afterwards along with the amount enlargement of catalytic agent. The acid number maximizes at the point of weight fraction of catalytic agent 0.3%. Crude oil viscosity has the opposite change tendency. It may be reason that amount enlargement of catalytic agent leads to oil acid condensation reaction and increase the molecular chain of heavy oil. So thus, 0.3% seems to be the optimum weight fraction of catalytic agent.

Table 1. Flues gas composition of catalytic oxidation reaction.

Flues gas composition, %	
N_2	71.23
O_2	7.95
CO_2	0.07
Gaseous hydrocarbon	20.75

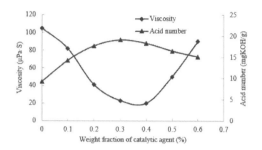

Figure 1. Weight fraction of catalytic agent versus viscosity and acid number.

2.3.2 The impact of temperature on the catalytic oxidation reaction

Temperature is an important factor for the catalytic oxidation reaction. Higher the temperature is, stronger the reaction is. 50 g crude oil, 50 ml configured water and 0.3% weight fraction of catalytic agent were in the reaction kettle. Figure 2 demonstrates that the acid number and crude oil viscosity lower rapidly before the temperature reaches 220°C, whereas, they lower modestly when the temperature is beyond 220°C. Therefore, the optimum temperature should be 220°C.

2.3.3 The impact of injection pressure on the catalytic oxidation reaction

50 g crude oil, 50 ml configured water and 0.3% weight fraction of catalytic agent were in the reaction kettle, and keep a constant temperature at 220°C, injected air under different pressure. The consequences were demonstrated in Figure 3. The results show that the acid number decreases firstly, and increases afterwards along with the injection pressure, the crude oil viscosity always decreases. Comprehensive consideration, the optimum injection pressure is 0.6 Mpa.

2.3.4 The impact of injection approach on the catalytic oxidation reaction

Single tube continuous displacement experiments were carried out, as shown in Figure 4. Firstly, sand tube was taken out to vacuum sate and saturated configured water, according to the absorbing capacity of water, computing the pore volume and porosity. Secondly, sand tube was saturated oil; the exit of model would be closed when it reached the state of bound water. After all of these, different experiments of steam flooding assisted by air could be carried out. To optimize injection approach, the paper designs three single tube displacement experiments. 1. Continuously injecting steam; 2. Slug flooding by firstly injecting catalytic agent and air, then injecting steam; 3. Injecting catalytic agent, air and steam at the same time. Table 2 demonstrates that slug flooding is the optimum injection approach.

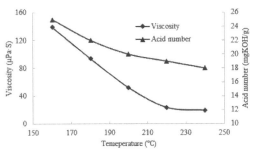

Figure 2. Temperature versus viscosity and acid number.

Figure 3. Injection pressure versus viscosity.

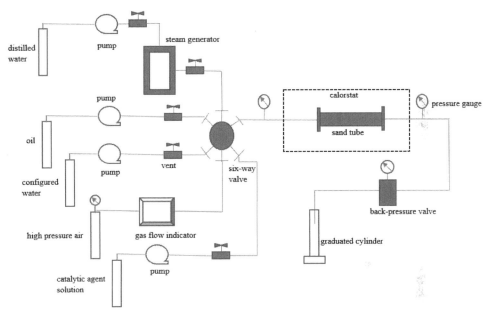

Figure 4. Experimental flow diagram.

Table 2. Displacement efficiency of different approach.

Injection approach	Displacement efficiency, %	Increased displacement efficiency, %
1#	75.56	/
2#	90.6	15.04
3#	87.78	12.22

3 PILOT TEST

The target block started cyclic steam stimulation in 1987 which is a heavy oil reservoir. After nearly 20 years, this block converted to steam flooding successively since 2007. Though it produced well at the beginning of the conversion displacement, now the oil rate decreases obviously, the water cut has been beyond 93%, and the oil steam ratio has dropped to 0.12 or so. In order to improve the recovery, In June 2014, 45000 m3 air and 7.3 t high-temperature catalytic agent were injected into well 23-k033, 51000 m3 air and 7.6 t high-temperature catalytic agent were injected into well 19–030. The production shows that the oil rate is almost doubled and the water cut decreases rapidly by 4.3% after air injection, all of these indicate that the steam channeling are restrained effectively, so the sweep efficiency and the development efficiency are improved.

4 CONCLUSIONS

Laboratory experiments demonstrate that injected air can react with crude oil, the reaction not only reduces the oil viscosity, but also generates surfactant and flue gas, all of these improve the development efficiency.

The results of laboratory experiment show that the optimum weight fraction of catalytic agent is 0.3%, the optimum reaction temperature is 220°C, the appropriate injection pressure is 0.6 Mpa, and the reasonable injection approach is slug injection.

Steam flooding assisted by air was tested in the target block, the production shows that the oil rate is increased and the water cut is decreased, it can serve as a kind of technology applied in other similar reservoirs.

REFERENCES

[1] Alvarado, V. and Manrique, E. 2010. Enhanced Oil Recovery: An Update Review. Energies 3(9): 1529–1575.

[2] Burger, J.G. and Sahuquet, B.C. 1972. Chemical Aspects of In-Situ Combustion-Heat of Combustion and Kinetics. SPE J. 12(5): 410–422.

[3] Clara, C., Durandeau, M., Quenault, G. et al. 2000. Laboratory Studies for Light-oil Air Injection Projects: Potential Application in Handil Field. SPE Res Eval & Eng 3(3): 239–248. SPE-64272-PA.

[4] Greaves, M., Ren, S.R., Rathbone, R.R. 1998. Air Injection Technique (LTO Process) for IOR from Light Oil Reservoirs: Oxidation Rate and Displacement Studies. Presented at the SPE/DOE Improved Oil Recovery Symposium, Tulsa, 19–22 April. SPE-40062-MS.

[5] Greaves, M., Ren, S.R., Rathbone, R.R. et al. 2000. Improved Residual Light Oil Recovery by Air Injection (LTO process). J can Pet Technol 39(1): 57–61. JCPT Paper No. 00–01–05.

[6] Li, D.W. 1991. Current Situation of EOR of GAS Flooding in Soviet Union. World Petroleum Science 42(5): 58–63.

[7] Liu, Y., Sun, L., Sun, L.T. et al. 2008. Study on Low Temperature Oxidation Kinetics for Light-Oil Air Injection. Appl. Chemical Industry 37(8): 899–902.

[8] Manrique, E.J., Thomas, C.P., Ravikiran, R. et al. 2010. EOR: Current Status and Opportunities. Presented at the SPE Improved Oil Recovery Symposium, Tulsa, 24–28 April. SPE-130113-PA.

[9] Moore, R.G., Belgrave, J.D.M., Mehta, R. et al. 1992. Some Insights Into the Low—Temperature and High-Temperature In-Situ Combustion Kinetics. Presented at the SPE/DOE Enhanced Oil Recovery Symposium, Tulsa, 22–24 April. SPE-24174-MS.

[10] Moore, R.G., Laureshen, C.J., Ursenbach, M.G. et al. 1999. A Canadian Perspective on In Situ Combustion. J Can Pet Technol 38(13): 1–8. JCPT Paper No. 99–13–35.

[11] Ren, S.R., Greaves, M., and Rathbone, R.R. 2002. Air Injection LTO Process: An IOR Temperature for Light-Oil Reservoir. SPE J. 7(1): 90–99. SPE-57005-PA.

[12] Sarma, H.K., and Das, S.C. 2009 Air Injection Potential in Kenmore Oil field in Eromanga Basin, Australia: A Screening Study Through Thermogravimetric and Calorimetric Analyses. Presented at the SPE Middle East Oil and Gas Show and Conference, Bahrain, 15–18 March. SPE-120595-MS.

[13] Teramoto, T., Uematsu, H., Takabayashi, K. et al. 2006. Air-Injection EOR in Highly Water-Saturated Light-Oil Reservoir. Presented at the SPE Europec/EAGHE Annual Conference and Exhibition, Vienna, Autria, 12–15 June. SPE-100215-MS.

Advances in Future Manufacturing Engineering – Yang (Ed.)
© 2015 Taylor & Francis Group, London, ISBN 978-1-138-02817-3

Research on fault diagnosis system with causal relationship in equipment technical manual's deep knowledge

L. Wang, Y.L. Qian, T.F. Xu & Y.B. Liu
Laboratory of Science and Technology on Integrated Logistics Support,
National University of Defense Technology, Changsha, Hunan, China

ABSTRACT: This article focus on the circumstances of progress of science and technology and increasing complexity of modern equipment' structure, which make the development of fault diagnosis towards the direction of knowledge and maintenance staff have constantly increasing training subjects with higher complexity. A method of using equipment' technology manuals as the carrier of knowledge to extract deep knowledge of equipment structure and principle in manual by hand or computer is put forward, then build a fault diagnosis system based on causal relationship in order to reduce maintenance staff's mental labor. At first, the article analyzes the fault diagnosis based on knowledge and the effect of deep knowledge on fault diagnosis; secondly, discusses the possibility of applying causality model to express the equipment principle and reasoning fault; then introduces the basic structure of fault diagnosis system with causal relationship based on deep knowledge in manual and briefly introduces the operating steps of the system; finally using VE distribution pump as the research object to illustrate how the system work in fault diagnosis.

Keywords: fault diagnosis; deep knowledge; causal relationship

1 INTRODUCTION

With the rapid development of science and technology, the development of equipment continues to accelerate, the structure and function of the equipment is becoming more and more complex, while the informatization of the society is emboldening. Especially modern equipment, which include and apply many subject knowledge and high technology group, is no longer monosomic and single professional military equipment in tradition but the integrated system with light, mechanics, electricity, liquid, gas, software and many other technical fields[1]. For example, self-propelled antiaircraft gun system consists of fire system, photoelectric fire control system, single shot search radar system, chassis system and each directly under monomer, whether structure complexity or interdisciplinary degree of related knowledge of the equipment are greatly improved. Therefore, fault diagnosis, the basic work of maintenance support of equipment, changes to be more knowledgeable.

At the same time, the gymnastics and their complexity of maintenance staff is increasing, which means greater challenges to them. In this case, the diagnostic ability of maintenance staff depends on their knowledge level of equipment structure and working mechanism. At present, the main way to obtain domain knowledge is searching technical manuals[2,3]. But most content of technical manuals is about the structure, function and working process of equipment, which reflects the mechanism of failure occurrence and belongs to deep knowledge[4,5], which could be used to determine fault location when it occurs. However, these information consist of massive unstructured words and graphs, which is impossible for maintenance staff to absorb in a short period of time. Therefore, if a computer system could be built, which is able to change above-mentioned knowledge to diagnosis knowledge and has ability of reasoning, the efficiency of fault diagnosis would be greatly increased.

2 FAULT DIAGNOSIS BASED ON KNOWLEDGE

2.1 *Fault diagnosis based on knowledge/ Knowledge-based fault diagnosis*

Knowledge-based fault diagnosis has developed for long time, including ten years' research combined with artificial intelligence. It does not need precise mathematical models, but obtain diagnostic information automatically from existing technical manuals and expert experiences.

From the perspective of cognition theory, fault diagnosis knowledge could be divided into three parts:

1. Experiential knowledge

   Experiential knowledge is obtained through long time practice to diagnostic objects by experts. It is often given in the form of "premise—diagnostic conclusion", but premise does not have strong causal relationship with diagnostic conclusion.

2. Causal knowledge

   Causal knowledge describes parts and their relations of the diagnostic object, which reflects knowledge of inside structure and function. Compared with Experiential knowledge, causal knowledge has potential theory evidence, which gives causal relation between premises and conclusions.

3. The first law knowledge

   The first law knowledge means the knowledge related to physical properties, functions and principles of diagnostic objects, including theory, law, formula, and rule. The first law knowledge is based on specific scientific theories, having attribute of universality and generality.

Among these knowledge, experiential knowledge is regarded as shallow knowledge, while causal knowledge and the first law knowledge is deep knowledge[6]. Low level external cognition of domain experts to diagnostic objects belongs to shallow knowledge. In terms of questions about fault diagnosis, whose efficiency could be increased due to comprehensiveness and high generality of shallow knowledge. The lack of a thorough understanding of shallow knowledge of diagnostic object would lead to degradation of diagnosis, which could not provide perfect interpretation of diagnostic results. The content about structure and principle of equipment in service manuals describes diagnostic objects' information concerning the structure, function and working process under normal circumstances[2,3], gives internal laws, including internal mechanism of faults. Therefore, the content is deep knowledge of fault diagnosis[4,5] and its form is like Figure 1[7]:

From Figure 1, we can know the content reflects designers' description of working principle of every part, which matters very much. But its form is not clear, so it is often connected with the structure in Figure 2.

The content in Figure 2 is unstructured document for individuals[8], and it does not explain the fault diagnosis knowledge explicitly. In fact, only the full use of this deep knowledge could help maintenance staff remove faults in time by analogy and overcome lack of diagnostic experience, difficulty of acquiring solving knowledge and identification of unexpected faults. However, due to the difference between documents' quality and per-

> **The theory of VE distribution pump stop oil by electromagnetic valve:**
> Electromagnetic valve consists of coil, spring, valve etc. When the coil energized, it will generate magnetic force to suck up the valve, then fuel from the pump cavity can flow to the plunger head, oil can be supplied. When coil magnetic force disappears, the spring force press the valve on the oil inlet orifice of the plunger, the oil in pump cavity can not flow to the plunger head, fuel injection pump just stop feeding.

Figure 1.   Describes the way to stop oil feeding.

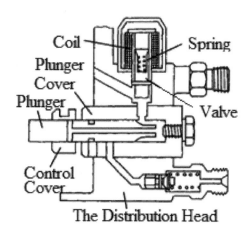

Figure 2.   Sketch map on stop oil feeding.

sonal understanding ability, the deep knowledge which a person could master about fault diagnosis is local and limited, which limits the development of knowledge about equipment's structure and working principles. So if maintenance staff could master these information, then the maintenance and repair efficiency would improve enormously.

3   CAUSAL RELATIONSHIP MODEL

3.1   Outline of causal model

Causal relationship model has been studied many times at home and abroad, and at present the

312

related results, often combined with a specific area, are mainly applied in the relevant areas of artificial intelligence[9,10]. Causal model is often embedded in an expert system or knowledge-based systems as a relatively independent sub-reasoning model. Causal model can also be used to explain some physical phenomena in qualitative reasoning of physical system.

According to the characteristics of the equipment technical manual, description of functional principle can be considered as an event, and the event is generally considered one of the most common causal relationships between entities. An event must involve some subject and the subject is parts and some objects (other parts), as well as the attribute of these objects (attribute usually defined by requirements). Occurrence of an event is usually accompanied by some changes in attribute value from one state to another, and these changes are usually obtained by a suitable occurrence of movement of a subject. Here the appropriate action of parts in mechanical system is called function possessed by parts. That is to say, a function of one or some parts in equipment structure will lead to changes of related state of one or several parts, which will lead to appearing new operation (function). Causal model is applied in diagnostic reasoning based on knowledge, especially based on deep knowledge of the equipment technical manual. The description of operational connection among parts in the mechanical system can be considered as such a link relationship.

3.2 Instance of causal model

In this paper, we present a basic model according to the above mentioned relationship with characteristics of equipment technical manual. Then the contents in Figure 1 can be converted into Figure 3.

In computer, the model in Figure 3 is stored in the form of an XML file, as shown in Figure 4.

In Figure 3, the block expresses an event which represents a section of text in Figure 1. Block contains the basic elements of an event: subject, object, and the relationship between them. They are the main function of an event. Obviously, there are more than one property of an object. The reader can add it as you need. In this paper, each object is only to write a major attribute intended to explain the causal relationship. The block is connected by single arrow in the figure. And we need to know that the complexity of modern mechanical system structure is far proportion

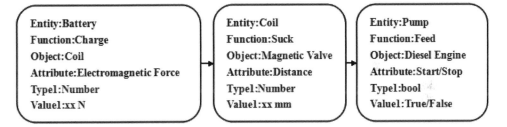

Figure 3.   Magnetic valve stop/supply principle model.

```
<?xml version="1.0" encoding="utf-8" standalone="yes" ?>
- <Events>
  - <Event Event_Description="When the coil energized"
          As_Because_GUID_1="37e3a570-0e15-455f-aac7-7b1dc7620205"
          As_So_GUID_1="" As_So_GUID_2="" As_So_GUID_3="">
      <Part Part_ID="0" Entity="Battery" Function="Charge" Object="Coil"
            Attribute="Electromagnetic Force" Attribute_Value="N"
            Attribute_Type="Number" />
    </Event>
  - <Event Event_Description="The coil will generate magnetic force to suck up the valve"
          As_Because_GUID_1="41d665d3-be01-4a78-9d2e-182163830298"
          As_So_GUID_1="37e3a570-0e15-455f-aac7-7b1dc7620205" As_So_GUID_2="" As_So_GUID_3="">
      <Part Part_ID="0" Entity="Coil" Function="Suck" Object="Magnetic Valve"
            Attribute="Distance" Attribute_Value="mm" Attribute_Type="Number" />
    </Event>
  - <Event Event_Description="Oil can be supplied"
          As_Because_GUID_1="3d8e6e83-72d6-4109-85b4-ca46d90b1b48"
          As_So_GUID_1="41d665d3-be01-4a78-9d2e-182163830298" As_So_GUID_2="" As_So_GUID_3="">
      <Part Part_ID="0" Entity="Pump" Function="Feed" Object="Diesel Engine"
            Attribute="Start/Stop" Attribute_Value="True/False" Attribute_Type="bool" />
    </Event>
  </Events>
```

Figure 4.   The causality model document.

Table 1. The explanation of nodes and attributes in causality document.

Name	Explanation
Event	Event node, include all the information about event.
Event_Description	Event description, description of event in manual text.
Entity	Named entity appears in manual event.
Function	The function of named entity.
Object	The main subject effected by entity
Attribute	Object attribute, object parameter variations, such as the property of magnetic valve is distance.
Attribute_Value	The value of the property, and the corresponding properties.
Attribute_Type	The type of attribute, the variable type of program.
As_Because_GUID_1	Causal marker, represent the part of reason, is also the main code of the event node.
As_So_GUID	Causal marker, represent the part of result, the same result caused by multiple reasons is not only, to be differentiated by adding digital in the rear.

to the example, but it does not affect the expression of the content.

In Figure 3, the operation principle of the VE dispensing pump is so clear. More importantly, the link between the boxes reflects the principle of causality exactly. Because, assuming the function of a particular box is not to achieve, then the function after it will not work correctly. Thereby, these relationships can not only expressing the structure and the operation principle of mechanical systems, but also finding out the one or more components which causing the malfunction of entire system by finding the causal chain of reasoning.

In Figure 4 each node and attribute explained in Table 1.

4 FAULT DIAGNOSIS SYSTEM

According to section second of the description, this document use the method of manual extraction to extract deep knowledge about diesel engine VE distribution pump in the literature [7], the causal relationship generated the XML document like Figure 4. And based on the establishment of the causal relationship mode, the fault diagnosis system can be build, and the system interface is shown in Figure 5.

Step 1, detecting the possible fault component.

In the "Select Type" click "Function Failure". At this point, all the parts and their functions will appear in the "Information" text box. Then select the most likely failure component in the "Start Part" down list according to the functions of these components. For example: the failure is that engine cannot stop the oil, and the function of oil switch is stopping the oil, so selecting "switch the oil" and checking the corresponding components of the equipment. If there is a problem on the oil switch, you can click "No" in the "Function". If there is

no problem on "switch the oil", it means that the initial parts selection is wrong, and needing to re-select according to function.

Step 2, searching the fault source according to the search path.

Click on "Child". At this moment, all the components and their functions, which have direct causal relationship between the starting components, will be displayed in the "Information" text box. Then selecting the name of fault component in the text box directly and clicking "Select" button. And now, the names and functions of the corresponding components will appear in the "Current Components" text box. Then the following operation is same as the step 1. If it still has fault after checking that the second step is repeated until the search to the bottom node. If all nodes on this search path are no problems, it needs to search other path. This time, you can through the "Parent" button to return. If all nodes on a layer have no problem, but there is a problem of its parent node, you can also find the faulty component. For example in the step 1, if the oil switch is no problem in the parent node, it can be concluded that there is a problem on the oil switch. After finding the cause, you can indicate how to deal with in the notes, and provide a reference for the next repair.

Step 3, recording the troubleshooting process and results.

Click on "Detailed Investigation". At this moment, all of the investigation process will be displayed in the "Information" text box, you can make the appropriate changes to the details, and then click on the "Enter" button, the contents of the text box will be saved. Alternatively, you can also click on "Generate Case", then the system will generate and store the cases with the malfunction and the appropriate approach for reference.

Through above description of the procedure, even if the maintenance staff does not know the

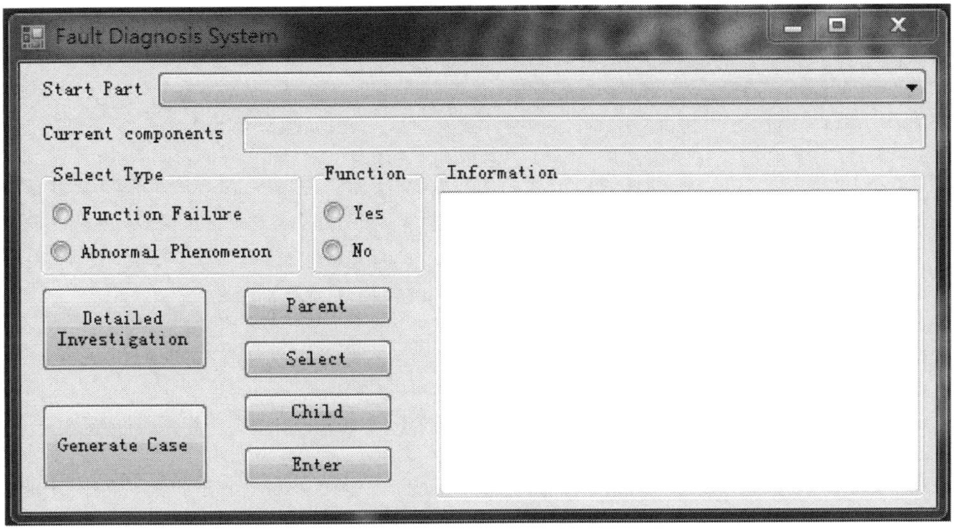

Figure 5. Fault diagnosis system.

original structure of the equipment well, he only needs to follow the operation process provided by the system, and he also can basically check out the functional faults and the abnormal phenomenon. Obviously, the system can also be extended to other functions. Such as increasing related information of the maintenance procedures, providing structures illustrated, etc. But these did not affect the idea of this paper.

5 TEST

Here a fault case of Iveco 40 L10 automotive diesel engine system, which is difficult to start, is taken into account using the system introduced in preceding section.

The circumstances of failing to start engine means that the engine itself may have problems, or some problems of one or more systems which have direct causal relationship with the engine. Aiming at this kind of situation, the fault diagnosis system will put forward the troubleshooting solutions according to the fault diagnosis network stored early. See Table 2.

Check sequence in the plan is according to the size the maintenance staff set early based on possibility of parts happened to failure, which can be changed.

If the engine is fine, the maintenance staff can check other components according to the prompt of the system which have a direct causal relationship the engine, the investigation order based on the size of the possibility of a list, either. The check sequence listed below is an important component

Table 2. The inspection plan of engine.

No.	Mechanism	Check project
1	Fuel tank	Oil or not?
2	Channel for oiling	Clear or not? Leak or not? Air or not?
3	Preheating plug	Failure or not? Preheating time achieve?
4	Oil extraction	Is front angle correct?
5	Fuel injector	Is needle valve stuck? Is nozzle blocked? Is atomizer OK? Is opening pressure normal?
6	Cylinder	Is compression pressure normal?
7	Start machine	...

of VE distribution pump, which effects engine start, see Table 3.

It can be found that the investigation scheme of fault diagnosis system is closely related to the underlying causal relationship network. If maintenance staff find the reason, the system will be based on the stored information about previous record to provide the corresponding treatment methods, for example: when the oil solenoid valve failure, the system will prompt: all parts on VE dispensing pump removed and thoroughly cleaned, refitted VE distribution pump and check. According to the prompt, the maintenance staff's operation can be carried out immediately, which can improve the overall efficiency of maintenance.

Table 3. Inspection plan of VE distribution pump.

No.	Mechanism	Check project
1	Oil pump	Weather start oil meets the requirements?
2	Return oil screw	Whether confusion with the oil inlet screw?
3	Oil inlet screw	Whether confusion with the return oil screw?
4	Electromagnetic valve	Move or not?
5	Battery	Is voltage enough or not?
6	Head plug	Is seal intact?
7	Oil outlet valve	Is model correct?
8	Governor	Is fly hammer rotating?
		Whether to move the sliding cover?
		Is starting rod flexible?

6 SUMMARY

This article focus on the present situation in which modern fault diagnosis increasingly relies on knowledge and technology manuals play an irreplaceable role in maintenance actions as the carrier of using and repairing knowledge. Especially the manuals, which introduce equipment' structure and working principle, become important foundation for maintenance and technical staff at all levels to understand equipment deeply and develop maintenance ability. Put forward new fault diagnosis system based on deep knowledge in manuals, and use it to make unstructured implicit knowledge as support to diagnosis and maintenance work.

Overall, the method put forward in this article is able to assist maintenance staff in diagnosis, overcome the lack of experience, and identify come unexpected fault. Moreover, the method provides new thought and way to expand the application of technical manuals, especially the use of Interactive Electronic Technical Manual.

REFERENCES

[1] Liu Fujun, Qian Yanling, Yin Xiaohu. The construction of "equipment support cloud", achieve "knowledge center" for equipment support [C]. 2013 integrated equipment support theory and technology conference. Changsha: Hunan, 2013: 495–500.

[2] GJB 6600 1-2008, Equipment interactive electronic technical manuals[S].

[3] ASD & AIA & ATA S1000D. International specification for technical publications utilizing a common source database (Issue 4.1) [S], 2013.

[4] Chen Aixiang, Chen Qingliang, Pan Jiuhui. Fast diagnosis algorithm through the diagnosis chart analysis[J]. Chinese Journal of computers, 2009, 32 (8): 1470–1485.

[5] Chandrasekaran B, Smith J W, Sticklen J. Deep models and their relation to diagnosis [J]. Artificial Intelligence in Medicine, 1998, 1(1):29–40.

[6] Yang Jun, Feng Zhensheng, Huang Cowley. Equipment intelligent fault diagnosis technology [M]. Beijing: National Defence Industry Press, 2004:24–25.

[7] Du Xianping. The diesel engine VE distribution pump structure and maintenance of graphic [M]. Beijing: National Defence Industry Press, 2004.

[8] Wei Moji, Yu Tao. The professional document semantic annotation method based on domain ontology. Computer Applications, 2011, 31 (8): 2138–2142.

[9] Li Jing, Chen Hao, Chen Zhicheng. Study on the current ship's maintenance management system of knowledge and the construction of [J]. Value Engineering, 2012, 31 (28): 284–285.

[10] Zheng Shaohua, Qian Yanling, Extraction product functional knowledge based on causal ordering theory, Japan advanced equipment management and maintenance technology conference [C], 2012: 65–7.

Advances in Future Manufacturing Engineering – Yang (Ed.)
© 2015 Taylor & Francis Group, London, ISBN 978-1-138-02817-3

Design and implementation of remote workpiece detection system

Wenwen Gu, Mengyou Huo & Yuming Liu
Research Institute of Digital Technology, Shandong University, Jinan, China

ABSTRACT: Expansion of production scale and economic development has put forward new demands for the quality of the workpiece and workpiece detection technology. The traditional artificial detection methods are inefficient, low-accuracy and labor-intensive, so designing of a new remote workpiece dimension detection system is particularly important.

Keywords: dimension measurement; freeRTOS; image processing

1   INTRODUCTION

After thirty years of rapid economic development, China has become the second largest economy country in the world. The manufacturing sector constitutes the main body of the national economy, leading the development of China's economy. Workpiece detection is an essential part of manufacturing, and intelligence and automation level of detection technology is especially important for improving the overall competitiveness of the manufacturing sector. However, with the continuous expansion of production scales, utilization of machine workpiece detection method is not high enough. The original manual contact detection methods are more common, but human eye recognition cannot reach the machine measurement accuracy, thus affecting the production efficiency and accuracy. In some high-speed production processes, artificial is unable to complete the real-time detection. So for large quantities of the workpiece, people can only use sampling approach instead of bulk detection, which will result in uneven quality. In addition, environment of many factory floors is so harsh, and inspectors are in the face of dust and harmful gas all the year round. With site operation at high temperatures, especially at night, inspectors are easy to produce negligence due to excessive fatigue, resulting larger deviation in detection, so the accuracy is not guaranteed, and may even lead to an industrial accident.

Therefore, using unattended mode in the production workshop, to make use of automation equipment to realize remote detection will has a realistic significance. Workers through computers to observe production situation in real-time, and be respond promptly to the scene. For the steel workpiece to be detected, computers carry out a mathematical operation by means of image processing, to obtain the image details and complete testing. In this way, it can do real-time analysis and real-time control, eliminating the need for manual measuring, and increasing the accuracy of detection.

In this paper, we design a new remote workpiece detection system based on embedded system. The embedded technology, image compression technology, network technology, machine vision technology is applied in the workpiece dimension detecting system. Based on the study of application status at home and abroad work detection technology, we put forward the overall design scheme of remote detection system, and complete hardware and software design of embedded WEB server.

2   OVERALL DESIGN OF DETECTION SYSTEM

2.1   *Overall structural design*

Remote workpiece detection system consists of embedded WEB server, client computers, LAN equipment or WAN equipment. Embedded WEB server is the embedded terminal to achieve WEB server functionality, responsible for on-site signal acquisition and data exchange; Client GUI interface is responsible for receive images and image processing. Overall structure framework of remote workpiece detection system is shown as Figure 1.

2.2   *Embedded WEB server hardware architecture*

Embedded WEB server hardware mainly includes central processor module, Ethernet module, image sensor module, SD card module, memory module, power module, download module, etc. Embedded WEB server hardware architecture is shown as Figure 2.

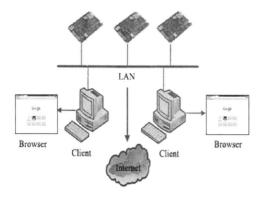

Figure 1. Overall structure framework of detection system.

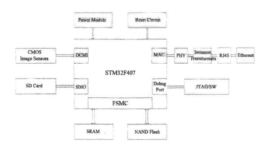

Figure 2. Embedded WEB server hardware architecture.

2.2.1 Central processor module

This system uses a single processor solution, with characteristics of digital, networking, low power, low cost and small size. Processor is STM32F407 produced by ST Company, is a high-performance and low-cost processor based on ARM Cortex-M4 and 32-bit DSP cores. STM32F407 uses a unique ART accelerator to realize memory code zero-wait execution, with multi-channel DMA controller and 7 re-AHB bus matrixes, and supports parallel processing mode, accelerating the rate of data transfer and the program runs.

Cortex-M4 is a high-effective processor core, based on the Harvard architecture. As the same as the microcontroller, Cortex-m4 have the characteristic of multi-function, multi-role, controllability and powerful digital processing capabilities of digital signal controller.

2.2.2 Ethernet module

Physical layer device in this system is DP83848, which is a low power physical layer chip with intelligent power management features. It supports high-speed Ethernet communications, containing various sub-layers and interfaces to code, encode and connect data. This system uses RMII interface mode, pins number needed is a lot less than the MII interface mode, signal line reduced from 16 to 8, data sending and receiving form 4 bit to 2 bit, which reduce the cost of the system and save the port.

2.2.3 Image sensor module

STM32F407 processor provides specific image sensor interface (DCMI) to connect digital image sensor. Digital camera interface is a synchronous parallel interface, receiving YCbCr422/RGB565 progressive scan image. Image receiving module comprises burst and snap, supporting 8 bit, 10 bit, 12 bit and 14 bit connections. It can receive high-speed data streams up to 54 Mbyte/s. Using 8 bit data lines, I2C bus data and clock lines, pixel clock line, horizontal synchronizing signal line and vertical synchronizing signal line.

2.2.4 SD card module

SD card supports SDIO interface and SPI interface modes, while this system uses SDIO interface mode. SDIO interface is a kind of high-speed SD card interface, faster than SPI mode, achieving STM32F407 processor speed communication via APB2 peripheral bus and SD card. SDIO interface via four data lines, command line, clock line pin inserting into detection line to connect SD card, and respectively connect to the resistor to ensure the SD card module to work.

2.2.5 Memory module

SRAM in inside STM32F407 processor is only 192 K, which cannot meet the storage requirements, so SRAM need to be expanded. Embedded web page requires a lot of storage space, so by expanding NAND Flash to meet the requirement of storage. There are two memory modules in this system, which connect SRAM and NAND FLASH by FSMC. FSMC can control different types of external memory at the same time, such as SRAM, PSRAM, ROM, NAND Flash, etc. External memory shared address, data and control signal by FSMC.

3 SOFTWARE DESIGN OF DETECTION SYSTEM

3.1 Software architecture

Embedded server requires multi-task scheduling of Free RTOS to achieve the corresponding functions, including image task and network task. In order to capture the real-time image, detection system requires JPEG compression encoding to reduce the amount of data, thus increase the transmission rate in the case of finite bandwidth. Software architecture of detection system is shown as Figure 3.

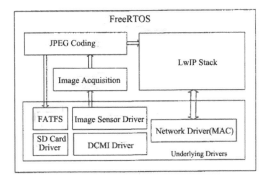

Figure 3. Software architecture of detection system.

3.2 FreeRTOS

Real-time operating system runs on an embedded device can call resources to complete real-time control tasks, making full use of resources and performance of the processor. FreeRTOS is a real lightweight and open source real-time operating system, with good readability, maintainability and portability. The main functions of FreeRTOS are task management, queue management, interrupts management and resource management.

3.3 Image acquisition

Image acquisition tasks are achieving by DCMI interface. Since the DMA transfers data register DMA SxNDTR only 16 bites available, and the maximum number of transmission DMA channel is 65535, similarly, maximum of DMA_BufferSize is 65535.

This system uses VGA 640 * 480 images. Since one frame of image data is beyond the maximum limit on the number of DMA transfer, it needs to execute multiple DMA interrupt. Change the image data storage address in the interruption, to store the image data in an external SRAM, and re-initialize the DMA channel before image reacquisition.

3.4 Image processing

3.4.1 Image preprocessing

a. Color mode conversion. Image sensor to collect images of the system is set in RGB format, which needs to be converted to YCbCr color mode, so as to be supported by JPEG compression coding.

b. Sampling. In the YCbCr color mode, Y component is more important than Cb component and Cr component, so sampling ratio of these three components is set to 4:1:1.

c. Block. The sampled data is divided into blocks. The data for each component is divided into

8*8 data sub-blocks, which will be taken as a unit to complete the discrete cosine transform in the subsequent data processing.

3.4.2 Discrete Cosine Transform

Discrete Cosine Transform (DST) separates the image data into high-frequency signal and low-frequency signal. JPEG compression coding adopts forward DCT, while decompression adopts backward DCT. JPEG compression adopts two-dimensional DCT. The DCT expression is shown as Formula (1):

$$F(a,b) = u(a)v(b)\frac{2}{N}\sum_{x=0}^{N-1}\sum_{y=0}^{N-1}f(x,y)$$
$$\times \cos\left(\frac{2x+1}{2N}a\pi\right)\cos\left(\frac{2y+1}{2N}b\pi\right) \quad (1)$$

3.4.3 Quantify

Quantization is used to reduce data amount of 8*8 data sub-block, which is a lossy compression method. Quantization process requires two quantization tables. Divide the coefficients by the value in corresponding position of the quantization table. Although the processor STM32F406 supports hardware divide, in order to reduce the amount of computation, we can adopt the form of multiplied by the reciprocal.

3.4.4 Entropy coding

Entropy coding is a lossless compression technology, realized by Huffman coding in JPEG coding. Quantized 8*8 data sub-block is divided into two parts: the maximum value in the first column and the first row is low-frequency component, also called DC component; the remaining 63 coefficient is high-frequency components, also called AC component. Other than low-frequency component processing, most of the rest of the coefficient is zero or close to zero. Huffman coding includes DC component of luminance coding, AC component of luminance coding, DC component color code, and AC component color code. DC component calculation is shown as Formula (2):

$$F(0,0) = u(0)v(0)\frac{1}{4}\sum_{x=0}^{7}\sum_{y=0}^{7}f(x,y)\cos\left(\frac{2x+1}{16}0\pi\right)$$
$$\times \cos\left(\frac{2y+1}{16}0\pi\right) = \frac{1}{8}\sum_{x=0}^{7}\sum_{y=0}^{7}f(x,y)$$
$$\quad (2)$$

3.4.5 JPEG format package

For final compressed data, package JPEG file header, file tail and data segment. Among them,

JPEG file header is the beginning of JPEG data, which value is 0xFFD8; JPEG file tail is the end of JPEG data, which value is 0xFFD9. JPEG file is organized in unit of data segment. Data segment includes 2 bytes segment identifier, 2 bytes length identifier and specific content. Segment identifier is used to indicate the purpose of the data segment. After package, the data will become standard JPEG format files.

REFERENCES

[1] Milano A., Perotti F., Serpico S.B., et al. Recent Issues in Pattern Analysis and Recognition [M]. Springer Berlin Heidelberg, 1989:324–337.
[2] Thomas C.E. Acoustic detection of contact between cutting tool and workpiece [J]. The Journal of the Acoustical Society of America, 1987, (5):1858.
[3] H. C, X. C, K.T.V. G, et al. Automatic micro dimension measurement using image processing methods [J]. Measurement, 2002, (2):71–76.
[4] Bulcke J.V.D, Acker J.V., Stevens M. Image processing as a tool for assessment and analysis of blue stain discolouration of coated wood [J]. International Biodeterioration & Biodegradation, 2005, 56:178–187.
[5] Adyanthaya S.K., A. Vijaya P. Emerging Research in Electronics, Computer Science and Technology[M]. Springer India, 2014:685–693.
[6] Liu Y.X., Zhou P. Application of Machine Vision System in the Precision Inspection [J]. Advanced Materials Research, 2012:1382–1386.

Advances in Future Manufacturing Engineering – Yang (Ed.)
© 2015 Taylor & Francis Group, London, ISBN 978-1-138-02817-3

Analysis on the water resource balance in project "water conservation and provision increase" of Jarud, Tongliao

Zhiyong Wang, Huiling Jiang & Jianwen Zong
Tongliao City Water Conservancy Planning Design and Research Institute, Tongliao, China

Fuxiang Sun
Jarud Water Authority, Inner Mongolia, China

ABSTRACT: The balance between the water resource supply and demand in different hydrologic years in the project areas was obtained through comparatively analyzing the total water consumption and the available water supply in different years in the targeted Jarud villages. The balance result analysis showed that the supply and demand of each targeted village were balanced if $P = 75\%$. The analysis on the water resource balance in project "water conservation and provision increase" found that it is necessary to improve the existing extensive agricultural irrigation mode in the project areas and construct a high-standard water-conserving irrigation mode for guaranteeing the good agricultural harvest in drought years and the sustainable utilization of groundwater resources.

Keywords: Jarud; water conservation and provision increase; water resources; balance analysis

1 INTRODUCTION

Jarud is located in the northwest of Tongliao and crossed by the water flows of Great Khingan Mountains. Restricted by factors such as physical geographical and landform conditions, water resources are relatively scarce in hilly regions. The bulletin of Tongliao's 2010 water resources and the perennial underground hydrologic monitoring data showed that the underground water level tends to decline year by year, because the underground consumption increases in Jarud in recent years, the shortage of groundwater level increasingly rises, wells are blindly dug and also unreasonably arranged, and the water yield of the source wells is insufficient. To alleviate the water shortage situation, it is very necessary and urgent to vigorously develop the water-conserving agriculture and rationally adjust the number of wells and layouts through efficient water conservation engineering measures.

2 THE GENERAL SITUATION OF WATER RESOURCES

2.1 *The general situation of the water resources in Jarud*

2.1.1 *The total water resources*
According to the summaries of the water resources of all countries in the Comprehensive Planning Report on the Water Resources of the Inner Mongolia Autonomous Region [1] and the report about the Research and Appraisal on the Development and Utilization of Inner Mongolia Water Resources [2], the multiple-year surface water resource quantity at an average was 402, 68 million m³ in Jarud, the perennial groundwater resource quantity at an average was 517, 9736 million m³, and the perennial water resource quantity at an average was 825, 6375 million m³ after the repeated computation (95, 0166 million m³) for the mutual transformation between the above two quantities was deducted.

2.1.2 *The available water resources*
The available surface-water and underground-water resources quantities were determined according to the results of Inner Mongolia and also by referring to Jarud's Water Resources Evaluation and the Reasonable Water Resources Configuration based on Ecological Economy [3], The Sustainable Utilization Plan for the Water Resources of Tongliao [4], the Bulletin of Tongliao's 2010 water resources, and Inner Mongolia's 2000 plan for allocating the surface water of western Liaohe River Basin.

In Jarud, there are 9 main rivers: Huolin River, A Ri Kun Du Cold River, Wu Bu Kun Du Cold River, Wu Li Ji Mu Ren River, Ba Yan Ta La River, Bai Yin Ju Liu River, Lubei River, Qianjin River, and Ailin River.

The groundwater in the area is mainly replenished from atmospheric precipitation. The groundwater in the plain mainly sources from the lateral runoff of the northern mountainous area and Wu Li Ji Mu Ren River, in addition to receiving atmospheric precipitation. The available perennial groundwater quantity at an average is 354.9042 million m³.

The available surface water resource quantity was 294, 3573 million m³ (the available transitioning water quantity was 74.12 million m³, and the available self-made water resource quantity was 220.2373 million m³), and the available perennial groundwater exploitation quantity at an average was 354.9042 million m³. After the repeated computation (47.7562 million m³) was deducted, the total available water resource quantity was 601.5053 million m³ in Jarud.

2.2 *The general situation of the water resources in the project areas*

The bulletin of Tongliao's 2010 water resources and the perennial underground water-level monitoring data showed that the underground water level tends to decline year by year, because the underground consumption increases in recent years, the shortage of groundwater level increasingly rises, wells are blindly dug and also unreasonably arranged, and the water yield of the source wells is insufficient in Jarud. To alleviate the water shortage situation, it is very necessary and urgent to vigorously develop the water-conserving agriculture and rationally adjust the number of wells and layouts through efficient water conservation engineering measures.

2.2.1 *The total water resources*
According to the summaries of the water resources of all countries in the Comprehensive Planning Report on the Water Resources of the Inner Mongolia Autonomous Region and the report about the Research and Appraisal on the Development and Utilization of Inner Mongolia Water Resources, Jarud's administrative area covered an area of 17, 061 km², the total groundwater resource quantity was 517, 97 million m³, the available groundwater exploitation was 354.9042 million m³, the comprehensive groundwater supply modulus was 30, 400 m³/a · km², and the available exploitation modulus was 20, 800 m³/a · km². However, the north of Jarud is crossed by the runoffs of Great Khingan Mountains, the south is actually the northwest edge of western Liaohe fluvial plain, the north-central is low hilly land, and the northwest is high mountains, because of the Topographical, hydrological and geological characteristics. Therefore, the available exploitation module is determined according to the evaluation on a small range of water resources in the targeted water-conserving farmlands and the actual situation of the local place.

According to the Sustainable Utilization Plan for the Water Resources of Tongliao (March 2003), the groundwater was evaluated from terrain, river system, and administrative partition, and Jarud was divided into three parts: the upstream water supply area of Huolin River, the tributary water supply area of Wu Li Ji Mu Ren River, and the water supply area of Wu Li Ji Mu Ren River. The comprehensive water supply from the upstream water supply area of Huolin River was 97.2 million m³, the supply modulus was 23, 000 m³/a · km², the available exploitation quantity was 48.6 million m³, and the available exploitation modulus was 11, 500 m³/a · km²; the comprehensive water supply from the tributary water supply area of Wu Li Ji Mu Ren River was 174.6 million m³, the supply modulus was 80, 200 m³/a · km², the available exploitation quantity was 69.8 million m³, the available exploitation modulus was 32, 100 m³/a · km²; the comprehensive water supply from the water supply area of Wu Li Ji Mu Ren River was 166.5 million m³, the supply modulus was 92, 000 m³/a · km², the available exploitation quantity was 91.6 million m³, and the available exploitation modulus was 50,600 m³/a · km². The comprehensive groundwater supply for Jarud was 438.3 million m³ according to the plan; the average comprehensive supply modulus was 25, 500 m³/a · km², inclined to be lower compared with the supply modulus (30,400 m³/a · km²) calculated in the summaries. The groundwater characteristic parameters corresponded to the local topographical conditions. In this study, the actual water resources divisions of the local place were seriously considered when the results of Inner Mongolia were designed. In the project areas, Ba Yan Bao Li Gao Village, Yongle Village, Xiangshan Village, and Hanshan Village of Xiangshan Town belong to the tributary water supply area of Wu Li Ji Mu Ren River, and other villages were distributed in the water supply area of Wu Li Ji Mu Ren River; the supply modulus was 80200 m³/a · km² in the tributary water supply area of Wu Li Ji Mu Ren River and 92, 000 m³/a · km² in the water supply area of Wu Li Ji Mu Ren River, respectively; the available exploitation modulus was 32, 100 m³/a · km² in the tributary water supply area of Wu Li Ji Mu Ren River and 50,600 m³/a · km² in the water supply area of Wu Li Ji Mu Ren River. After analysis and calculation, the perennial groundwater resource quantity at an average was 90.4982 million m³ in the project areas, and the available exploitation quantity was 48.1743 million m³. The calculation results are shown in Table 1.

2.2.2 *The available water resource*
The only available water resource in the project areas was groundwater and its available exploitation quantity was 48.1743 million m³.

Table 1. The results of calculating the groundwater resource quantity in the project areas.

County	Town	Village	Area (km²)	Perennial comprehensive water supply modulus at an average (m³/a·km²)	Available exploitation modulus (m³/a·km²)	Comprehensive underground water supply modulus (m³/a·km²)	Available exploitation (m³/a)
Jarud	Lubei Town	Xianfeng Village	18	9.20	5.06	165.60	91.08
		Zhaojiapu	31.5	9.20	5.06	289.80	159.39
		Xinsheng Village	24.67	9.20	5.06	226.96	124.83
	Dao Lao Du Town	Bei Bao Li Gao Village	580	9.20	5.06	5336.00	2934.80
	Xiangshan Town	Ba Yan Bao Li Gao Village	40	8.02	3.21	320.80	128.40
		Central Village	40.67	9.20	5.06	374.16	205.79
		Yongle Village	25.4	8.02	3.21	203.71	81.53
		Xiaoshan Village	52.33	8.02	3.21	419.69	167.98
		Hanshan Village	15.47	8.02	3.21	124.07	49.66
		Wuduandi Village	14.8	9.20	5.06	136.16	74.89
		Wudaojingzi Village	40.93	9.20	5.06	376.56	207.11
	Xiangshan Farmland	Branch 1	19	9.20	5.06	174.80	96.14
		Branch 2	33.33	9.20	5.06	306.64	168.65
		Branch 3	28	9.20	5.06	257.60	141.68
		Branch 5	13.33	9.20	5.06	122.64	67.45
		Daoziba branch	23.33	9.20	5.06	214.64	118.05
In total			1000.76			9049.82	4817.43

3 THE CURRENT AVAILABLE WATER CONDITIONS IN THE PROJECT AREAS

3.1 *The current water supply conditions*

Based on the field survey and investigation on the current conditions of the project areas, there were 387 agricultural electro-mechanical wells to normally work, the water yield of the wells was 30 m³/h, and the water supply capacity of the existing water source during the crop growth period in a normal year was 11.1456 million m³; there were 16 drinking water wells, and the water supply capacity was 1.6352 million m³. The total groundwater supply capacity was 32.3072 million m³, less than 48.1743 million m³, the perennial available groundwater exploitation quantity at an average. Thus, an available resource was still t available to explore in the project areas.

According to the Technical Specification for Water Resources Supply and Demand Forecast Analysis SL429-2008, the available water supply quantity for the project was the minimum among the local available groundwater exploitation quantity, electro-mechanical well water lifting capacity, and user water demand. The above calculation showed that the available water supply quantity was 26.4145 million m³ in the targeted villages if $P = 75\%$.

3.2 *The water consumption condition*

In the project areas, the total area of land was 1000.76 km²; the area of cultivated land was 32,1400 mu; the area of existing irrigable land was 116, 309 mu; the traditional irrigation mode was applied; a total of 26600 people lived there; a total of 81,400 domestic animals were breed, in which 12, 776 were big and 68, 651 were small. The total water consumption quantity of the targeted villages was 26.4145 million m³ in the project areas, the water quantity drunk by people and animals was 826,500 m³, and farmland irrigation consumed 25.588 million m³.

4 ANALYSIS ON THE BALANCE BETWEEN WATER SUPPLY AND DEMAND IN THE PROJECT AREAS

Balance result analysis showed that the water supply and demand was balanced in all the targeted villages if $P = 75\%$. The calculation results in the balance analysis are as shown in Table 2.

4.1 *The available potential for the development and utilization of water resources in the project areas*

From the above analysis, the available groundwater quantity was 48.1743 million m³, the total water consumption quantity was 26.4145 million m³, and the groundwater exploitation quantity was 54.83% of the total available groundwater resource. The groundwater was abundant, indicating a certain

Table 2. The balance between water supply and demand in all the targeted villages in different hydrologic years (Unit: m³).

Town	Project village Zai Village	P = 75% Total water consumption quantity	P = 75% Available water supply for project	P = 75% Supply and demand balance analysis
Lubei town	Xianfeng Village	50.41	50.41	0.00
	Zhaojiapu	74.91	74.91	0.00
	Xinsheng Village	81.20	81.20	0.00
Daolaodu town	Bei Bao Li Gao Village	1316.77	1316.77	0.00
Xiangshan town	Ba Yan Bao Li Gao Village	118.26	118.26	0.00
	Central Village	125.39	125.39	0.00
	Yongle Village	25.17	25.17	0.00
	Xiangshan Village	166.59	166.59	0.00
	Hanshan Village	47.77	47.77	0.00
	Wuduandi Village	46.91	46.91	0.00
	Wudaojizi Village	140.91	140.91	0.00
Xiangshan Farmland	Branch 1	73.61	73.61	0.00
	Branch 2	157.46	157.46	0.00
	Branch 3	93.77	93.77	0.00
	Branch 5	25.81	25.81	0.00
	Daoziba Branch	96.51	96.51	0.00
In total		2641.45	2641.45	0.00

potential was available for the development of water resources in the project areas. It could meet the needs of the designed irrigation area.

4.2 The balance between water supply and demand

The irrigated land area was 116,309 mu in the targeted villages. In the project, the improved water-conserving irrigation area was 40,124 mu; drip irrigation under mulch was applied to the designed area (40124 mu); the water-supply source was groundwater; the water demand in the designed irrigation area (40124 mu) was 4.7972 million m³; the water demand by people and animals was 1.1177 million m³; the total water demand was 22.6756 million m³, accounting for 47.07% of the total available exploitation quantity. The available groundwater exploitation quantity in the project could totally meet the water demand.

5 CONCLUSION

The analysis of the water supply and demand in the project found that the available groundwater exploitation quantity and the available water supply capacity for the project could meet the drinking water demand of people and livestock and the agricultural irrigation water demand in the normal

hydrologic years, but the available groundwater exploitation quantity in drought years could not meet the drinking water demand of people and livestock and the agricultural irrigation water demand. To ensure a good harvest and a sustainable utilization of groundwater resources in the project areas, it is necessary to improve the existing extensive agricultural irrigation mode and construct a high-standard water-conserving irrigation mode.

REFERENCES

[1] Inner Mongolia Water Conservancy and Hydropower Survey and Design Institute. The Comprehensive Planning Report on the Water Resources of the Inner Mongolia Autonomous Region. 2005, 9.
[2] Inner Mongolia Water Conservancy and Hydropower Survey and Design Institute, Inner Mongolia Water Hydrology Bureau, and Inner Mongolia Water Management Center. Research and Appraisal on the Development and Utilization of Inner Mongolia Water Resources. April, 2008.
[3] Inner Mongolia Agricultural University. Jarud's Water Resources Evaluation and the Reasonable Water Resources Configuration based on Ecological Economy. 2007.
[4] Tongliao Water Supplies Bureau. The Sustainable Utilization Plan for the Water Resources of Tongliao. February, 2003.

Advances in Future Manufacturing Engineering – Yang (Ed.)

Study on the application of finite element method and CAE technology in modern mechanical engineering

Lie Yang

Yunnan Open University, Kunming, Yunnan, China

ABSTRACT: Analysis and discussion are launched for the application of CAE technology, with the finite element method as the typical representative, in the mechanical engineering field from the perspective of mechanical engineering. The current status of CAE application in mechanical engineering field is summarized, and the general thoughts of finite element and application advantages are discussed. Finally, the main contents of the application of finite element method are illustrated, hoping to draw attention from all related parties.

Keywords: finite element method; CAE technology; mechanical engineering; advantages; application

1 INTRODUCTION

The finite element method was born in the middle of the 20th century, and with the development of computer technology and computer method, it has already developed to be the most effective calculation method in the computational mechanics and Calculation engineering field. Many engineering analysis problems, such as the displacement field and stress field analysis in solid mechanics, electromagnetic field analysis in the electromagnetism, vibration characteristics analysis, temperature field analysis in thermo physics, and flow field analysis in hydromechanics, etc. can be summarized as the problem of working out the control equation in the given boundary condition. The occurrence of finite element technology provides forceful tools to the manufacture and design of mechanical engineering structure, and it can also solve many problems that cannot be solved by the previous manual calculation, which has brought tremendous economic and social benefits for the enterprise. In order to design excellent and reliable mechanical products in the modern mechanical industry, it is impossible to realize it if the calculation and auxiliary design analysis is not applied. Therefore, production and design departments pay lots of attention to the application of advanced calculation technology in the manufacturing and designing process. With the finite element method, it can further improve and optimize the quality of product design in the modern mechanical engineering field.

2 APPLICATION STATUS OF CAE TECHNOLOGY IN MECHANICAL ENGINEERING FIELD

In the traditional mechanical engineering product manufacturing field, the operation steps are relatively tedious. The expert mainly designs the product preliminarily according to the individual experience, and forms the model machine. And then, it will obtain the related data through prototype testing, and the test data will be further analyzed. Eventually, the prototype shall be optimized reasonably according to the analysis result. And the above mentioned steps shall be repeated till it reaches a satisfactory design effect. In the whole set of processing flows, it is difficult in operation, and it also consumes much time, and it is really difficult to satisfy the demands proposed by the current market for the product performance in the mechanical engineering field.

With the rapid development of computer technology at present, the application of the technology in the mechanical engineering field also presents remarkable development trend. The computer-aided engineering based on the virtual prototype and its related technologies are closely combined to the mechanical engineering field. Moreover, it is fully reflected in several links, such as the engineering analysis and design. Till now, the superiority of CAE technology in the mechanical engineering field is irreplaceable, meaning that CAE technology, with finite element method as the typical representative, becomes one of the key tools for enterprises engaging in the mechanical engineering field to win in the market competition.

It is considered in related studies that through the comprehensive application of CAE technology, it can reduce or eliminate the prototype manufacturing and test links, so as to reach the product development and preparation circle. Besides, CAE technology is also of great value in the control of product manufacturing cost. Typical CAE technology includes the finite element method, the finite difference method, as well as the boundary element method, etc. among which, the finite element method is widely applied in the mechanical engineering field.

3 GENERAL THOUGHT OF FINITE ELEMENT METHOD

Finite element is actually an analysis method which is applied in the ordinary differential and partial differential equation solving. In other words, as long as it can be included into the category of differential equation solving, it can be analyzed with finite element method. Therefore, in the structure field, thermoelectricity, fluid field, acoustics field, as well as the mechanical field, the finite element method has accurate application value.

Seen from the application features of the mechanical engineering field, during the process of applying finite element method in the analysis of related problems, it shall divide the region into several units connected by several nodes with the discrete method, for constructing a complete element stiffness matrix. And then, several independent element stiffness matrixes shall be combined to form a complete stiffness matrix. On this basis, equivalent transformation shall be applied to transfer the non-node load to form the nodal load. Meanwhile, the element load array shall be applied to form an overall load matrix through integration. Through the above mentioned treatments, it can obtain a total stiffness equation corresponding to the structure, and meanwhile, the boundary condition can be introduced for working out the equation. Eventually, a solution similar to the structure can be made.

4 MAIN ADVANTAGES OF FINITE ELEMENT METHOD

In the modern mechanical engineering field, there are many advantages in the application of finite element method, which are mainly reflected in the following aspects: firstly, in the process of analyzing the mechanical structure problem with finite element method, element may form the set through different connection methods, and the shape of the element is not restrained clearly, for structures with relatively complicated geometrical shape, finite element method also has strong adaptability, and the differences between the approximate solution and practical solution is quite small. Secondly, in the finite element analysis process for the mechanical structure, the solving steps required shall be characterized by systematization and standard features. Therefore, it can support the development of computer system operating procedures, and it also has good universal advantages. Thirdly, in the process of analyzing the mechanical problems with the finite element method, the general thought is to introduce the boundary condition based on the construction of overall stiffness equation. Therefore, no matter for the boundary condition or for the structural model, it has relatively good independent advantages. Fourthly, there is no need to be applicable for the corresponding interpolation function of the entire structure during the analysis of mechanical structural problems with finite element method, and each element shall correspond to its own interpolation function. Therefore, it is relatively easy form the perspective of mathematical treatment, especially for the analysis of complicated structure, and it can simplify the solving steps to certain degree. Fifthly, for the uneven continuous medium that cannot be treated by CAE technology, it is relatively easy and accurate to treat with the finite element method. Sixthly, the finite element method is applicable for the analysis of mechanical structure coupled fields, and meanwhile, it has precise value for the solving of nonlinear problems. Seventhly, the finite element method can support the optimization of product design in the entire mechanical structure, and it can be combined to the analysis method of structure reliability, so that it can give full play to the advantages of two analysis methods.

5 APPLICATION OF FINITE ELEMENT METHOD IN MECHANICAL ENGINEERING FIELD

In the support of traditional technical condition, the product design and manufacture in mechanical project have certain particularity. Firstly, the production lot size is relatively huge, so it can accumulate experience in the product manufacturing and operating process, and meanwhile, it can transform and optimize. Secondly, the prototype testing cost of some products is relatively low in the mechanical engineering field, and its reliability is higher than that of the computer simulation test. Thirdly the majority of products and spare parts of the mechanical engineering field refers to the similar products and spare parts, and on that basis, it is obtained through improvement or approximation design. Therefore, the performance is also related to the performance of spare parts. Just because of the above features, the finite element method cannot be widely applied in the mechanical engineering field, and it is also applied in the analysis of some key devices. However, with the further intensification of competitive relationship in the mechanical engineering field, and in the context of constantly improving economic and social environmental benefit, the application of finite element method in the mechanical engineering field grows more and more mature. Generally speaking, the mainly application points are reflected in the following aspects:

① Statics analysis method

Statics analysis method refers to the analysis of stress and strain of 2D or 3D mechanical structure, and it is the main analysis mode in the application of finite element method. When the load on the mechanical structure does not vary or vary with time slowly, workers shall conduct in-depth analysis with the statics analysis method.

② Modal analysis method

Modal analysis method is one of the typical plans of the kinetic analysis. By using this method, it can further the study on the inherent frequency of mechanical structure, and it can also achieve the typical vibration performance of the structure (such as the natural vibration, etc.). It shall pay attention to this: in the analysis process, the load on the mechanical structure can only be the pre-stress load or displacement load.

③ Thermal-stress analysis

Thermal-stress analysis is also one of the typical finite element analysis methods in the mechanical engineering field. Workers can study the corresponding temperature stress relation when there are differences between the working temperature and installation temperature (or under normal working state, there are differences in the internal temperature) with this method.

④ Contact analysis method

Contact analysis method is a sate nonlinear analysis method, applicable for the analysis of corresponding state and normal force in the contact reaction of two works. From the perspective of the mechanical engineering field, since the force transmission between structures is realized on the basis of contacts, the application value is quite outstanding. But, it is relatively limited in calculation ability, and as a result, it is seldom applied in the mechanical engineering field.

⑤ Buckling analysis method

Buckling analysis method is a kind of nonlinear geometric analysis, applied for confirming the corresponding critical load and buckling mode of the structure when it gets unstable. For instance, the pressure lever stability problem shall belong to the research category of buckling analysis.

6 CONCLUSION

In the context of rapid modern scientific technology development, CAE technology has always been in the dynamic development and exploration process, and the mechanical engineering field can also try to apply the CAE technology comprehensively to reach the purpose of improving the product design efficiency. Meanwhile, it can also consider about the proper control over the development cycle of mechanical products through applying the finite element method. Therefore, more and more enterprises and technicians start to realize the production potential and value of CAE technology. And apply various CAE technologies to the finite element method as the representative in the development of related products in mechanical field, receiving perfect effect. Therefore, it is considered that with the constant maturity of the finite element method application, its value in the mechanical engineering field will be further developed.

REFERENCES

[1] Zhao Zhong, Peng Xudong, Sheng Song'en, et al. Numerical Analysis of Laser Textured Mechanical Seals with a Porous Sector Face [J]. CIESC Journal, 2009, 60 (4): 965–971.

[2] Xu Jijin, Chen Ligong, Ni Chunzhen, et al. Effect of Mechanical Stress Relieving Method on Welding Residual Stress [J]. Journal of Mechanical Engineering, 2009, 45 (9): 291–295.

[3] Zhou Guobin, Lu Yanlin, Universal Weight Function Method for Mechanical and Thermal Loading and Its Application [J]. Journal of Mechanical Engineering, 2007, 43 (6): 116–121.

[4] Su Xiongbo, Yang Jun, Hou Shunli, et al. Analysis on mechanical property of lab equipment of multilayered drying section based on finite element method [J]. China Pulp & Paper Industry, 2012, 33 (4): 34–38.

[5] Ying Yinqiong, Jiang Hong, Li Cheng, et al. Structural Analysis of PSD Mechanical System Based on FEA [J]. Electric Drive for Locomotives, 2012, (2): 75–78.

[6] Hu Yanjuan, Wang Zhanli, Dong Chao, et al. Integrated Symmetry Fuzzy Number and Cutting Force Prediction for Finite Element Method [J]. Journal of Vibration, Measurement & Diagnosis, 2014, 34 (4): 673–679.

[7] Liu Yushan, Xu Anping, Fu Dawei, et al. Pre-offset-based Approach to Optimizing and Compensating Weld-line of Tailor-welded Blanks [J]. Journal of Mechanical Engineering, 2012, 48 (16): 59–63.

[8] Zhang Jun, Zhao Wenzhong, Zhang Weiying, et al. Research on Acoustic-structure Sensitivity Based on FEM and BEM [J]. Journal of Vibration Engineering, 2005, 18 (3): 366–370.

Advances in Future Manufacturing Engineering – Yang (Ed.)
© 2015 Taylor & Francis Group, London, ISBN 978-1-138-02817-3

Manuscript digital recognition system of automobile centre console

Guxiong Li
Guangzhou College, South China University of Technology, Guangzhou, Guangdong, China

Kai Huang
South China National Centre of Metrology, Guangzhou, China

ABSTRACT: One automobile centre console is designed for automotive users who prefer to input characters or digits by handwriting on a touch screen. Because it is one of the most natural and convenient ways for human-machine interaction, without the distraction of driver's attention, and it would greatly improve traffic safety. This paper proposes a method which can recognize handwritten digits by using Discrete Hopfield Neural Network. The experimental results show that the proposed algorithm is practical and reliable.

Keywords: digital recognition; automobile centre console; neural network

1 INTRODUCTION

The degree of car automation is increasing with the development of science and technology. One automobile centre console is designed for automotive users who prefer to input characters or digits by handwriting on a touch screen. Because it is one of the most natural and convenient way for human-machine interaction, without the distraction of driver's attention, and it would greatly improve traffic safety.

This problem has a strong engineering representation. In daily life, we often encounter problems with noise symbol recognition. Such as recognition of automobile license plate in the transportation system, digital recognition in automatic classification line of mail, parcels, express, and factory logistics.

As a new technology of artificial intelligence, artificial neural network reflects many basic features of human brain function. It simplifies, abstracts and simulates the human nerve system. Artificial neural network is one of the most promising subjects, which has broad application prospects in the fields of meet and optimize the restrictive conditions, data compression, prediction, judgment of danger, control, pattern recognition, diagnose of content, addressable memory and data fusion of multi sensors. To solve the combinatorial optimization problem by using artificial neural network is a shortcut. Hopfield network is a kind of artificial neural network.

There are several ways to carry out character recognition, respectively is the use of artificial neural network, probability statistics, fuzzy algorithm. Traditional digital recognition methods are not good for numeral recognition in case of disturbances. Discrete Hopfield neural network has the function of associative memory, which can achieve satisfactory results and the calculative convergence is very quick for digital identification.

This paper proposed a method to solve identification problems by using Hopfield network.

Neural networks are parallel, distributed, adaptive information-processing systems so the calculation will not trigger exponential explosion with the increase of dimension. It is particularly effective for the high-speed computation of recognition.

2 THE STRUCTURE OF HOPFIELD NEURAL NETWORK

Hopfield neural network is an interconnection network [1]. It introduces the concept of energy function, which is similar to the Lyapunov function. The topological structure of neural network (represented by the connection weight matrix) corresponds to the problem (described by the objective function), and that convert it into an evolution problem of neural network dynamical system. The evolution process is a nonlinear dynamic system, which can be described by a set of nonlinear difference equations (discrete) or differential equations (continuous). The stability of the system can be analyzed by using energy function. If conditions were met, energy of certain energy function would decrease continuously in the process of network operation, and tend to the stable equilibrium state finally.

For a nonlinear dynamic system, the state of a system from an initial value may have the following evolutional results: asymptotic stability of sub (attractor), limit cycles, chaos, state divergence [2].

The transformation function of the artificial neural network is a bounded function, so state of the system does not diverge. Artificial neural networks often solve certain problems by using asymptotic stability point. Treating the stable point of the system as a memory, the evolution from the initial state to a stable point is a process of searching for memory. Considering the stable point of the system as a minimum of functions and treating energy function as the objective function of an optimization problem, the evolution from initial state to stable point is a process of solving the optimization problem. Therefore, evolution of Hopfield neural network is a calculation of associative memory or a process of solving optimization problem. Do not need to calculate actually, you constitute a feedback neural network firstly, and the appropriate design of its connection weights and input can achieve this goal [3].

Hopfield neural network is divided into discrete type and continuous type.

3 DISCRETE HOPFIELD NEURAL NETWORK

As to Discrete Hopfield neural network, the output of each neuron node can have only two states: 1 or −1. The activation function of neural network is generally represented by threshold function. 1 indicates that neurons in the activated state, and −1 expressed neuron in the inhibitory state. It is a single layer network [4]. The structure of discrete Hopfield neural network, which is composed of 3 neurons, is shown in Figure 1.

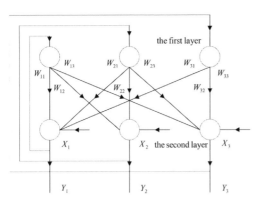

Figure 1. Structure of discrete hopfield neural network.

In Figure 1, the first layer just acts as the input of the network. It is not a real neuron, without calculation function. The second layer is neuron, perform summation of product of the input information and the weight coefficient, and produce the output information after treatment of a nonlinear function. F is a simple threshold function. If the output of neural is greater than the threshold, then the output value of neurons is 1. If the output of neural is smaller than the threshold, then the output value of neurons is −1 [5].

For a neuron of two values, its formula is as follows:

$$U_j = \sum_i W_{ij}Y_i + X_j \qquad (1)$$

X_j is the external input, and

$$\begin{cases} Y_j = 1, & u_j \ge \theta_j \\ Y_j = -1, & u_j < \theta_j \end{cases} \qquad (2)$$

The state of the network is information collection of the output neurons. For a network whose output layer composed of N neurons, the state of t time is an n-dimensional vector.

$$Y(t) = [Y_1(t), Y_2(t), ..., Y_n(t)]^T \qquad (3)$$

$Y_i(t)(i = 1, 2, ..., n)$ can be taken as 1 or, so $Y(t)$, an n-dimensional vector, has 2^N states. Therefore, there is 2^N state of network.

Considering the general state of the node, $Y_j(t)$ represents the j neuron, the state of node j at t time. The state of nodes in the next moment $(t+1)$ is

$$Y_j(t+1) = F[U_j(t)] = \begin{cases} 1, U_j(t) \ge 0 \\ -1, U_j(t) < 0 \end{cases} \qquad (4)$$

$$U_j(t) = \sum_{i=1}^{n} W_{ij}Y_i(t) + X_j - \theta_j \qquad (5)$$

If $W_{ij} = 0$ and $i = j$, it mean that an output of neuron does not feedback to its input. Then, this discrete Hopfield neural network is called non-self-feedback network. If $W_{ij} \ne 0$ and $i = j$, it mean that an output of neuron will feedback to its input. Then, this discrete Hopfield neural network is called self-feedback network.

4 OPERATION MODE OF NETWORK

Hopfield neural network runs in a dynamic mode, and its working process is the evolution of the neuron state. Evolution is in reducing direction of Lyapunov energy function from the initial state, until a steady state is reached. The steady state is the output of network [6].

There are two main working mode of Hopfield neural network.

1. serial (asynchronous) working mode. At any moment t, only one neuron i (random or deterministic) changes according to equation 4 and 5, while other neurons state is invariant.
2. parallel (synchronous) work mode. At any moment t, the state of part or all of the neurons change simultaneously.

Let us illustrate operation procedure with serial (asynchronous) working examples.

Step 1, initializing network.
Step 2, selecting neurons i randomly from the network.
Step 3, calculating the input $u_i(t)$ of neuron i.
Step 4, calculating the output $v_i(t+1)$ of neuron i, while the output of other neurons remain unchanged.
Step 5, determining whether the network reaches a stable state. If the condition is met, program will terminate the operation. Otherwise, continue with step 2.

The network stable state is defined here as: the network state will no longer change since a moment later.

$$v(t + \Delta t) = v(t) \qquad (6)$$

Here $\Delta t > 0$.

5 NETWORK STABILITY

As to the stability of continuous Hopfield neural network, it can be derived by definition of energy function [7].

The definition of energy function is as follows:

$$E(t) = -\frac{1}{2}\sum_{j=1}^{n}\sum_{i=1}^{n}w_{ij}V_i(t)V_j(t) - \sum_{j=1}^{n}V_j(t)I_j$$
$$+ \sum_{j=1}^{n}\frac{1}{R_j}\int_0^{V_j(t)}g^{-1}(V)dV \qquad (7)$$

In the formula, $g^{-1}(V)$ is the inverse function of $V_j(t) = g_j(U_j(t))$.

The time derivative of the energy function is

$$\frac{dE(t)}{dt} = -\frac{1}{2}\sum_{j=1}^{n}\sum_{i=1}^{n}\left[w_{ij}\frac{dV_i(t)}{dt}V_j(t) + w_{ij}V_i(t)\frac{dV_j(t)}{dt}\right]$$
$$- \sum_{j=1}^{n}\frac{dV_j(t)}{dt}I_j + \sum_{j=1}^{n}\frac{U_j(t)}{R_j}\times\frac{dV_j(t)}{dt} \qquad (8)$$

If $w_{ij} = w_{ji}$, the above formula can be written as

$$\frac{dE(t)}{dt} = \sum_{j=1}^{n}\sum_{i=1}^{n}\left[w_{ij}\frac{dV_j(t)}{dt}V_i(t) - \sum_{j=1}^{n}\frac{dV_j(t)}{dt}I_j\right.$$
$$\left. + \sum_{j=1}^{n}\frac{U_j(t)}{R_j}\times\frac{dV_j(t)}{dt}\right]$$
$$= -\sum_{j=1}^{n}\frac{dV_j(t)}{dt}\sum_{j=1}^{n}\left[w_{ij}V_i(t) + I_j - \frac{U_j(t)}{R_j}\right] \qquad (9)$$

Put the dynamic equation into the formula (9),

$$\frac{dE(t)}{dt} = \sum_{j=1}^{n}\frac{dV_j(t)}{dt}\times C_j\frac{dU_j(t)}{dt} \qquad (10)$$

$V_j(t) = g_j(U_j(t))$, so $U_j(t) = g_j^{-1}(V_j(t))$, formula (10) can be rewritten as

$$\frac{dE(t)}{dt} = -\sum_{j=1}^{n}\frac{dV_j(t)}{dt}\times C_j\frac{d[g^{-1}(V_j(t))]}{dt}$$
$$= -\sum_{j=1}^{n}\frac{dV_j(t)}{dt}\times C_j\frac{d[g^{-1}(V_j(t))]}{dt}\times\frac{dV_j(t)}{dt}$$
$$= -\sum_{j=1}^{n}\left[\frac{dV_j(t)}{dt}\right]^2\times C_j\times[g^{-1}(V_j(t))]' \qquad (11)$$

If $g(u)$ is monotonically increasing bounded continuous function. Its inverse function is also monotone increasing function. So its derivative must be greater than 0, $[g^{-1}(V_j(t))]' > 0$. We can know that $C_j > 0$, $dV_j(t)/dt \geq 0$, then $dE(t)/dt \leq 0$. $dE(t)/dt = 0$ is true only when $dV_j(t)/dt = 0$.

As can be seen from the above proof:

1. When the transfer function is monotonically increasing and the weight coefficient matrix is symmetric, the network energy will decrease or change over time;
2. The network energy is constant only when the output of neuron will not change over time.

6 METHOD OF MODEL DESIGN

Design a discrete Hopfield neural network with associative memory function, which is required to identify 10 numbers from 1 to 9 correctly. When the numbers have noise interference, this network still has a good recognition effect.

We assume that the network consists of 10 steady states, from 0 to 9. A 10×10 matrix represents each steady state. These matrixes describe the simulation of digital Arabias. Number 1 means an effective coverage, number -1 means a blank

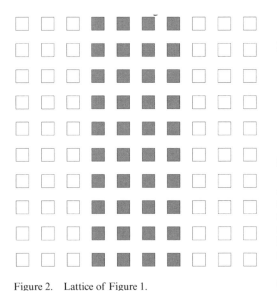

Figure 2. Lattice of Figure 1.

section. Here, green boxe represents 1, colourless block mean –1, as shown in Figure 2. When the network is input with noisy digital lattice, the network output can be close to target vector so as to achieve the correct recognition result.

7 DESIGN STEPS

The steps are as follows: designing digital matrix—constructing a discrete Hopfield neural network—generating digital lattice with noise—digital identification test—result analysis.

1. Designing digital matrix
   Number 1 means an effective coverage. Number –1 means a blank section. Here, green boxe represents 1, colourless block mean –1, as shown in Figure 2.
2. Generating digital lattice with noise
   Noisy digital lattice mean the value change of some lattice position. There are many ways to generate noisy digital lattice. Here we introduce two, fixed noise method and random noise method.
   Fixed noise method is named artificial generation method. It is the method of artificial modification to change digital lattice value of certain position. If demanding different noisy digital lattices, we need to do artificial modifications many times. But surely, this is the question that really matters. Relatively, random noise method is more convenient.
   Random noise method determines the lattice position need modified by using the random number generation method. Because the

value of digital matrix is only 1 and –1, 1 will be replacing –1, or –1 will be replacing 1.
3. Digital identification test
   The noisy digital lattice is input to prepare Hopfield neural network. The output of network is target vector which is most close to the digital matrix, i.e., a digit from 0 to 9. That is a technique to achieve the function of associative memory.

8 EXPERIMENTAL RESULT

As shown in Figure 3 and Figure 4, with the use of associative memory, Hopfield neural network can correctly identify noisy digital lattice. Figure 3 is recognition result as noise intensity = 0.1 (10% digital lattice position have changed), it shows that the formula works rather well.

Further research found that the recognition effect gradually decreased as the noise intensity increases. Figure 4 is recognition result as noise intensity = 0.2, Figure 5 is recognition result as noise

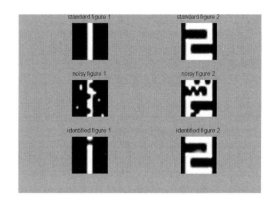

Figure 3. Recognition result as noise intensity = 0.1.

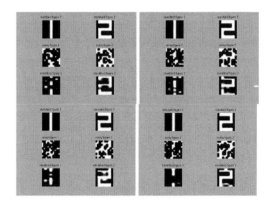

Figure 4. Recognition result as noise intensity = 0.2.

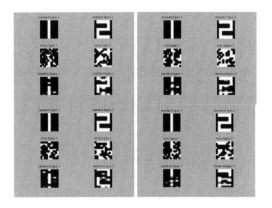

Figure 5.   Recognition result as noise intensity = 0.3.

intensity = 0.3. It is not difficult to see that, when the noise intensity is equal to 0.3, Hopfield neural network feel painful for numeral recognition.

9   CONCLUSIONS

Discrete Hopfield neural network has the function of associative memory. In recent years, many researchers have tried to apply it to various fields, such as water quality assessment, the fault diagnosis of generator, etc. The traditional methods could not solve these problems well. If we combine some optimization algorithms with Hopfield neural network, that would enhance associative memory and highlight its application effect.

REFERENCES

[1] Zhang Liangjun. Practical tutorial of neural network [M]. Beijing:National Defence Industry Press, 2013.
[2] Cui Yonghua. Water quality evaluation based on Hopfield neural network [J]. Water Resources Protection, 2013, 23(3):13–16.
[3] Shi Hui. Principle and application of artificial neural network [M]. Beijing: Science Press, 2013.
[4] Song Tao. Analysis of the project risk based on Hopfield neural network [J]. Statistics and Decision, 2013, 3:23–26.
[5] Sun Shouyu. An improved algorithm of Hopfield network to solve the TSP [J]. Chinese Journal of Electronics, 2013, 23 (1): 77–81.
[6] Feisi science and technology R & D center. Implementation of neural network [M]. Beijing: Electronic Industry Press, 2013.
[7] Zhu Shuangqi. The application of neural network [M]. Shenyang: Northeastern University Press, 2013.

Computer science and applications

Advances in Future Manufacturing Engineering – Yang (Ed.)
© *2015 Taylor & Francis Group, London, ISBN 978-1-138-02817-3*

A novel speech recognition method based on LLE and SVM

Jiabao Gao

Center of Modern Education Technology, Hechi University, Guangxi, Yizhou, China

ABSTRACT: To effectively recognize speech and improve its classification and recognition accuracy, it is needed to perform nonlinear dimensionality reduction for speech feature data lying on a nonlinear manifold, embedded in high-dimensional acoustic space, and conduct classification well. This paper proposes a novel speech recognition method, based on Locally Linear Embedding (LLE) and Support Vector Machine (SVM), which enhances the discriminating power of low-dimensional embedded data and possesses the optimal generalisation ability. The proposed algorithm is used to conduct multi-class data and nonlinear high-dimensional speech emotional feature data. Experimental results on the natural speech emotional database show that the proposed algorithm obtains higher classification and recognition accuracy compared to other popular methods.

Keywords: speech emotion recognition; Locally Linear Embedding (LLE); Support Vector Machine (SVM)

1 INTRODUCTION

Since speech, one of the most important communication mediums for humans; it carries sufficient emotional information from speakers, it creates research, which is about analysis and extracting affective characteristics in speech signals so that computers can identify the affective state of speakers automatically, which is especially important— and it has a great application values [1] in the neighbourhood human-computer interaction, call centers, Intelligent robot, etc.

In recent years, the two typical manifold learning algorithms for nonlinear dimensionality are, Local Linear Embedding (LLE) [2] and Isometric Mapping (Isomap) [3]. Although these two algorithms can effectively achieve low-dimensional visualisation for speech data, they are ineffective for speech recognition, and are even worse than traditional Principal Component Analysis (PCA) [4]. The reason for this is that these two manifold learning algorithms are in the unsupervised way for dimensionality reduction, which means they don't consider the class information among the existing data points, which is helpful for classifying things. To overcome the weakness of unsupervised manifold learning algorithms in recognition, we propose a novel speech recognition method based on Locally Linear Embedding (LLE) and Support Vector Machine (SVM), which enhances the discriminating power of low-dimensional embedded data and possesses the optimal generalisation ability. The proposed algorithm is used to conduct multi-class data and nonlinear high-dimensional

speech emotional feature data. Experimental results on the natural speech emotional database show that the proposed algorithm obtains higher classification and recognition accuracy compared to other popular methods.

2 OUR PROPOSED METHOD

We propose a novel speech recognition method based on Locally Linear Embedding (LLE) and Support Vector Machine (SVM), namely, SVM-LLE. First, we introduce the LLE and its improved version. Set the N input data points of D-dimension equal to X_i ($X_i \in R^D$, $i \in [1, N]$), type number equal to L_i, embedded output dimension (d, $d \le D$). N data points equal to Y_i ($Y_i \in R^d$, $i \in [1, N]$). LLE can be divided into 3 steps:

Step1. Find out each data point's k neighbourhood points by calculating the nonlinear supervised distance between data points.

Step2. Calculate the partial reconstruction W matrix by each data point's neighbourhood points. When it's calculating optimum reconstruction W matrix W_{ij}, it needs to minimise cost function:

$$\varepsilon(W) = \sum_{i=1}^{N} X_i - \sum_{j=1}^{N} W_{ij} X_j \tag{1}$$

It requires $\sum_{j=1}^{N} W_{ij} = 1$; when X_j is not X_i's neighbourhood point, $W_{ij} = 0$.

Step3. Calculate the data point's output value by its partial reconstruction WV matrix and neighbourhood points. Calculating the embedded image of data points lying on low-dimensional space to minimise low-dimensional reconstruction error needs to minimizse cost function:

$$\phi(Y) = \sum_{i=1}^{N} Y_i - \sum_{j=1}^{N} W_{ij} Y_j = tr(YMY^T) \tag{2}$$

where $M = (I-W)^T (I-W)$ and Y consists of feature vector corresponding to the Matrix M's d minimum nonzero feature value.

In the Step1 of LLE mentioned above, the nonlinear supervised distance formula (improved version) is:

$$\Delta' = \begin{cases} \sqrt{1 - e^{-\Delta^2/\beta}}, & L_i = L_j \\ \sqrt{e^{-\Delta^2/\beta}} - \alpha, & L_i \neq L_j \end{cases} \tag{3}$$

In which Δ' is the distance calculated from data point class information, while Δ is the original Euclidean distance which neglects the data point class information. Vector β is used to prevent Δ in exponential function from growing excessively rapid, and it is more obvious when Δ is large. So β is related to data-intensive degree of data concentration and generally takes the Euclidean distance average value of paired data points, while $\alpha(0 \leq \alpha \leq 1)$ is a constant factor used to control the distance of different class data points, and it is close to, or less than that, of the same class data points on a certain probability.

Based on the LLE, we integrate the Support Vector Machine (SVM) with the aim to possess the optimal generalisation ability. SVM is first put forward [5]. At first, we are given a set of training data $\{(x_1, y_1), ..., (x_m, y_m)\}$, x_i is a vector in n dimension. y_i represent the class the point x_i belongs. $V(y, f(x))$ is the error of a given x. $f(x)$ is a predicted value. The purpose is to find a function f which minimises the expected error.

$$\int V(y, f(x)) P(x,y) dx \, dy), \tag{4}$$

For a given set of points, we can find enormous classifiers which can separate the data. Firstly, we need to set a standard to judge. Firstly, let us observe the following classifier that is a better classifier. Thus we consider it as a better classifier. However, what we need is a quantitative determination. Consider a binary classification problem with a train set consists of n sample. Each is defined as a two-tuples (x_i, y_i) $(i = 1, 2, ...n)$, $x_i = (x_{i1}, x_{i2}, ..., x_{in})$ $y_i \in (-1,1)$ represent the class the sample belongs to. A classifier can be describesd as

$$wx + b = 0, \tag{5}$$

where, w and b are the parameters of the classification model.

For the cause of simplicity, we adjust w and b so that $min |wx_i + b|$ is larger than or equals to1, the vector that fulfills the requirement $|wx_i + b| = 1$ is called support vector. Margin d is defined as follows.

$$d = \frac{2}{\| w \|} \tag{6}$$

What we need do now is to find the w and b which minimise d. Thus the problem can be transformed to a constrained optimisation problem.

$$\min \frac{1}{2} \| w \|^2$$
$$s.t. y_i(w * xi + b) \geq 1, \ i = 1, 2, ... n \tag{7}$$

This is called the primal problem. We can solve the problem using the method of Lagrange functions. The corresponding Langrage function is as follows.

$$J(w,b,\partial) = \frac{1}{2} w \cdot w - \sum_{i=1}^{m} \partial_i \{ y_i(w \cdot x_i - b) - 1 \} \tag{8}$$

where, ∂_i is Lagrange multiplier. The solution is determined by the saddle poking. After a series of transformations, we can get the following dual problem.

$$\min Q(\partial) = \sum_i \partial_i - \frac{1}{2} \sum_i \sum_j \partial_i \partial_j y_i y_j x_i x_i$$
$$s.t. \sum_i \partial_i y_i = 0$$
$$\partial \geq 0 \tag{9}$$

We can get the optimum Langrange multipliers which is denoted by ∂_i^*. Finally, we can compute the optimum weight vector w^*.

$$w^* = \sum_i \partial_i^* y_i x_i \tag{10}$$

3 EXPERIMENTAL RESULTS

To test the speech emotion recognition ability of SVM-LLE and LLE [2] will be used to extract dimensionality reduction for speech feature data; then we compare the speech emotion recognition results of these five algorithms in different dimensions.

All feature parameters should be unified to [0,1] and use the KNN classifier when the experiment

338

Table 1. Recognition description.

Emotion type	Anger	Happiness	Sadness	Neutral
Anger	166	24	10	0
Happiness	36	140	20	4
Sadness	16	40	123	21
Neutral	0	20	8	172

Table 2. Optimised average value of SVM-LLE's constant factor α in different dimension.

D	2	3	4	5	6	7	8	9	10	11
α	0.16	0.27	0.33	0.42	0.25	0.36	0.21	0.59	0.12	0.24
Error rate %	52.95	48.18	44.32	32.23	35.68	33.86	31.23	30.85	27.50	28.06

Table 3. Optimised average value of LLE's constant factor α in different dimension.

D	2	3	4	5	6	7	8	9	10	11
α	0.62	0.35	0.41	0.72	0.44	0.76	0.53	0.35	0.78	0.24
Error rate %	54.09	46.04	43.86	38.63	35.82	34.14	33.50	29.54	30.90	28.61

is running. *KNN* has a better speech emotion discriminating performance when using a nearest neighbour training mode. The 10-times cross-check technology will be used to improve the reliability of recognition result—that is, all sentences be divided equally to 10, then 9 of them will be used to train and the remaining 1 will be used to test. Repeat the experiment 10 times and get the average value to be the recognition results. The automatic optimised algorithm of constant factor α should also be done once in every cross-check.

Experiment: Don't perform any dimensionality reduction on the extracted original 48D speech feature data, and do the emotion recognition experiment directly. The recognition results are shown in Table 1.

We can know that the recognition results about anger and neutral are more satisfactory from Table 2, and the correct recognition rate is up to 83% and 86%. The average correct recognition rate of 4 kinds of emotion is 75.13%. While the correct recognition rate of happiness and sadness is lower, in which the one about happiness is 70% and that about sadness is 61.5%. The main reason leading to this is that the rhythm features of happiness are similar to that of anger and the voice quality features of sadness are similar to that of anger. So that we mix the two paired emotions (happiness and sadness, sadness and anger) easily.

From Tables 2 and 3, we can know that the optimised average value of SVM-SLLE's constant factor in different dimension is very stable that its value is generally less than 0.5. But the optimised average value of LLE's constant factor α changes obviously. It's because that there's a parameter β existing in nonlinear supervised distance that Improved-SLLE is using. β's value equals to the average value of all paired data points' Euclidean distance, so that β has equilibrium activity on the change of α.

4 CONCLUSIONS

This paper proposes a novel speech recognition method based on Locally Linear Embedding (LLE) and Support Vector Machine (SVM), which enhances the discriminating power of low-dimensional embedded data and possesses the optimal generalisation ability. We can conduct nonlinear dimensionality reduction for 48-dimensional speech emotional features, extracting low-dimensional embedded discriminating feature for speech emotion recognition.

ACKNOWLEDGMENTS

This work is supported by scientific research project directed by Education Department of Guangxi Province (201010 LX454).

339

REFERENCES

[1] R. Picard. Affective computing [J]. MIT Press, Cambridge, MA, pp. 1–24, 1997.

[2] S.T. Roweis and L.K. Saul. Nonlinear dimensionality reduction by locally linear embedding [J]. Science, vol. 290, no. 5500, pp. 2323–2326, 2000.

[3] J.B. Tenenbaum, V. Silva and J.C. Langford. A global geometric framework for nonlinear dimensionality reduction [J]. Science, vol. 290, no. 5500, pp. 2319–2323, 2000. 2319–2323.

[4] I.T. Jolliffe. Principal component analysis [J]. New York: Springer, pp. 150–165, 2002.

[5] Z.Y. Liu. Research on Algorithms and Application of Support Vector Regression [J]. Lecture Notes in Computer Science, vol. 2720, pp. 145–151, 2010.

Advances in Future Manufacturing Engineering – Yang (Ed.)
© *2015 Taylor & Francis Group, London, ISBN 978-1-138-02817-3*

A novel identification method of two phase flow based on LDA feature extraction and ELM in ERT system

Yanjun Zhang
The School of Information Science and Technology, Heilongjiang University, Harbin, China
Instrument Science and Technology, Post-Doctoral Research Station, Harbin University of Science and Technology,
Harbin, Heilongjiang, China

ABSTRACT: Flow regime is a very important parameter in two phase flow measurements. It can provide critical priori-information for real-time, on-line control of industrial processes. This paper proposes a new identification approach for common two phase flow regimes using LDA feature extraction and Extreme Learning Machine based on Electrical Tomography measurement. Simulation was carried out for typical flow regimes using the approach. The results show its feasibility. Especially, the results indicate that the extreme learning machine can provide the best generalization performance at extremely fast learning speed.

Keywords: Electrical Resistance Tomography; flow regime identification; Linear Discriminate Analysis; Extreme Learning Machine

1   INTRODUCTION

Two-phase flow is a mixed-flow pattern widely found in nature, especially in the chemical, petroleum, electricity, nuclear power and metallurgical industries[1]. Flow regime is a very important parameter in two phase flow measurements. Two-phase flow regime identification plays an increasingly important role in the automation process of energy industry. It can provide valuable information for a rapid and dynamic response which facilitates the real-time, on-line control of processes including fault detection and system malfunction. Two-phase flow regimes not only affect the two-phase flow characteristics and mass transfer, heat transfer performance, but also affect the system operation efficiency and reliability. While other parameters of the two-phase flow measurements have a great impact. Therefore, on-line identification of two-phase flow regimes is important for oil mixed transportation systems.

2   THE FEATURE EXTRACTION BASED ON LDA

This paper uses 12-electrode sensor array, the incentive is in any two adjacent electrodes between the one electrode to the excitation current of the input and one for output, then cycle all the two adjacent electrodes detect electrical potential difference (in addition to two other excitation electrode) can be nine voltage data; is to inspire the

next cycle of the electrode, changing the incentives electrode, and can detect the 9 voltage data; to any two adjacent electrodes are incentives once a week to encourage voltage data obtained 12 groups, so that each image acquisition to have 12 * 9 = 108 elements, the number of independent elements is

$$V = \{v_1, v_2, \dots v_n\} \tag{1}$$

where, n is the index of a flow regime sample, v_i is composed of 12 vectors, each vector have 9 voltage values.

In this paper, core flow, laminar flow, circulation flow and trickle flow are selected to be tested.

The first key step before training and recognition is feature extraction. Linear Discriminate Analysis is employed in this paper. LDA method is widely used in pattern recognition and machine learning to find a linear combination of features which characterizes or separates two or more classes of objects or events[2]. The resulting combination may be used as a linear classifier, or more commonly for dimensionality reduction before later classification. LDA is also closely related to Principle Component Analysis (PCA) and factor analysis in that they both look for linear combinations of variables which beat explain the data[3] LDA attempts to model the difference between the classes of data. PCA on the other hand does not take into account any difference in class, and factor analysis builds the feature combinations based on differences rather than similarities.

It is supposed that there are M classes, each class has a mean μ_i and the same covariance, then between class variability may be defined by the sample covariance of the class means

$$S_b = \sum_{i=1}^{C} P_i(\mu_i - \mu)(\mu_i - \mu)^T \qquad (2)$$

where, μ_i is the mean of C_i class means, μ is the mean of all samples, P_i is priori information.

$$S_w = \sum_{i=1}^{C} P_i S_i \qquad (3)$$

where,

$$S_i = E[(v - \mu_i)(v - \mu_i)^T | x \in C_i]$$

This paper used Fisher's linear discriminate and LDA interchangeably. According to Fisher Criterion Function

$$J(W_{opt}) = \arg\max_w \frac{|W^T S_b W|}{|W^T S_w W|} \qquad (4)$$

where, W_{opt} satisfies

$$S_b W_i = \lambda_i S_w W_i \qquad (5)$$

where, $i = 1, 2, \dots m$.

The class means and covariances are not known. They can, however, be estimated from the training set. Either the maximum likelihood estimate or the maximum a posteriori estimate may be used in place of the exact value in the above equations. Although the estimates of the covariance may be considered optimal in some sense, this does not mean that the resulting discriminant obtained by substituting these values is optimal in any sense, even if the assumption of normally distributed classes is correct.

3 EXTREME LEARNING MACHINE CLASSIFICATION

To satisfy the on-line identification, it is necessary to reduce the time consumed in training and recognition phase. Some Single–hidden Layer Feed Neural networks (SLFNs)[4] models have been used in flow regimes classification such as BP neural networks. However, it has been a major bottleneck due to the behind two reasons: the gradient-based traditional learning methods may needs large number of iterations in the training models, and

the numbers of models are decided by iterative tuning. An extreme learning machine is proposed by Huang GB. The main objective of the methods is to tend to provide the best generalization performance at extremely fast learning speed. The algorithm randomly chooses the input weights and analytically determines the output weights of SLFNs. The significant advantage of the algorithm is to reduce the amount of time needed to train a Neural Network.

In the paper, consider the minimum norm least-squares solution of a general linear system in Euclidean space

$$Ay = O \qquad (6)$$

where $A \in R^{m \times n}$ and $O \in R^m$.

For N arbitrary distinct samples y_i, where $X_i = [x_{i1}, x_{i2} \cdots x_{in}]^T \in R^n$.

The standard SLFNs with $\hat{N}$ hidden neurons and activation function $g(x)$ can approximate these N samples with zero error means that

$$\sum_{i=1}^{\hat{N}} \beta_i g(w_i \cdot y_j + b_i) = O_j, \ j = 1, \dots, N \qquad (7)$$

where, $w_i = [w_{i1}, w_{i2} \cdots w_{in}]^T$ is the weight vector connecting the ith hidden neuron and the input neurons, and $\beta_i = [\beta_{i1}, \beta_{i2} \dots \beta_{in}]^T$ is the weight vector connecting the ith hidden neuron and the output neurons, and b_i is the threshold of the ith hidden neuron. The above equations can be written acompactly as

$$H\beta = O \qquad (8)$$

where,

$$H(w_1, \dots, w_{\hat{N}}, b_1 \dots, b_{\hat{N}}, y_1, \dots, y_n)$$
$$= \begin{bmatrix} g(w_1 \cdot y_1 + b_1) & \cdots & g(w_{\hat{N}} \cdot y_1 + b_{\hat{N}}) \\ \vdots & \cdots & \vdots \\ g(w_1 \cdot y_N + b_1) & \cdots & g(w_{\hat{N}} \cdot y_N + b_{\hat{N}}) \end{bmatrix}_{N \times \hat{N}} \qquad (9)$$

$$\beta = \begin{bmatrix} \beta_1^T \\ \vdots \\ \beta_{\hat{N}}^T \end{bmatrix} \text{ and } O = \begin{bmatrix} o_1^T \\ \vdots \\ o_{\hat{N}}^T \end{bmatrix} \qquad (10)$$

As named in Huang[5], Huang and Babri[6], H is called the hidden layer output matrix of the neural network.

Through the simulations and theories, it can be achieved that The input weights and the hidden layer biases of SLFNs need not be adjusted at all and can be arbitrarily given . Thus, for fixed input weights and the hidden layer biases, to train an SLFN is simply equicalent to finding a least-squares solution $\hat{\beta}$ of the linear system $H\beta = O$.

According the theory: Let there exist a matrix G such that Go is a minimum norm least-squares solution of a linear system $Ay = O$. Then it is necessary and sufficient that $G = A^{\dagger}$, the moore-penrose generalized inverse of matrix A.

The smallest norm least-squares solution of $H\beta = O$ is

$$\beta = H^{\dagger}O \tag{11}$$

4 EXPERIMENTS AND CONCLUSIONS

The approach based on LDA and ELM identifier is tested on laminar flow, trickle flow, the core flow and circulation flow, Samples of flow regimes used in the experiments are as follows:

1. The core flow, laminar flow and circulation Each flow regime was divided into three groups according to the fluid phase holdup. The core flow were classified into three groups according to fluid contents holdup 20%, 50%, 80% as the figure. 60 samples is trained for each group respectively.
2. Trickle Each flow regime was divided into according to the different phase holdup has also been divided into three groups, which contains 5% water, 10% and 20%, 60 samples were trained.

The simulations are carried out in MATLAB 6.5 environment running in a Intel Core 2, 2.93 GHZ CPU. Experiments results are as Table 1.

The average accuracy rate is 96.93%.It is noted that the accuracy rate of identification of trickle flow is significantly lower than other flow regimes; it can be explained by the complexity of trickle

Table 1. The identification accuracy rate.

Flow regime	TC for FA/ms	TC for T/R/ms	Accuracy rate
Core 20%	29	1036/30	98.3%
Core 50%	29	1039/32	98.3%
Core 80%	31	1030/32	98.3%
Circulation 20%	28	1042/33	98.3%
Circulation 50%	29	1041/32	96.7%
Circulation 80%	29	1037/32	98.3%
Laminar 20%	29	1032/33	98.3%
Laminar 50%	30	1035/34	96.7%
Laminar 80%	32	1039/34	96.7%
Trickle 5%	29	1051/33	95%
Trickle 15%	30	1053/34	95%
Trickle 20%	31	1051/35	93.3%

Table 2. The comparison between BP, SVM and ELM.

Alogrithm	Time consuming training/ms	Time consuming recogniton/ms
BP	21532	102
SVM	5648	114
ELM	1040	34

flow. It is observed that the fluid phase holdup is bigger; identification accuracy is lower for core flow, circulation flow, laminar flow and trickle flow. The average time-consuming of recognition is smaller than 34 micro-second. The speed can satisfied with the real time control. In order to prove its advantage in improving speed, running time using back propagation and support vector machine algorithms are compared with extreme learning machine. The experiments results indicate that this method significant reduces time consumed in the training phase.

ACKNOWLEDGEMENT

This work is partially funded by Natural Science Foundation of Heilongjiang Province (F201019).

REFERENCES

[1] Yanjun Zhang, Yu Chen. A Novel PCA-SVM Flow pattern Identification Algorithm For Electrical Resistance Tomography System, In Proceeding(s) of the Advances in Intelligent and Soft Computing, Volume 160, 2012.
[2] Yu, H., Yang, J. A direct LDA algorithm for high-dimensional data—with application to face recognition, Pattern Recognition, Volume 34 (10), pp. 2067–2069.
[3] Huang G-B. Learning capability and storage capacity of two hidden-layer feedforward networks. IEEE Transactions on Neural Networks, 2003,14(2): 274–281.
[4] G. Henkelman, G. Johannesson and H. Jónsson, in: Theoretical Methods in Condencsed Phase Chemistry, edited by S.D. Schwartz, volume 5 of Progress in Theoretical Chemistry and Physics, chapter, 10, Kluwer Academic Publishers (2000).
[5] Huang G-B, Zhu Q-Y, Siew C-K. Extreme learning Machine: Theory and Applications. Neurocomputing, 2006, 70(1–3): 489–501R.J. Ong, J.T. Dawley and P.G. Clem: submitted to Journal of Materials Research (2003).
[6] Huang G-B, Bari H.A, Upper Bounds on the number of hidden neurons in feedforward networks with arbitrary bounded nonlinear activation functions. IEEE transactions on neural Networks, 1998, 9(1):224–2.

Advances in Future Manufacturing Engineering – Yang (Ed.)
© 2015 Taylor & Francis Group, London, ISBN 978-1-138-02817-3

Vulnerability analysis and improvement of RFID grouping-proof protocols

Xin Wei, Zhiwei Shi & Yongle Hao
China Information Technology Security Evaluation Center, Beijing, China

ABSTRACT: A grouping-proof in RFID system is evidence that two or more tags are scanned simultaneously. Juels introduced the first grouping-proof guaranteeing a pair of tags have been simultaneously scanned namely yoking-proof. Then, yoking-proof is generalized to the proof for a larger number of tags called grouping-proof. In this paper, we first introduce some existing RFID grouping-proof protocols and show the vulnerabilities in these protocols. Then we present a novel grouping-proof protocol with an untrusted reader.

Keywords: RFID; grouping-proof protocols; security and privacy

1 INTRODUCTION

Radio Frequency Identification (RFID) technology represents a fundamental change in the information technology infrastructure. A grouping-proof is evidence that two or more RFID tags were scanned simultaneously by a reader within its broadcast range. For instance, pharmaceutical distributors may want to guarantee a medicine has been sold with its prescription or with the instructions leaflet. In a meeting scenario, we can leverage grouping proof to make sure that a group of people equipped with RFID tags are present at a specific location.

Ari Juels introduced a new RFID application in 2004 called yoking-proof [1] that involves generating evidence of simultaneous presence of two tags in the range of an RFID reader. Yoking-proof has been extended to grouping-proof in which groups of tags prove simultaneous presence in the range of an RFID reader. We refer to yoking-proof under the circumstance that the protocol deals with exclusively two tags and grouping-proof when the protocol involves more than two tags.

2 ATTACKS

Typically, a RFID system consists of three main entities: tags, readers and a verifier. The tags are resource limited devices allowing only simple operations such as XOR, pseudorandom number generator etc. The readers communicate wirelessly with the tags, who present its identification number or other stored information to the reader upon request. A back-end server is denoted as a verifier, which is a trusted entity that maintains a database containing the information needed to identify tags.

Grouping-proof includes a set of tags being scanned in the same session by a RFID reader. The reader links those tags of a group and makes the tags to generate a proof of simultaneous presence within its broadcast range. The proof should be verifiable by the verifier.

As for security and privacy issues in grouping-proof system, we mainly consider the following attacks:

- **Replay attacks:** The adversary impersonates valid tags by replaying the messages previously sniffed from the tags.
- **Multi-proof session attacks:** The adversary combines several incomplete proofs to generate a valid proof.
- **Concurrency attacks:** The adversary currently executes several grouping proof protocols with different tags to generate a valid proof. The tags claimed by this proof are not involved in the same grouping proof protocol. For example, the adversaries execute two protocols with tag A and tag B, and tag C and tag D. However, the final proof claims to be tag B and tag C.
- **Multiple impersonation attacks:** The adversary eavesdrops on the public messages among tags set $S = \{T_1 \dots T_n\}$ during the execution of a grouping-proof. Then, the adversary replayed the captured messages to the tag T_x which is not

included in S to build a counterfeit proof of tags $\{T_1, ..., T_n, T_x\}$.

- **Privacy attacks:** There are two notions of privacy in RFID grouping-proof system: the first one is commonly known as tag anonymity. That is, an adversary A should not be able to disclose the identities of tags he reads from or writes into. The second notion is called untraceability: an adversary A should not be able to trace or track the person (product) attached with tags by a fake reader. There is a special untraceability named forward untraceability (also known as forward privacy), that is, an adversary capturing the tag's secret information cannot correlate the tag with its responses before the last complete protocol run with a valid reader.
- **Forged proof attacks:** The adversary is able to impersonate one or more tags to generate a grouping-proof.

3 RELATED WORK

The first attempt of multi-tag scheme in RFID systems is due to Juels et al. [1] in 2004, they proposed the concept of a yoking-proof, namely a proof that a pair of RFID tags has been scanned simultaneously. In their protocol, there are two tags T_A and T_B present in the range of a reader. The tags have secret keys x_A, x_B, known to the verifier but not the reader.

The protocol steps are summarized in Figure 1 and described as follows:

1. The reader sends a query to T_A, T_A then generates a random number r_A, and returns r_A and the identity of T_A as A to the reader.
2. The reader relays r_A to T_B.
3. T_B generates a random number r_B and computes $m_B = MACx_B[r_A]$ and responses to the reader r_B, m_B and the identity of T_B as B.
4. The reader sends r_B to T_A.
5. T_A computes $m_A = MACx_A[r_B]$ and sends m_A to the reader.

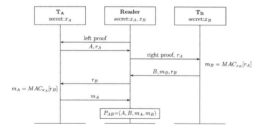

Figure 1. Juels' yoking-proof protocol.

6. The reader sends m_A and m_B to the verifier. The verifier can verify that T_A and T_B were scanned simultaneously.

Saito and Sakurai later showed that Juels' proof is vulnerable to replay attacks, and proposed an improved yoking-proof using time stamp and also extended their proposal to plural RFID tags. However, Piramuthu [2] demonstrated that Saito's new proof for plural tags is also vulnerable to replay attacks.

Bolotnyy and Robins [3] proposed a new solution for the grouping-proof and addressed the requirements on privacy. The new protocol is called anonymous yoking and each tag is supposed to compute a keyed hash function and a MAC. The main drawback of the scheme is the computational complexity on the side of the verifier being $O(n^2)$. Peris-Lopez et al. [4] later proposed an improvement, that the verification takes $O(n)$ steps.

Burmester et al. [5] present a security model based on the universal composability framework for this so-called group-scanning problem.

Lien et al. [6] proposed an order-independent protocol, which should improve the efficiency and reduce the failure rates. The reason for improved efficiency is the fact that there is no requirement on predefined reading order. Lien's grouping-proof is vulnerable to privacy attacks due to disclosing the real identities of the tags.

Batina et al. [7] proposed a yoking-proof system that is based on public key cryptography. The security of Batina's protocol is claimed to be related to the security of the Schnorr identification protocol, however, the adversary can forge a valid grouping proof that T_a and T_b were scanned together.

4 PROPOSED PROTOCOLS

Although we have reviewed some yoking-proof protocols just for two tags, actually these schemes can easily extend to grouping-proof ones. For better understanding of whole process of grouping-proof, we will illustrate our protocols in the form of grouping-proof rather than yoking-proof. Our new protocol on the tag is based on Ma's proposal [8] using only pseudorandom number generator.

Generally speaking, tags do not contain clocks and cannot communicate with each other directly, they communicate via a reader. As in grouping-proof system, the reader is considered untrusted. That is to say, any reader can provide a grouping-proof to verifier proving a group of tags were scanned simultaneously. This important property is often neglected by researchers which lead to many problems. We mainly take our efforts on designing an order dependent grouping protocol. Order dependent implies that the input of a tag T_i should be exclusively dependent on a value generated by T_{i-1} or $\{T_1 \ldots T_{i-1}\}$. According to Peris-Lopez [9], the computation of T_i should rely on $\{T_1 \ldots T_{i-1}\}$ rather than T_{i-1} only, if so, the computation and communication cost will be a big issue. The reason why Peris-Lopez believe so is that he assumed all the tags are identical, the order dependent grouping-proof process will look like Figure 2. The reader first sends a message m_{r0} to T_1, after processing m_{r0}, T_1 sends m_{t1} to the reader, the reader then forwards and updates the message to the next tag T_2, the steps repeat until T_n sends m_{tn} to the reader. The structure is just like a one-way linked list, the reader is easy to add another tag T_{n+1} to forge a n+1 grouping-proof, which is how multiple impersonation attacks work. Both Juels' and Burmester's protocols were in such case. We believe a grouping-proof protocol ought to have a master tag mixed with auxiliary tags, which look like a circular linked list. If the adversary wants to add another tag T_j into a grouping-proof between T_i and T_{i+1}, he must guarantee that the input of T_j must be same to output of Tj, the probability is negligible.

We denote the first tag as master tag, and leverage Ma's authentication to achieve untraceability, the protocol steps are shown in Figure 3 and described as follows:

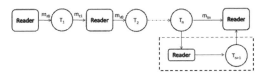

Figure 2. The general process of grouping-proof.

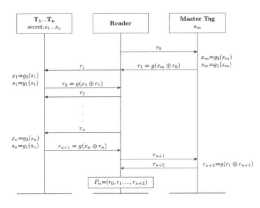

Figure 3. Our protocol.

1. Reader R generates a random number r and sends it to the master tag.
2. Upon receiving r, the master tag computes $x_m = g_2(s_m)$, $s_m = g_1(s_m)$ and responses with $r_1 = g(x_m \oplus r)$.
3. R then forwards r_1 to T_1, T_1 computes $x_1 = g_2(s_1)$, $s_1 = g_1(s_1)$ and responses with $r_2 = g(x_2 \oplus r_1)$, the steps repeated (T_i receives r_i and sends back r_{i+1}) until T_n computes $x_n = g_2(s_n)$, $s_n = g_1(s_n)$ and responses to R the value $r_{n+1} = g(x_n \oplus r_n)$.
4. R forwards r_{n+1} to the master tag, the master tag computes $r_{n+2} = g(x_m \oplus r_{n+1})$ and returns to R.
5. R generates the grouping proof $P_n = (r_0, r_1, \ldots, r_{n+2})$ and sends P_n to the verifier.
6. The verifier V checks each pair of r_i, r_{i+1} according to Ma's authentication protocol, if V recognizes each tag in set $T_1, \ldots, T_n$, V will believe that all the tags in are simultaneously present.

5 PERFORMANCE AND SECURITY ANALYSIS

In this section, we mainly discuss some attacks that are commonly used by adversaries against RFID grouping-proof systems and present how our protocols to resist these attacks.

- **Replay attacks:** In our protocol, a random number is generated by each tag, thus our protocol is immune to replay attacks.
- **Privacy attacks:** Privacy attack includes anonymity and traceability. Since Ma's authentication protocol is an anonymous and untraceable protocol, our protocol can resist privacy attacks.
- **Multiple impersonation attacks:** In our protocol, if the adversary wants to add another tag T_j into a grouping-proof between T_i and T_{i+1}, he must make sure that the output of T_j is the same to input of T_{i+1} when it is queried by the output of T_i. The probability is negligible.
- **Forged proof attacks:** The adversary is not able to impersonate one tag or unless he can break the security of Ma's protocol.

We compare our two proposed protocols with other schemes in terms of the attacks as mentioned above and communication cost, as some proposals are just for two tags, we extend these schemes into grouping-proofs by adding auxiliary tags for better parallel. We assume that the number of the tags is n including the master tag. The result is shown in Table 1. From the comparison table, we can see that our protocols can resist all the listed attacks and are more efficient in communication cost.

Table 1. Protocols comparison.

Schemes	Replay	Anonymity	Traceability	Multiple impersonation	Forged proof	Communication cost
Juels	X	X	X	X	O	$2n+2$
Saito	X	O	O	X	O	$3n-1$
Bolotnyy	O	O	O	X	O	$2n+2$
Lien	O	X	X	X	O	$7n-5$
Burmester	O	O	O	X	O	$5n+1$
Batina	O	O	O	X	X	$2n+2$
Ours	O	O	O	O	O	$2n+2$

6 CONCLUSION

In this paper, we propose a novel RFID grouping-proof protocols with an untrusted reader to address security and privacy attacks related to them. Our analysis shows that our proposed protocols provide stronger security robustness and are efficient at communication cost.

ACKNOWLEDGMENT

This work is partially supported by by National High Technology Research and Development Program of China under Grant No. 2012AA012903.

REFERENCES

[1] Juels, "yoking-proofs" for rfid tags," in PerCom Workshops, 2004, pp. 138–143.

[2] S. Piramuthu, "On Existence Proofs for Multiple RFID Tags," in IEEE International Conference on Pervasive Services, Workshop on Security, Privacy and Trust in Pervasive and Ubiquitous Computing—SecPerU 2006, IEEE. Lyon, France: IEEE Computer Society, June 2006.

[3] L. Bolotnyy and G. Robins, "Generalized yoking-proofs for a group of rfid tags," in Annual International Conference on Mobile and Ubiquitous Systems, 2006, pp. 1–4.

[4] P. Peris-Lopez, J.C. Hernandez-Castro, J.M. Estevez-Tapiador. "Solving the Simultaneous Scanning Problem Anonymously: Clumping Proofs for RFID Tags," in IEEE International Conference on Pervasive Services, 2007, pp. 55–60.

[5] M. Burmester, B. de Medeiros, and R. Motta, "Provably Secure Grouping-Proofs for RFID Tags," in Proceedings of the 8th Smart Card Research and Advanced Applications—CARDIS 2008, ser. Lecture Notes in Computer Science, G. Grimaud and F.-X. Standaert, Eds., vol. 5189. Royal Holloway University of London, United Kingdom: Springer, September 2008, pp. 176–190.

[6] Y. Lien, X. Leng, K. Mayes, and J.-H. Chiu, "Reading order independent grouping proof for rfid tags," in ISI, 2008, pp. 128–136.

[7] L. Batina, Y.K. Lee, S. Seys, D. Singel´ee, and I. Verbauwhede, "Extending ecc-based rfid authentication protocols to privacy-preserving multiparty grouping proofs," Personal and Ubiquitous Computing, vol. 16, no. 3, pp. 323–335, 2012.

[8] Ma, "Low cost rfid authentication protocol with forward privacy," Chinese Journal of Computers, vol. 34, no. 8, 2011.

[9] P. Peris-Lopez, A. Orfila, J.C. Hernandez-Castro, and J.C.A. van der Lubbe, "Flaws on rfid grouping-proofs. Guidelines for future sound protocols," J. Network and Computer Applications, vol. 34, no. 3, pp. 833–845, 2011.

Advances in Future Manufacturing Engineering – Yang (Ed.)
© 2015 Taylor & Francis Group, London, ISBN 978-1-138-02817-3

Computer aided design of continuous Extractive Distillation processes for separating compound m-Diisopropylbenzene and p-Diisopropylbenzene with Dinonyl Phthalate

Chunyong Zhang, Jiehong Cheng & Chunzhi Zheng
Jiangsu Key Laboratory of Precious Metal Chemistry and Technology, Jiangsu University of Technology, Changzhou, Jiangsu, China

ABSTRACT: A new process for separating compound m-Diisopropylbenzene (m-DIPB) and p-Diisopropylbenzene (p-DIPB) with high purity and yield has been put forward in which the Extractive Distillation (ED) with Dinonyl Phthalate (DP) as extractant is employed by feasibility studies and rigorous simulation using Aspen software, and the simulation result was verified by experiment. A process with Double Column System (DCS) was designed for separating of the mixture. The position of the stage of extractant (n_{ex}), the position of the stage of raw material (n_{DI}), solvent to feeding ratio (mole ratio, S/F) and reflux ratio (R) were adjusted in order to get optimal results for extractive distillation. The extractive distillation process of m-DIPB separation from the mixture of m-DIPB and m-DIPB was simulated with a tower with 60 theoretical plates. The simulation result indicated that the mole fraction (x) and the recovery (y) of m-DIPB could be up to 98% and 96% with the conditions as follows: the mixture was fed at 31th plate, the extractant was fed at 2th plate, the solvent ratio was 5 and the reflux ratio was 6. The extractive distillation simulation result was verified by the experiment. The simulation process is reasonable and reliable.

Keywords: m-Diisopropylbenzene; p-Diisopropylbenzene; extractive distillation; aspen

1 INTRODUCTION

Distillation is the separation method most frequently applied in the chemical industry, which is based on the difference of volatility of the components of a liquid mixture [1–3]. Because of its high demand of energy the optimal design and operation of the distillation columns is a very important issue. For the separation of the two components (A and B) forming an azeotrope a special distillation method must be applied such as Pressure Swing Distillation (PSD), extractive and heteroazeotropic distillation [4]. In the case of Extractive Distillation (ED) a third, heavy component (solvent, E) is added to the system, which breaks the azeotropic composition by changing the liquid phase activity efficient [5]. ED is an important separation method the petroleum and petrochemical industries, and is used to separate compounds with similar boiling points or compounds [1,6].

Recently with the rapid increase of demand for resorcinol derivatives, the production of m-Diisopropylbenzene (m-DIPB) has become more and more important, which mainly can be oxidized to resorcinol. Since most of the Diisopropylbenzene (DIPB) comes from the byproduct during manufacturing cumene in the manufacture of phenoal-acetone plants by alkylation of benzene with propylene, the separation of m-DIPB from a mixture mainly composed of m-DIPB and p-Diisopropylbenzene (p-DIPB) has great commercial significance. However, the boiling points of the isomers of m-DIPB and p-DIPB are very close over a large range of pressure: m-DIPB 203.2 °C, p-DIPB 210.3 °C. Therefore, it is difficult to separate by ordinary distillation due to the very high reflux ratio and the very large number of theoretical trays [7]. ED is an expensive and time-consuming task, because of the large number of parameters involved [8].

Aiming at the disadvantages of ordinary distillation such as high reflux ratio and large number of theoretical tray, this paper is therefore focused on the separation of m-DIPB and p-DIPB and intended to optimize the process by adding solvent with the help of available simulation programs Aspen.

2 EXPERIMENT

2.1 *Materials and instrumentation*

m-DIPB, p-DIPB and dinonyl phthalate (DP) were of analytical grade and purchased from Guo Yao Chemical Company, China. All chemicals were

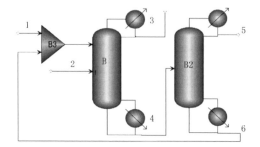

Figure 1.   Flow sheet of ED.

used without further treatment. For the quantification of extractive product, analyses were performed by a HP6890 gas chromatograph with a FID at a carrier gas (N₂) flow rate of 20 ml/min, a column temperature of 200 °C and a sample injector temperature of 250 °C.

2.1.1   The process of ED

A methodology covering the complete design framework for the ED with a Double Column System (DCS) is described in detail. The flow sheet of ED is showed in Figure 1. The mixture of feed (mole ratio, m-DIPB/p-DIPB = 3) was added into tower B1 (ED tower) at above part. DP (1) is a high boiling solvent hence it is being added into B1 on a stage higher than the feed stage of the m-DIPB and p-DIPB mixture (1). B1 was heated by reboiler, high purity of 98% m-DIPB (3) can be obtained through a cooler in B1. The bottom product–a mixture of p-DIPB and DP–is being fed to the second distillation tower B2 (solvent recovery tower). 98% (mole fraction) p-DIPB (5) and 100% DP can be obtained as top distillate and bottom product of B2, respectively. DP can be recovered by feeding back to mixer B3.

3   SIMULATING RESULTS AND EXPERIMENTAL VERIFICATION

Once a distillate objective in terms of purity and recovery is set, the range of operating parameters (the position of the stage of extractant (n_{ex}), the position of the stage of raw material (n_{DI}), solvent to feeding ratio (mole ratio, S/F) and reflux ratio (R)) is determined for each operating step so as to match a general feasibility criterion [9, 10]. In our case, we would like to achieve a minimum mole fraction (x) of 98% and the recovery (y) of 96% m-DIPB at the top of tower B1 (ED tower). The ED column was simulated using the Aspen Plus.

3.1   Effects of nex on ED

The segregated-feed setup is that DIPB and *DP* were fed into the column, respectively. DIPB was introduced into the lower section because it is less volatile than DP. The effect of n_{ex} on the mole fraction (x) and the recovery (y) of m-DIPB could can be seen in Figure 2. This finding can be ascribed to that DIPB and DP can contact in countercurrent by adding DIPB (light component) from the top and DP (heavy component) from the bottom separately into the tower which increases the contact time of DIPB and DP, leading to a high x and y. From this figure, optimal n_{ex} 2th can be read. At this point the mole fraction and the recovery of m-DIPB is 98.25% and 96.61%, respectively.

3.2   Effects of nDI on ED

Figure 3 shows the effects of n_{DI} on the mole fraction (x) and the recovery (y) of m-DIPB. The feed location is the only parameter that recovery ratio and the mole fraction (x) of m-DIPB exhibit an extremum. When the feed location of DIPB increased from 1 to 31, leading to a increase in the mole fraction (x) of m-DIPB from 83.61 to

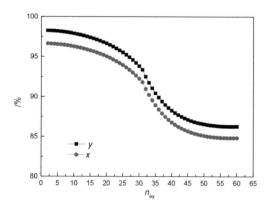

Figure 2.   Effects of n_{ex} on ED.

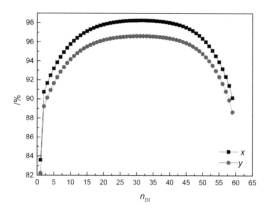

Figure 3.   Effects of n_{DI} on ED.

98.25%, the recovery (y) of m-DIPB accumulated at the bottom of the tower from 82.21% to 96.61%, respectively. From this figure, optimal feed location 31th can be read.

3.3 Effect of S/F on ED

DP and DIPB were added into the tower from the 2th tray and the 31th tray of the distillation column, respectively, with the mole ratio of DP to DIPB changed. The profiles of the mole fraction (x) and the recovery (y) of m-DIPB at different S/F are presented in Figure 4. The mole fraction (x) and the recovery (y) of m-DIPB can be increased by increasing the amount of DP. Therefore, the mole fraction (x) and the recovery (y) of m-DIPB increases from 30.7 to 38.9% and from 30.7 to 38.9% with the solvent ratio increased from 0.47 to 3.45. However, a large amount of excess DP will coexist with the product. Aiming at the 98% mole fraction and the 96% recovery of m-DIPB, S/F is 5.

3.4 Effect of R on ED

To study the effect of R in the process, it was changed while keeping the rest of the variables constant. Figure 5 shows how R affects the product purity and the recovery of m-DIPB. The recovery of m-DIPB increases with the increase of reflux ratio, which can be attributed to that increasing the R, the refluxing of p-DIPB can take part in the interaction with DP once again. The increase of reflux can advance the product purity and the recovery (y) of m-DIPB at the cost of increasing energy consumption. Increasing R improves the operating lines in the various sections of the column, increasing the separation and, nevertheless, the energy consumption [11–13]. When the optimal economic benefit is considered, lower reflux is preferred in the production at given product purity.

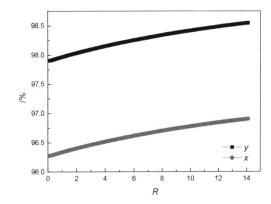

Figure 5. Effect of R on ED.

Aimed at the m-DIPB mole fraction of 98% in the products and the m-DIPB recovery of 96%, the reflux ratio is 6.

The simulation result was verified by experiment. The m-DIPB mole fraction is of 98.33% and the m-DIPB recovery of 96.24% in the experimental products. It can be seen that the simulating results and the experimental data agree very well and the product purity and the product purity.

4 SUMMARY

Continuous ED processes for separating compound m-DIPB and p-DIPB with DP was demonstrated for optimum extractive conditions using commercial software program Aspen software as an aid for process development. The DCS proved to be easy method, and it has produced satisfactory results.

ACKNOWLEDGMENT

This research is financially supported by the National Natural Science Foundation of China (Grant no. 21307009), the University Science Research Project of Jiangsu Province (Grant no. 13KJB530004) and the Innovative Research Team Development Program in Jiangsu University of Technology (Ma Di).

CORRESPONDING AUTHOR

Chunyong Zhang, Master, lecturer, Jiangsu Key Laboratory of Precious Metal Chemistry and Technology, Jiangsu University of Technology, Changzhou 213001, P. R. China. Tel: 0086-519-86989073, E-mail: zhangcy@jsut.edu.cn.

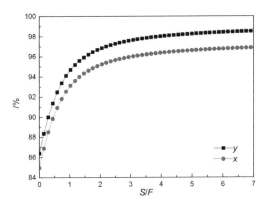

Figure 4. Effect of S/F on ED.

REFERENCES

[1] Roberto Gutierrez-Guerra, Juan Gabriel Segovia-Hernandez and Salvador Hernandez: Chemical Engineering Research and Design Vol. 87 (2009), p. 145.

[2] Emhamed A.M., Czuczai B., Rev E. and Lelkes Z.: Ind Eng Chem Res Vol. 47 (2008), p. 9983.

[3] Gil I.D., Botia D.C., Ortiz P. and Sanchez O.F.: Ind Eng Chem Res Vol. 48 (2009), p. 4858.

[4] Modla G., Kopasz A. and Lang P.: Comput Aided Chem Eng Vol. 27 (2009), p. 1017.

[5] I.M. Mujtaba: Trans IChemE Vol. 77 (1999), p. 588.

[6] Jie He, Baoyun Xu, Weijiang Zhang, Cuifang Zhou and Xuejia Chen: Chemical Engineering and Processing Vol. 49 (2010), p. 132.

[7] J.M. Resa, M.A. Betolaza, C. Gonzalez and A. Ruiz: Fluid Phase Equilibria Vol. 110 (1995), p. 205.

[8] R. Munoz, J.B. Monton, M.C. Burguet and J. de la Torre: Separation and Purification Technology Vol. 50 (2006), p. 175.

[9] Csaba Steger, Endre Rev, Laszlo Horvath, Zsolt Fonyo, Michel Meyer and Zoltan Lelkes: Separation and Purification Technology Vol. 52 (2006), p. 343.

[10] Hilal N, Yousef G and Langston P: Chem Eng Process Vol. 41 (2002), p. 673.

[11] Kossack S., Kraemera K., Gani R. and Marquardt W.: Chem Eng Res Des Vol. 86 (2008), p. 781.

[12] Ligero E.L., Ravagnani T.M.K.: Chem Eng Process: Process Intensif Vol. 42 (2003), p. 543.

[13] Marcella Feitosa de Figueiredo, Brenda Pontual Guedes, Joao Manzi Monteiro de Araujo, Luis Gonzaga Sales Vasconcelos and Romildo Pereira Brito: Chemical Engineering Research and Design Vol. 89 (2011), p. 341.

Advances in Future Manufacturing Engineering – Yang (Ed.)
© 2015 Taylor & Francis Group, London, ISBN 978-1-138-02817-3

Decision model in wine producer marketing based on analytical hierarchy process

Tongkui Gu, Zhiyi Huang & Xiaomin Lv
Department of Management Science and Engineering, Zhejiang Gongshang University, Zhejiang, China

ABSTRACT: Since its reform and opening-up, China's economy has experienced rapid growth and its per capita income as well as consumption has also expanded, which has brought tremendous opportunities and potential markets to its wine industry. Hence, it becomes the primary focus for wine producers to initiate rapid development for higher profits. This paper analyses the main factors affecting wine sales, based on AHP. We then weigh each factor which could help wine producers chalk out their marketing strategies.

Keywords: wine; AHP; weigh; marketing strategies

1   INTRODUCTION

So far, white wine and beer are more popular in China, whereas wine consumption in general is not so strong. However, with the growing per capita income, wine is gaining a huge potential market due to its beneficial effects on our health. A big problem rises for small and medium sized wine producers on how to find a place in the fierce wine market. AHP, with its mathematical background, strict logical thinking and strong operability, provided us a scientific and practical method for multiple targets. Based on the AHP method, this paper comprehensively analyses the package, taste and price of wine in terms of both qualitative and quantitative routes. Thus it proposes a marketing plan to aidr wine producers.

2   AHP

AHP was put forward by T. L. Satty in the 1970s. It is an important multi-objective method in Operational Research with a combination of both qualitative and quantitative analysis. We can take following steps when using AHP.

2.1   *To construct a judgment matrix*

By comparing the importance of an element it is constructed under a layer of the related factors of a judgment matrix. The judgment matrix usually adopts "1–9 and the countdown nine scale methods". The proportion of the scale is shown in "Table 1".

The established $A = (a_{ij})$ according to the results of pair is called the judgment matrix. Obviously, $a_{ii} = 1$, $a_{ji} = 1/a_{ij}$ the matrix A is a positive reciprocal matrix.

2.2   *Calculate weight coefficient*

Weight calculation: when the judgment matrix A is established, we should calculate the maximum

Table 1.   1–9 Scale meaning.

Scale	Meaning
1	Compare c_i with c_j, the two are equally important
3	Compare c_i with c_j, the former is slightly more important than the latter
5	Compare c_i with c_j, the former is more important than the latter
7	Compare c_i with c_j, the former is obviously more important than the latter
9	Compare c_i with c_j, the former is extremely important than the latter.
2, 4, 6, 8	Compare c_i with c_j, the importance of ratio is in the middle of the two adjacent levels.
$\frac{1}{2}, \frac{1}{3}, ..., \frac{1}{9}$	Compare c_i with c_j, The importance of the ratio is the reciprocal of the scale.

Table 2.

n	1	2	3	4	5	6	7	8	9
RI	0	0	0.58	0.90	1.12	1.24	1.32	1.41	1.45

eigenvalue ρ_{max} of A and its corresponding eigenvectors ω_i, and give ω_i for the normalization process to sort the weight of each factor. In order to judge if the degree of inconsistency of the matrix A is in the allowable range, it needs to check the consistency of it. The steps are as follows:

2.3 Calculate the consistency index CI

$$CI = \frac{\rho_{max} - n}{n-1} \qquad (1)$$

Find the corresponding random consistency index RI in "Table 2".

2.4 Calculating the consistency ratio CR

$$CR = \frac{CI}{RI} \qquad (2)$$

When CR < 0.1, then the judgment matrix can pass the consistency test. Otherwise adjust the matrix A until it passes the consistency test.

3 ESTABLISH WINE MARKETING DECISION MODEL

Ae survey of the wine sales market, reveals that the main factors effecting the wine sales are wine package, taste, price, advertising and packaging, quality, durability, pure degrees, etc. A wine marketing decision hierarchy model was established dependent on the interrelation and the subordinate relations between various factors "Figure 1".

According to the hierarchical structure of the Wine marketing decision model, it uses an expert scoring method and yaahp software. The weights of each factor indicators obtained are shown in Tables 3–16.

Table 3. Marketing decisions: Judgment matrix consistency ratio CR = 0.0072 < 0.1 $\rho_{max} = 4.0492$.

a_i	b_1	b_2	b_3	b_4	ω_i
b_1	1	2	3	7	0.4901
b_2	1/2	1	2	5	0.2879
b_3	1/3	1/2	1	3	0.1619
b_4	1/7	1/5	1/3	1	0.0601

Table 4. Price: Judgment matrix consistency ratio CR = 0.0000 < 0.1 $\rho_{max} = 2.0000$.

b_1	c_1	c_2	ω_i
c_1	1	5	0.8333
c_2	1/5	1	0.1667

Table 5. Taste: Judgment matrix consistency ratio CR = 0.0370 < 0.1 $\rho_{max} = 3.0385$.

b_2	c_3	c_4	c_5	ω_i
c_3	1	5	3	0.6370
c_4	1/5	1	1/3	0.1047
c_5	1/3	3	1	0.2583

Table 6. Advertisements: Judgment matrix consistency ratio CR = 0.0000 < 0.1 $\rho_{max} = 2.0000$.

b_3	c_6	c_7	ω_i
c_6	1	3	0.7500
c_7	1/3	1	0.2500

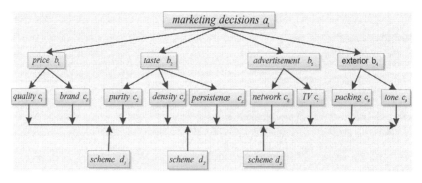

Figure 1. Wine marketing decisions hierarchy model.

Table 7. Exterior: Judgment matrix consistency ratio CR = 0.0000 < 0.1 ρ_{max} = 2.0000.

b_4	c_8	c_9	ω_i
c_8	1	3	0.7500
c_9	1/3	1	0.2500

Table 8. Quality: Judgment matrix consistency ratio CR = 0.0624 < 0.1 ρ_{max} = 3.00649.

c_1	d_1	d_2	d_3	ω_i
d_1	1	7	3	0.6491
d_2	1/7	1	1/5	0.0719
d_3	1/3	5	1	0.2790

Table 9. Brand: Judgment matrix consistency ratio CR = 0.0121 < 0.1 ρ_{max} = 3.0126.

c_2	d_1	d_2	d_3	ω_i
d_1	1	1/5	1	0.1336
d_2	5	1	7	0.7471
d_3	1	1/7	1	0.1194

Table 10. Purity: Judgment matrix consistency ratio CR = 0.0279 < 0.1 ρ_{max} = 3.0291.

c_3	d_1	d_2	d_3	ω_i
d_1	1	5	1/3	0.2654
d_2	1/5	1	1/7	0.0629
d_3	3	7	1	0.6716

Table 11. Density: Judgment matrix consistency ratio CR = 0.0624 < 0.1 ρ_{max} = 3.0649.

c_4	d_1	d_2	d_3	ω_i
d_1	1	5	1/3	0.2790
d_2	1/5	1	1/7	0.0719
d_3	3	7	1	0.6491

Table 12. Persistence: Judgment matrix consistency ratio CR = 0.0068 < 0.1 ρ_{max} = 3.0070.

c_5	d_1	d_2	d_3	ω_i
d_1	1	3	1/3	0.2426
d_2	1/3	1	1/7	0.0879
d_3	3	7	1	0.6694

Table 13. Network: Judgment matrix consistency ratio CR = 0.0036 < 0.1 ρ_{max} = 3.0037.

c_6	d_1	d_2	d_3	ω_i
d_1	1	1/5	1/3	0.1095
d_2	5	1	2	0.5816
d_3	3	1/2	1	0.3090

Table 14. Tone: Judgment matrix consistency ratio CR = 0.0036 < 0.1 ρ_{max} = 3.0037.

c_9	d_1	d_2	d_3	ω_i
d_1	1	2	1/3	0.2297
d_2	1/2	1	1/5	0.1220
d_3	3	5	1	0.6483

Table 15. Sub-criterion layers of each factors relative to target layer weight.

a_1								
c_1	c_2	c_3	c_4	c_5	c_6	c_7	c_8	c_9
0.4084	0.0817	0.1834	0.0301	0.0744	0.1214	0.0405	0.0451	0.0150

Table 16. The final weight of scheme layer relative to target layer.

a_1		
d_1	d_2	d_3
0.4006	0.2237	0.3757

4 THE RESULT ANALYSIS

Judged by the matrix structure, scheme d_1 pays more attention to wine price, scheme d_2 to wine brand and advertising and scheme d_3 to the taste and quality. According to the final weight of the scheme layer to the relative target layer we can judge that the weight in schemed_1 is the largest among all the schemes. Therefore, we should choose the scheme d_1 as the main marketing decision scheme. Comparing scheme d_1 with actual life, we find that people focus more on the commodity prices in wine consumption, which shows that the hierarchical analysis decision method is feasible. But if we observe the data carefully, we find that the proportion of price in marketing decision is less than 50%. Besides wine quality and advertising is also closely subject to people's attention.

First of all, according to the "Table 15" we can see that enterprises should strengthen the wine brand publicity. Let more consumers understand the enterprise wine brand, increase the wine brand visibility and improve consumer's confidence for the wine brand.

Secondly, according to "Table 15" we can find that wine enterprises should focus on wine quality improvement and promotion for the high income groups. Enterprises can start from the purity of the wine, concentration and persistence, and increased high-grade. Wine consumer groups should expand the market share of high-end wine.

Thirdly, for ordinary wine consumer groups, enterprises should focus on innovating the wine package design; increase the allure of the wine to consumers. At the same time, enterprises should conduct product innovation, in order to meet the personalised needs of ordinary wine consumer groups.

On the top of this, by constructing the judgment matrix and Tables 3–16, it can be concluded that the price factor is important for wine consumption, so enterprises should, according to market demand for the production of low, medium and high-grade wine, should be met for different consumer groups.

Therefore we should not only consider the price when making marketing decisions, but also pay attention to the quality of the wine and brand promotion. To enhance their image of wine enterprises should strengthen their brand position and expand their market share.

5 CONCLUSIONS

When consulting relevant literature, we can find that the current research on wine market mainly concentrates too much on the qualitative analysis of the wine market, but lacks a relatively scientific qualitative analysis method. We also find that analytical hierarchy process can be used to solve complex problems. AHP combines qualitative analysis with quantitative analysis, which not only makes full use of expert experience but also effectively controls influence on subjective factors. Through the consistency check of the judgment matrix, we can ensure the rationality of the calculation results. Furthermore, the logical reasoning process of the AHP method is clear and convenient in calculation, so that the calculation results really stand out. Getting the best marketing decision is in line with people's needs, which shows that using AHP in the decisions for marketing wine is feasible.

However, the AHP method also has its shortcomings. It has a certain subjectivity because the importance of each factor is presented by experts. Besides,

it is also difficult to consider all the relationships between factors while making a marketing decision. In the actual application, you can refer to the method in this article and make adjustments accordingly, on the influencing factors of marketing decisions and the the importance of each factor, so as to increase the accuracy and feasibility of the marketing decision.

ACKNOWLEDGMENT

In this paper, this research was sponsored by The National Key Technology R&D Program of China (Grant 2012BAI34B01-5), Major Project for Science and Technology Plan of Hangzhou (No. 20122511A26), Natural Science Foundation of Zhejiang Province (Grant LY14F020002), This research was sponsored by the Contemporary Business and Trade Research Center of Zhejiang Gongshang University which is the Key Research Institutes of Social Sciences and Humanities Ministry of Education (Grant 14SMXY04YB). The authors also gratefully acknowledge the helpful comments and suggestions of the reviewers, which have improved the presentation.

ABOUT THE AUTHOR

Tongkui Gumajors has majored in management science and engineering from Zhejiang Gongshang University, Zhejiang, China. His research interests are data mining and electronic commerce. His mobile phone number is 15700060942, and the e-mail address is 15700060942@139.com.

Zhiyi Huang majors in management science and engineering of Zhejiang Gongshang University, zhejiang, china. His research interests are data mining and music information retrieval. His mobile phone number is 13157172925, and the e-mail address is 1203105280@qq.com.

REFERENCES

[1] ZekiAyağ, R.G.Özdemir: Journal of Intelligent Manufacturing. Forum Vol. 179–190(2006), p. 17.
[2] Ludovic-Alexandre Vidal, Evren Sahin, Nicolas Martelli, Malik Berhoune Brigitte Bonan: Expert Systems With Applications, Forum Vol. 1528–1534(2009), p. 37.
[3] Michele Bernasconi, Christine Choirat, Raffaello Seri: Journal of Mathematical Psychology. Forum Vol. 152–165(2010), p. 55.
[4] S.Perez-Vega, Senior Peter, A.Nieva-de la Hidalga, P.N. Sharratt: Process Safety and Environmental Protection. Forum Vol. 261–267(2011), p. 89.
[5] ThadsinKhamkanya, George Heaney Stanley McGreal: Journal of Corporate Real Estate, Forum Vol. 80–93(2012), p. 14.
[6] Pravin Kumar, Rajesh K Rajesh: Journal of Modelling in Management. Forum Vol. 287–303(2012), p. 7.

Advances in Future Manufacturing Engineering – Yang (Ed.)
© 2015 Taylor & Francis Group, London, ISBN 978-1-138-02817-3

IDEF and UML-based enterprise modeling

Feng Yang
Jilin Teacher's Institute of Engineering and Technology, Jilin, China

Youquan Cheng
Changchun Institute of Engineering Technology, Changchun, Jilin Province, China

Lidan Fan
Jilin Teacher's Institute of Engineering and Technology, Jilin, China

ABSTRACT: The purpose of enterprise modeling is to describe the architecture of enterprise, so as to build a series of models that express the status of the enterprise. Enterprise Modeling, as the decision support tools methods of the collection of enterprise information, is the key to information system development. Currently, there are two kinds of modeling methods. One is the structure-based modeling methodology; the other is the object-oriented modeling methodology. Although similar characteristic lie in both, there are some differences between them. In this paper, we product a new modeling method that integrates IDEF and UML. After establishing a series of models with one example, the modeling steps are analysed. The method can help modeling engineers in various fields to build and test specialised models of their own, based on integrated IDEF models.

Keywords: enterprise modeling; IDEF; UML

1 INTRODUCTION

Enterprise modeling is the use of appropriate expressions from one or more angles (such as business processes, applications, information, organization, resources and so on) describing the architecture of the enterprise, in order to produce a series of enterprise modeling representing the actual situation of enterprises [1]. Enterprise Modeling for the Enterprise Information Integration of decision support tools and methods of collection is the key of information system development [2]. Enterprise modeling is currently the main method of structured modeling methodology and object-oriented modeling methodology, the former represented to IDEF (Integrated DEFinition), ARIS, GRAI-GIM, CIM-OSA, [3], the latter is represented in a uniform modeling language (Unified Modeling Language, UML). Usually structured modeling methodology is aimed at engineering and object-oriented modeling methodology for software engineering major [4], but there is an intersection between the two, although there are different characteristics in them as well. Therefore, some scholars have tried to integrate the advantages of both modeling methodologies to achieve the entire life cycle of enterprise modeling and implementation purposes [5][6]. To complete this work, this paper based on the above study was based on IDEF and UML modeling language modeling combining method, and established a relevant model as example.

2 IDEF AND UML OVERVIEW

IDEF (Integrated Computer-Aided Manufacturing, ICAM) is used to describe a set of internal operations modeling. IDEF was invented by the U.S. Air Force, and is now based on knowledge base system development. It was originally only used in the manufacturing industry, and thereafter its use has become widespread in transformation, and for general software development. IDEF is the is integrated computer aided manufacturing. IDEF is the first project built in the U.S. Air Force ICAM [7], and originally developed 3 methods: Functional Modeling (IDEF0), information modeling (IDEF1), dynamic modeling (IDEF2), later, information systems were developed, as was the development of the IDEF family. From IDEF0 to IDEF14 (including IDEF1X included) a total of 16 sets of methods, each method is routed through the modeling program to obtain a specific kind of information. The IDEF method is used to create images of various systems to analyse the expression system modules, to create the best version of the system and help convert things from different systems [8].

UML (Unified Modeling Language, which is short for the Unified Modeling Language, and was created by Grady Booch, James Rumbaugh and Ivar Jacobson, and then standardised by the Object Management Group (OMG) in 1997 [9], is a universal visual modeling language used to describe the various components of this complex software system—visual processing, construction and building models of software systems, and the establishment of software documentation [10].

It records constructing a system of rules and understanding of the system which can be used for understanding, design, navigation, configuration, maintenance, and information control. UML, which is the object-oriented development of a common, unified graphic modeling language, is the object of a modern software engineering environment and an important tool for analysis and design. UML provides a standard model of 3 basic building blocks: things, links and graphics.

Model building blocks of all the essential links have 4 characteristics [11]: Dependency, Association, Generalisation and Realisation. Model building blocks and contact combination can be constructed out of a well-structured system model (Well-formed Model). UML model elements of graphics are a visual representation of the collection. It is from the static and dynamic modelst one can describe both the system model and static model, also called structural model, with an emphasis on object structure in a system—including their classes, interfaces, attributes and relationships. Dynamic models, also known as behavioral models, emphasise the behavior of the system objects, including their methods, interaction, collaboration, and state changes [12]. These models can be at different stages of software development from different perspectives and different levels of description of the development of the project, which is very suitable for the structure and behavior of mock objects.

3 INTEGRATED IDEF AND UML MODELING

UML, through the establishment of the UML activity diagram, collaboration diagram, component diagram, deployment diagram, sequence diagram, state diagram, static structure diagrams, use case diagram of several graphics—fully embodies the object-oriented design, the development of standardization of the system and is highly explorative. But hinders its readability and understanding, as with many modeling notations the UML modeling language model built complex requires a higher level of domain experts to understand—thus affecting the business of the enterprise project management decision-making process in the early stages. The IDEF method, from the manufacturing information systems modeling, object-oriented design perspective, needs knowledge representation and software development to improve further. The UML approach from the field of object-oriented software development, business process modeling needs further expansion. Therefore, the large-scale integrated information management system in the development process, in the simple use of IDEF and UML method cannot be simple, to clear and to build flexible, scalable, reusable, management information systems. Integration of these two types of modeling language can be used to the greatest extent to overcomethis shortage. A combination of both enterprise information system modeling is shown in Figure 1.

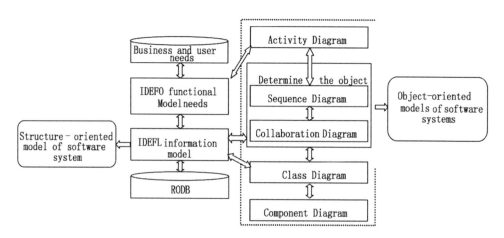

Figure 1.   IDEF and UML modeling method of combining.

4 INTEGRATED IDEF AND UML MODELING EXAMPLE

When we design a system by using the Integration of IDEF and UML languages, we must first know the demand for the characteristics of the system, analyse the system requirements, and then create the static model, based on system needs to construct the structure of the system, and then finally, describe the behavior of the system and establish the system dynamic model.

5 SYSTEM REQUIREMENTS ANALYSIS

The main task of requirements analysis is to determine the software system requirements. Namely: to define user requirements for the software to solve a problem or achieve an objective of capacity and also refer to a system or system components which must have the ability. System requirements analysis results can be a use case (Use Case) model expression. UML use case analysis is based on object-oriented modeling process, a significant feature in the modeling process based on UML by using the cases in a central location. Any software project development process is driven by the use cases, which runs through the system analysis, system design, system implementation, testing and the deployment of the entire process. This has become the current object-oriented analysis and design methods of the mainstream. In UML, a use case model consists of a number of use case diagrams describe the main elements of use case diagram is a system-wide (System Scope), use cases (Use Case), actors (actor), and the links between them. Here's inventory management system Use Case, as shown in Figure 2.

6 STATIC MODEL

System Static model is mainly expressed by the static object model Class Diagram and Object Diagram. Object is the most basic model elements of object-oriented modeling. The object class diagram expresses a set of classes and their links. In the object class diagram, it on the one hand describes the attributes of each object class itself, operations, and constraints and on the other hand, describes the system between the various classes of objects in static contact.

Object represents a specific instance of an object class. Object diagram is an instance of an object class diagram. Object Graph represents an object at a time for the state of the system, the state of the link between the object and the object's state of the static behavior. The classification defined in this system is: Storage, Stock, InStock, OutStock Product Position, LP Position, Stock Detail and Handled Operator. To clearly organize these classes, while establishing a static structure model, the class closely linked together will be into the package, as shown in Figure 3 and Figure 4.

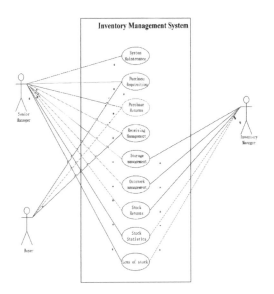

Figure 2. Inventory management system use-case diagram.

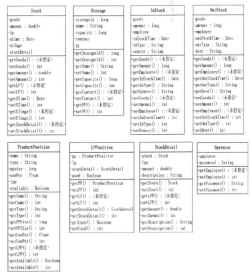

Figure 3. Parts inventory management system object graph.

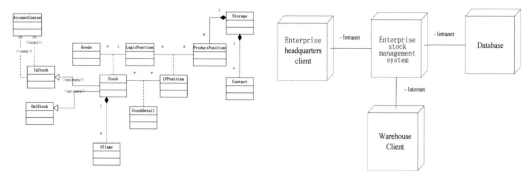

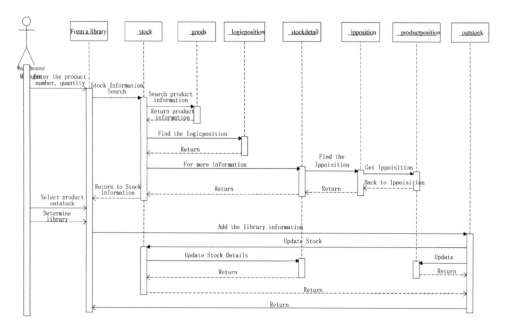

Figure 4. Inventory management system use-case diagram.

Figure 5. Parts inventory management system object graph.

Figure 6. Products out of sequence diagram library.

7 DYNAMIC MODEL

When the objects and their relationship to change the situation at any time are concerned, the system, a dynamic model with UML describing its visualization must be dynamically analyzed. There are three dynamic models: Interaction Diagram expressing objects timing diagram and State Diagram expressing object state changing the state diagram as well as the system's business process workflow and concurrent activity diagram (Activity Diagram). In the system analyzing and designing, class mapping of these graphics should be made for the main use cases and object in order to analyze the dynamic and behavior. The following sequence diagram shows parts of the system, whose objects can be seen from the interaction between the sequence shown in Figure 6.

8 PHYSICAL MODEL

In constructing an object-oriented software system, the system must be considered from the logical and physical aspects. The physical model of the system design, and intranet-based central database three-tier C/S structure, system, and as shown in Figure 5.

9 CONCLUSIONS

IDEF is used to describe a set of internal operations modeling. IDEF was invented by the U.S. Air Force, now based on knowledge base system development. It was originally only used in the manufacturing industry, and after-use has become widespread transformation, and for general software development.

UML is a well-defined, easily expressed, powerful and universal modeling language, which is integrated into the software engineering field of new ideas, new methods and new technologies. Its scope is not only support for object-oriented analysis and design, but also supports requirements analysis starting from the whole process of software development. This language established the combination of IDEF and UML modeling method Inventory management through business process reengineering example shows the feasibility of modeling, and discussed the combination of methods and procedures. The method can help the engineers of modeling in various fields build and test the specialized models of their own fields based on an integrated IDEF models.

ACKNOWLEDGMENTS

This paper is supported by "Twelfth five-year" science and technology research project of department of education of Jilin province, funds No. 2013482.

CORRESPONDING AUTHOR

Corresponding author is Feng Yang, Email: aolinpike0808@163.com.

REFERENCES

[1] M.C. Panis, B. Pokrzywa. Deploying a System-wide Requirements Process within a Commercial Engineering Organization[J]. 15th IEEE International Requirements Engineering Conference (RE 2007), 2007, 295–300.

[2] Aranda J., Easterbrook S., Wilson G. Requirements in the wild: How small companies do it[J]. 15th IEEE International Requirements Engineering Conference (RE 2007), pp. 39–48.

[3] Jacobson I. Object-Oriented Software Engineering[M]. Addison Wesley Professional, 1 edition (June 30, 1992), pp. 1–552.

[4] Kim C.H., Weston R.H., Hodgson A., et al. The complementary use of IDEF and UML modeling approaches[J]. Computers in Industry, 2003, 50(1): 35–56.

[5] Noran, Ovidiu Sever. UML vs IDEF: An Ontology-oriented Comparative Study in View of Business Modelling[J]. In Proceedings of INSTICC2009.204-212.

[6] Darius Silingas, Rimantas Butleris. UML-Intensive Framework for Modeling Software Requirements[J]. Proceedings of the 14th International Conference on Information and Software Technologies, IT 2008, Kaunas, Lithuania, April 24–25, 2008, p. 334–342.

[7] Glinz M. Problems and Deficiencies of UML as a Requirements Specification Language[J]. 10th International Workshop on Software Specification and Design, 2000, p. 11–22.

[8] Šilingas D., Vitiutinas R. Towards UML-Intensive Framework for Model-Driven Development. Proceedings of CEE-SET 2007, Springer: Lecture Notes in Computer Science Volume 5082, 2008, pp. 116–128.

[9] Castro J., Kolp M., and Mylopoulos J. Towards Requirements-Driven Information Systems Engineering: The Tropos Project[J]. Information Systems, Elsevier, 2002, 27 (6): 365–389.

[10] Engels G., Heckel R., and Sauer S. UML—A Universal Modeling Language? In M. Nielsen, D. Simpson (Eds.): ICATPN2000, LNCS 1825, pp. 24–38, 2000.

[11] K. Kapocius, R. Butleris. Repository for Business Rules Based IS requirements[J]. Informatica, Vol. 17, No. 4, 2006, 503–518.

[12] D. Silingas, R. Butleris. IDEF-Intensive Framework for Modeling Software Requirements[J]. Proceedings of International Conference on Information Technologies (IT 2008), Kaunas, 2008, 334–342.

Advances in Future Manufacturing Engineering – Yang (Ed.)
© 2015 Taylor & Francis Group, London, ISBN 978-1-138-02817-3

The research on the general and efficient method of network download

Wentao Liu
School of Mathematics and Computer Science, Wuhan Polytechnic University, Wuhan, Hubei Province, China

ABSTRACT: The network download component is often used in many systems such as PC client software, mobile APP and some web server systems. It is also the important part in the web page crawler and the internet search engine and they need a high efficient method to download the massive web pages. Some methods of realizing the function are introduced and include the WinINet and WinHTTP. The advantages and disadvantages of them are pointed out and some technologies related are discussed including the GET and POST style and how to get and set the Cookies. In this paper, a general method of web page download is developed and it uses the socket to download web page. The method pays more attention to the efficiency and it realizes the address redirect processing and the timeout control. According to the different style of web page data, the method provides the module of dealing with the gzip encoding and chunked data. The method contains the characters of high performance, adaptive download and portability.

Keywords: HTTP; socket; chunked; gzip; timeout control

1 INTRODUCTION

The network download is a very frequently used function with the rapid development of the network and the popularization of network applications. Some applications pay more attention to the network download component especially network applications such as professional network download software, network game and a variety of network clients. In the mobile platform, the network download has been widely used in the APP software. It can also be used in the network update upgrade module of various software applications. The network download component which can adapt to a verity of the network environment is very important. The HTTP protocol download is the most widely downloaded form and it is directly used more applications such as network search engine, network data collection system, web crawler system for various uses. Many network stations provide professional application program interface for a variety of special features. The aim of the HTTP download is to get the data from the server and it often uses the GET and POST method to extract data. There are many problems to be solved in the HTTP download progress. The download component must deal with the content encoding, cookies, chunked, timeout and so on. The commonly used encoding is gzip and the download needs to unzip the raw data from the server. If the server doesn't know the length of response information, it can provide the data in chunked format. The client must resolve the chunked data to get complete information. If the client needs to keep the cookies in the total session between the client and the server, it needs to set the cookies in the download progress. The download progress must be timely and effective, so the timeout must be taken seriously in order to deal with a variety of complex network environments.

2 THE DOWNLOAD METHOD

The WinINet firstly uses the InternetOpen to make an HINTERNET handle to indicate the session [1–2]. The simple method to open the URL is to use the InternetOpenUrl and it will make a new HINTERNET to state the connection. When the action is fulfilled, the WinINet can use the InternetReadFile to read the response data from the server in a loop statement with the second handle. When all the data has been received, it requires close the all handles. The InternetOpenUrl method cannot deploy the http header data according to the customer needs. It can use the InternetConnect to make a new connection handle and use the HttpOpenRequest to make a request handle with the connection handle. It can use the HttpAddRequestHeaders to add the http header field to the data. The data can be sent by the HttpSendRequest and it contains the http header data. The response data is received by use of the InternetReadFile to get the all data. It can use the HttpQueryInfo to get the information of HTTP response such as status code, cookie, and content-coding and so on.

The WinINet has the fault which cannot set the timeout correctly and it can use the asynchronous mode to solve this problem. The asynchronous WinInet uses the InternetOpen with the last parameter of INTERNET_FLAG_ASYNC. It uses the InternetSetStatusCallback to set the callback to keep watch over the progress of every operation. It uses the InternetConnect to get the connection handle and set the connection mask to the callback so the callback function call can understand the connection operation. The HttpOpenRequest also uses the other different mask to mark the request operation. The request data is sent by the HttpSendRequest and the response data is received by the InternetReadFileEx. In the callback function, the WinInet can judge the action according to the state value such as handle created, request sent and so on. If the user wants to wait for the completion of some actions, it can use the event synchronization mechanism to send the message to the main progress.

The WinHttp uses the HINTERNET handle to describe the structure of connection and operation [3]. It uses the WinHttpOpen to make a session handle of HINTERNET firstly and the function can set up the property of proxy and asynchronous features. Use the WinHttpConnect to make a connect handle and it uses the session handle and set the server IP and port. If the connection is finished successfully, the WinHttpOpenRequest can make a request handle with the connection handle. It can set the type of request and the URL. When the request connection is made, it can use the WinHttpSendRequest to send the HTTP data to the server with the request handle. It can make more property such as additional http headers and request data. If the request operation is finished, it can use the WinHttpReceiveResponse to get the data from the server. The data is received by a loop and it uses the WinHttpQueryDataAvailable to get the size of data and it is the mark of break of loop. The real data is received by the WinHttpReadData and it uses the buffer to store the data. When all operations are finished, it uses the WinHttpCloseHandle to close the handle of session, request and connection.

When the WinHttp uses the POST method to send the HTTP, it uses the POST mask in the WinHttpOpenRequest. It uses the WinHttpAddRequestHeaders to set the http header data such as Content-type and so on. The post data is sent by the WinHttpSendRequest with the fourth parameter. The other operation is same with the GET method. The WinHttp uses the WinHttpSetTimeouts to set the timeout and it can set the resolve timeout, connection timeout, send timeout and receive timeout with the session handle.

The cookie can contain many important information of server. The client can send the cookie information to the server and let the server maintain some status. The HTTP uses the Cookie field to describe the cookies and the server uses the Set-Cookie filed to return the cookie of the session. The IE browser can save the cookie information to the cookie files. When the browser makes a new connection to the server with the same URL, the browser will read the cookies from the local cookie files. The HTTP session will send the cookie information to the server. The cookies use the domain and path to mark the correct URL. The Set-Cookie uses expire to set the expiration time and if the cookie has expired over the specified time, it will delete the cookie. The WinINet can use the cookie of IE by default to send the server and the HttpOpenRequest can use the INTERNET_FLAG_NO_COOKIES mark to delete the cookie of IE. But sometimes the default cookies can solve many problems correctly. When you want to get the information in the user logged on, the cookie can make the download to get the private data. The mechanism makes the user to download the sensitive information. The WinHTTP cannot use the cookie of the IE by default and it must set the cookie information in the header by hand. It can use the InternetGetCookie of WinINet to get the cookie of one URL. But it cannot get the cookie which has the HTTPONLY property. Thus cookies can be got by the InternetGetCookieEx. If the WinInet want to modify the cookie, it can use the InternetSetCookie or InternetSetCookieEx to set the cookie.

The download method realized by the pure socket directly is defined in Figure 1. It does not depend on the IE control and can get more flexibility with the asynchronous name resolves, timeout control, redirect handling, chunked data analysis and gzip data processing.

The first step is to parse the provided URL to the server name and address. According the type of request data, the header of HTTP request must be made with the rule of HTTP protocol rules. For example, the following http headers are sent to server to get one web page.

GET /s/blog_48874cec0102edz8.html HTTP/1.1
Accept:*/*
Accept-Encoding: gzip
User-Agent: TestAgent
Host: blog.sina.com.cn
Connection: Keep-Alive
Referer: http://blog.sina.com.cn/s/blog_48874ce c0102edz8.html.

The Host field and the GET field are got from the module of parsing URL. The Accept-Encoding field is gzip and the server can send the response data in gzip format. In order to avoid the prohibited

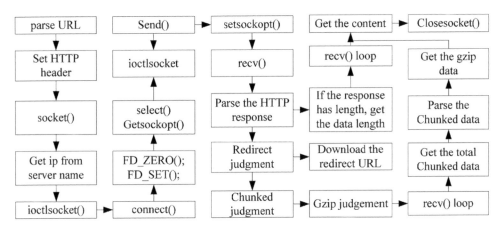

Figure 1.   The download progress based on socket.

external chain in some web server, the download component uses the URL in the Referer field. This method can achieve the purpose of deception server. Use the socket function to make a handle of socket and the domain is AF_INET. The type is SOCK_STREAM and the protocol is IPPROTO_ TCP because the HTTP is based on the connected TCP. The server name must be converted into IP address using the DNS server. The traditional gethostbyname may get the timeout defect when the network state have some problems so the asynchronous name resolves library c-ares can be used in the step [4]. Use the connect function to link the server with the socket handle. Use the select method to set the timeout of connection because the connection can cost more time especially the server broken the connection. Firstly use the ioctlsocket to set the socket is non-blocking mode. The connect function will return immediately and use the select function and getsockopt to detect the timeout. When the connection progress is finished, the socket will be set into the blocking mode with the ioctlsocket.

Use the send function to send the http header data and the recv function to get the response. Use the setsockopt to set timeout of getting data from the server with the SO_RCVTIMEO mark. The data maybe cannot get totally by one recv function and it can use the loop to get the total data from the server. Pare the response data to get the HTTP response with the HTTP end mark. Analyze the HTTP response to get the response code and make the further step according to the response code.

The part of the response data from the server is shown as following.

HTTP/1.1 200 OK
Server: nginx/1.2.8
Content-Type: text/html
Transfer-Encoding: chunked

Connection: keep-alive
Vary: Accept-Encoding
Cache-Control: no-cache
Content-Encoding: gzip.

When the HTTP response has the Transfer-Encoding field and its value is chunked, the HTTP data is organized by the chunked format [5]. Chunked data series are composed by a number of Chunks and its end flag is marked by a chunk with zero length. Each Chunk block is divided into two parts which are the head and body. The head of content specifies the total character number of the under paragraph text. The body part is the actual content with the length. Two parts are separated by the CRLF mark. The length of last chunk is zero and it was as a sign of the end of data. Because the server cannot determine the size of the data returned and the HTTP response header has no the fiend of Content-Length, it can return information using chunk data and the data can be produced dynamically with the server. In this download component, the response code can deal with the 200, 301 and 302. If the code is 301 or 302, the component can get the location redirection URL and use the same download component to download the URL. The Transfer-Encoding field expresses that the response data is chunked. So the further step is to get the total chunked data and resolves the chunked data to get the real response data. If the response the data is gzip by the Content-Encoding fiend, the data must be unzip with the gzip format and it can be realized by the zlib [6].

3   SUMMARY

The download module is required in many software and it must be able to adapt to a variety of

365

network situations. It must require high efficiency and contain the character of generality. There are many technologies to implement the module and various techniques have their advantages and disadvantages. The WinINet is widely used and it can realize the function with simple steps. It has more functions and can support for multiple protocols with the same calling interface so it is easy to be used for the programmer. It has the flaw in terms of handling timeouts and cannot properly handle the timeout setting and some bugs can affect download performance [7–8]. The program consumes a lot of resources in the case of multi-threaded. It needs more time and total components relatively bloated and it relies on the IE component. Another advantage is the ability to retain cookie information and it is convenient to download the webpage with the necessary cookie information. The WinHTTP is focused on programming for the HTTP protocol and it doesn't contain the compatible cookie information so it needs to set the cookie manually. The socket can be directly used to realize the download in order to make an efficient method and it can get more flexible implementation. The method can resolve the chunked data and unzip the gzip correctly with the capability of data timeout handling in the download progress.

REFERENCES

[1] WinINet Functions, http://msdn.microsoft.com/en-us/library/aa385473.aspx.
[2] WinINet Reference, http://msdn.microsoft.com/en-us/library/aa385483.aspx.
[3] WinHTTP, http://msdn.microsoft.com/zh-cn/library/aa382925(v=vs.85).aspx.
[4] c-ares: library for asynchronous name resolves, http://c-ares.haxx.se/.
[5] Hypertext Transfer Protocol—HTTP/1.1, http://tools.ietf.org/html/rfc2616, 1999.
[6] zlib: A Massively Spiffy Yet Delicately Unobtrusive Compression Library, http://www.zlib.net/.
[7] BUG: WinInet Application Stops Responding and Shows 100-Percent CPU Utilization, http://support.microsoft.com/kb/263754/en-us.
[8] BUG: InternetSetOption, http://support.microsoft.com/kb/176420/zh-cn.

Advances in Future Manufacturing Engineering – Yang (Ed.)
© 2015 Taylor & Francis Group, London, ISBN 978-1-138-02817-3

Numerical simulation research about offshore wind lifting ship structural damage in the process of lifting

Zhenqiu Yao, Ya Shu & Linxin Bu
School of Naval Architecture and Ocean Engineering, Jiangsu University of Science and Technology, Zhenjiang, China

ABSTRACT: For accidents during the process of lifting, the structure damage of offshore wind lifting vessels has been researched by this paper—choosing 3.6 MW nacelle as the object of the study, and using MSC. Dytran does the numerical simulation research. The result shows that the collision will produce greater iMPact force because of the great weight of the nacelle. Common ships cannot even sustain the collision falling from a low height—or the collision may produce severe damage to the ship structure. So these kind of accidents need be completely eradicated.

Keywords: offshore wind lifting vessel; falling; numerical simulation

1 INTRODUCTION

In recent years, as the offshore wind power technology gets more progressive, it has become the world's latest direction in the field of energy development[1]. And it needs large offshore wind lifting ships to complete the installation of the wind turbines. The offshore wind lifting ship is a new offshore engineering ship which has the functions of transforming, installing, lifting and so on, to complete the entire installation of the offshore wind turbine. In the process of installing the wind turbine, it may have some potential safety hazards, caused by a series of emergencies, such as suddenly a generator falling down accidently, and which can cause serious damage to the structure of the vessel.

We can calculate the probability of an object falling at each lifting, based on the statistical data of accidents from 1980 to 1986, provided by the British Department of Energy[2]. During this period, the entire number of lifting was about 370000—and a crane lifts 4500 times in each year. And the probability of falling is 2.2e-5 at each lifting. For the object, whose mass is over 20 tons, the probability of falling is about 3e-5 at each lifting: 70% of them may fall onto the platform deck, and about 30% of these objects may fall into the sea. The main section on the vessel carrying the components of the wind generator is the deck, which is also the platform for completing the assembling of the wind generator. It needs a crane to assist during the process of assembling. An object has the probability falling because of accidents or errors in work—so this emergency can cause serious damage to the structure and equipment. We need, therefore, to do some research about the structure security of the ship because this is something quite new and we have little design experience.

This paper will take the 3.6 MW wind turbine as an example and use the software MSC. Dytran will research the main component, nacelle. During the process of lifting, it will use different falling speeds to research the damage which the object causes to the structure, so that we can get the safety limit of the structure of the vessel, which can then be used to guide the structure design of the vessel, and also can provide some measures to avoid collision.

2 DEFINITION OF CONTACT

In the analysis of structural dynamic problems, the interaction between different objects can be obtained through the contact calculation. The contact module is provided by MSC. Dytran is an effective way to solve this kind of problem. The using of this contact module can not only simulate the continuous contact between the variable body and rigid body, but also allow the contact objects to make long-distance relative moves.

Ship collision[3] is a kind of complex nonlinear dynamic response process when the hull structures cave under a huge iMPact load, in a very short time (about tenths of a second to a few second). After decades of development, ship collision is mainly divided into three methods—simplified analytical method, experimental method and numerical simulation calculation method[4].

So this paper chooses the accident during nacelle lifting process, as the object of its study, to carry out the hull structure damage numerical simulation research because. there is a lot of equipment in the nacelle, and each collision may cause damage to the nacelle and inner equipment. This paper will mainly study the damage conditions of the hull to simplify the problem and the study will also use personal computers to assist the calculation. In order to simplify the simulation model, we can simplify the nacelle as a solid element according to the integrity of the nacelle, the definition and calculation of main-minor surface contact—and we can also make an assumption that the intensity of nacelle is stronger than the hull For the ship the consideration is the hull damage strength of the central part, which is the main platform used for assembly operations and is also the most prone to accidents. At the same time, when the nacelle contacts the hull, the hull is also needed to define the adaptive contact itself.

Table 1. The parameters of wind turbine.

Name	Dimension (m)	Weight (t)
Blade	$\Phi 2.8 \times 58.5 \times 4.2$	100
Turbine	$18 \times 5 \times 5$	140
Tower	$\Phi 4.5 \times 90$	90

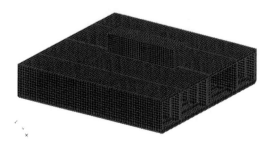

Figure 1. The finite element model of simplified icebreaker.

3 COLLISION MODEL

According to the structural damage caused by the collision of objects, the DNV specification presents two kinds of calculation analysis methods. One is the method of considering the energy condition, based on simple elastic-plastic to analysis, and the other, is to use finite element software based on nonlinear dynamics for simulation. But the first one is a simplified analytical method, and it omits many related content, which cannot be calculated conveniently and thus leads to imprecise results. And the second takes all the influencing factors of the collision into account, so it can calculate the relationship between structure deformation and stress effect more precisely[5]. The common contact calculation methods are a penalty function method, the Lagrange multiplier method and the Hertz contact force method. The Lagrange multiplier method is an accurate method to calculate these things.

Wind Power Installation and Wind Turbine. This paper chooses a large offshore wind lifting ship as the object of its study. The main dimensions: Overall Length 132.34 m, Length between Perpendiculars 121 m, Breadth 39 m, Depth 9 m deep, Design Draft 5.8 m. The ship is similar to a square box type, and has a single bottom and a single deck; and there are three continuous longitudinal bulkheads. The platform can carry up to 10 sets of Siemens-type SWT-3.6-120 3.6 MW wind turbines to the specified area to be installed. The specific parameters of wind turbines are shown in Table 1.

This paper will choose the heaviest main component, wind nacelle, as the object of the study to

research the structure damage numerical simulation. And the paper will also mainly focus on the damage conditions of the hull, to simplify the problem and use personal computers to assist in the calculation. In order to simplify the simulation model, we can simplify the nacelle as a solid element according to the integrity of nacelle, definition and the calculation of main-minor surface contact; we can also make an assumption that the intensity of nacelle is stronger than the hull. For the ship the consideration is the hull damage strength of the central part, which is the main platform used for assembly operations and is also the most prone to accidents. Using the most commonly used Plastic Kinematic Model as the computational model, we will consider the Strain hardening at the same time. Once the calculated unit reaches the maximum plastic failure strain value which is assumed before, it is identified that the unit has been broken.

It uses the QUAD 4 element to build the model of the main structure in the middle of the hull. There are 36024 elements and 53762 nodes in this model. Because of the high strength, the offshore wind turbine is simplified as a solid element and its dimension is $5 \times 5 \times 18$; in this model, there are 3600 solid elements and 4477 nodes. The numerical simulation of the whole model is shown in Figure 1.

Collision Parameters. The hull body parts are made of NV-D36 high strength steel which has a sensitive strain rate. So it needs to consider the relationship between its yield stress and strain rate. The formula 1–1 can be applied in the strain rate model of hull material which is called is the Cowper-Symonds model[6,7]. This formula can

Table 2. Hull material model parameters.

Density of material [kg/m³]	Modulus of elasticity [N/m²]	Yield stress [MPa]	Hardening modulus [Gpa]
7850.0	2.1×10^{11}	355.0	1.18
Strain rate parameters C	Strain rate parameters P	Poisson ratio	Plastic deformation
40.40	5.00	0.30	0.30

Table 3. Turbine material model parameters.

Total of material [kg]	Modulus of elasticity [N/m²]	Yield stress [MPa]	Poisson ratio
140000	2.1×10^{11}	345.0	0.30

reflect the influence of strain rate on the yield stress, and the calculated data tallies with the experimental data.

$$\frac{\sigma_0'}{\sigma_0} = 1 + \left[\frac{\dot{\varepsilon}}{C}\right]^{\frac{1}{p}}. \tag{1}$$

In the formula: σ_0'—Dynamic yield stress when the plastic strain rate is ε
σ_0—the corresponding static yield stress
C—Constant, for ship steel, C = 40.4
P—Constant, for ship steel, P = 5.
We can consider the wind turbine as homogeneous material to simplify the calculated factors.

4 NUMERICAL SIMULATION EXAMPLE

In order to research the accident occurrences during the process of lifting, as well as the failure mode, and dynamic characteristics after the hull was iMPacted, we make an assumption that the speed is 2 m/s and 5 m/s when the wind nacelle falls onto the hull deck. We keep observing the deformation or damage of the hull until the hull element is destroyed. We can fix the two sides of the midship rigidly to do the Numerical Simulation calculation, because the weight of the wind turbine is 140 tons and the ship is 23000 tons. The collision time is 5 s for the sake of saving time.

Wind Nacelle Falls onto the Deck with the Speed of 2 m/s. The function of the deck is to sustain the static load, so we don't consider the shock load when designing. The weight of the wind nacelle is huge, and there is no possibility for the deck to sustain the large kinetic energy iMPact, so we use a lower speed to do the simulation analysis.

We can get all kinds of curve diagrams about energy and speed via collision analogue

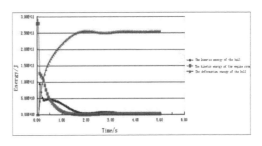

Figure 2. Curve of the internal energy at the speed of 2m/s.

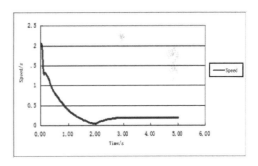

Figure 3. Velocity-time curve at the speed of 2m/s.

simulations; also we get Figure 2 and Figure 3. We can get information from these figures that the initial energy of the whole system is the kinetic energy of the nacelle. Then the kinetic energy mainly turns into the hull's deformation energy, after the collision, because the parallel middle body is restricted. When the nacelle hits the hull at about 0.1 s, the kinetic energy of the nacelle falls sharply, at the same time the deformation energy of the hull increases quickly. The collision will finish after about 1.8 s and the curve figures tend to be gentle. The initial kinetic energy of the nacelle is 2.8e11J,

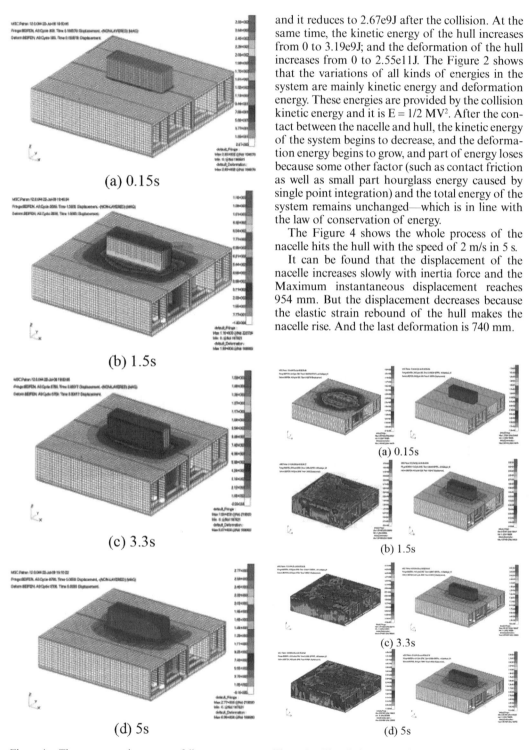

(a) 0.15s

(b) 1.5s

(c) 3.3s

(d) 5s

Figure 4. The process engine room to fall.

(a) 0.15s

(b) 1.5s

(c) 3.3s

(d) 5s

Figure 5. The whole process of stress and strain diagram.

and it reduces to 2.67e9J after the collision. At the same time, the kinetic energy of the hull increases from 0 to 3.19e9J; and the deformation of the hull increases from 0 to 2.55e11J. The Figure 2 shows that the variations of all kinds of energies in the system are mainly kinetic energy and deformation energy. These energies are provided by the collision kinetic energy and it is $E = 1/2\ MV^2$. After the contact between the nacelle and hull, the kinetic energy of the system begins to decrease, and the deformation energy begins to grow, and part of energy loses because some other factor (such as contact friction as well as small part hourglass energy caused by single point integration) and the total energy of the system remains unchanged—which is in line with the law of conservation of energy.

The Figure 4 shows the whole process of the nacelle hits the hull with the speed of 2 m/s in 5 s.

It can be found that the displacement of the nacelle increases slowly with inertia force and the Maximum instantaneous displacement reaches 954 mm. But the displacement decreases because the elastic strain rebound of the hull makes the nacelle rise. And the last deformation is 740 mm.

Figure 5 shows stress and strain changes of the whole process.

It shows in Figure 5, that in 0.15 s the stress quickly surpasses the yield stress of the steel for ship—355 MPa to reach 484 MPa in a very short period when the nacelle falls onto the hull. Then the steel becomes the plastic state. The stress is always greater than the yield stress though the stress has decreased. About 2 s later, the nacelle begins to rebound, and 5 s later the stress becomes 485 MPa. In the process of the fall, the plastic strain at most reaches 0.29, and it doesn't reach the largest plastic failure strain 0.30, so there are no breaking elements. We can find that the acting force has far

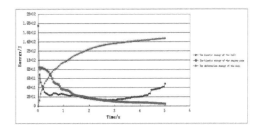

Figure 7. Curve of the internal energy at the speed of 5m/s.

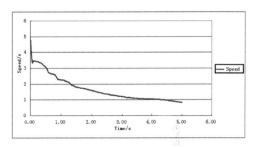

Figure 8. Velocity-time curve at the speed of 5m/s.

surpassed the allowed stress; the structure has plastic deformation and some parts are broken.

Wind Nacelle Falls onto the Deck with the Speed of 5 m/s. We continue to enlarge the fallen speed, to research the failure mode and dynamic characteristics of the hull after the collision. The model and method used here are the same as above. The calculated results include factors such as displacement diagram, stress and strain diagram, energy diagram and speed curve diagram which are shown as diagrams 6–8.

It can be found that the nacelle breaks the deck completely after 5 s, and keeps falling down to the bottom—even has the tendency to break the bottom. It shows from the stress and strain diagram that when the nacelle falls the iMPact force has reached 485 MPa, which has far surpassed the allowed stress, and the maximum stress is double the time of the static yield limit. The strain produced by the collision has already surpassed the maximum plastic strain 0.3, so some parts of the structure elements have been broken. Nacelle's falling speed is 5 m/s, corresponding to free fall from the height of 1.28 m. The middle section of the hull sustains great iMPact force from the view of energy. The kinetic energy of nacelle decreased sharply from the initial energy 1.75e12J, but the hull structure cannot stop the nacelle immediately, so the kinetic energy produced by speed is enough to break the ship structure, and the structure

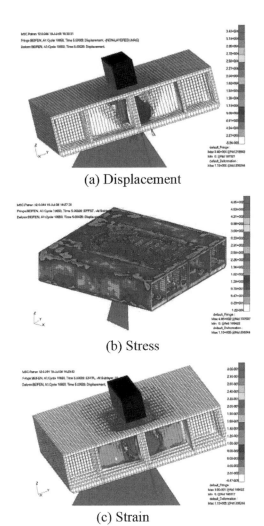

(a) Displacement

(b) Stress

(c) Strain

Figure 6. The whole process of displacement stress and strain diagram.

assimilates the deformation energy which seriously makes the structure out of shape. On the aspect of speed, the nacelle keeps falling, and the hull only reduces the nacelle's speed, but doesn't stop it after hitting the hull and even makes the nacelle rebound.

5 CONCLUSION

This paper takes the accident during the process of lifting nacelle as object of its study, uses numerical method to simulate the damage condition of the strong structure of the hull when the nacelle falls onto it with different speeds—and as a result, gets the stress-strain diagram and speed-time diagram. We can draw the conclusions below via the analysis the time-distance diagram of collision.

1. The collision will lead to the deck damage, longitudinal deformation even though there is longitudinal bulkhead under the deck at the collision position when the fallen speed reaches 5 m/s (corresponding to the height of 1.28 m). After the nacelle contacts the hull, the collision phenomenon is more obvious. As the nacelle keeps falling, the ship structure will go out of shape and even get broken. So the nonlinear characteristics are very obvious.
2. The energy provided by the collision is not enough to break the hull structure, but can also produce plastic deformation when the falling speed is 2 m/s. When the iMPact force reaches its maximum, the displacement of the ship remains unchanged, and the iMPact force decreases sharply.
3. For the structure of offshore wind lifting ship, which doesn't consider the iMPact force, it may result in the structure deformation or damage once the accident occurs during the process of lifting. And the strength of ship structure cannot sustain the collision., so we must completely eradicate this kind of accident.

ACKNOWLEDGEMENTS

Project2013 AA050602 supported by the National High-tech R&D Program of China.

REFERENCES

[1] Zhang Taiji, Wang Zhangtang. The development of offshore wind power equipment installation vessel [J] SHIP&BOAT, 2009, (5), pp. 38–40.
[2] DNV 1996b. Det Norske Veritas, 1996.
[3] Paik J.K., Petersen P.T. Modelling of the Mechnics in Ship. Collisions [J]. Ocean Engineering, 1996(02), Vol. 23, No. 2, pp. 107–142.
[4] Hu Zhiqiang, Cui Weicheng. Research summarize of ship collision mechanism and crashworthiness structure design. 2005,04, VOL.9, pp. 131–142.
[5] Zhou Guobao, Wang Lin. Numerical simulation research about offshore platform under the iMPact force [J]. China Offshore Platform, 2007, 22(2), pp. 18–21+27.
[6] Minorsky V.U. An Analysis of Ship Collision to Protection of Nuclear Powered Plant [J]. Journal of Ship Research, 1959, 3(2):1–4.
[7] Wang Ren, Xiong Zhuhua, Huang Wenbin. Plastic mechanics Foundation [M]. Beijing, Science Press, 1982, 03, pp. 1–598.

Advances in Future Manufacturing Engineering – Yang (Ed.)
© 2015 Taylor & Francis Group, London, ISBN 978-1-138-02817-3

Study on navigation radar image fusion denoising algorithm based on multiple wavelet packet bases

Wenjie Lang
National Electrical and Electronic Laboratory Teaching Exemplary Center, North University of China, Taiyuan, Shanxi, China

Guoguang Chen, Xiaoli Tian & Changfan Xin
School of Mechatronics Engineering, North University of China, Taiyuan, Shanxi, China

ABSTRACT: To reduce navigation radar image's noise and clutter interference, an image fusion denoising algorithm based on multiple wavelet packet bases is proposed in this paper, which is in the light of the fact that wavelet packet bases with different features can effectively lay different importance on image. Navigation radar image is decomposed by wavelet packet bases with different features. The processed coefficient is fused to get a new set of fused sub-images. Finally the inverted transformation of wavelet packet is reconstructed to obtain the new denoised image. Experimental results show that this method can achieve better denoising effect on navigation radar image.

Keywords: wavelet packet transform; image fusion; image denoising

1 INTRODUCTION

Navigation radar is an important tool for vessels' navigation and collision avoidance. With the development of science and technology, its function is greatly improved with much more superiority and practicability. However, navigation radar image is often affected by noise and clutter in the process of image collection, transmission and display, resulting in the phenomenon of low Signal-to-Noise Ratio (SNR) and less precision. The digital preprocessing of the radar image is the basis of target tracking, intelligent early-warning, thus is of vital significance.

Wavelet transform, with the characteristic of multi-resolution analysis, can focus on any details of the analyzed object, hence has been widely applied in image processing, especially in the field of image noise reduction. Time-frequency transformation based on wavelet packet can provide signal with a more precise approach, which can divide the frequency band to multiple levels. The high frequency and low frequency of the image can be further divided to keep the original image information better and can effectively remove the noise. In wavelet decomposition, different wavelets bases have generally different time-frequency characteristics to show signals' different parts or features. In general, the effect of signal denoising is affected

by coefficient distribution of each dimension after wavelet decomposition, which is, to some extent, affected by wavelet base. Therefore, image denoising effect is associated with the selection of wavelet base [1]. This article puts forward the image fusion denoising algorithm of navigation radar reduction on the basis of multiple wavelet base packets, which can make the de-noised image improve its Peak Signal-to-Noise Ratio (PSNR) and subjective visual effect greatly.

2 IMAGE WAVELET PACKET TRANSFORM [2]

If unify dimension subspace V_j and wavelet subspaces W_j with a new subspace U_j^n and suppose:

$$\begin{cases} U_j^0 = V_j \\ U_j^1 = W_j \end{cases} (j \in Z).$$ (1)

Orthogonal decomposition of Hilbert space $V_{j+1} = V_j \oplus W_j$ can be decomposed with U_j^n.

$$U_{j+1}^0 = U_j^0 \oplus U_j^1 \quad (j \in Z).$$ (2)

Subspace U_j^n is defined as closure space of function $u_n(t)$, while U_j^{2n} as closure space of function

$u_{2n}(t)$. $u_n(t)$ is also made to meet the following dual scaling equation,

$$\begin{cases} u_{2n}(t) = \sqrt{2} \sum_{k \in Z} h(k) u_n(2t-k) \\ u_{2n+1}(t) = \sqrt{2} \sum_{k \in Z} g(k) u_n(2t-k). \end{cases} \tag{3}$$

in which $h(k)$ and $g(k)$ are the low and high pass filter coefficients of filter bank respectively.

$$g(k) = (-1)^k h(1-k) \tag{4}$$

That is to say, these two filter coefficient has orthogonal relation, esp, when $n=0$, eq. 3 can be changed to the following:

$$\begin{cases} u_0(t) = \sqrt{2} \sum_{k \in Z} h(k) u_0(2t-k) \\ u_1(t) = \sqrt{2} \sum_{k \in Z} g(k) u_0(2t-k). \end{cases} \tag{5}$$

From comparison, $u_0(t)$ and $u_1(t)$ are degraded as the scale function $\phi(t)$ and wavelet base function $\psi(t)$ respectively.

$$\begin{cases} \phi(t) = \sqrt{2} \sum_{k \in Z} h(k) \phi(2t-k) \\ \psi(t) = \sqrt{2} \sum_{k \in Z} g(k) \phi(2t-k). \end{cases} \tag{6}$$

Sequence $\{u_n(t)(n \in Z_+)\}$ constructed from Eq. 6 is considered as orthogonal wavelet packet determined by basis function $u_0(t) = \phi(t)$.

If $g_j^n(t) \in U_j^n$, $g_j^n(t)$ can be expressed as:

$$g_j^n(t) = \sum_l d_l^{j,n} u_n(2^j t - l). \tag{7}$$

Wavelet packet decomposition algorithm is to get $\{d_l^{j,2n}\}$ and $\{d_l^{j,2n+1}\}$ from $\{d_l^{j+1,n}\}$ as follows:

$$\begin{cases} d_l^{j,2n} = \sum_k a_{k-2l} d_k^{j+1,n} \\ d_l^{j,2n+1} = \sum_k b_{k-2l} d_k^{j+1,n}. \end{cases} \tag{8}$$

Wavelet packet reconstruction algorithm is to get $\{d_l^{j+1,n}\}$ from $\{d_l^{j,2n}\}$ and $\{d_l^{j,2n+1}\}$,

$$\{d_l^{j+1,n}\} = \sum_k \left[h_{l-2k} d_k^{j,2n} + g_{l-2k} d_k^{j,2n+1} \right]. \tag{9}$$

in which $\{a_k\}$ and $\{b_k\}$ are decomposition filter, while $\{h_k\}$ and $\{g_k\}$ are reconstruction filter [3].

3 IMAGE FUSION DENOISING BASED ON MULTIPLE WAVELET PACKET BASE

3.1 Image fusion

Image fusion [4] is a field of data fusion technology research. Its advantages are mainly to improve image quality, increase the geometric registration accuracy or Signal-to-Noise Ratio (SNR), generate a three-dimensional effect, realize real-time or quasi real-time dynamic observation, overcome the incomplete of image data in target extraction and recognition, and expand the space-time range of sensors and so on. In this paper, image fusion denoising based on multiple wavelet packet bases is adopted to fuse the denoised wavelet packet coefficient before reconstruction. In the process of fusion, the choice of fusion principle is the key factor in determining the wavelet fusion effect. In this paper the wavelet packet coefficient are calculated with weighted average fusion algorithm.

3.2 Method and implementation procedures

When denoising image with wavelet transforms, the key problem is how to distinguish noise coefficient, restrain and even remove it. The approach to minimize the influence on effective signal coefficient when restraining and eliminating noise coefficient has been discussed a lot by other researchers. In this paper, the image fusion denoising method based on multiple wavelet packet base is to use wavelet packet base with different features to decompose noise-contained image, threshold processing wavelet packet coefficient on different levels. Then weighted average fusion algorithm is adopted to process the low frequency and high frequency components to get a new set of the fused sub-images. Finally the inverted transformation of wavelet packet is reconstructed to get the new denoised image. The basic process is shown in Figure 1.

Implementation procedures:

1. Choose n different wavelet packet bases with different qualities, and decompose the noise-contained image. Thus the multi-resolution image sequence can be gotten.
2. Process wavelet packet coefficient on each frequency band with soft threshold method.
3. Obtain a new set of fused sub-images from the fusion of the low and high frequency of the image based on given fusion rules.
4. Reconstruct wavelet packet of the processed coefficient.

Short support wavelet packet base db3 and symmetry biorthogonal wavelet packet base bior4.4 are

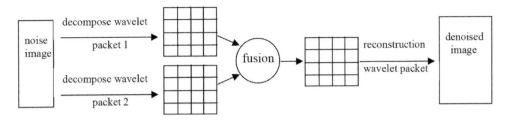

Figure 1. Image fusion denoising based on multiple wavelet packets bases.

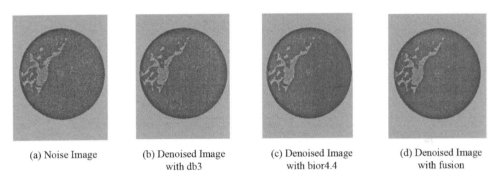

| (a) Noise Image | (b) Denoised Image with db3 | (c) Denoised Image with bior4.4 | (d) Denoised Image with fusion |

Figure 2. The result of image denoising.

fused to decompose noise-contained image, which can respectively get better denoising effect in the smooth and edge areas. Image is decomposed as three layers, Bayes Shrink threshold is adopted to process wavelet coefficient. The low and high frequency components of image are fused with weighted average image fusion method. If the fusing images are A and B, their image sizes are $M \times N$, F is the fused image. Then the pixel gray value weighted average fusion process is expressed as following:

$$F(m,n) = w_1 A(m,n) + w_2 B(m,n). \qquad (10)$$

in which $m = 1, 2 \ldots M$, $n = 1, 2 \ldots N$, w_1, w_2 are weighted coefficient. $w_1 + w_2 = 1$. Here we choose $w_1 = w_2 = 0.5$.

3.3 Experiment and analysis

To illustrate the effectiveness of the proposed method, the virtual radar image containing radar clutter or white noise is denoised, the noise variance σ^2 is 15 as shown in Figure 2.

Data from the denoised image with method of db3, bior4.4 and the new method respectively are shown as Table 1.

It can be seen that this algorithm can better remove noise. In addition, the denoised image is

Table 1. The results of PNSR, SNR, MSE after wavelet packet denoising.

[type]	[Noise Image]	[db3 denoising]	[bior4.4 denoising]	[fusion Denoising]
[PSNR]	[24.7012]	[30.2190]	[30.5134]	[30.8639]
[SNR]	[22.2181]	[27.8952]	[27.9123]	[28.0916]
[MSE]	[0.05060]	[0.00318]	[0.00302]	[0.00268]

clearer and has better visual effect. To show the superiority of this method, the Signal-to-Noise Ratio (SNR) and minimum mean square error are used as evaluation criteria.

Table 1 indicates that under the same noise condition, PSNR of the denoised image with wavelet packet base db3 and bior4.4 is slightly higher compared with that with the single wavelet packet base.

4 CONCLUSION

This paper introduces the basic theory of wavelet packet, and proposes a new image fusion denoising algorithm of navigation radar on the basis of different characteristics of wavelet packet base.

The contrast experiment with traditional wavelet packet threshold denoising method shows that this new algorithm can improve the subjective visual effect with higher Peak Signal to Noise Ratio (PSNR), which lays some foundation for further image processing.

CORRESPONDING AUTHOR

Name: Wen-jie Lang, Email: langwenj@163.com, Mobile phone: 13753116738.

REFERENCES

[1] Cai Dun-hu, Yi Xu-ming. Influence of Selection of Wavelet Packet Base on Image Denoising [J]. *Mathematics Journal*, Vol. 25 (2005), pp. 185–190.
[2] Ge Zhe-xue, in: *Javert Wavelet Analysis Theory and MATLABR2007 Implementation* [M]. edited by Electronic Industry Publications, Beijing (2007).
[3] Xu Chen, Zhao Rui-zhen, in: *Wavelet Analysis. Applied Algorithm* [M]. Science Publications, Beijing (2004).
[4] Gao Shao-shu, Jin Wei-qi, Wang Ling-xue, et al. Objection Quality Assessment of Image Fusion [J]. *Journal of Applied Optics*. Vol. 32 (2011), pp.672–677.

Advances in Future Manufacturing Engineering – Yang (Ed.)
© 2015 Taylor & Francis Group, London, ISBN 978-1-138-02817-3

Simulation analysis on contact stress of new type energy-saving conductor with different structure

Jian-cheng Wan, Long Liu, Jia-jun Si & Liang-hu Jiang
China Electric Power Research Institute, Beijing, China

ABSTRACT: In this paper, medium strength aluminum alloy conductor, one kind of new type energy-saving conductor, has been chosen as the simulation subject to analyze the conductor stress distribution by macroscopic and microscopic analysis. And the finite element model of assemble gather of conductor and suspension clamp has been built. The contact stress on conductors with different wire number of same cross section has been compared. By the analysis result, the anti-fatigue performance of JLHA3-845-91 is superior to JLHA3-850-61.

Keywords: contact stress; new type energy-saving conductor; conductor structure; anti-fatigue performance

1 INTRODUCTION

For the past few years, with the development of power grid construction in China, domestic design and manufacture technology of electric conductor developed rapidly and several kinds of new type energy-saving conductor [1–5] have been designed. As one part of the research on the new type energy-saving conductor, the stress in overhead transmission conductor is very complex. The bending of the conductor will make the stress between single wires more complex. As a result, analysis on contact stress of energy-saving conductor with different structure becomes important to the application of energy-saving conductor. In this paper, medium

strength aluminum alloy conductor, one kind of new type energy-saving conductor, has been chosen as the simulation subject to analyze the conductor stress distribution [6–7] by macroscopic and microscopic analysis. And the varying pattern of contact stress has been obtained.

2 THE FINITE ELEMENT MODEL

The macroscopic analysis focuses on the contact stress between the conductor surface and the clamp port. With the finite element model of assemble gather of conductor and suspension clamp, as shown in Figure 1, the distribution of

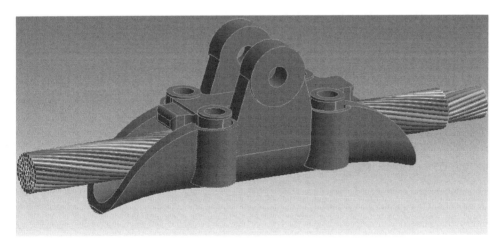

Figure 1. The finite element model of assemble gather of conductor and suspension clamp.

Table 1. Parameters of medium strength aluminum alloy conductor.

Conductor type	Electric conductivity	Number of layer	Nominal cross section/ mm²	Diameter of aluminum alloy wire/ mm	Twisted incremental	Wire strength/ MPa	Elastic modulus/ GPa	Thermal coefficient of expansion
JLHA3-850-61	58.5%IACS	5	850	4.21	1.0237	230	55	23×10^{-6}
JLHA3-845-91	58.5%IACS	6	845	3.44	1.0243	240	55	23×10^{-6}

Table 2. Parameters of medium strength aluminum alloy conductor.

Conductor type	Aluminum alloy area/ mm²	Resistivity/ (nΩ·m)	Conductor diameter/mm	Mass per unit length/ (kg/km)	RTS/kN	Direct-current resistance (20°C)/(Ω/km)
JLHA3-850-61	849.15	29.472	37.89	2349.6	195.30	0.0355
JLHA3-845-91	845.76	29.472	37.84	2341.6	192.83	0.0357

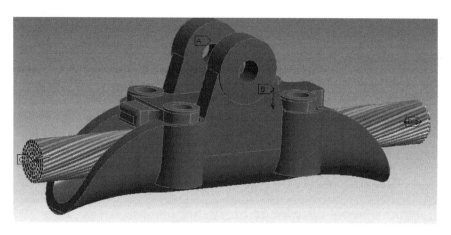

Figure 2. The computational boundary condition.

contact stress [8–9] of conductor on defined tension has been analyzed. And the microscopic analysis focuses on the contact stress between single wires. The distribution of contact pressure stress and friction shear stress has been analyzed with the finite element model. The conductor parameters are shown in Table 1 and Table 2.

3 COMPUTATIONAL BOUNDARY CONDITION

The computational boundary condition is shown as follows:

1. The fixed constraint is imposed on the bolt hole of suspension clamp;

2. Load of 40 kN is imposed on the cover plate of suspension clamp;
3. Load of 25% RTS is imposed on both ends of conductor;
4. The calculation model of JLHA3-850-61 contains 774 contact pairs while the calculation model of JLHA3-845-91 contains 1585 contact pairs.

The computational boundary condition of finite element model is shown in Figure 2.

4 CALCULATION RESULTS

The calculation results are shown as Table 3 and Table 4. With the analysis of the result, the

Table 3. Contact stress between the conductor surface and the clamp port.

JLHA3-850-61	JLHA3-845-91
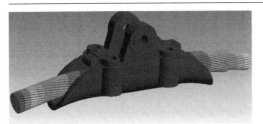	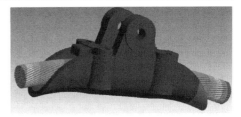
Maximum displacement: 1.84 mm	Maximum displacement: 0.93 mm

Table 4. Contact stress between single wires.

JLHA3-850-61	JLHA3-845-91
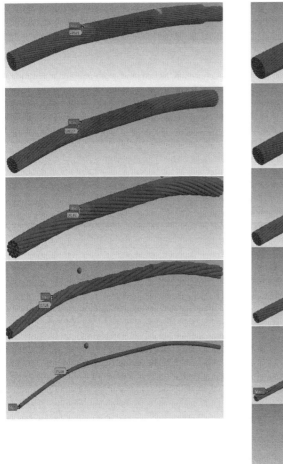	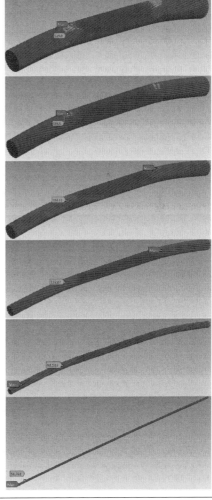

conclusion of distribution of contact stress is shown as following:

1. The maximum contact stress of JLHA3-850-61 and JLHA3-845-91 located in the clamp port and the contact stress in conductor decrease from the outer layer to the inner layer, as well as the maximum contact stress and the average contact stress;
2. Compared with JLHA3-845-91, the results of JLHA3-850-61 shows bigger deformed displacement and the maximum contact stress and the average contact stress of each layer is bigger;
3. The stress concentration phenomenon appears on the place where maximum stress locates on JLHA3-850-61 outer surface while the maximum stress of JLHA3-845-91 spread out along each adjacent wire. And the increase of wire number helps the stress diffusion.

5 CONCLUSIONS

With the same tension (25%RTs), the contact stress of JLHA3-850-61 is bigger than JLHA3-845-91. And the stress concentration phenomenon is more serious. To a certain extent, the stress concentration will aggravate the conductor fretting damage caused by conductor vibration and speed up the crack extension. As a result, the anti-fatigue performance of JLHA3-845-91 is superior to JLHA3-850-61.

REFERENCES

[1] Wan Jian-cheng, Application of Aluminum Alloy Cored Aluminum Stranded Wire for UHVDC Transmission Line with Large Capacity[J], Electric Power Construction, 2013, 34(8): 105–111.
[2] Huang Peng. Technical-Economic Analysis on Application of Energy-Saving Conductors in Transmission Lines[J]. Electric Power, 2013, 46(7): 153–160.
[3] Zhang Rui-yong, Polarization and Application of the New-type Energy-saving Conductors in Transmission Lines[J], Electric Power Construction, 2012, 33(6): 89–92.
[4] Ding Guang-xin, Analysis on Application of Energy-Saving Conductors in Transmission Lines[J]. Power System Technology, 2012, 25(2): 101–104.
[5] Ye Hong-sheng. Application of Moderate strength All Aluminum Alloy Conductor in Transmission Line[J]. Electric Power Construction, 2010, 31(12): 14–19.
[6] Diana G., Falco M. On the forces transmitted to a vibrating cylinder by a blowing fluid. Mechanica. 1971, 6(l): 9–22.
[7] Sinha N.K., Hagedorn P. Wind-excited over head transmission lines: Estimation of connection stresses at junctions. Journal of sound and Vibration. 2007, 301(1–2): 400–409.
[8] Kong D.Y., LI L., etal. Analysis of Aeolian vibration of UHV transmission conductor by finite element method. Journal of Vibration and Shock. 2007, 6(8): 64–67.
[9] Poffenberger J.C., Swart R.L. Different displacement and dynamic conductor strain. IEEE Transaction Paper. 1965, Vol. Pas-85(4): 564–587.

Advances in Future Manufacturing Engineering – Yang (Ed.)
© 2015 Taylor & Francis Group, London, ISBN 978-1-138-02817-3

SC-cylinder parameterization design based on Pro/Engineer

Yongchuan Lin
Doctor, Associate Professor, Guangxi University, Nanning City, Guangxi, China

Fawen He
Postgraduate, Guangxi University, Nanning City, Guangxi, China

ABSTRACT: The most distinctive feature of Pro/Engineer system is the parameter. This paper mainly discusses the design of the SC cylinder with parameters based on Pro/Engineer. Parametric design is the use of parts of various parameters to regulate the family size and design components of a design method. The design principle is to change a few specific parameters of the parts simply, which will be able to serve the needs of a variety of component models easily. By parametric design, with high accuracy and convenient advantages, there will be plenty of promotional value.

Keywords: Pro/Engineer; SC-cylinder; parametric design

1 INTRODUCTION

The SC-cylinder belongs to a precision instrument which may have many inevitable deficiencies designed by human hand, for example, difficulty to realise optimal design, time-consuming designing and large consumption of material. According to personal experience, the design can only be an estimation, rather a than calculation. Besides, by personal design, the accuracy and reliability of the product are limited. Apart from that, during the designing, the designers have to spend much time and energy in the complicated and repeated hand calculation, drawing and table-drawing. Furthermore, during designing, since it is hard to get an accurate calculation, a comparatively large safety factor has to be selected. With the rapid development of economy, technology and society, the market has various requirements for products; the turnover for updating of products is very fast and products are manufactured for the trend of of multiple-varieties and in small batches. The corporate must strive to provide the best products with the minimum development time and minimum cost— and provide the best after-sales service. Besides, the company should research and develop new products constantly so that it can survive and develop in the fierce competition between the domestic and foreign markets in high and advanced technology. The application of CAD is one prerequisite for realising this objective. Therefore, the design for overall dimensions of the SC-cylinder should be with the help of CAD tools.

1.1 *Ideas for the parametric design of SC-cylinder*

Since the SC-cylinder is commonly used, its features and dimensions vary according to different requirements. In order to be parameter-driven, the design methods manifested by the control features have to be adopted so as to establish the cylinder components with different dimensions and remain parameter-driven; therefore, the parametric design of the SC-cylinder in some products can be realised. The parametric design can mainly be divided into the following five steps:

1.2 *Create a solid model: According to the requirements of dimension and shapes, create the three dimension solid model*

1. Sever the parts and find out the feature size of each part: according to the function of the each part and the features for establishing the model, sever the parts. In order to make the established model to better reflect the basic features of the parts, the features should be typical and conceptual.
2. Define the characteristic parameter: after creating the model, the designing list of **PROGRAM** can automatically list all the components of the model. The defined dimensions of the parts will automatically be named as D1, D2, D3 ... according to the created sequence by the system.
3. Input the characteristic parameter: input the defined parameter into the "input part" of the list of designed parts and define the corresponding relation between the dimensions of each part in the

"relation defined part". Besides, the constrained relation between the same parts should be defined as well in the "relation defined part".

4. Revise the characteristic parameter: there are altogether two methods to revise the parameter, in which one is to choose the name of each parameter and change it one by one according to the prompting of the specific system in the software; the other is to put all the parameters that require revision into a data file and revise all the parameters once by reading the files.

2 THE PROCESS OF THE PARAMETRIC DESIGN FOR THE STANDARD SC-CYLINDER BY PRO/E

Before designing, the basic methods should be described. The basic methods need to create the three dimensions model by interactive mode; create the design parameter according to the parameter function of Pro/E; and then retrieve the parameter symbols by the applications of Pro/E itself and provide the "edit function" of the parameter so as to create the three dimensions model by the new design parameter. Its flow chart is shown in Figure 1.

2.1 To define the characteristic parameter

The shape of the SC-cylinder is mainly composed of two end covers, four screw bolts and one cylinder tube. To create the solid model of the cylinder one needs to draw two end covers firstly and the other parts subsequently. To draw the solid models of the two end covers is easy, which can be obtained only by pulling the mirror. Finally, the modeling of the standard SC-cylinder is as shown in Figure 2.

Then, omit the unimportant or those without common details (such as chamfering, relief groove) to reduce the workload of parametrization and improve the efficiency of designing—which means severing the parts and finding the feature size of the key parts. Based on the contents of this paper, the

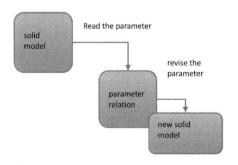

Figure 1.

data of all the SC-cylinder data should be acquired—and then designed according to the data. Figure 1 is all the acquired data; then the parametric design of this project should be finished by these data.

In order to be well-prepared for the subsequent parametrisation of the dimension, various features of parametric symbols for each size should be obtained. The concrete operation is as follows: click the corresponding solid model—highlight the model tree—right-click—display the value of the dimension in the figure—information—change size—change the value of the dimension into parametric symbol. As is shown in Figure 3, the dimension length is displayed in the figure rather than the letter "B" in Figure 4. Therefore, it should be transferred into parametric symbol by the above-mentioned methods and then revise its name.

2.2 The parametric procedures for creating the SC-cylinder

The model of the SC-cylinder is determined by the diameter of the cylinder and the process. In other words, the input parameter in the created parametric procedures in the following is the diameter of the cylinder process. Create the program and operate it; then the model of the cylinder is determined after inputing the value. The following is the concrete operation, after adding the input parameter: click "Tools" menu—"Application" option—choose the sub-option of "Edit" in the options of "Application" on the popped dialog—input the parameter between the INPUT and ENDINPUT in the popped notepad.

```
INPUT
ID NUMBER; cylinder diameter
TRAVEL NUMBER; process of the cylinder
D NUMBER
E NUMBER
F NUMBER
G NUMBER
N NUMBER
T NUMBER
S NUMBER
V NUMBER; diameter of the small column
END INPUT
```

The above mentioned processes are the program that needs to be edited. After operating the program, input the value of different ID and TRAVEL—then the different models of the SC-cylinder will be obtained. Figure 5 is the flow diagram of the generated new solid model.

```
RELATIONS
D25 = ID
D24 = TRAVEL
```

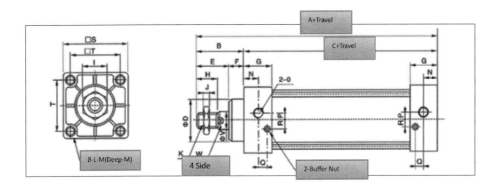

Table 1.

Bore ID	Travel	A	B	C	D	E	F	G	H	I	J	K
32	86	140	43	93	22	34	13	16	22	17	7	M16 × 1.25
40	108	142	54	93	28	43	16	20	22	17	7	M16 × 1.25
50	135	150	68	93	35	54	20	25	22	17	7	M16 × 1.25

Bore ID	Travel	L	M	N	O	P	Q	R	S	T	V	W
32	86	M6 × 1	9.5	8	G1/8	6	8.2	9	48	38.4	4.8	14
40	108	M6 × 1	9.5	10	G1/4	6	8.2	9	60	48	6	14
50	135	M6 × 1	9.5	12.5	G1/4	6	8.2	9	75	60	7.5	14

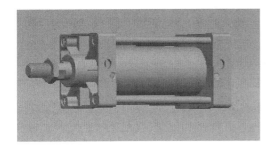

Figure 2.

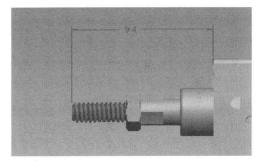

Figure 3.

D27 = D
D295 = E
D26 = F
D0 = G
D296 = N
D5 = S
D293 = T
D10 = V
D27 = 0.7*D25
AD297 = D24
D9 = D24
D295 = 0.4*D24
D26 = 0.15*D24
D0 = 0.5*D25

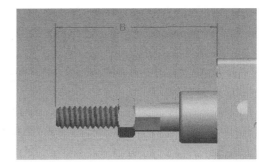

Figure 4.

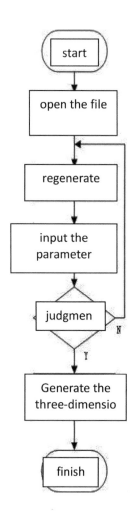

Figure 5.

D15 = 0.5*D25
D296 = 0.25*D25
D5 = 1.5*D25
D293 = 1.2*D25
D10 = 0.15*D25
END RELATIONS

All the parameters according to the requirements of the design should be considered in the main program, so that all the following unknown parameters are from the above mentioned figures. Adding the equation for the parametric design is similar to the operation of input diameter and the

difference lies in that the former is added between the RELATIONS and END RELATIONS between the open notepad—the concrete process is as shown in Figure 5.

3 CONCLUSION

The topic in this paper is beneficial to modernisation construction and follows the pace of development in the science of technology of the times—which is not only helpful for the manufacturing industry of the SC-cylinder, but will also make some contributions to advanced technology.

ACKNOWLEDGEMENT

Project: supported by Guangxi Natural Science Foundation (No: 2014GXNSFAA118347); Municipality & University S&T Cooperation Project of Yulin and Guangxi University (No. 201402801); supported by Guangxi Key Laboratory of Manufacturing System & Advanced Manufacturing Technology (No.:12-071-11S08).

REFERENCES

[1] Sun Jiaguang. Basic Technology for CAD [M]. Beijing: Tsinghua University Press, 2000:112–118.
[2] Fang Xin. Mechanical CAD/CAM Technology[M]. Xian:Xian Electronic Science & Technology University Press, 2004:4–7.
[3] Huang Kai, Li Lei, Liu Jie. Pro/E Parametric Design[M]. Beijing: Chemistry Industry Press, 2008:1–16.
[4] Wang Yongmei, Hao Anlin, Wang Weiping. Introduction to Mastery of Pro/E Wildfire4.0 (Chinese Version). Beijing: Publishing House of Electronics Industry, 2008:127–1321.
[5] Lin Anqing. Basic Component Design for Pro/E Wildfire4.0 (Chinese Version) [M]. Beijing: Publishing House of Electronics Industry, 2009.
[6] Li Shiguo, Li Qiang. Sample Courses for the Chinese Version of Pro/engineer wildfire. China Machine Press, 2004:132.
[7] Tian Ling, Tong Bingqu, Liu Xiong. Product modeling methods based on the intelligent CAD system[J]. Mechanical Science and Technology, 1998, III, Vol. 17, pp. 130–132.
[8] Shan Quan et al. Introduction to Mastery for the parametric design of Pro/ENGINEER Wildfire 4.0 in Chinese Version [M/CD]. Beijing: China Machine Press, 2008: 80–83.

Advances in Future Manufacturing Engineering – Yang (Ed.)
© 2015 Taylor & Francis Group, London, ISBN 978-1-138-02817-3

Design of inverter system based on PID control

Yuan Cao
Nanyang Institute of Technology, Nanyang, Henan, China

ABSTRACT: This paper introduces a small power inverter system based on STM8. The system mainly consists of a push-pull DC booster circuit, the single-phase full bridge inverter circuit, the filter circuit and a control module. The control module uses the two STM8 single chip microcomputers. One piece serves as the main control chip used in the inverter output terminal and the other piece serves as the auxiliary chip used in the DC boost end, and adopts the PID control algorithm. The experimental results show that: the inverter can stabilise output valid 220 V values under the power rating; sine wave alternating current frequency is 50 HZ; the output voltage harmonic content is small; the reliability is high; and it has a good application prospect in the field of photovoltaic power generation.

Keywords: PID; STM8; inverter; push-pull boost

1 INTRODUCTION

The global shortage energy and environmental pollution have become two important factors that restrict the sustainable development of human society, and vigorously developing new alternative energy sources have become the pressing matter of the moment. Solar energy, wind energy and other green energy generation options are new energy production avenues. They are safe, and without pollution. Resources are abundant, widely distributed. These features show good development space and application prospects. However, the new energy through directly generated electricity cannot be used directly, so the inverter with generating has new demand for energy increases. In view of the energy shortage and environmental pollution problems, and especially the national policy on the photovoltaic industry, a series of different power invertershave emergede at this historic moment, more and more in-depth research is being conducted.

2 THE OVERALL DESIGN OF THE SYSTEM

The core circuit is two pieces of STM8 serving as the control module. The STM8 is an 8 bit microcontroller, CPU performance up to 20 MIPS; 10 bit A/D converter with less than 3 microseconds up to 16 channels, 16 bit timer advanced, and can be used for motor control, capture/compare and PWM function.

In the system, STM8 (2) through the output of 4 pairs of complementary PWM wave drives 4 pairs of power push-pull circuit tube, to implement DC-DC boost output of 4 transformers, in order to realise high pressure from 24 V boost to approximately 330 V, it serves as the inverter input; STM8 (1) output 2 channel SPWM without unipolar complementary tube and two 50 Hz square wave drives the full bridge inverter tube, so as to realise the DC-AC transform function. The control circuit control STM8 (2) output of 4 pairs of complement duty ratio of PWM wave according to the voltage of an inverter AC output value. So as to form a closed loop control to make the inverter that will output voltage effective value at 220 V, the whole system is shown in Figure 1.

3 THE CONTROL PRINCIPLE OF SPWM WAVEFORM

In the sampling control theory, there is an important conclusion: in the impulse of equal and different shapes of narrow pulse in inertia link, the effect is basically the same [3]. Impulse refers to the narrow pulse area. The area of the equivalence

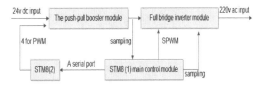

Figure 1. The overall structure block diagram of inverter.

principle above, a half sine wave is averagely divided into N equal parts (as shown in Fig. 1 as shown, N = 6), and then surrounded by the sine curve and the horizontal axis of each equal area is one and the same area of high rectangular pulse to replace, if the midpoint rectangular pulse and sine wave every equal area center point (as shown in Fig. 1), which is composed of N the amplitude but not the rectangular pulse width of waveform and half sine wave equivalent. The negative half cycle sine wave is also the same way to equivalent, and is the improved equal area algorithm. Obviously this series of pulse width and switching time can be rigorously calculated with mathematical method.

A sinusoidal signal is $U_r = U_m\sin t$. U_m is the sine wave amplitude of sinusoidal signal. The positive half cycle is divided into N equal parts (N is a multiple of 3), each PI radians. Figure 2 known that/N pulse height is $U_s/2$, as shown in Figure 3, a pulse width is W_K, pulse midpoint to theta K. by K sine wave area and corresponding to the K SPWM pulse area equal, so:

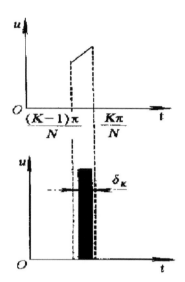

Figure 3. Equal area algorithm of monopole SPWM.

$$W_k U_s/2 = \left(\int_{(K-1)\pi IN}^{K\pi IN} U_m \sin t \cdot dt \right) \tag{1}$$

$$W_k = (2U_m / U_s)(\cos((K-1)\pi IN)) - \cos(K\pi IN)) \tag{2}$$

And because of θ_K that is the center of every sin wave, get:

$$\int_{(K-1)\pi IN}^{\theta_K} U_m \sin t \cdot dt = \int_{\theta_K}^{K\pi IN} U_m \sin t \cdot dt \tag{3}$$

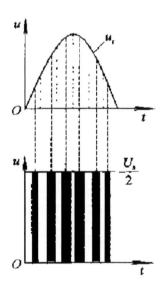

Figure 2. Improved phase voltage waveform of monopole SPWM.

The solution:

$$\theta_K = ar\cos[(1/2)(\cos(K-1)\pi IN) + \cos(K)\pi IN] \tag{4}$$

And because the NO.k $\theta_{on(K)}$ and $\theta_{off(K)}$ meet to:

$$\theta_{on(K)} = \theta_K + W_K/2, \tag{5}$$

$$\theta_{off(K)} = \theta_K + W_K/2, \tag{6}$$

Get:

$$\theta_{on(K)} = \arccos \frac{1}{2}\left[\cos\left[\frac{K-1}{N^\pi}\right] + \cos\left[\frac{K}{N^\pi}\right] \right]$$
$$- \frac{U_m}{U_s}\left[\cos\left[\frac{K-1}{N^\pi}\right] - \cos\left[\frac{K}{N^\pi}\right] \right], \tag{7}$$

$$\theta_{off(K)} = \arccos \frac{1}{2}\left[\cos\left[\frac{K-1}{N^\pi}\right] + \cos\left[\frac{K}{N^\pi}\right] \right]$$
$$+ \frac{U_m}{U_s}\left[\cos\left[\frac{K-1}{N^\pi}\right] - \cos\left[\frac{K}{N^\pi}\right] \right]. \tag{8}$$

4 THE IMPROVED ALGORITHM OF PID CONTROL

As shown in Figure 4, the output voltage U_0 is compared with the reference voltage of $UreF$ after the voltage feedback, the error after having

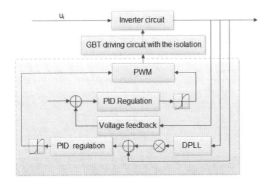

Figure 4. Voltage current negative feedback.

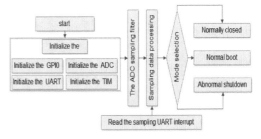

Figure 5. System flow chart of the main program.

anti windup function of **PID** regulator, the output adjustment data set table. Getting the **SPWM** drive signal, the feedback loop can realise the system voltage stabilizing function. In addition, the output voltage after DPLL to generate the reference current, frequency, phase and magnitude of the reference current by IM, gets the reference current Iref, Iref and IO compared to the output stream error signal. The signal is with the antiwindup function of the PID regulator, so as to adjust the data table, and get **SPWM** drive signal. This feedback loop ensures the output current and output voltage with the same frequency and phase. When the output voltage or current is over voltage and current limit, we close the **PWM** output in a timely manner; protect the **IGBT** inside the inverter device.

5 SOFTWARE DESIGN

The system has two single chips. One piece only has the sample, 1 pair of **PWM** wave complementary output and serial communication function. Function procedure that we only analysis is the main machine; the main control machine must first configure two timers, arranged to the power PWM band complementary output mode to drive the full bridge inverter tube. In order to output SPWM wave, we need to put the sine function values to store up form. Then we need to configure the AD sampling module, serial communication module and system main program flow chart is as shown in Figure 5.

6 THE EXPERIMENTAL RESULTS

The PID controller is verified in the application of grid photovoltaic inverter. In different load conditions, the output waveform harmonic content is

small, the output voltage is low, the inverter can stabilisze output valid values for the 220 V at rated power conditions, and frequency of sine wave alternating current is 50 HZ (oscilloscope probe attenuation is 10 times).

7 CONCLUSIONS

Currently a number of photovoltaic inverters on the market are mainly controlled by the control chip DSP. Complex circuit is complex, high cost. This paper mainly introduces a kind of photovoltaic inverter system based on the STM8 single chip microcomputer. The system has the advantages of a simple circuit structure. Cost is low and the efficiency is high. The design has practical value for medium and small power photovoltaic inverter.

REFERENCES

[1] Hu Xinb liu, Peng Xiao-bing, Mu Xin-hua. Single polarity SPWM inverter power supply control method [J]. Mechanical and Electrical Engineering, 2004 (1):38–42.
[2] Liu Zhi-fu, Li Xue-yong, Gao Si-yuan. The development method of a push -pull inverter with high frequency transformer [J]. Modern Electronic Technology, 2008 (19):70–73.
[3] Yu Y.J, Yang B, Wu J.D, et al. Design and implementation of photovoltaic grid-connected inverter with digital control [J]. Mechanical and Electrical Engineering, 2008(10):40–45.
[4] Pan Wen-cheng. Single polarity SPWM Inverter Bridge drive mode research [J]. Journal of Zhejiang Institute of Science and Technology, 2012(3):200–204.

Advances in Future Manufacturing Engineering – Yang (Ed.)
© 2015 Taylor & Francis Group, London, ISBN 978-1-138-02817-3

The design and realization of security model for P2P network

Zhimin Wang
Zhangjiakou University, Zhangjiakou, Hebei, China

ABSTRACT: Through the comparative analysis of the various popular characteristics for current P2P network structure, the paper puts forward the design idea of a kind of hybrid network structure with more security and flexibility. The paper analyzes the JXTA platform technology, and discusses the design and realization method of each function module of the security model, according to the design process; we design and realize the creating, the privilege of members, the node searching, node joining, and node communication part of the contents of the security model.

Keywords: P2P; JXTA; secure peer group; network model

1 INTRODUCTION

P2P network is widely concerned by people in recent years. It has been applied in many fields, and has large development space in the future. However, the process of the development of P2P network also produces a series of urgent needs to study and solve problems. Security is one of the most important problems. Due to the characteristics of P2P network, to solve the security problem in P2P network environment is very complex. P2P network application early basic does not consider the safety factor; the later development of some of the P2P network application system, despite its own security mechanisms, is not universal and limited security features; some present good development platforms for the development of application system security are only to provide certain support. Therefore, more need to improve the design of the security model.

The appearance of the P2P technology makes the user nodes on the network can communicate with each other freely. Each user peer could respond to other users' requirements, can also search for the resources they need or services to other users on the network, to achieve the widely shared network computer resources. As in the P2P network, each peer user may dynamically join or leave the network at any time, so than traditional network exists greater randomness and uncertainty. A method to improve this problem is to create a secure peer group. A secure peer group can bring a relatively stable security environment, and realize the feasibility of local management of network supervision, cost reduction and so on the advantages to become a good solution to the problem of network security P2P.

Because JXTA has outstanding performance in support and secure part of the P2P network, so the network security model is designed and realized on JXTA platform. Classified by function implementation, security model can be divided into the peer group part and the node part. Through the JXTA platform, the peer group member part of creating module, authentication module, module, search method of node part get peer group membership module and intra group node communication protocol module and other parts can get the design and implementation of more efficiency. Through the interaction of each module, and create a safe environment which is relatively stable in the network.

2 THE ARCHITECTURE OF P2P NETWORK STRUCTURE

The basic structure of the SAP2P is as shown in Figure 1. In P2P network, SAP2P is a trust relation layer, application program is established on the SAP2P. In P2P network, all nodes have the same SAP2P code; it provides the safe operation of the interface to. In the SAP2P structure, the communication infrastructure is not limited to the TCP/1P network layer. In fact, since the transmission independence of JXTA, any network can provide reliable data transmission in the network layer can be used in JXTA based on the realization of SAP2P.

In the figure, the above layer is P2P network application. In a specific application, P2P applications are the same when run on different peers. The second layer components are trust and policy management. In this layer trust model will be given specific definitions. In the specific design and

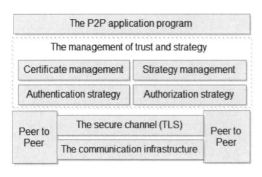

Figure 1.　The hierarchy schematic diagram of SAP2P.

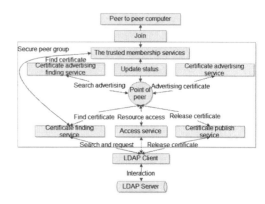

Figure 2.　Component module of security model.

implementation is used in the SPKI/SDSI system, of course, the user can also choose to use other trust management systems, such as PKI, PGP or reputation system etc. In the actual execution process, data structure is used in the management system for the SPKI certificate, and the certificate is placed in the certificate server. Strategy management is to provide authentication and authorization strategy.

Authentication and authorization strategies include the user's identity and authority for verification service. Login credentials provided by the user includes: password, X.509 public key certificate or other such as login credentials like an IC card. The model is to identify the user holding the public key of the identity. SAP2P according to the authentication method is verified on these documents, confirmed in the effective cases, produce identification certificate for the user. In the authorization, SAP2P according to the unified security credentials submitted by the users, add a description of the local security policy, legal analysis. Finally draw a conclusion whether authorized results.

The secure channel is responsible for providing encryption to protect data in transit, which provides tamper, replay protection active or passive attacks to the use. It can easily use TLS which is a mature security protocol based on JXTA platform.

3　THE OVERALL ARCHITECTURE OF THE SECURITY MODEL

As shown in Figure 2, a peer is the model in the active entity. A peer can call service based on peer group as the boundary, that is to say, a peer in the peer groups can call the group to provide services, for securing peer group. It provides important peers the supply of services: advertising trust membership service, discovery service, advertising services, access service certificate, certificate and certificate issued a service discovery service.

Peer point that be outside the peer secure peer group, only interact with the group membership service to apply to join the group, through the verification, the peer will obtain a within a group ID. After joining the group, peer through membership services to update their status, can also provide various service calls within the group.

The LDAP server in the model is mainly used to store all kinds of SPKI certificates. The visit must be carried out by a LDAP client.

Peer group can obtain the certificate from the LDAP server or publish certificate to obtain certificate discovery service and certificate publish service.

An application to join the group for the equivalence point, by the group membership service finds the peer related certificates according to the peer identity information through the certificate discovery service, to verify the identity and authority.

Peer within a group can also through advertising means to discover and release certificate.

Within group peer access to a shared resource through the access service to each, and the use of the service and access to SPKI certificate related resources for access control. In general, interactive range access service is only within the group, so the relevant SPKI certificate can be obtained by way of advertising, and do not have to interact with the LDAP server. But in order to improve the reliability, retain the interaction through the LDAP client and LDAP server interface.

4　THE REALIZATION OF P2P NETWORK SECURITY MODEL

4.1　To create a secure peer group

Process to create secure peer group is as shown in Figure 3.

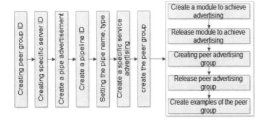

Figure 3. Creation process of secure peer group with extended service.

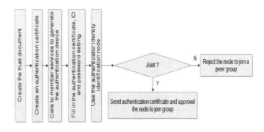

Figure 4. The process of member that obtains privileges.

If we create peer group using a fixed ID to identify, so no matter what, the peer group is created once, will use the same ID. This is a very important point. It guarantees that no nodes are for the peer group to create a different ID.

SpecPeerGroupID = (PeerGroupID) IDFactory. fromURL(new URL("run", "", "jxta:uuid—DEAD BEEFDE AFBABAFEED BABE000000010206")).

In order to add an allocated specific ID for a peer group which needs to expand the service, the ID must be shared by all instances of this reactor group.

SpecSrvID = (ModuleSpecID) IDFactory. fromURL (new URL ("run", "", "jxta: uuid-3371 DAF8A7B043E7A8718B102B62055A614CAC047 AD240B8910ABDE677D12FA659")).

4.2 Member privileges

Because the secure peer group banned group members using the group shared resources and use the group services, so the nodes necessary must first put forward to join the peer group requirements. Node to request to join peer groups using the membership protocol for certification, certification through the nodes can be added to the after service of peer group. Authentication process of peer group to join the group of nodes is as shown in Figure 4.

1. Create the trust documents.
   StructuredDocument credDoc = null;
   Trust the document credDoc is set in the initialization for null values, obtained within group information to facilitate, create authentication certificate.
2. Create authentication certificate.
   AuthenticationCredential autCrd = new AuthenticationCredential (peerGrp, null, credDoc);
   Get information from joining the secure peer group in peerGrp, on the basis of trust certificate credDoc, create autCrd authentication certificate.
3. Calls to member services to generate the authentication device.

MembershipService memSrv = (MembershipService) (peerGrp.getMembershipService);
   Authenticator autor = memSrv.apply (autCrd);
4. Complete authentication certificate, identity and password setting.
   PasswdMembershipService.PasswdAuthenticator pwdAutor = (PasswdMembershipService. PasswdAuthenticator) autor;
   PwdAutor.setAuth1Identity (MYPEERGRP_ID); pwdAutor.setAuth2_Password (MYPEERGRP_PWD).

4.3 The process to join peer group

Peer-to-peer users into secure peer group processes are as shown in Figure 5.

1. Create a node trust document.
   StructuredDocument peerCredDoc = null;
   Create a node trust document peerCredDoc, initialized to an empty.
2. Create a node authentication certificate.
   AuthenticationCredential peerAutCrd = new AuthenticationCredential (peerGrp, null, peerCredDoc);
   Get information from peer groups, based on the node trust document peerCredDoc, create a node authentication certificate, stored in the peerAutCrd.
3. Calls to member services, to generate the authentication device.
   MembershipService memSrv = (Membership Service)
   PeerGrp.getMembershipService ();
   Authenticator autor = memSrv.apply (peerAutCrd);
   The use of the standard peer group getMembershipService () method calls to member services, and member services to generate the authentication device autor.
4. Fill in the user ID and password.
   String peerId = PEER_ID;
   String peerPwd = PEER_PWD;
   PwAutor.setAuth1Identity (peerId);

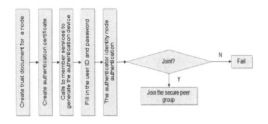

Figure 5. Process of joining a peer group.

PwAutor.setAuth2_Password (peerPwd);
PEER_ID and PEER_PWD are equivalent user fill in the user ID and password, password is pwAutor certification certification is created by autor, respectively, the user ID and password populated.

5. The authenticator node identity authentication. Call isReadyForJoin () authenticator node identity authentication method. Through the authentication is joining a peer group, add failed authentication failure.

6. Add secure peer group.
If through authentication, then call the member service to join () method, to join the secure peer group.

5 CONCLUSIONS

Because JXTA has outstanding performance in support and secure part of the P2P network, so the network security model is designed and realized on the JXTA platform. Classified by function implementation, security model can be divided into the peer group part and the node part. Through the JXTA platform, creating module, authentication module, module, search method of node part get peer group membership module and intra group node communication protocol module and other parts of the peer group member can get more efficient design and implementation. Through the interaction of each module, we create a relatively stable and safe environment in the network.

REFERENCES

[1] Tang Hui, Zhang Guojie, Huang Jianhua, Li Zupeng. Research and design of a hybrid P2P network model [J]. Computer application, 2005, 25 (3): 521~535.

[2] Shi Yanfen, Sui G.H. P2P network information security model based on JXTA [J]. Microcomputer and application, 2005, 3: 29~31.

[3] Zhang Chunrui, Jiang Fan, Xu Ke. Trust and reputation model for peer to peer network application [J]. Journal of Tsinghua University, 2005, 45 (10): 1436~1440.

[4] Liu Baoxu, Li Xueying, Xu et al. Peer to peer network technology and Application [J]. Computer engineering and application of 2003, 39 (18).1–3,12.

[5] Li Zhenwu, Yang Jian, Bai Ying Cai. Research on peer to peer networks and its challenges [J]. Computer software and application of. 2004, 21 (2).54-55110.

Advances in Future Manufacturing Engineering – Yang (Ed.)
© 2015 Taylor & Francis Group, London, ISBN 978-1-138-02817-3

The design and implementation of PHP vulnerability protection in e-commerce website

Jinzhao Yang
Tangshan Polytechnic College, China

ABSTRACT: This article mainly aims at the several common network security holes on PHP, with the actual needs for application. To achieve the specific preventive measures in the PHP e-commerce website, and give the specific code to prevent mode, which especially gives the user registration and login, user business form, and concrete the idea of safety aspects to solve the info problem of the goods for the PHP in e-commerce site.

Keywords: network security; PHP; e-commerce; vulnerability protection

1 INTRODUCTION

In recent years, with the rapid development of network technology, the vigorous development of the Internet industry, terminals are becoming more and more intelligent; the function of the network is more and more abundant; the topological structure of the network is more and more complex; the requirements of network security are higher and higher, which makes the network security protection and management issue even more important. Especially, since the mobile and wireless access to the customer has the ability to more powerful and has greater access mode, which exacerbates the problem of network security. Network security in its essence is the security of network information, and generally speaking, it all relates to the network of information confidentiality, integrity, availability, reliability and related technology that can be controlled; the theory belongs to the field of network security research.

Information security and network security also involve the public communication network, internet network and many other aspects. Therefore, the enterprise, relates to the planning, procurement, strengthening, maintenance and business and other departments; network security also involves industry management, support units, a large amount of third party partners, and also involves general public users. Network information system and information security play a very important role.

As everyone knows, Professional Hypertext Preprocessor (PHP) has now become one of several network programming languages, and is the most popular in the world. It ensures a large and busy network operation and provides a variety of efficient network services. At the same time, it also brings about many issues of concern, such as performance, maintainability, reliability—among which the most important is safety. Some of the features and languages, such as the conditional expression, branch structure, circulation structure, and security are the more abstract part in language. In fact, the security is more like characteristics of developers, not linguistic characteristics. Any language cannot avoid the security problem of the code; although in some language features the security conscious developer effect is not so obvious.

PHP language builds on driving in high performance database and dynamic flexibility Web2.0 technology. It has become the most popular website development tool, especially in e-commerce, and is especially prominent in the perfect combination of it and other open source softwares such as MySql, database and Apache server. However, as more and more sites start using PHP development, they also become vulnerable to attack targets. The developer must initiate attack preparations. With the increase in attack frequency and the changing mode of attack, security and protection measures have become major concerns; this mainly includes: preventing cross site scripting attacks vulnerability, preventing a SQL injection attack, session hijacking etc.

2 THE ANALYSIS OF PHP NETWORK SECURITY LOOPHOLE

In general, network security and threat have several forms:

The first is the "system security threat". It refers to the use of resources in the network or computer without consent, for non-normal use of the

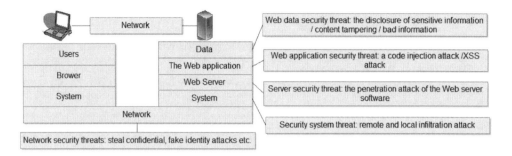

Figure 1. Common form of network security threats.

network equipment and resources, or unauthorised expansion of permission, unauthorised access information;

The second is "Web data security threat". It refers to the sensitive data being leaked out consciously or unconsciously—such as the information in the transmission loss or leakage, loss or leakage in the storage medium; then the attacker can steal sensitive information through the establishment of a concealed tunnel etc.;

The third is "Web application security threat". It refers to the attacker by illegal means of the right to the use of stolen data, insert, modify, delete some important information or retransmission, response to obtain the beneficial to the attacker, and malicious add, modify data, to interfere with normal users;

The fourth is "service security threat". It refers to the ongoing interference on the network service system, to change its normal work flow, the slow independent program execution system response or even paralyse, affecting the normal operation of users. Even legitimate users of rejection cannot enter into the computer network system or can't get the corresponding service;

The fifth is "network security threat". It refers to the attacker's fake legal identity disruption of normal work, and through the network transmission of the virus. Its destruction is much higher than that of the single computer system, and the user is difficult to guard against [1].

PHP network security loophole is as shown in Figure 1.

3 APPLICATION OF LOOPHOLE PROTECTION IN ELECTRONIC COMMERCE WEBSITE

3.1 The global variable protection

User registration in the member management module can be the prevention and optimisation of code level for the global variables. Here is a combination of a database module, and modification of the variable characteristics of GET mode, use of the COOKIE technology to carry on the protection of global variables.

```
if(isset($_GET[login]))                              url=administratorManager.php");
{                                                    }
    require_once("dbconnect.php");                   else
    $sql="select * from users where yhm='$_         {
    GET [yhm]'andpwd='$_ GET                             echo    "<script>alert('   Your
    [pwd]'";                                             username or password is incorrect
    $rs=mysql_query($sql);                              ')</script>";
    $row=mysql_num_rows($rs);                           }
    if($row==1)                                      }
    {                                                else
        $rec=mysql_fetch_array($rs);                 {
        if($rec[power]==1)                               echo "<script>alert(' Your username or
        {                                            password is incorrect ')</script>";
            $_SESSION[admin]=$_ GET[yhm];            }
            header("refresh:0                        }
```

3.2 Remote file protection

In the E-shop e-commerce system, remote file protection is mainly reflected in the system information portal platform, the information portal platform; the remote file request is required to view from the document tree to the corresponding file system drive and the specific file. In the security request remote files in the process, we need to create a separate thread to obtain the file the size of the corresponding file, and, depending on the size of the file open, download the thread to enter the corresponding file processing; the remote file requests and processes is shown in Figure 2.

3.3 Inject protective of SQL statement

SQL injection refers to the insert of the SQL metacharacter via the input area of Internet, (special character's 'represents some data and instruction), manipulate and perform back-end SQL query technology. These attacks are targeted at other organizations WEB server. CSS attack is by inserting the script tag in the URL, then induced to trust their users to click on them, to ensure that malicious Javascript code runs on the victim's machine.

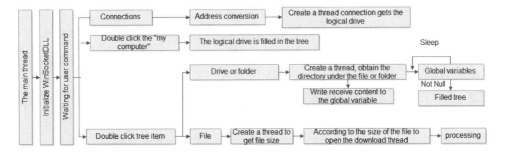

Figure 2. Remote file requests and processing map.

These attacks exploit the trust relationship between users and servers. In fact input and output info are not detected by servers, which does not refuse to JavaScript code.

Commodity table is more important table in electronic commerce, and it records the goods_id product number, goods_name: name of commodity name, provider_name: commercial providers, shop_price: shop price, market market_price: retail price of other information.

The code:

```
<?
$sql="select  *  from  phpe_goods,invoice
where
phpe_goods.goods_id=invoicploy.no and
(enddatetime-getdate())>=0";
$rs=mysql_query($sql);

$row=mysql_num_rows($rs);
for($i=0;$i<$row;$i++)
{
$rec=mysql_fetch_array($rs);
echo "<tr>";
}
```

The User table records the user_id user number closely related with the user information, the username is the user name and password information filled in the password registered users and on important information, in the table; we need the user registration and login function to make the following code prevention measures, specific code:

```
<?
if(isset($_GET[newyhm]))
{
$flag=0;
if($_GET[newyhm]!=$_SESSION[admin])
{
    $sql="select*  from  phpe_admin_user
    where
    user_name=$_GET[newyhm]'";
    $rs=mysql_query($sql);
    $row=mysql_num_rows($rs);
    if($row==1)
    {
        echo " This account has been forbidden

to visit ";
        $flag=1;
    }
}
$sql="select money from phpe_admin_user
where
user_name='$_SESSION[yhm]'";
$rs=mysql_query($sql);
$rec=mysql_fetch_row($rs);
$mymoney=$rec[0];
$rs=mysql_query($sql);
}
```

Pay table records the user some important payment and so on information; here we combined JavaScript, AJAX and other technical advance to check on the front end; the basic steps of the process are shown in Figure 3. The user of the entire transaction payment process is realized by the AJAX technology—that is, the entire transaction process is asynchronous; the page does not have

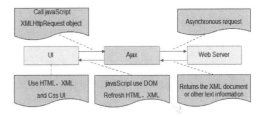

Figure 3. The AJAX response flow chart.

the overall refreshing, and so the attacker using GET mode of SQL injection finds it difficult.

We give only background PHP program based on AJAX for the state table phpe_stats:

```
<?
$tsql="select range from phpe_stats where
[no]=$rec[pid] and
(enddatetime-getdate())>=0";
$trs=mysql_query($tsql);
$trec=mysql_fetch_row($trs);
if($trec[0]=='')
{
$hyprice=$hyprice+$rec[hyprice];

}
else
{
$hyprice=$hyprice+$rec[hyprice]*(float)($rec[
0])*0.1;
} echo "<span id='ss$rec[pid]'
style='display:none'>$trec[0]</span></td></t
r>";
?>
```

Although the state table is set in the database, it is not the most critical, but we can also find some worthy of injection safety from the table to solve the problems in the SQL.

4 CONCLUSIONS

PHP language is established on driving in high performance database and dynamic flexibility Web2.0 technology. It has become one of the most popular web development tools, and is especially prominent in the e-commerce. It can also be perfectly combined with open source softwares such as MySql database and Apache server. However, as more and more sites are using PHP development, they are also become vulnerable targets for attack. The developer must initiate attack preparations. With the increase of attack frequencyand change in methods of attack—security and protection measures have become a major concern—mainly: to prevent cross site scripting attack vulnerability, to prevent SQL injection attack, session hijacking etc.

REFERENCES

[1] Song Shangping Li Xingbao. To realize the configuration and application of PHP Smarty template engine [M]. 2007.01, 45–97.

[2] Pan Kaihua, Zou Tiansi. PHP programming book [M]. Beijing: Tsinghua University press, 2010, 25–67.

[3] Chen Xiangyang, Chen Guoyi. The development and design of PHP5+MySQL Webpage system [M]. Beijing: Publishing House of electronics industry, 2007, 45–57.

[4] Song Shangping, Li Xingbao. The configuration and application realization of PHP Smarty template engine [J]. modern education technology, 2007, 11–55.

[5] Zou Tiansi, Sun Peng. PHP from entry to the master [M]. published: Beijing: Tsinghua University press, 2008, 32–57.

Advances in Future Manufacturing Engineering – Yang (Ed.)
© 2015 Taylor & Francis Group, London, ISBN 978-1-138-02817-3

The design of the wireless communication system based on voice and text

Qingsong Zhou

The Department of Computer Science, PU'ER University, Pu'er, Yunnan, China

ABSTRACT: In this system, the information acquisition and the wireless communication of short message service are implemented using the S3C2410X microprocessor, the JXARM9-2410 as the main hardware platform, and connecting to the sensor, microcontroller, microprocessor and wireless communication modules. The system is stable and timely, and is easy to expand and promote.

Keywords: sensor; ARM; embedded system; GPRS; AT command

1 INTRODUCTION

ARM is mainly used in the current wireless communications equipment, and the demand for the application of GPRS is also very extensive in the era of 3 g and 4 g. GPRS is a data business to develop based on GSM and has the features of high resource utilisation rate, rapid transmission speed, short access time, etc.

GPRS contains abundant data services such as PTP (Point to Point), PTM-M (Point to Multipoint), broadcasting data service, PTM-G (Point to Multipoint–Group), IP-M radio service, and GSM-based voice broadcasting and voice group call service. It can be widely used in wireless pay stations, remote monitoring, car navigation, vehicle scheduling, intelligent transportation, remote monitoring, banking and finance, and other fields.

In the following, the ARM-based GPRS wireless voice and text communication are mainly discussed. Then, users can directly communicate with voice and text and maske wireless-free transfers of the monitored and collected information.

2 DESIGN FUNCTIONAL SPECIFICATION

In the system, the collection of external information such as temperature, humidity, and altitude, longitude and latitude is completed by the sensor. The emphasis is laid on the study of communication function.

2.1 *Information collection*

In the system, the collection of information such as temperature and GPS is completed mainly by a corresponding sensor. The collected information can be directly sent to the storage unit or to the target terminal equipment, in the form of text message.

Temperature information collection is taken for example: DS-18B20 digital temperature sensor features wear and collision resistance, is small in size, easy to use, and comes in various package forms, so it is applicable to the digital temperature test control field of small space equipment [1].

It has a unique one-way wire interface. DS18B20, when connecting with the microprocessor, only needs one line for implementing the two-way

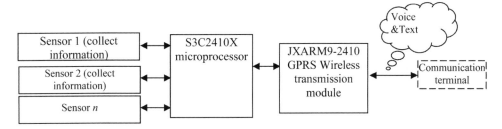

Figure 1. Hardware functional interface.

communication of microprocessor and DS18B20; the range of testing temperature is in −55°C~+125°C, and the inherent temperature measuring resolution is 0.5°C; working power supply is in 3~5V/DC.

2.2 *GPRS voice communication function*

1. GPRS dialing function

The system, after starting up, will initialise GPRS module and get into a state of standby once it acquires network support. In the standby mode, the user can dial with the keyboard. Users dial with key Enter after inputting the phone number, and can hear the ring if getting contact with the called successfully, or return to BUSY or NOCARRIER if unsuccessful.

2. GPRS incoming call function

The system, after starting up, will initialise GPRS module and get into a state of standby once it acquires network support. In the standby mode, the user can answer the phone by pressing Enter once there is a telephone call; the incoming call is controlled also with a simple state machine and the flow is divided into the idle state, answering state, and busy state, so the incoming call function is well completed.

2.3 *GPRS text communications functions*

1. Text sending function

For sending SMS, PDU Mode is applied, which is supported by all mobile phones and can send any characters set. When GPRS sends a short message, a state machine is used for the control, and the whole flow is divided into the idle state, number inputting state, and sending state.

2. Text receiver function

When the system starts and completely initialises GPRS module as well as the corresponding hardware module, it will get into a state of standby. In the standby mode, it prompts the arrival of a new message if there is an incoming short message, and then analyses and prints the message by pressing Enter to display the detailed contents.

3   HARDWARE ANALYSIS

From the perspective of application, the JXARM9-2410 ARM embedded experimental system is used as the hardware platform. The S3C2410X microprocessor is used for the JXARM9-2410 target processor; ADT inheriting development environment of Wuhan CVTech Info &Tech CO., LTD is used as the development platform [2].

The S3C2410X microprocessor features low power consumption and high integration, which is a design based on ARM920T by Samsung for

handheld devices. To reduce the total cost of the system and reduce peripheral devices, the following parts are integrated in the chip: 16 KB instruction Cache, 16 KB data Cache, MMU, external storage controller, LCD controller, NAND Flash controller, 4 PWM timers and 1 internal timer, general I/O port, real-time clock, 8-channel 10-bit ADC and touchscreen interface, USB master-slave, SD/MMC card interface [3].

The JXARM9-2410 is the main hardware platform, and contains the following interfaces: RS232/RS485 serial interface, Ethernet interface, USB interface, CF card interface, IDE interface, MMC/SD card interface, IIS interface, CAN bus interface, A/D and D/A conversion interface, standard computer print interface, colour LCD display plus touch screen, GPRS wireless communication module, keyboard module, and 5.7" colour display touch screen module [4].

4   ANALYSIS ON THE GPRS SYSTEM STRUCTURE

The GPRS network is implemented by adding GGSN and SGSN into the existing GSM network, so users can send or receive data in the end-to-end grouping way. GPRS grouping is used to send a message from the base Station to the GPRS Support Node (SGSN), but not to connect to the voice network via the Mobile Switching Center (MSC). SGSN communicates with the GPRS Gateway Support Node (GGSN); GGSN accordingly processes the grouping data and then sends it to the target network such as PDN or the Internet. IP packs, marked with the Internet and containing the mobile station's address, are received by GGSN and then transferred to SGSN and sent to the mobile platform.

A Public Land Mobile Network (PLMN) supporting GPRS may involve any other networks when it operates GPRS, and then a demand for a mutual network communication emerges.

The mobile terminal with GPRS business functions comes with the address provided by GSM and GPRS business operators, and thus the public data network terminal can directly send data to the GPRS terminal, using the data network identification number. In addition, GPRS supports the network communication based on IP and provides TCP/IP header compression, if datagram is used in TCP connection.

5   SOFTWARE DESIGN

5.1 *AT command program design*

The JXARM9-2410, using the integrated GPRS wireless communication module, provides an interface supporting RS232. The module can be driven directly by the computer serial port via S3C2410X UART1 interface. In the following, some AT commands commonly used in the system are listed and also programmed [5].

In editing AT commands, the following two points should be focused:

1. The instruction characters, constant, and PDU data packets containing AT command are transmitted in the form of ASCII codes. For example, "A" is coded with 41H, "T" is with 54H, and "0" is with 30H, according to ASCII codes.
2. The work mode of text information communication is set to the format of PDU and the command is AT+CMGF = 0.

   1. Main voice communication control instructions
   ATD: dial; ATH: hang up; ATA: answer the phone; AT+CEER: check the call failure causes; AT+VTD: set DTMF (dual tone multiple frequency) voice length; AT+VTS: send DTMF voice; ATDL: redial the last number; ATS0: set up automatic reply; AT+VGR: adjust the voice receiving gain; AT+VGT: adjust the sending voice gain; AT+CMUT: set up microphone mute; AT+SPEAKER: microphone option; AT+ECHO: set up echo cancellation; AT+SIDET: set up side tone correction; AT+VIP: restore the voice by default.
   2. Network service instruction
   AT+CSQ: check network signal quality; AT+COPS: select service provider; AT+CREG: check current network registration status; AT+WOPN: display network provider in the form of text; AT+CPOL: check the preferred network list.
   3. Short message communication control instruction
   AT+CSMS: choose SMS; AT+CNMA: confirm and answer new message; AT+CPMS: choose text storage area; AT+CMGF: choose SMS format; AT+CSAS: set up parameters for short message storage; AT+CRES: set up parameters of equipment to recover the stored short message; AT+CSDH: display the parameters in the model of TEXT message; AT+CNMI: choose how to receive short message; AT+CMGR: read short message; AT+CMGL: list the stored information as required; AT+CMGS: send short message; AT+CMGW: write a short information in storage area; AT+CMSS: send the short messages stored in the storage area; AT+CSMP: set up parameters for TEXT message mode; AT+CMGD: delete short messages; AT+CSCA: set up the address of

the SMS service center; AT+CSCB: choose the broadcast information types in the area; AT+WCBM: check the broadcast message identifier in the area; AT+WMSC: modify the message status; AT+WMGO: cover a short message; AT+WUSS: keep the SMS status unchanged.

There are three ways to send and receive short messages: Block Mode; Text Mode based on AT command; PDU (Protocol Data Unit) Mode based on AT command. Block Mode requires mobile phone manufacturers to provide driver support; Text Mode can make codes send and receive SMS simple and is also easy to implement: directly sending the original text or non-ASCII contents is allowed, but this function needs a mobile phone to support for properly displaying SMS, so the application of SMS in Chinese is restricted. PDU mode can not only support both English and Chinese messages. Therefore, PDU mode is applied in this design.

Three types of coding can be used in PDU mode to send and receive short messages: 7-bit, 8-bit, and UCS2 codes. The 7-bit code is used to send ordinary ASCII characters; 8-bit codes are usually used to send data messages; UCS2 codes are used to send Unicode characters. To implement Chinese short messages, GB2312 encoding in Chinese is required to be converted to Unicode. The PDU data format is mainly composed of short message center address, file header byte, the types of information, the destination address, protocol identification, data coding scheme, the period of validity, the user data length, and user data.

5.2 *Voice communications instance*

Basic calls include dialing and incoming call.

Dialing: ATD "15125699999"; when the instruction is sent out, connection succeeds and the call is available, if "OK" returns.

Incoming call: There is an incoming call if "RING" signal is received; "ATA" instruction is sent if user is willing to communicate; connection succeeds and call can start, if "OK" returns.

5.3 *Text information communication*

Text message management includes operations such as "write short message ", "send a short message", and "receive short message".

1. Send message
   Message mode is set to PDU mode first and command AT+CMGF = 0 is completed, "OK" returns after the setting-up. The short message center number is set up and the command is AT+CSCA = "8613808790500", and the number of Yunan Pu'er mobile

communication center is set. "How are you!" is sent to the mobile terminal and the command is AT+CMGS = "15125699995"<CR> How you! <Ctrl-Z>. It is sent successfully, if "+ CMGS: OK" returns.

2. Receive message

If AT+CMGF = 0, the message mode is set to PDU mode, and "OK" returns after the setting-up. Command AT+CNMI = 2,1,0,0,0, sets message display mode to "+CMTI:<mem>, <index>" (i.e.the name of the memory and the message number), and "OK" returns after the setting-up. New message is stored in the fifth record of SIM card, if +CMTI: "SM",5 returns. The message can be read using command AT+CMGR = 0.

AT+CMGR = 1 means one message has already been read, and AT+CMGR = 0 means one unread message (new message) is read.

5.4 *Program function specification*

5.4.1 *The main functions of GPRS*
The functions in the GPRS module mainly manage GPRS module initialisation, data reading, data sending, etc. The functions below are mainly included:

void gprs_init( ) //*Initialise a part of GPRS module*

void gprs_send_cmd(char *cmdstring) //*Send command AT to GPRS module*

int gprs_recv_cmd(char *cmd) //*Receive the data of GPRS module*

void gprs_print_msg(SM_PARAM* pMsg) //*Print a short message*

BOOL gprsSendMessage(const SM_PARAM* pSrc) //*Send short message*

int gprsEncodePdu(const SM_PARAM* pSrc, char* pDst) //*Analyze short message*

void gprs_tel_call_in( ) //*Incoming call function*

void gprs_tel_call_out( ) //*Dialing function*

6 CONCLUSIONS

In this design, wireless communication technology is combined with embedded technology, greatly improving the practicality and expansion of the system. The system runs stably, but its features are not expanded and studied further. The timely voice and short message communication functions are mainly implemented in the system.

ACKNOWLEDGEMENT

Scientific Research Project: The Scientific Research Project of PUER University (No. 201317).

REFERENCES

[1] Jun Lv, Hui-qiang Tang. ARM-based Wireless Meteorological Data Communication System Design [J]. Journal of Electronic Design Engineering, 2012.9, pp.184–186.

[2] Chengpeng Ru. ARM-based Beidou Navigation Satellite System/GPRS/GIS Vehicle Monitoring System Design [J]. Journal of Railway Transportation and Economy, 2013.3, pp.84–88.

[3] Ze Chen. ARM9 Embedded Technology and Senior Practice Linux Tutorial [M]. Beijing: Beijing University of Aeronautics and Astronautics Press, 2005.

[4] Yaowu Guan, Zongde Yang, et al. ARM9 Embedded Wireless Communication System Development and Instances [M]. Beijing: Electronic Industry Press, 2006.

[5] Dong Su. Mainstream ARM Embedded System Design Technology and Instances [M]. Beijing: Electronic Industry Press, 2007.

[6] Yafeng Li. ARM Embedded Linux System Development from the Basic to the Master [M]. Beijing: Tsinghua University Press, 2007.

Advances in Future Manufacturing Engineering – Yang (Ed.)
© 2015 Taylor & Francis Group, London, ISBN 978-1-138-02817-3

Study on the online monitoring system of instrument and equipment based on wireless multimedia sensor networks

Lin Bao

Dalian Ocean University, Dalian, Liaoning, China

ABSTRACT: At present, the sensor network technology has been developed to be a wireless multimedia sensor network technology. It is a product of the integration of many modern technologies such as wireless communications, multimedia, sensor, computer, and microcomputer point. In this paper, an online monitoring system of instrument and equipment, including three task modules (adaptive intelligent decision module, interactive operation module, and emergency alarm module) is constructed using wireless multimedia sensor network based on the performance parameters and natural environment parameters of instrument and equipment. With the system, the fundamental shift from making amends, after the emergence of failure to taking preventive measures before the emergence of failure, so that the maintenance work becomes timely, effective, economical, and convenient. And the management work is gradually specialised, programmed, and automatic. Thus, the stable operation of the precise instrument and equipment is secured and protected.

Keywords: wireless multimedia sensor network; instrument and equipment; online monitoring

1 INTRODUCTION

In many fields such as military, scientific research, teaching, and finance, many instruments and equipment are playing a very important role. Whether instruments and equipment can work constantly efficiently and safely and the maintenance can be implemented on time, and is available or not, has a close link with the competitiveness of an institution, a company, and even a nation [1]. Therefore, users expect to fully maintain the use value and economic benefits of instruments and equipment, and they require accurate data, safe and stable operation, and need to save the maintenance cost and prolong its service life.

The complex problems such as parameters and high precision should be considered, because, the system composition, product performance, working conditions and technical indexes, in all kinds of instruments and equipment are not identical. For this reason, online monitoring precise instrument and equipment with Wireless Multimedia Sensor Networks (WMSNs) has become the first choice of management and maintenance personnel. Compared with the traditional wireless sensor networks, WMSNs can perceive audio, video, images and other multimedia information and implement the fine-grained, accurate environmental monitoring [2]—so it possesses an incomparable advantage.

2 AN INTRODUCTION TO THE SYSTEM

2.1 *Structural topology and hardware specifications*

The structural topology of the system is as shown in Figure 1.

In wireless multimedia sensor networks, the Multimedia Sensor is responsible for connecting the information about a variety of parameters and then transferring it to Sink node for coding; it is then transmitted to the Server in the control network via wired or wireless network. The control management network is composed of the Server and Controller.

Regulatory Systems are a series of regulating equipment, generally including air filter, cooler, dehumidifier, and controller, etc. The controller receives the Server's open and close instructions, or adjusts the running of the regulating equipment.

2.2 *Operating mechanism*

The basic working principle can be divided into information collection, information transmission, and information processing.

In information collection, tiny sensors are installed to correctly reflect the environmental factors and the performance of instruments and equipment, so as to implement comprehensive, fine-grained, highly-precise, and real-time

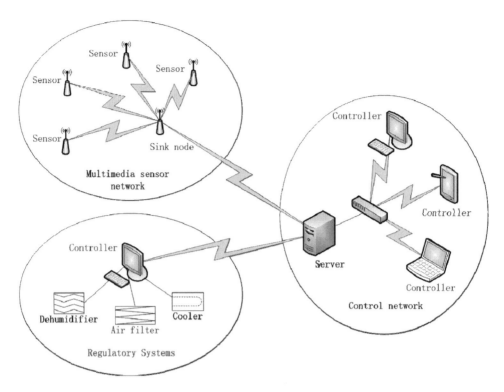

Figure 1. The structural topology of the system.

monitoring on all kinds of monitoring objects in the network distribution area.

In information transmission, there are many sensors and also the data collected by a sensor is simple and scattered, so the linkage between multiple sensor nodes should be implemented using the wireless communication mode according to the different communication protocol, in order to ensure the integrity and rationality of the information. As a result, the compression and transmission of multimedia information is completed.

In information processing, the processing system running in the Server will calculate, integrate and analyse the received information resources, and also operate the corresponding task modules in the backend according to manually or automatically set parameter information and indicators, so as to implement the next decision-making or warning.

2.3 *Performance analysis*

In the design, WMSNs are applied to the remote monitoring system of instruments and equipment, so the remote communication and the acquisition and transmission of many types of scientific teaching research and equipment parameters, and instruments and equipment parameters, are

implanted. Meanwhile, a variety of complex information such as pictures, video, and audio is identified, compressed and consolidated—and a fast, reliable, easy-to-interactive, and complete processing system, with automatic processing ability, is constructed according to the monitoring objects.

The remote monitoring system of college and university scientific teaching research instruments and equipment, based on WMSNs, not only meets the requirements for a wide coverage and a high monitoring quality, but also makes the grade in many performance indicators such as resource acquisition, transmission distance, and processing ability. Therefore, it is of great significance to open up a new field for the application of wireless multimedia sensor networks and make a new plan for the scientific teaching research instruments and equipment maintenance and management system of colleges and universities.

3 PARAMETER DESIGN

3.1 *The performance parameters of instruments and equipment*

The performance parameters of instruments and equipment refer to those parameters reflecting the performance and function of instruments and equipment, and they are usually described with concrete numbers and units in details. There are many factors to cause the malfunction or damage of instruments and equipment; the reasons and characteristics are varied, in which the decision basis is impossible to get rid off from the effect of the performance indicators. Therefore, it is important and necessary to monitor the performance parameters of instruments and equipment, using wireless sensors as the basic system data.

3.2 *Natural environment parameters*

The natural environment parameters include temperature, humidity, salinity, dust, light intensity and other natural environment factors. It is an environment for placing and operating instruments and equipment and always affects the state of instruments and equipment. The effect on precise instruments and equipment should be never ignored.

4 TASK MODULE

Task modules are divided into three sub-modules (adaptive intelligent decision module, interactive operation module and emergency alarm module). The three can supplement each other and provide a full range of support for the management of instruments and equipment.

4.1 *Adaptive intelligent decision module*

In adaptive intelligent decision module, there are many decision rules with adaptive capacity, and a corresponding decision task will be immediately activated to take decisions once the change of the perceived parameters matches with the set of rules within a certain range. In short, the adaptive intelligent decision module is a sub-routine, allowing the subject of decision-making to automatically choose different strategies, according to the changes of the environment.

4.2 *Interactive operation module*

The Interactive operation module is an information processing program for the interaction between the user and the system. First, users set up the default and pre-warning value for the parameters within a secured range. The system will send a warning and also make the intervention by starting up the corresponding equipment through the adaptive decision system, if it judges the state of equipment to deviate from the default and there may be equipment failure or operation error. Once pre-warning value exceeds, the system will prompt the user for a

potentially dangerous state with lights, alarms, and other forms of information, and also will guide the user to reduce the corresponding load, or terminate the operations, etc.

4.3 *Emergency alarm module*

The Emergency alarm module appropriately classifies the emergencies according to the performance parameters of instruments and equipment and the collected state information through the state judgment methods within it; it also automatically sends the alarm information to the specified host. The host, after receiving the alarm signal, triggers different alarm sounds according to the different warning levels, and also achieves the linkage between the users and the maintainers.

4.4 *Application example*

With the online monitoring system, the occurrence of emergency can be prevented.

The system hardware settings are as follows:

1. The main monitoring parameters of sensor networks: projector's internal temperature, internal air speed, environmental temperature, environment humidity, environmental light intensity, environmental dust dispersion, and sound and video monitoring
2. Control management network: remote server and computer on control site
3. Regulatory systems: air conditioning, dehumidifier, air filter and electric curtains.

The operation process of the system is as shown in Figure 2.

A ventilation system is located in the projector and the internal air flow takes away heat through the internal fan—but the dust in the air will often be attached to the air entrance of the cooling system, so the projector's internal temperature has a very close relationship with environmental temperature and environmental dust scattering degree. Inside the projector, therefore, a temperature sensor and air velocity transducer are also set up. Additionally temperature, humidity, luminosity and dust sensors are also set up in its working environment.

1. The operational mode of the adaptive module: It is an automatic cycle module. It will start a corresponding regulating equipment to intervene once the temperature, humidity and dust parameters in the environment are abnormal, and will close the regulating equipment until parameters are back to normal.
2. The operational mode of the interactive module.

403

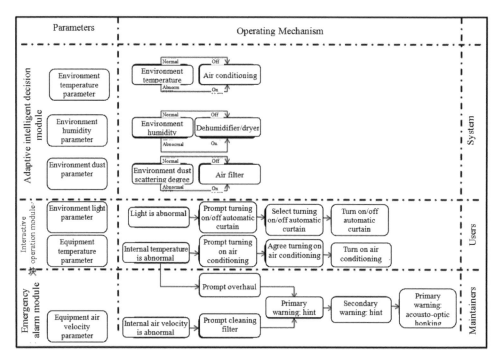

Figure 2. The operation process of the system.

3. The operational mode of the emergency warning module: If internal temperature of the projector among the equipment parameters is higher than the pre-set, the control end will send a question to the manager about whether air conditioning or fan is turned on for cooling, and simultaneously issue a prompt message for maintenance. The pre-warning system will start the corresponding warning according to the temperature from low to high, if the temperature is higher, because managers do not take any measures, or, the taken measures are ineffective.

5 CONCLUSION AND PROSPECT

Unblocked communication channels can ensure information—accurate, comprehensive and timely; so they are the important technical guarantees for the smooth implementation of this system. However, static and moving obstacles and other wireless communication devices will seriously affect the quality of the communication between sensor nodes in wireless multimedia sensor networks, so that the data packets frequently lose and have errors in the network [4] and become the biggest bottleneck to restrict the online monitoring

system of instruments and equipment. Therefore, how to optimise system performance, effectively improve the accuracy and rate of information— and ensure the communication quality of instruments and the equipment's remote monitoring system, have become an important issue and severe challenge to the field of wireless multimedia sensor network applications.

REFERENCES

[1] Guang Meng. Review on Condition-based Equipment Residual Life Prediction and Preventive Maintenance Scheduling. Doctoral Dissertation, Shanghai Jiaotong University.

[2] Huadong MA, Dan Tao. Multimedia Sensor Network and Its Research Progresses [J]. Journal of Software, 2006, 17 (9): 2013–2028.

[3] Deji Wang. Study on the Adaptive Intelligent Decision Support System under a Complicated Environment [J]. Journal of University of Science and Technology of China, 2007 (5).

[4] Tingpei Huang, Haiming Chen, Zhaoliang Zhang, Li Cui. Easipled: An Approach to Discriminate the Causes of Packet Losses and Errors for Wireless Sensor Networks based on Supervised Learning Theory [J]. Journal of Computer Science, 2013 (3): 471–484.

Advances in Future Manufacturing Engineering – Yang (Ed.)
© 2015 Taylor & Francis Group, London, ISBN 978-1-138-02817-3

Design of a planar MIMO antenna with high isolation for LTE handsets

Yutian Zhang & Jinfang Wen

National Engineering Research Center of Mobile Communication, China

ABSTRACT: A Dual-Element Multi-Input Multi-Output (MIMO) antenna with high isolation operating in the band 2.6 GHz is designed for Long Term Evolution (LTE). The proposed MIMO antenna system covering from 2570 MHz to 2650 MHz is based on meandered lines. With the introduction of two arms and a ground split of GND plane, the isolation better than 25 dB can be obtained. The peak gain is 2.08 dBi while the efficiency is 77.3%. The size of the proposed antenna occupies 46×110 mm^2 which corresponds to the size of the LTE handsets. The planar structure and high isolation, along with the Omni-directional radiation pattern, make it very suitable for LTE terminals.

Keywords: LTE; MIMO; meandered lines

1 INTRODUCTION

Long Term Evolution (LTE) is the term that is given to 4G in mobile communication networks. LTE will be Internet Protocol (IP) based and will provide high throughput, broader bandwidth and better handoff capabilities compared with current 3G networks. Several key enabling technologies are used in LTE such as Multiple-Input Multiple-Output (MIMO) systems, Adaptive Modulation and Coding (AMC) and adaptive beam forming. The use of these technologies will enhance the system spectral efficiency to about 15 bits/Hz with the use of MIMO within the 20 MHz bandwidth allocated per channel. This is six times higher efficiency than current 3G based networks. LTE will have a new air interface that is based on Orthogonal Frequency Division Multiple Access (OFDMA). MIMO technology offers parallel data transfer via multiple antennas. The best MIMO performance is achieved when data channels are uncorrelated to each other. In reality, however, it is impossible to avoid the channel correlation due to multi-path fading along the propagation channel and antenna correlation between antennas. While several adaptive signal processing techniques exist to compensate such impairments, managing low antenna correlation is rather challenging, especially for MIMO antennas implemented in a limited space. In general, the mutual coupling between two closely placed antennas is mainly caused by induced currents due to the sharing of common ground and near field coupling [1]. This adverse impact can degrade the radiation performance of the antenna. Significant research efforts to reduce the mutual coupling or achieve good isolation between closely placed antenna elements have been made [2–7]. High isolation was achieved by etching slits into the ground plane [8]. Ground branches were applied in [9] to achieve low mutual coupling within a narrow frequency band. Parasitic elements were added to improve the port isolation of a MIMO antenna [10].

In this work, the design and simulation of a compact dual element MIMO antenna system that operates in the 2.6 GHz band of the LTE specification is presented. It consists of two meander line antennas that cover the frequency band from 2570 MHz to 2650 MHz, with a center frequency of about 2600 MHz. The antenna system corresponds with approximate 77.3% efficiency. The isolation is more than 25 dB between the two elements in the operating band, which is a good metric for diversity systems.

2 DESIGN AND ANALYSIS OF PROPOSED MIMO ANTENNA

The proposed MIMO antenna consists of two meandered line elements which are bent opposite to each other to weaken mutual coupling. This antenna is built on an h = 0.51 mm FR-4 substrate with a dimension of 46 mm × 110 mm, a relative permittivity of $\varepsilon_r = 4.4$ and a dielectric loss tangent of 0.02. In order to miniaturize the size of the antenna for the application of the portable devices, 50Ω microstrip lines are used to excite the printed antenna. The width of the feed line is 1.5 mm. The dimensions of the designed antenna after optimization are as follow: L = 110 mm, L0 = 90 mm, Lg = 24 mm, W = 46 mm, W1 = 15.5 mm W2 = 1.05 mm, W3 = 1.05 mm, W4 = 1.6 mm, h = 0.8 mm, S = 2 mm. In addition, a GND split is introduced in ground plane along with two shorted parasitic elements to enhance the isolation and reduce the mutual coupling between two antennas.

Several isolation enhancement technologies have been investigated in many works. In [11–13] GND plane cuts or introducing parasitic elements between two antenna elements are introduced. In this paper, the two technologies are combined in a new structure.

The radiation pattern, radiation efficiency, correlation coefficient are usually calculated to evaluate the performance of the proposed MIMO antenna. The envelope correlation coefficient ρ of a two antennas system can be determined using the following equation:

$$\rho = \frac{|S_{11}^* S_{12} + S_{21}^* S_{22}|^2}{(1-|S_{11}|^2-|S_{21}|^2)*(1-|S_{22}|^2-|S_{12}|^2)}$$

Figure 2 depicts the reflection coefficient of the two antenna elements. The proposed MIMO

Table 1. The simulated S parameters of the proposed MIMO antenna.

	Frequency (2570 MHz~2650 MHz)
S11 (dB)	<−20
S22 (dB)	<−20
S12 (dB)	<−25
S21 (dB)	<−25

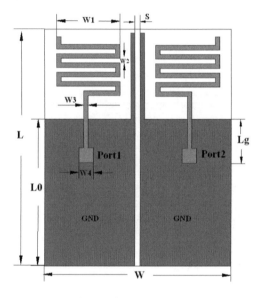

Figure 1. The layout of the proposed MIMO antenna.

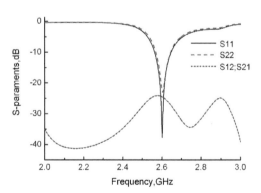

Figure 2. Reflection coefficient and isolation curves simulated by HFSS.

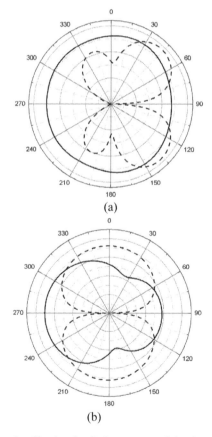

(a)

(b)

Figure 3. Simulated radiation patterns of the element 1 at 2.6 GHz on (a) xz-plane (b) yz-plane, Gainθ (solid) and GainΦ (dashed).

antenna centered at 2.6 GHz covers the frequency band from 2570 to 2650 MHz which can be used for LTE handsets, and the reflection coefficient is more than –20 dB. Owing to mutual coupling, different antenna elements have different reflection coefficients. Apparently, with the introduction of GND split and the two arms of ground plane, the isolation between port 1 and port 2 is better than 25 dB which could offer good performance for LTE systems.

Figure 3 and Figure 4 depict the simulated radiation patterns of the MIMO antenna system. The radiation patterns of the two elements both follow the same trends. The MIMO system shows Omni-direction approximately which makes it ideal for today's wireless communication devices. The peak gain of the MIMO antenna system is 2.08 dBi and the radiation efficiency of the system is 77.3%. The electric filed distributed on each element at 2.6 GHz is shown in Figure 5, from which we can see that resonance occurs at the vertical direction of the meandered lines.

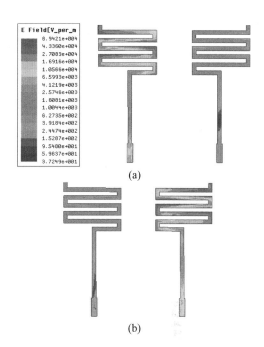

Figure 5. The electric field distribution of MIMO system at 2.6 GHz on (a) element 1 (b) element 2.

3 CONCLUSION

In this paper, a MIMO antenna system composed of two opposite meandered line elements is designed for mobile LTE handsets. This antenna is built on an h = 0.51 mm FR-4 substrate with a relative permittivity of $\varepsilon_r = 4.4$ and a dielectric loss tangent of 0.02. The MIMO system covers 2570 MHz to 2650 MHz band. The peak gain of the system is 2.08 dBi while the efficiency is 77.3%. With the introduction of the two arms and ground slot on GND plane, isolation can be improved to 25 dB at communication band. In addition, omni-directional radiation pattern makes it very ideal for LTE devices.

ACKNOWLEDGEMENT

This work was supported by the National Natural Science Foundation of China (51172034) and the Fundamental Research Funds for the Central University of UESTC (ZYGX2010 J120).

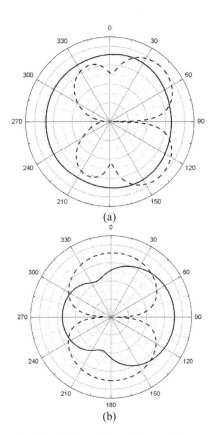

Figure 4. Simulated radiation patterns of the element 2 at 2.6 GHz on (a) xz-plane (b) yz-plane, Gainθ (solid) and GainΦ (dashed).

REFERENCES

[1] D. Pozar, "Input impedance and mutual coupling of rectangular microstrip antennas," IEEE Trans. Antenna and Propag.,Vol.30, pp.

[2] D. Ahn, J.-S. Park, C.-S. Kim, J. Kim, Y. Qian, and T. Itoh, "A design of the low-pass filter using the novel microstrip defected ground structure," IEEE Trans. Microw. Theory Tech., vol. 49, pp. 86–93, 2001.

[3] S. Zhang, Z. Ying, J.Xiong and S. He, "Ultrawideband MIMO/Diversity Antennas With a Tree-Like Structure to Enhance Wideband Isolation," IEEE Antennas Wireless Propag. Lett., Vol.8, pp. 1279–1282, 2009.

[4] Y.Ding, Z. Du, K.Gong and Z. Feng, "A Novel Dual-Band Printed Diversity Antenna for Mobile Terminals," IEEE Trans. Antenna and Propag., Vol.55, pp.2088–2096, 2007.

[5] S. Zhang, S.N. Khan, and S. He, "Reducing Mutual Coupling for an Extremely Closely-Packed Tunable Dual-Element PIFA Array Through a Resonant Slot Antenna Formed In-Between," IEEE Trans. Antenna and Propag., vol. 58, pp. 2771–2776, 2010.

[6] S-C. Chen, Y-S. Wang and S-J. Chung, "A Decoupling Technique for Increasing the Port Isolation Between Two Strongly Coupled Antennas," IEEE Trans. Antenna and Propag., vol. 56, pp. 3650–3658, 2008.

[7] G. Dadashzadeh, A. Dadgarpour, F. Jolani and B.S. Virdee, "Mutual coupling suppression in closely spaced antennas," IET Microw. Antennas Propag., Vol.5, pp. 113–125, 2011.

[8] F.G. Zhu, J.D. Xu and Q. Xu, "Reduction of mutual coupling between closely-packed antenna elements using defected ground structure," Electronics Lett.4., vol. 45, no. 12, Jun. 2009.

[9] Y. Ding, Z.W. Du, K. Gong, and Z. H Feng, "A novel dual-band printed diversity antenna for mobile terminals," IEEE Trans. Antennas Propag., vol. 55, no. 7, pp. 2088–2096, Jul 2007.

[10] Z.Y. Li, Z.W. Du, and K. Gong; "A dual-slot diversity antenna with isolation enhancement using parasitic elements for mobile handsets." Asia Pacific Microwave Conference, 2009:1821–1824.

[11] H.Li, J.Xiong,andS.He,"Acompact planarMIMOantenna systemof four elements with similar radiation characteristics and isolation structure," IEEE Antennas Wireless Propag. Lett., vol. 8, pp. 1107–1110, 2009.

[12] C.Y. Chiu et al., "Reduction of mutual coupling between closely packed antenna elements," IEEE Trans. Antennas Propag., vol. 55, no.6, pp. 1732–1738, Jun. 2007.

[13] K.S. Min, D.J. Kim, and Y.M. Moon, "Improved MIMO antenna by mutual coupling suppression between elements," in Proc. Eur. Conf. on Wireless Technology, Paris, France, Oct. 2005, pp. 125–128.

Data processing and information management

Advances in Future Manufacturing Engineering – Yang (Ed.)
© 2015 Taylor & Francis Group, London, ISBN 978-1-138-02817-3

A kind of QT-Bézier curves and its applications

Jin Xie
Institute of Scientific Computing, Hefei University, Hefei, China

Xiaoyan Liu
Department of Mathematics, University of La Verne, La Verne, CA, USA

Le Zou
Institute of Scientific Computing, Hefei University, Hefei, China

ABSTRACT: A kind of quadratic trigonometric Bézier curves with two shape parameters, briefly QT-Bézier curves, are presented in this paper. The trigonometric curves retain the main superiority of cubic Bézier curves. The shape of the curve can be adjusted by altering the values of shape parameters while the control polygon is kept unchanged. Shape parameter has the property of geometry, the larger is the parameter, and the better approach is the curves to the control polygon respectively. The ellipses can be represented exactly using quadratic trigonometric Bézier curves.

Keywords: QT-Bézier curve; shape parameter; shape control; ellipse; parabola

1 INTRODUCTION

It is well known that Bézier curves, particular the quadratic and cubic Bézier curves, have gained widespread application. However, the position of these curves is fixed relative to their control polygon. The control polygon must be changed if the shapes of the curves need to be adjusted. Although the weight numbers in the cubic rational Bézier curves, see [4], have an influence on adjustment of the shape of the curves. For given control points, changing the weight numbers to adjust the shape of a curve is quite opaque to the user. Recently, in geometric modeling and computer graphics, in order to improve the shape of a curve and adjust the extent to which a curve approaches its control polygon, some methods of generating curves are presented by using tension parameters; see [2–12].

On the other hand, since the Bernstein basis is a basis of algebraic polynomials it has many shortcomings especially in representing transcendental curves, such as the arcs and ellipses. In order to avoid the inconveniences, many bases are presented in other new spaces (Zhang, 1996; Peña, 1997; Walz, 1997; Sánchez-Reyes, 1998; Chen and Wang 2003). But the shapes of these curves based on the above methods can be adjusted with a global shape parameter.

The purpose of this work is to present practical quadratic trigonometric Bézier curves, analogous to the cubic Bézier curves, with two shape parameters.

The present paper is organized as follows. In Section 2, the basis functions, with two shape parameters, of the quadratic trigonometric Bézier curves are established and the properties of the basis functions are shown. In Section 3, the quadratic trigonometric Bézier curves are given and some properties are discussed. The figures of the curves affected by the shape parameters are given. And the representation of ellipses and parabola are also illustrated in the Section 4. At the end, the conclusions are given.

2 QUADRATIC TRIGONOMETRIC BÉZIER BASIS FUNCTION

Firstly, the definition of quadratic trigonometric Bézier (i.e. QT-Bézier) is given as follows.

Definition 2.1 For two arbitrarily selected real values of λ and μ, where $\lambda, \mu \in [-1, 1]$, the following four functions of $t \in [0, \pi/2]$ are defined as quadratic trigonometric Bézier (i.e. QT-Bézier) basis functions with two shape parameters λ and μ:

$$\begin{cases} b_{0,j}(t) = (1-\sin t)(1-\lambda \sin t), \\ b_{1,j}(t) = (1+\lambda)\sin t(1-\sin t), \\ b_{2,j}(t) = (1+\mu)\cos t(1-\cos t), \\ b_{3,j}(t) = (1-\cos t)(1-\mu \cos t). \end{cases} \tag{1}$$

Obviously, the functions $b_{i,j}(t)(i=0,1,2,3)$ are the quadratic trigonometric Bézier basis functions with a shape parameter when $\lambda = \mu$, see [9].

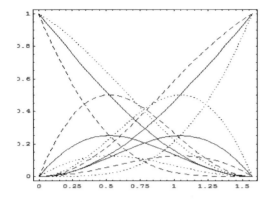

Figure 1. The quadratic trigonometric basis function.

Theorem 2.1 The basis functions (2.1) have the following properties:

a. Non-negativity: $b_{i,j}(t) \geq 0$, $i = 0, 1, 2, 3$.
b. Partition of unity: $\sum_{i=0}^{3} b_{i,j}(t) = 1$.
c. Monotonicity: For a given parameter t, $b_{0,j}(t)$ and $b_{3,j}(t)$ are monotonically decreasing for shape parameters λ and μ, and $b_{1,j}(t)$ and $b_{2,j}(t)$ are monotonically increasing for shape parameters λ and μ respectively.
d. Symmetry: $b_{i,j}(t,\lambda,\mu) = b_{3-i,j}(\pi/2-t,\mu,\lambda)$ for $i = 0,1,2,3$.

Proof: (a) For $t \in [0,\pi/2]$ and $\lambda, \mu \in [-1, 1]$, then $1 - sint \geq 0$, $1 - cost \geq 0$, $sint \geq 0$, $1 - \lambda cost \geq 0$, $1 - \lambda sint \geq 0$;
$1 - \mu cost \geq 0$, $1 + \lambda \geq 0$ and $1 + \mu \geq 0$.
It is obvious that $b_{i,j}(t) \geq 0$, $i = 0, 1, 2, 3$.
(b) For $b_{0,j}(t) + b_{1,j}(t) = \cos^2 t$ and $b_{2,j}(t) + b_{3,j}(t) = \sin^2 t$, then $\sum_{i=0}^{3} b_{i,j}(t) = 1$
The remaining cases follow obviously.
Figure 1 shows the curves of the trigonometric basis functions for $\lambda = 1$, $\mu = -0.5$ (dotted lines) for $\lambda = 0$, $\mu = 0$ (solid lines) and for $\lambda = -0.5$, $\mu = 1$ (dashed lines).

3   QT-BÉZIER CURVES

3.1   *Construction of the QT-Bézier curve*

Definition 3.1 Given points $P_i (i = 0,1,2,3)$ in R^2 or R^3, then is called a quadratic trigonometric Bézier curve with two shape parameters. From the definition of the basis function, some properties of the T-Bézier curve can be obtained as follows:

$$r(t) = \sum_{i=0}^{3} P_i b_{i,j}(t), t \in [0,\pi/2], \ \lambda, \mu \in [-1,1] \quad (2)$$

Theorem 3.1 The QT-Bézier curves (3.1) have the following properties:

a. Boundary properties:

$$\begin{cases} r(0) = P_0, r(\pi/2) = P_3 \\ r'(0) = (1+\lambda)(P_1 - P_0), r'(\pi/2) = (1+\mu)(P_3 - P_2) \end{cases}$$
$$(3)$$

b. Symmetry: P_0, P_1, P_2, P_3 and P_3, P_2, P_1, P_0 define the same QT-Bézier curve in different parameterizations, i.e.

$$r(t, \lambda, \mu; P_0, P_1, P_2, P_3) = r(\pi/2-t, \lambda, \mu; P_3, P_2, P_1, P_0),$$
$$t \in [0, \pi/2], \lambda, \mu \in [-1, 1] \quad (4)$$

c. Geometric invariance: The shape of a QT-Bézier curve is independent of the choice of coordinates, i.e. (3.1) satisfies the following two equations:

$$r(t, \lambda, \mu; P_0 + q, P_1 + q, P_2 + q, P_3 + q)$$
$$= r(\pi/2-t, \lambda, \mu; P_3, P_2, P_1, P_0) + q \quad (5)$$
$$r(t, \lambda, \mu; P_0 * T, P_1 * T, P_2 * T, P_3 * T)$$
$$= r(\pi/2-t, \lambda, \mu; P_3, P_2, P_1, P_0) \quad (6)$$
$$*T, t \in [0, \pi/2] \lambda, \mu \in [-1, 1]$$

where q is an arbitrary vector in R^2 or R^3, and T is an arbitrary $d \times d$ matrix, $d = 2$ or 3.

d. Convex hull property: The entire QT-Bézier curve segment must lie inside its control polygon spanned by P_0, P_1, P_2, P_3.

3.2   *Shape control of the QT-Bézier curve*

For $t \in [0, \pi/2]$, we rewrite (3.1) as follows:

$$r(t) = \sum_{i=0}^{3} P_i c_{i,j}(t) + \lambda sint(1-sint)(P_1 - P_0)$$
$$-\mu cost(1-cost)(P_3 - P_2) \quad (7)$$

where

$$c_{0,j}(t) = 1 - sint; c_{1,j}(t) = sint(1-sint);$$
$$c_{2,j}(t) = cost(1-cost); c_{3,j}(t) = 1-cost;$$

Obviously, shape parameters λ and μ only affect curves on the control edge $P_1 - P_0$ and $P_3 - P_2$ respectively. In fact, from (3.6), we can also predict the following behavior of the curves.

The shape parameters λ and μ serve to effect local control in the curves: as λ increases, the curve moves in the direction of the edge $P_1 - P_0$; as λ decreases, the curve moves in the opposite direction to the edge $P_1 - P_0$. The same effects on the edge

412

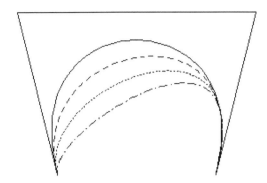

Figure 2. The effect on the shape of T-Bézier curves of altering the value of λ.

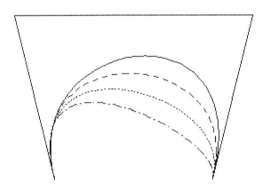

Figure 3. The effect on the shape of T-Bézier curves of altering the value of μ.

$P_3 - P_2$ are produced by the shape parameter μ. As the shape parameters $\lambda = \mu$, the curve moves in the direction of or the opposite direction to the edge $P_2 - P_1$ when λ increases or decreases. The effect of the shape parameters of the QT-Bézier curves (3.1) is clear. Figures 2 and 3 show the effects on shape of the curve of altering the values of λ and μ respectively.

In Figure 2, the curves are generated by changing $\lambda = 0$ to $\lambda = 1$ (solid lines), $\lambda = 0.5$ (dashed lines), $\lambda = 0$ (dotted lines), and $\lambda = -0.5$ (dash–dotted lines), and setting $\mu = 1$.

In Figure 3, the curves are generated by changing $\mu = 0$ to $\mu = 1$ (solid lines), $\mu = 0.5$ (dashed lines), $\mu = 0$ (dotted lines), and $\mu = -0.5$ (dash–dotted lines), and setting $\lambda = 0.5$.

4 THE APPLICATIONS OF THE CURVES

4.1 The representation of ellipses

Theorem 4.1 Let P_0, P_1, P_2 and P_3 be four control points. by proper selection of coordinates,

their coordinates can be written in the form $P_0 = (-a,-b), P_1 = P_2 = (0,b), P_3 = (a,-b)$, then the corresponding QT-Bézier curve with the shape parameters $\lambda = \mu = 0$ and local domain $t \in [t_0,t_1]$ represents an arc of an ellipse.

Proof If we take $P_0 = (-a,-b)$, $P_1 = P_2 = (0,b)$, $P_3 = (a,-b)$ into (1), then the coordinates of the QT-Bézier curve with the shape parameters $\lambda = \mu = 0$ is

$$\begin{cases} x(t) = a(\sin t - \cos t), \\ y(t) = -3b + 2b(\sin t + \cos t), \end{cases} \tag{8}$$

where $0 \le t_0, t_1 \le 2\pi$. This gives the intrinsic equation

$$\frac{x^2}{a^2} + \frac{(y+3b)^2}{4b^2} = 1. \tag{9}$$

It is an equation of an ellipse.

When a = 2b, the ellipse becomes the circle.

Theorem 4.2 By proper selection of coordinates, let four control points P_0, P_1, P_2 and P_3 be in the form $P_0 = (2a,0), P_1 = (a,0), P_2 = P_3 = (0,b)$, then the corresponding QT-Bézier curve with the shape parameters $\lambda = \mu = 1$ and local domain $t \in [t_0,t_1]$ represents an arc of a parabola.

Proof If we take $P_0 = (2a,0)$, $P_1 = (a,0)$, $P_2 = P_3 = (0,b)$ into (1), then the coordinates of the QT-Bézier curve with the shape parameters $\lambda = \mu = 1$ is as follows

$$\begin{cases} x(t) = 2a(1 - \sin t), \\ y(t) = b\sin^2 t, \end{cases} \tag{10}$$

which is the parametric equation of a parabola.

5 CONCLUSIONS

As above mentioned, QT-Bézier curves have all the properties that cubic Bézier curves have. However, they can deal precisely with circular arcs, cylinders, ellipses and cones, etc., which can only be approximated by cubic curves. Also, because there is nearly no difference in structure between a QT-Bézier curve and a cubic Bézier curve, it is not difficult to adapt a QT-Bézier curve to a CAD/CAM system that already uses the cubic Bézier curves.

ACKNOWLEDGMENTS

The work was funded the Key Project Foundation of Scientific Research, Education Department of

Anhui Province under Grant no. KJ2014ZD30, and the Key Construction Disciplines Foundation of Hefei University under Grant no. 2014XK08.

REFERENCES

[1] J.W. Zhang. C-curves: An extension of cubic curves[J]. Computer Aided Geometric Design, 13(1996.):199–217.

[2] J.M. Peña. Shape preserving representations for trigonometric polynomial curves[J]. Computer Aided Geometric Design, 14(1997.):5–11.

[3] G. Walz. Trigonometric Bézier and Stancu polynomials over intervals and triangles[J]. Computer Aided Geometric Design 14(1997.):393–397.

[4] Sanchez-Reyes, J. Harmonic rational Bézier curves, p-Béziercurves and trigonometric polynomials[J]. Computer Aided Geometric Design, 15(1998):909–923.

[5] X. Han, Quadratic trigonometric polynomial curves with a shape parameter[J]. Computer Aided Geom. Design 21(2004):535–548.

[6] P.E. Koch, Multivariate trigonometric B-splines[J]. Journal of Approximation Theory, 54(1988): 162–168.

[7] P.E. Koch, T. Lyche, M. Neamtu, L.L. Schumaker. Control curves and knot insertion for trigonometric splines[J]. Advance Computational Mathematics. 3(1995): 405–424.

[8] T. Lyche, L.L. Schumaker, Quasi-interpolants based on trigonometric splines[J]. Journal of Approximation Theory, 95(1998): 280–309.

[9] X.Q. Wu, X.L. Han, S.M. Luo. Quadratic Trigonometric Polynomial Bézier Curves with a Shape Parameter[J]. Journal of Engineering Graphics, 1(2008):84–77.

[10] I.J. Schoenberg. On trigonometric spline interpolation[J]. Journal of Mathematical Fluid Mechanics. 13(1964):95–825.

[11] G. Walz, Some identities for trigonometric B-splines with application to curve design[J]. BIT, 37 (1997): 189–201.

[12] Q.Y Chen, G.Z. Wang. A class of Bézier-like curves[J]. Computer Aided Geometric Design, 20(2003):29–39.

Advances in Future Manufacturing Engineering – Yang (Ed.)
© 2015 Taylor & Francis Group, London, ISBN 978-1-138-02817-3

The application analysis on share management of supply chain information collaboration by using triangular fuzzy TOPSIS decision method

Xidong Deng & Weiping Cao
School of Economics and Management, China Three Gorges University, Yichang, Hubei, P.R. China

ABSTRACT: The management of sharing information in a collaborative supply chain has become an increasingly important issue. This paper puts forward a new decision-making method, namely—fuzzy TOPSIS, to evaluate whether the management of sharing information in a collaborative supply chain achieves the maximum performance. Then we use this method to do a case study, which involves five factors including cost, usability, reliability, responding ability and predicting ability—and then find out the most suitable solution for managing information-sharing in a collaborative supply chain, for the analysed company, by calculating the scores and final ranking with fuzzy weight and fuzzy threshold.

Keywords: collaborative supply chain; information sharing; fuzzy TOPSIS

1 INTRODUCTION

In the past twenty years, the management of the supply chain has been regarded as a powerful commercial tool which is used in competitive marketing. Many manufacturers and shopkeepers have already realised that can reduce the primary cost and increase income through the effective management of the supply chain. In the entire management of the supply chain, more and more researchers have started to pay attention to the discovery of the collaborative supply chain field.

The management of the collaborative supply chain, constitutionally speaking, exists in the plenty of cooperators, which are complex and dynamic [1]. In order to face these challenges, the management of the collaborative supply chain can help a company to manage their cooperators effectively. Thus it helps them go a step further to build a long-term partnership [2]. The management of the collaborative supply chain can be defined as a complementary flowing tool which manages material, information and finance, available between company and cooperator. In this machinery, there are two vital problems: resource sharing and collaborative working [3].

Firstly, information errors are caused by the inconsistencies between upstream partners and downstream partners [4]. The cooperators of the supply chain may predict the marketing demand based on incomplete information. This will cause the increase in production cost—the well named "The bullwhip effect". A lot of research about the management of the collaborative supply chain,

at present, emphasizes on the management of collaborative information. Secondly, the essence of collaborative supply chain behaviour with a favourable definition, is constructed on the participants' collective decision, based on achieving a collective aim. Constitutionally speaking, this provides them with clear data about future demand, before exploiting a practical plan to meet the demand, while coordinating the interrelated activities to finish this task by using the method of systematisation. This is an effective driving force practice of the supply chain. However, in order to make the cooperation propitious, the management pattern of the collaborative supply chain is required to be highly effective, which can not only promote the goods flowing in the supply chain, but also help to predict demand more effectively.

Although it can be seen from the research at present, the effective management of supply chain may reduce the upstream members in supply chain, especially the production order suppliers. It can bring effectively available data processing which can improve the supply chain property. Thus producing more profit [5].

This paper analyses the management of the collaborative supply chain and information-sharing by discussing, while using fuzy TOPSIS, the management of the collaborative supply chain information. The second part of essay reviews the internal and external research of its development. The third part introduces and analyses fuzzy TOPSIS. The fourth part combines data with cases and useage method.

2 THE INFORMATION-SHARING MANAGEMENT OF THE COLLABORATIVE SUPPLY CHAIN IN INTERNAL AND EXTERNAL RESEARCH

Collaborative information is the essential precondition to bring about the working process of the supply chain achieving the dynamics of high visibility and agility. The collaborative information flow is the essential precondition of the collaborative logistics in the supply chain and cash flow [6]. Information is the inner-core of the collaborative supply chain that is high importance and deep value.

In the research of domestic scholars, ZhangQing [7] discussed the relationship between information source sharing and the collaborative information technology in the supply chain system of enterprises and pointed out the difference. This shows that information resources sharing is the primary condition to achieve collaborative information technology, which may reach its optimum result in the supply chain system of the enterprise only through the accurate information sharing. Haohao, Xia Jianming [8] and others summarised the main factors, including customer response effect, competitive advantage, system's total holding cost, outsourcing risks and others, in the collaborative supply chain management. Then they first pointed out the pattern basis in the collaborative supply chain management system. In the research of the collaborative supply chain, Li lingual [9] found that it could mainly include three parts: the collaborative purchasing information and selling information between the core enterprise and suppliers and core enterprise internal information.

However, in the research of foreign scholars, Indicator research was focused more on the basis of the information-sharing in the collaborative supply chain. Rai et al [10] considered that the three types of information sharing are operation, tactics and strategy. This information collaboration constitutes a large sharing investment cost and undertakes the corresponding risks, but it can also receive a lot of complementary resources. When doing research on the ability of the information-sharing in the collaborative supply, Zacharia found that supply chain knowledge, the ability of information utilisation and cooperative ability in the process of the supply chain, have a great influence in a periodic collaborative supply chain, which can effectively promote the level of collaboration and relationship among enterprises [11]. The information-sharing and collaboration amongst some partners mainly concentrates on the degree of communication, trust and interdependence. They are willing to work jointly, which

can lead to more stable trading and increase the uncertainty of the market. Some studies have also pointed out that the social exchange beliefs, such as trust, commitment, and interdependence, are the critical factors in the information-sharing and collaboration of the supply chain.

In this paper, according to the information-sharing management of the collaborative supply chain, we use five important indexes: primary cost, usability, reliability, reactive potency, predicted ability. These indexes evaluate it objectively from different aspects, including qualitative and quantitative analysis. The definitions of five indexes are listed in the chart one.

3 METHODOLOGIES

The previous studies on supply chain collaboration mainly adopt three research methods: weight analysis (e.g. AHP), mathematical programming (e.g. linear programming) and probability statistic (e.g. DEA). AHP is devoted to the comprehensive evaluation of problems with an hierarchical structure. It decomposes the whole problem layer by layer into multiple evaluation problems of a single criterion, and then makes the comprehensive evaluation on the basis of evaluation results of the sub-problems. However, AHP applies the mean value method to calculate eigenvalue, which would cause systematic deviation for some ill-conditioned matrix. The linear programming method is mainly used to analyse quantitative factors, while it cannot deal with qualitative factors in most practical cases. DEA is adopted to evaluate the production performance, mainly according to the input-output ratio, though it cannot obtain very accurate measurement because of the absence of exact ratio.

The idea of fuzzy TOPSIS is to find out the solution which is nearest to the fuzzy positive ideal solution and furthest from the fuzzy negative ideal solution. The definition of fuzzy TOPSIS is like an index, seeking the solution which is similar to positive ideal ones and deviated to the negative ideal ones. Then, it can be implemented according to nearest solution of the positive ideal solution. Researchers cannot evaluate the accurate performance in the studies. The advantage of fuzzy method is to adopt the fuzzy number, but not accurate number, to make the objective analysis in the fuzzy environment—hence it is more suitable for the real world. This method is very useful for the group decision-making problem in the fuzzy environment. Therefore, this paper adopts fuzzy TOPSIS method to study the problem of the management of information-sharing in the collaborative supply chain.

This paper applies the fuzzy TOPSIS method to the information-sharing management of the collaborative supply chain. We firstly put forward three methods to make a fuzzy TOPSIS analysis. Through building a fuzzy matrix, we respectively figure out the distance from the fuzzy positive ideal solution and fuzzy negative ideal solution. Lastly, we can obtain the best solution for the management of the collaborative supply chain. The following is a simple review of the fuzzy theory and the development of fuzzy TOPSIS decision-making.

Step one: determine the indicators of the studied problem and the evaluation criterion. The indicators considered in this paper are shown in Table 1.

Step two: build the fuzzy decision-making matrix and select the proper language level. Figure 1 demonstrates fuzzy decision-making matrix as an example.

$$\tilde{D} = \begin{array}{c} A_1 \\ A_2 \\ \vdots \\ A_M \end{array} \begin{bmatrix} \tilde{x}_{11} & \tilde{x}_{12} & \tilde{x}_{1n} \\ \tilde{x}_{21} & \tilde{x}_{22} & \cdots & \tilde{x}_{2n} \\ \vdots & \vdots & \ddots & \cdots & \vdots \\ \tilde{x}_{m1} & & \cdots & \tilde{x}_{mn} \end{bmatrix}$$

$i=1,2,\ldots,m; \ j=1,2,\ldots,n$
$W = [\tilde{w}_1, \tilde{w}_2, \ldots \tilde{w}_n]$
$C_1 \ C_2 \ C_N$

Figure 1. Fuzzy decision-making matrix.

$\tilde{x}_{ij}$ is the evaluation value corresponding to the criterion of c_j under the solution A_i. $\tilde{W}_j$ refers to the weight in such a case, which can reveal the importance of solution A_i under the criterion of c_j.

Step three: normalise the fuzzy decision-making matrix. If $\tilde{R}$ is the normalised form of the fuzzy decision-making matrix, then we can get Formula 1.

Table 1. The indexes in the information sharing management of the collaborative supply chain.

Cost (C1)	It is the cost about relevant data gotten from other supply chains cooperative partners
Usability (C2)	It includes the process of using information related to the management of supply chain. In another word, it represents the importance of information in the collaborative supply chain
Reliability (C3)	It means the precision of information
Reactive potency (C4)	It means the reactive pace the company makes when it gets information
Predicted ability (C5)	It means the biggest benefit obtained when using specific and effective information

$$\tilde{R} = [\tilde{r}_{ij}]_{m \times n}, i = 1, 2, \ldots, m; \ j = 1, 2, \ldots n \quad (1)$$

In Formula 1, the fuzzy number is expressed as the triangular fuzzy function (a_{ij}, b_{ij}, c_{ij}); hence the normalised value of the matrix under the condition of benefit and cost can be received as follows.

$$\tilde{r}_{ij} = \left(\frac{a_j^-}{a_{ij}}, \frac{a_j^-}{b_{ij}}, \frac{a_j^-}{c_{ij}} \right)$$

where

$$a_j^- = min_i a_{ij} \quad (2)$$

The normalised $\tilde{r}_{ij}$ is still the triangular fuzzy function. Likewise, the normalisation of the trapezoidal fuzzy number can be done in the same manner. The weighted fuzzy normalised decision-making matrix is matrix $\tilde{V}$, which can be developed as follows:

$$\tilde{V} = [\tilde{v}_{ij}]_{m \times n}, i = 1, 2, \ldots, m; \ j = 1, 2, \ldots, n \quad (3)$$

$$\tilde{v}_{ij} = \tilde{r}_{ij} \times \tilde{w}_j.$$

Step four: determine the fuzzy positive ideal solution and fuzzy negative ideal solution.

Through calculating the weighted normalised decision-making matrix, we know that the value of $\tilde{v}_{ij}$ is in the range of [0, 1], therefore the fuzzy positive ideal solution A^+ and fuzzy negative ideal solution A^- can be defined as:

$$A^+ = \left(\tilde{V}_1^+ \ldots \tilde{V}_j^+ \ldots \tilde{V}_n^+ \right)$$
$$= \left\{ \left(max_i \ v_{ij} \mid i = 1, 2, \ldots, m \right) \right\}, \ j = 1, 2, \ldots, n \quad (4)$$

$$A^- = \left(\tilde{V}_1^- \ldots \tilde{V}_j^- \ldots \tilde{V}_n^- \right)$$
$$= \left\{ \left(min_i \ v_{ij} \mid i = 1, 2, \ldots, m \right) \right\}, \ j = 1, 2, \ldots, n \quad (5)$$

where $v_j^- = (1,1,1) \times \tilde{w}_j = (xw_j, yw_j, zw_j)$, $v_j^- = (0,0,0)$, $j = 1, 2, \ldots n$.

Step five: compute the distance between every optional solution and fuzzy positive ideal solution and fuzzy negative ideal solution.

$$d_i^+ = \sum_{j=1}^n d_v(\tilde{v}_{ij}, \tilde{v}_j^+), i = 1, 2, \ldots, m \quad (6)$$

$$d_i^- = \sum_{j=1}^n d_v(\tilde{v}_{ij}, \tilde{v}_j^-), i = 1, 2, \ldots, m \quad (7)$$

Step six: lastly we can get the consistency coefficient.

$$\widetilde{CC}_1 = \frac{d_i^-}{d_i^- + d_i^+} = 1 - \frac{d_i^+}{d_i^- + d_i^+}, i = 1, 2, \ldots, m \quad (8)$$

We can get consistency coefficients through the above steps. The solution with the largest consistency coefficient has the best possibility of information-sharing management within the collaborative supply chain, and hence is the best option for the company.

4 CASE ANALYSES

In this paper, we assume three different ways for the management of the supply chain information collaboration—S1, S2, S3. These plans use Table 1, which proposes the standards of evaluation of the supply chain collaborative information-sharing management. There are three people involved in the decision-making process; P1, P2, and P3. Table 2 shows the language levels of rating table. Table 3 shows the determining level.

1. Compute the decision's fuzzy decision matrix. The aggregated fuzzy weights (C1) of w_{ji} are calculated as:

Table 2. Linguistic terms for alternatives ratings.

Linguistic term	Membership function
VP	(0.1.3)
P	(1.3.5)
F	(3.5.7)
G	(5.7.9)
VG	(7.9.10)

Table 3. Linguistic terms for criteria ratings.

Linguistic term	Membership function
VERYHIGH(VH)	(0.7;1;1)
HIGH(H)	(0.5;0.7;1)
LOW(L)	(0;0.3;0.5)
VERYLOW (VL)	(0;0;0.3)

$$w_{j1} = \min_k \{0.7, 0.5, 0\},$$

$$w_{j2} = \frac{1}{3}\sum_{k=1}^{3}(1+0.7+0.3), \quad w_{j3} = \max_k\{1,1,0.5\}$$

So $w_j = (0, 0.67, 1)$.

Likewise, we compute the aggregate weights for others in Table 4.

2. Compute the alternative's fuzzy decision matrix. Supplier A1 for criteria C1 using the rating given by the three decision makers is computed as follows:

$$a_{ij} = \min_k\{0,1,3\}, \quad b_{ij} = \frac{1}{3}\sum_{k=1}^{3}(1+3+5),$$

$$c_{ij} = \max_k\{3,5,7\}$$

So the $x_{\tilde{ij}} = (0, 3, 7)$.

Likewise, we compute the aggregate weights for others inTable 5 and 6.

3. In the next step, we perform normalisation of the fuzzy decision matrix of alternatives using Esq. for example, the normalised rating for alternative S1 for criteria C1 is given by:

$$c_j^+ = \max_i(7,10,7) = 10, \quad a_j^- = \min_i(0,1,1) = 0$$

Because we should consider maximising interests, so we choose the positive idea.

Other results as shown in Table 7. Based on $\tilde{v}_{ij} = r_{ij} \times \tilde{w}_j$, it is concluded to Table 8.

4. Then, we compute the distance of each alternative from the fuzzy positive ideal matrix (A+) and fuzzy negative deal matrix (A−). For example, for alternative S1 and criteria C1, the distances $d_v(S1,A+)$ and $d_v(S1,A−)$ are computed as follows:

$$d_v(S_1, A^-) = \sqrt{\frac{1}{3}\left[(0-0)^2 + (0.2-0)^2 + (0.7-0)^2\right]}$$

$$= 0.4204$$

$$d_v(S_1, A^+) = \sqrt{\frac{1}{3}\left[(0-1)^2 + (0.2-1)^2 + (0.7-1)^2\right]}$$

$$= 0.7594$$

Likewise, we compute the distances for the remaining criteria for the four alternatives. The results are shown in Table 9.

Table 4. Linguistic assessments for the 5 criteria.

Criteria	Decision makers			Aggregate fuzzy weight
	P1	P2	P3	
C1	VH(0.7, 1, 1)	H(0.5, 0.7, 1)	L(0, 0.3, 0.5)	(0,0.67,1)
C2	L(0, 0.3, 0.5)	VL(0, 0, 0.3)	H(0.5, 0.7, 1)	(0,0.33,1)
C3	H(0.5, 0.7, 1)	H(0.5, 0.7, 1)	VH(0.7, 1, 1)	(0.5,0.8,1)
C4	VH(0.7, 1, 1)	VH(0.7, 1, 1)	L(0, 0.3, 0.5)	(0,0.77,1)
C5	VL(0, 0, 0.3)	L(0, 0.3, 0.5)	L(0, 0.3, 0.5)	(0,0.2,0.5)

418

Table 5. Linguistic assessments for the three alternatives.

| | Alternatives | | | | | | | | |
| | S1 | | | S2 | | | S3 | | |
Criteria	P1	P2	P3	P1	P2	P3	P1	P2	P3
C1	VP	P	F	P	VG	P	F	F	P
C2	P	F	F	VP	P	P	F	F	VP
C3	F	P	VP	VG	P	P	G	VP	P
C4	G	VP	P	G	P	VP	VP	F	VG
C5	VG	F	G	F	VP	F	G	VG	G

Table 6. Aggregate fuzzy decision matrix for alternatives.

| | Alternatives | | |
Criteria	S1	S2	S3
C1	(0,3,7)	(1,5,10)	(1,4.33,7)
C2	(1,4.33,7)	(0,3.33,5)	(0,3.67,7)
C3	(0,3,7)	(1,5,10)	(0,3.67,9)
C4	(0,3.67,9)	(0,0.67,9)	(0,5,10)
C5	(3,7,10)	(0,3.67,7)	(5,7.67,10)

Table 7. Normalised fuzzy decision matrix for alternatives.

| | Alternatives | | |
Criteria	S1	S2	S3
C1	(0,0.3,0.7)	(0.1,0.5,1)	(0.1,0.433,0.7)
C2	(0.1,0.433,0.7)	(0,0.333,0.5)	(0,0.367,0.7)
C3	(0,0.3,0.7)	(0.1,0.5,1)	(0,0.367,0.9)
C4	(0,0.367,0.9)	(0,0.067,0.9)	(0,0.5,1)
C5	(0.3,0.7,1)	(0,0.367,0.7)	(0.5,0.767,1)

Table 8. Weighted normalised alternatives, FPIS and FNIS.

| | Alternatives | | | | |
Criteria	S1	S2	S3	FNIS (A−)	FPIS (A+)
C1	(0,0.2,0.7)	(0,0.34,1)	(0,0.29,0.7)	0	1
C2	(0.1,0.14,0.7)	(0,0.11,0.5)	(0,0.12,0.7)	0	0.7
C3	(0,0.24,0.7)	(0.05,0.4,1)	(0,0.29,0.9)	0	1
C4	(0,0.28,0.9)	(0,0.052,0.9)	(0,0.39,1)	0	1
C5	(0,0.14,0.5)	(0,0.073,0.35)	(0,0.15,0.5)	0	0.5

Table 9. Distance d_v (S1, A+) and d_v (S1, A−) for alternatives.

| | d_i^- | | | d_i^+ | | |
Criteria	S1	S2	S3	S1	S2	S3
C1	0.4204	0.6098	0.4375	0.7594	0.6917	0.7290
C2	0.4162	0.2956	0.3500	0.4748	0.5410	0.5250
C3	0.4274	0.6225	0.5759	0.7456	0.6487	0.7104
C4	0.5442	0.5205	0.6197	0.7138	0.7977	0.6763
C5	0.2998	0.2064	0.3014	0.3557	0.3807	0.3524
Σ	2.108	2.2548	2.2845	3.0493	3.0598	2.9931

419

Table 10. Closeness coefficient (CCi) for the four alternatives.

	Alternatives		
	S1	S2	S3
d_i^+	3.0493	3.0598	2.9931
d_i^-	2.108	2.2548	2.2845
cc_1	0.4075	0.4243	0.4329

5. Base on the Closeness coefficient (CCi) for the 3 alternatives, Table 10 shows the final ideal ranking. The Closeness coefficient of S1 is 0.4075.

$$\widetilde{CC_1} = \frac{\tilde{d}_1^-}{\tilde{d}_1^- + \tilde{d}_1^+} = 1 - \frac{\tilde{d}_1^+}{\tilde{d}_1^- + \tilde{d}_1^+} = 0.4075$$

By comparing the CCi values of the three alternatives, we find that S3 > S2 > S1. Therefore, we can say that supplier S3 has the best environmental performance according to share management of supply chain information collaboration. For S3, in the all shared management of the supply chain information collaboration. It pays more attention to the reaction ability and the ability to predict; it also suggests that the reactive potency and predicted ability are important to share the management of the supply chain information collaboration. But S2 pays more attention to the cost and reliability, and in the development of the whole supply chain, the cost-benefit principle is the most important, and the evaluation of response ability and the ability to predict are of less consideration.

So at the end of the consistency coefficient comparison, consistency coefficient value is less than the S3. And S1, takes into consideration availability, reliability and reactive potency, but in the case of the entire management of the supply chain information collaboration, didn't evaluate the prediction ability.

5 CONCLUSIONS

In this era of global competition, the modern company pays special attention to the development of the supply chain collaboration. Therefore, an effective model of development is very important for any company. This paper proposes a new method—the fuzzy TOPSIS method, to solve the problem of shared management of the supply chain information collaboration. In the past few years, some fuzzy TOPSIS methods were developed in different application fields. Chen and Mr.

Tsao use fuzzy TOPSIS method to analyse interval value fuzzy decision; Wang and Elhag put forward the fuzzy TOPSIS optimal selection, which is based on the alpha level, and created a kind of nonlinear programming solution.

The companies involved in the whole supply chain collaboration includer three dimensions: information synergy, goal congruence, and decision-making. And the information sharing of synergy can be proved in unstable conditions, by implementing information between customers, suppliers, and the ability of the full use of resources and knowledge, which can significantly improve business performance and promote the competitive edge of the supply chain. Information flow plays a key role in the supply chain collaboration; it works through the right channel and transmission mechanism, takes timely, efficient and accurate delivery, will create value for the company and lastly, enhance the overall competitiveness of the supply chain.

In the future development of research, we should pay attention to the benefits of shared management of the supply chain information collaboration, and the cost control, etc—Also to further help a company improve its overall efficiency and perform well.

REFERENCES

[1] Vijayasarathy, L.R., 2010. Supply integration: an investigation of stimuli—dimensionality and relational antecedents. International Journal of Production Economics 124(2), 489–505.

[2] Fynes, B., deBúrca, S., Mangan, J., 2008. The effect of relationship characteristics on relationship quality and performance. International Journal of Production Economics 111(1), 56–69.

[3] Smith, G.E., Watson, K.J., Baker, W.H., Pokorski, J.A., 2007. A critical balance: Collaboration and security in the IT-enabled supply chain. International Journal of Production Research 45 (11), 2595–2613.

[4] Prajogo, D., Olhager, J., 2012. supply chain integration and performance: the effects of long-term relationships, information technology and sharing, and Logistics integration. International Journal of Production Economics 135, 514–522.

[5] Katzenberg M.E, Rosenzweig E.D, Marucheck AE, Metters R.D. A framework for the value of information in inventory replenishment. European Journal of Operational Research 2006:1230–50.

[6] Soroor, J., Tarokh, M.J., Shemshadi, A. Theoretical and practical study of supply-chain coordination. Journal of Business Industrial Marketing, 2009, 24 (2):131–142.

[7] Zhangqing. Research on the framework and the essentials of information coordination in supply chain. Journal of intelligence, 2009(5):179–1.

[8] Haoahao, xiajiangming. Under the perspective of manufacturing outsourcing suppliers management model research, Productivity Research, 2008(8): 131–135.

[9] Lilingju. The information in the supply chain management information system synergistic effect analysis, Information Science, 2001, 24(1):101–103

[10] Rai, A., Patnayakuni, R., Seth, N., 2006. Firm performance impacts of digitally enabled supply chain integration capabilities. MIS Quarterly 30(2), 225–246.

[11] Acharia, Zach G., Nix, Nancy W., Lusch, Robert F. Capabilities that enhance outcomes of an episodic supply chain collaboration [J]. Journal of Operations Management, 2011, 29(6):591–603.

[12] Myhr, N., Spekman, R.E., 2005. Collaborative supply-chain partnerships built upon trust and electronically mediated exchange. Journal of Business and Industrial Marketing 20(4/5),179–186.

[13] Ramanathan U. Aligning supply chain collaboration using analytic hierarchy process [J]. Omega, 2013, 41(2):431–440.

[14] Hwang, C.L., & Yoon, K. (1981). Multiple attribute decision making: Methods and applications. Berlin: Springer.

[15] Sun C-C. A performance evaluation model by integrating fuzzy ahp and fuzzy topsis methods [J]. Expert Systems with Applications, 2010, 37(12): 7745–7754.

[16] Chen, T.Y., & Tsao, C.Y. (2007). The interval-valued fuzzy TOPSIS methods and experimental analysis. Fuzzy Sets and Systems.

[17] Wang, Y.M., & Elhag, T.M.S. (2006). Fuzzy TOPSIS method based on alpha level sets with an application to bridge risk assessment. Expert Systems with Applications, 31, 309–319.

[18] Liwei. The Research on performance evaluation of Supply Chain Information collaboration, Dalian University of Technology (2013), Master degree paper.

Advances in Future Manufacturing Engineering – Yang (Ed.)
© 2015 Taylor & Francis Group, London, ISBN 978-1-138-02817-3

Research of enterprise work safety information management system construction

Qiquan Wang
Department of Safety Engineering, China Institute of Industrial Relations, Beijing, China

ABSTRACT: Taking into account China's current safety information construction and work safety situation, the progress of China's safety information construction and its major problems are discussed. Then the basis, procedure, and frame of being is proposed. Then the work safety management information system frame is divided into four levels: data level, technology level, business level and interface level. The function of each level is analysed. The conclusion is: when constructing a safety information system, the objective analysis of enterprise safety management situation should be accounted for and the rationality of the work safety information management system aims should be assessed in order to prevent accidents and improve the management level of enterprise work safety management.

Keywords: safety management; information system; emergency; framework; construction

1 INTRODUCTION

Work safety is related to people's life and property security, and it is related to reform and development and to social stability. With the development of society and the economy, the number of enterprises has grown rapidly. With the rapid economy development of enterprises, the issue of safety at work is a very important one. Many enterprises generally have the problem of a weak and poor management of safety standards. In order to improve the level of safety management of enterprises, to curb possibility of accidents effectively, to adapt to the needs of the development of work safety, to truly achieve a people-centre development concept, an advanced scientific management approach needs to be brought in, to enhance the level of the safety management, and also achieve the goal of preventing accidents.

Presently enterprises have established a perfect work safety information management system in technology-developed countries. But in our country, our enterprise safety management is mostly on the level of propaganda, education, regular safety inspections, and mostly about qualitative or accident analysis. This management tool is passive and slow reacting. It is difficult to adapt to the requirements of modern production management. While numerous enterprises using of computer-aided management, due to the lack found in the safety production management information system based on theory, find that the construction and application of the system is not satisfactory [1]. With China's rapid economic development

and diversification of the production and business operation entity' property right and business management mode, an enterprise's production safety arrangements becomes increasingly complex; there is a dramatic increase in safety data; traditional means of work are difficult to meet the needs of the development under the new situation. By means of modern communications, electronic information network technical support and service of the enterprise safety production—to establish sensitive, responsive and reliable operation of the information system, one can understand the Safety Dynamics, and improve the level of safety production supervision, information management and productivity. It is imperative to comprehensively improve the informatization of enterprise safety production work [2]. Attaches great importance to and strengthen the construction work of the enterprise safety production management information system, promoting the results of research and development, promotion and application. It is significant in enhancing the level of safety and setting up a safety production long-efficiency mechanism.

The writer—starting from the construction of a safety production informatization platform, as well as the current situation of safety production informatization system research in our country, according to the related theory of enterprise security knowledge, relevant national laws, regulations, industry standards, technical specifications, and computer programming language—explains the basis of the connotation and establishment of a safety production management information system, to build an enterprise safety production management information system framework, for the enterprise to achieve an efficient and sensitive, responsive, reliable security information management, and provide the principle concept of safety information platform; also for enterprises to improve the level of safety production supervision, monitoring and emergency response; get traditional safety management from post-processing changes to advance prevention—and to get access to achieve the informatization.

2 THE SURVEY OF THE RESEARCH ON SAFETY PRODUCTION INFORMATION SYSTEM CONSTRUCTION

Global informatization was started in the 1960 of the 20th century, Japan began integration of information and communication technologies earlier than other countries and started the tenor of the informatization. Informatization began brewing in 1978 (1978–1983) in China, has gone through four major phases—the initial stage (1984–1992), advancing stages (1993–1997), accelerated phase (Since 1998); The 'Shanghai Forum' was discussed and adopted at the second session of the Forum on City Informatization in the Asia-Pacific region in June 2000—the proposition that uses information to solve basic management was put forward; after this, a large number of domestic service providers who were dedicated to enterprise information management were emerging; a large number of application systems, such as OA, logistics management, financial management, human resource management, and multiple "basic enterprise management" of informatization integration had been achieved, and began to move towards integration, depth, and professional increasingly [3].

China's production safety supervision and administration of information technology started in the early 21st century. After the establishment of the State Administration of Work Safety, safety supervision and monitoring of information basic construction work was gradually carried out; it included information network infrastructure, work safety, monitoring of applications and the underlying database construction. Because institutions at all levels are formed not for long, and have less outlay, parts of the construction of safety production informationization at all levels and level of application development are not balanced. "Accelerating the construction of safety information" began to "safety in information construction projects", the research and development of accident statistics system, monitoring of major hazard control systems, safety policies and regulations search expert database systems, security systems, safety supervision system were presented by 'National Safety

Plan (2004–2010) as stated' which was issued by the State administration of work safety in 2003 [4].

In the year of 2006, the fifth of the 9 key projects was the safety production information system construction project. The national safety production "Eleven-Five" project was issued by the General Office of the State Council. On government regulation, national major accidents, rail, water transport, air safety, airport safety, construction safety, the safety of special equipment, agricultural machinery of fishing vessel territory, started with improving the Government-level work safety management information system. In the same year, the National leading group for Informationization of national safety production information system, classified as the "Jinan" project, became the national information points over a period of time. 'The national safety production informationization "Eleven-Five" special planning' released the major assignment and significant project further by the State administration of work safety in 2007. China Petroleum and other key leading enterprises commenced constructing the enterprise safety production management information system.

At this point, under the industrial guidance and policy support, our country progressively developed a number of safety information industries specialising in the high-tech enterprises and institutions. Some Government and enterprise safety production management information system of safety production supervision and management information system service have been introduced—formed originally on the safety and production informationization industrial pattern, with enterprises as the main body; both the production, learning, and research was significant [5]. However, in China, mature enterprise safety production management information systems are few. They are unable to meet the dramatic increase in safety data, prevention, monitoring, emergency response, safety analysis, forecasting, integrated management of queries and service needs. Therefore, strengthening the development of the safety production of enterprise information resource integration, speeding up enterprise safety production management system development, is of great guiding significance, in order to improve the safety management level, and to be effective in preventing accidents.

3 THE BASIS OF THE CONNOTATION AND ESTABLISHMENT OF SAFETY PRODUCTION MANAGEMENT INFORMATION SYSTEM

3.1 *Connotation*

Safety production informationization is accompanied by a sense of safety information and communication technology; computer technology continues to progress in incorporating these technologies in production safety accident prevention and treatment, rescue and work safety management, changing the process of traditional production processes and structures, improving the efficiency of safety production management and reducing the possibility of occurrence of safety accidents. Safe production informatization is a process. It is achieved by computer data recording and storage, by LAN and Internet information delivery, by programs for data processing and feedback—to alter the traditional enterprise present status of safe production management in business.

Safety production management information system is the core of the safety management Informationization; it is a Core management tool which is composed of people, networks, computers and other peripheral devices—which are, able to carry out safety production management information collection, transmission, storage, processing, maintenance and use, using information flow direction to handle safety production—by also using historical data to predict the future, using real-time data to implement early warning, using data integration to achieve statistics—thus to support senior corporate decision-making, give the Government assistance to middle-level guidance on operating control, the grass-roots level in process operation, and help them to achieve its safety objective.

3.2 The basis of the establishment of safety production management information system

The Safety management information system construction transforms and optimises traditional safety production management business information, which is a large, complex system of engineering that refers to implantation of management thinking, historical data analysis, business process review management model reconstruction, construction of the hardware device, program development, training of key users and several other key work processes—all of which requires more than one aspect of the necessary support [6].

3.2.1 Business owners and management attention
To establish and promote safety production information systems, one must to do a lot of groundwork, and also invest a considerable amount of manpower and material resources—Especially during the preparatory phase of systems research and development, and also import data in initial operation stage, which will involve a great deal of analysis of integrated data migration. Consequently, business leaders must be given

attention and support to promote the project team, develop management programs, incentives, and assessment methods. Clear and definite safety informationization is the next inevitable direction of safety management, and enables employees to set up "technology gmelinii" safety concepts, actively participate in the development of new systems and applications.

3.2.2 Have professional technical delegate
Safety production management information system of construction is system engineering requires a strong technology and relates to a wider professional area. On one hand, it needs safety science and management professionals talent for support, on the basis of determining the safety management objectives, get safety production management subdivided into a number of module, such as Legal regulations, accident investigation, safety checks, emergency plan, the report analyses. Also ensure that the content of each module relates to the management of technology, the function they need and the goal needed to be achieved; On the other hand, it needs computer science and communications technology, as well as adequate technical support capability provided over long periods of system maintenance. So, when we build the system, we should consider not only the service provider's ability to develop the system, but also the safety management aspect of the legitimacy, integrity and service provider understanding of safety in business degree and the ability to continue to preserve and protect.

3.2.3 Great computer network environment
Network administration is the stage of production characteristic of management information; the management information system must be built on the basis of the enterprise's local area network or the Internet: a good network access environment is the base for implementation of safety management.

3.2.4 Sufficient funds and sufficient development cycle
Safety information systems involve a large amount of software and hardware facilities, which needs investment of sufficient funds. System software and services costs include information, system design and development costs, research costs, training costs and system maintenance costs, hardware costs, along with acquisition of servers, terminal addition and adjustment of the layout of the network costs. In addition, after the inputs, we also need a research and development cycle; generally speaking, a good safety management information system is the need for a year, before operating on the business.

4 PROGRAMS AND SYSTEM SETTINGS FRAMEWORK FOR ESTABLISHING SAFETY PRODUCTION MANAGEMENT INFORMATION SYSTEM

4.1 Establishment of safety management information system program

The process of the establishment of safety management information system [7] as shown in Figure 1.

4.2 Enterprise framework and functions of safety production management information system

Framework refers to the system of the building structure. A system framework is not easy to change; the safety production management information system framework is divided into 4 levels: layers of data, technical and operational level, the interface level, as shown in Figure 2. The feature is set thus.

4.2.1 Layers of data

It is the basis of the safety production management information system; a data bit yield, defining the safety data types, formats, requirements. According to the complexity of the system, the data tier includes several data tables or databases, to support the business of safety production management operations and systems' technology needs.

4.2.2 Layers of technical

Technology stack is technical, which means supporting the safety production information system—including two parts—the work safety management and computer technology; these technologies can improve the application of the system of advancement and adaptability, improving the usefulness of open systems, and advanced.

4.2.3 Operational level

Service platform is an actual embodiment of safety management in business; it is also a primary interface that interacts with the user. It is based on the work safety laws and regulations, internal operation management system in enterprises, the level and goal of enterprise safety management. According to the actual needs of the enterprise; classified by safety management modules, in the business layer construction we need to be aware of teasing out safety management processes, clear, concise and effective needs of analysis of business processes; research and construction of the service platform is the focus of the entire production information engineering.

4.2.4 Interface level

Interface layer is the system and the external data exchange interface; through it, the system can work

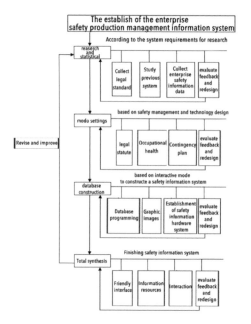

Figure 1. Safety production management information system program.

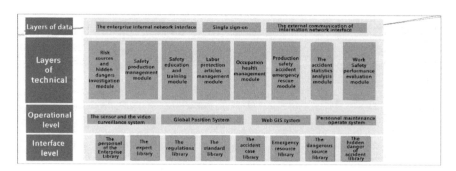

Figure 2. Typical safety production information system framework diagram.

426

with the government of safe production supervision and management system, the enterprise internal network system, financial, material supply system, communication system provided by the communication service provider, realising the sharing of data and extensive communication system.

5 CONCLUSION

Along with the rapid development of science and technology and social production, safety information management is becoming more and more important in safety management. Safety information and research on the safety production information management system, has become an important part of modern safety management. Realising effective integration of enterprise information management and enterprise management of production safety will be the advanced mode of leaping development of enterprise information. Enterprise safety informationization is an indispensable resource for management activities, and safety management information system is indispensable to the safety management system as a whole "nervous system".

Either from the current work safety needs, or from the long-term development of the science of work safety requirements, safety information management system of construction is imperative; but in this regard it started late in China; the process was slow, and also required relevant departments of the State and Governments at all levels to attach great importance to it, and also increase funding and organization, to change the current production safety conditions, in order to make a clean sweep of all kinds of hidden dangers for accident occurances, and to get safety production management in Chinese enterprises to take a qualitative leap [8].

Establishing a sound enterprise safety production information system is a complex engineering system. It is based on a lot of research. It is a matter of a multitude of complex systems, with diverse needs. The complexity of this technological system inevitably involves secrets in the business sector. Construction is very difficult, with a relatively long cycle. And it puts forward higher requirements for real-time updates of the data [9]. So, at the time of initially building the secure information systems, demand objective analysis on enterprise safety management status, assess the rationality of the goal of building a secure information system, and gradually improve the safety information system. An enterprise's safety production information management system could improve enterprise safety management efficiency and transparency. Share resources to provide the appropriate security decision services for enterprise leaders [10]. With the acceleration of enterprise informatization, strengthen the enterprise safety production management information system construction. The system is able to achieve dynamic updates, management, queries. Actualizing classification authorization management ensures accurate, confidential, secure and clear safety information.

REFERENCES

[1] Zhou Min-wen, Tan Hai-wen. Status and countermeasures analysis of safety production informationization in our country[J]. Opencast Mining Technology, 2005(6):36–38.

[2] Jiang ai-lin. Historical Changes of China's information technology development [J]. Information and Documentation Services, 2002(4):15–17.

[3] Sun Min-wen. Safety status and countermeasures analysis of Jiangsu Province information construction[J]. Informatization Construction, 2008(9):51–55.

[4] Production of national safey information, "Eleventh Five-Year" special plan[Z], 2007:23–24.

[5] Shi You-gang. Research on the information management system for safety inspection of the oil company[J]. Journal of Safety Science and Technology, 2010, (3):64–67.

[6] Wang Li-hong, Liu Chang. Analysis on Establishment of Petrochemical Enterprise Safety Production Management Information System[J]. Safety Environment Health, 2004(12):11–13.

[7] Wang Hao. Overview of Production Safety Information System [J]. Labor Protection, 2004(11):7–8.

[8] Wang Qi-Quan. Research on risk assessment information system for punching machinery injuries[J]. Journal of Safety Science and Technology, 2008, 18(5):102–108.

[9] Liu Lang, Cheng Yun-cai. Establishment of safety management information system in modern enterprises[J]. Journal of Safety Science and Technology, 2008, 18(5):133–137.

[10] Zhan Hong-chang, Chen Guo-hua. Probing into the development of safety management information system[J]. Industry safety and environmental protection, 2003, 29(3):38–40.

Advances in Future Manufacturing Engineering – Yang (Ed.)
© 2015 Taylor & Francis Group, London, ISBN 978-1-138-02817-3

Analysis of provincial domestic tourist arrivals in China: The role of national protected areas

Aiping Zhang

Institute of Geographic Sciences and Natural Resources, Chinese Academy of Sciences (CAS), Beijing, China
University of Chinese Academy of Sciences, Beijing, China

Linsheng Zhong & Yong Xu

Institute of Geographic Sciences and Natural Resources, Chinese Academy of Sciences (CAS), Beijing, China

ABSTRACT: This study analyses the effect of protected areas on tourist arrivals, to reveal the origin of the contradiction between tourism development and natural protection; the conclusions could provide supports for management of protected areas. Three types of protected areas are selected for analysis— they are National Natural Reserves, National Scenic Areas and National Geoparks. The main method for analysis is the Panel Data model, and the estimation method is Robust M-estimation. Major conclusions are the following: 1) Protected areas have a significantly positive influence on provincial tourist arrivals. 2) Protected areas show different effects on tourist arrivals associated with the different kind of protected areas—and the differences can be identified as: NGP > NSA > SCE > NNR. 3) The influence of protected areas presents a regional difference intervened by external factors—Eastern region is most affected, followed by Western and Central China.

Keywords: protected areas, tourist arrivals, tourism attractiveness, panel data

1 INTRODUCTION

Continuously improving for over 30 years, tourism has become one of important national economic industries in China. In terms of tourism development process, both international and domestic tourism market segments grew rapidly and steadily before the 21 century—which not only had great influence on social economy through incomes, employment, exports and taxes, but also raised China's status all over the world [1]. With its entry in the 21 century, the intrinsic drawbacks of tourism industry started showing up gradually. The international tourism market segment was deeply influenced by the global economic crisis, SARS and some other natural disasters, and inbound tourist arrivals dropped dramatically [2]. As for the domestic tourism market, tourist arrivals grew comparatively steadily from 2001 to 2011. In 2011, domestic tourist arrivals reached 2.64 billion, taking up more than 95% of the total tourism market in China. Thus, this paper pays attention to domestic tourism segment.

In order to caterto the development of domestic tourism market, conditions of tourism infrastructure and facilities in and around national protected areas, such as nature reserves, scenic areas and geological parks, have been greatly improved.

Both the construction of infrastructure and great tourist arrivals have had considerable impact on preserving natural resources and environmental protection. Tourism development and environmental protection has been a universal problem all over the world [3]. For instance, Banff National Park in Canada has suffered serious environmental deterioration as a result of the impact of tourism. Likewise, environmental problems associated with tourism urbanization—such as pollution, smog, crime, and overcrowded condition are prevalent in the Yellowstone region. From the aspect of a certain protected area, all these problems have risen from its attraction as a tourist hub. That is to say, the role of national protected areas on tourist arrivals decides the future negative impact on the protected areas [4]. In view of this, this paper will select three patterns of China's protected areas (National Natural Reserves, National Scenic Areas and National Geoparks) to reveal the origin of the negative impact on them.

2 METHODOLOGY AND DATA

2.1 *Modeling domestic tourism demand*

To investigate the determinants of provincial tourist arrivals, especially the effect on national protected areas, a tourism demand function is estimated in this study. The demand model is specified as

$$
\begin{aligned}
PDTA_{it} = {} & C + \alpha CPAS_{it} + \beta_1 SCE_{it} + \beta_2 SERV_{it} \\
& + \beta_3 CITY_{it} + \beta_4 INFRA_{it} \\
& + \beta_5 CPGDP_{it} + \mu_{it} + \varepsilon_{it}
\end{aligned} \quad (1)
$$

where *PDTA* is the quantity of provincial domestic tourist arrivals, and the subscripts i and t denote a certain province and time period. The *CPAS* is the total number of National Natural Reserve (*NNR*), National Scenic Areas (*NSA*) and National Geoparks (*NGP*) in one province. Meanwhile, the variables *SCE*, SERV, *CITY*, *INFRA* and *CPGDP* denote the quantities of provincial scenic spots, the level of tourism service, the level of city construction, the perfection of infrastructural facilities and the level of economic development, respectively. Besides, α is the influence coefficient of the main explanatory variables (*CPAS*) that we need to pay more attention to. Note that u_i is the unobserved variable that varies across provinces.

In order to identify the differentiation of tourism influence among *NNR*, *NSA* and *NGP*, the linear international tourism demand model is specified as

$$
\begin{aligned}
PDTA_{it} = {} & C + \alpha_1 NNR_{it} + \alpha_2 NSA_{it} \\
& + \alpha_3 NGP_{it} + \beta_1 SCE_{it} + \beta_2 SERV_{it} \\
& + \beta_3 CITY_{it} + \beta_4 INFRA_{it} \\
& + \beta_5 CPGDP_{it} + \mu_{it} + \varepsilon_{it}
\end{aligned} \quad (2)
$$

where *NNR*, *NSA* and *NGP* are the main explanatory variables, and α_1, α_2 and α_3 are relative variables of the three types of national protected areas.

The original source of each observed variable is as follows: the data on Provincial Domestic Tourist Arrivals (*PDTA*) comes from *The Yearbook of China Tourism Statistics*, while the data on the numbers of *NNR*, *NSA* and *NGP* come from *China Statistical Yearbook on Environment, China Urban Construction Statistical Yearbook* and national geopark list, respectively. In addition, the data on the other explanatory variables (*SCE*, *SERV*, *CITY*, *INFRA*, and *CPGDP*) are collected from provincial scenic spot list, *The Yearbook of China Tourism Statistics* and *China Statistical Yearbook*. Besides, it is found that there is some data has been missed in two provinces (Shanghai and Tibet) in the process of data collection. Therefore, the panel data comprise 29 provinces over the period from 2006 to 2011 with 174 observations after deducting the missing data for each variable.

2.2 Model test

It is generally recognised that Unit Root Test and Co-integration Test need to be carried out before the regression process of panel data model [5]. In fact, whether these tests are necessary or not depends on the structure of data itself. Baltagi (2005) argued that some problems, such as auto-correlation, unit root and co-integration, must be taken into account if the time period (*T*) is long enough. But if *T* is rather short, and number of cross section's observations is far larger than that of *T*, we need not consider these problems because the difference between groups is more considerable than that within each group [6]. In this research, *T* (6) is much less than *N* (29), and these is no need to conduct the Unit Root Test and Co-integration Test. But we still conduct these two tests, and five methods (Levin, Lin & Chu t* (LLC), Breitung t-stat, Im, Pesaran and Shin W-stat (IPS), ADF-Fisher Chi-square, PP-Fisher Chi-square) are used. The results show that all variables are not steady with original values, while most variables are steady with first order difference. Only *SERV*, *NRS* and *NGP* could not reject the null hypothesis in Breitung t-stat test and ADF-Fisher Chi-square test. Additionally, the result of KAO test indicates that this panel data could reject the null hypothesis.

2.3 Regression method

The rationality and effectiveness of estimation results for regression model not only needed to take care about R-square, but also needs to pay attention to the Robustness of the estimation process. If the robustness is lacking, estimation results are still invalid, despite the large value of R-square, because it could not reflect the factual influence of independent variables to the dependent variables [7]. During the process of estimation, estimators do not need to be sensitive to singular values. In view of this, this research draws a scatter diagram with the LOESS fitting method about explanatory variables and dependent variables to identify the influences of singular values in the first place (Fig. 1). The diagram and the LOESS curve

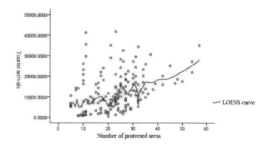

Figure 1. Scatter diagram about number of protected areas and domestic tourist arrivals.

show that there are a large number of singular values, and least-square estimation method is invalid because of the large residual error.

Huber (1964) proposed a regression method, namely M-estimation—the basic concept of which was that coefficients could be estimated through iterative weighted least-square [8]. In the process of this regression estimation, observations with different residual errors are endowed with different weights so that the influences of singular values could be smoothed. The target of iteration is to minimise the function value about residual values.

$$\sum_{i=1}^{n} \rho(\gamma_i/s) \to \min , \ \gamma_i = y_i - \sum_{j=1}^{n} X_{ij}c_j \qquad (3)$$

where γ_i denotes residual error, and s is the measure of spread. The process of minimising the function value is also a process of identifying the weights. In terms of selecting weight iteration method, the Huber-method is applied in this research, and the weighted function is specified as

$$\omega_i = \begin{cases} 1 & \text{if } |\mu_i| < c \\ \dfrac{c}{|\mu_i|} & \text{if } |\mu_i| > c \end{cases} , c = 2\sigma, \mu_i = \gamma_i/s \qquad (4)$$

where μ_i represents the standardisation index for residual errors. If the absolute value of μ_i is less than c, The Huber-method is actually the classical least-square method. Otherwise, the greater the absolute value of μ_i is, the smaller the weight is.

3 RESULTS

3.1 General influences of protected areas

The estimated results obtained from equation (1) are shown in Table 1. Through controlling the variable of tourism scenic spots step by step, Models (1), (2) and (3) could estimate the effects of protected areas and their differences from scenic spots. Results of the three models show that all explanatory variables have significantly positive influences. In terms of economic level, provincial tourist arrivals are positively affected by this economic factor, and economic development could improve the level of tourism product supply. Likewise, the level of city construction and the perfection of infrastructural facilities are also the methods to upgrade the tourism social environment and even transform a modernised city to be an excellent tourism destination.

As for the effects of protected areas, model (1) takes protected areas as provincial tourism

attractions individually. The coefficient associated with CPAS is positive and significant at the 1% statistical level, suggesting that the number of protected areas is actually one of the major driving forces of provincial tourist arrivals in China. That is, adding one protected area would on average increase the number of yearly provincial tourist arrivals by 2841.06 thousand after controlling other variables. In model (2), the influence of CPAS exhibits a moderate fall when the number of 5A—and 4A—class scenic spots (SCE) is included. The coefficient of variable SCE is positive and statistically significant, while the effect of CPAS remains significantly positive. In addition, the coefficient of CPAS is much larger than that of SCE, suggesting that protected areas is more attractive than scenic spots.

If a traveller plans to visit on an occasion, she (he) often formulates a linear travel route, implying that the combination of CPAS and SCE on the travel route might be considered together as the traveller makes her (his) travel plans. By adding the sum total of the CPAS and SCE as a new variable SITE, the estimate of model (3) shows that SITE has the expected positive effect on provincial tourist arrivals. That is, both adding one protected area and scenic spot would increase the amount of purchasing tourism products.

3.2 Different effects associated with protected area's types

The estimated results for equation (2) are shown in Table 2. In considering only protected areas, the estimate of model (4) shows that the coefficients of NNR, NSA and NGP are all positive and significant at the 5% statistical level. One point worth emphasizing is that the three coefficients are different from each other. Increasing the number of NNR by one would create an additional 1477.41 thousand tourist arrivals, and adding one NSA would create 2641.77 visitors, while one more NGP would increase provincial domestic tourist arrivals by 553.40. Model (5) estimates the effect of protected areas and scenic spots. We could observe that the coefficients of NNR, NSA, NGP and SCE are obviously different although they are all positive and significant at the 10% statistical level. Specifically, the coefficient of NGP is the largest, followed by NSA, SCE, while the effect of NGP is the weakest.

To sum up, national geological parks seem to be most favoured by domestic tourists and this coincides with the fact that geoparks have better aesthetics due to their geological spelndour. The aesthetics of nature reserves is relatively worse as a result of their less attractive biological landscapes. Nature reserves are suitable for a ecotourism

432

Table 1. Estimated result of domestic tourists arrivals (general influences).

Model	CPAS	SCE	SITE	C	SERV	CITY	INFRA	CPGDP	R-square
(1)	284.106***			−8909.068***	43.638***	4.639***	6093.927***	491.255**	0.856
(2)	215.322***	94.651***		−7902.939***	34.989***	4.409***	4189.975***	457.897***	0.874
(3)			139.067***	−6443.314***	32.096***	4.626***	3029.672***	361.328**	0.872

Table 2. Estimated results of domestic tourists arrivals (associated with protected area's types).

Model	NNR	NSA	NGP	SCE	C	SERV	CITY	INFRA	CPGDP	R-square
(4)	147.741**	264.177***	553.340***		−8182.266***	41.355***	4.760***	4856.203***	610.142**	0.863
(5)	82.732*	276.699***	338.056***	94.952***	−6959.876***	28.355***	4.770***	2824.972***	608.025*	0.878

experience, but it seems that domestic tourists are not interested at present. Due to several kinds of landscapes such as geological and cultural landscapes, the effect of the national scenic areas falls in between *NNR* and *NGP*. In view of this, the different effect of tourism attractive factors in terms of provinces can be identified as: *NGP > NSA > SCE > NNR*.

3.3 Comparison of regions in China

It is generally accepted that there are considerable differences as to the level of tourism development among the three regions (Eastern, Central and Western) in China. The development level of Eastern China is the highest while Western China is the lowest. On considering this regional difference, this study attempts to explore different influences of protected areas on provincial tourist arrivals in terms of the three regions. As for the regional division, *China Statistical Yearbook* provides a general reference. That is, both Eastern and Central China comprise 9 provinces respectively, and Western China comprise 10 provinces. The numbers of protected areas as to the three regions in 2011 are shown in Figure 2.

The estimated results are shown in Table 3. Comparatively, the statistical significance declines due to the less observations in each model. As for the coefficients, protected areas have positive effects on tourist arrivals in general. Model (6), (8) and (10) show that the coefficients as to Eastern, Central and Western China are 640.992, 132.862, 235.156, respectively. In considering the effect of *SCE* in model (7), (9) and (11), the coefficients of *CPAS* in each region exhibits a moderate fall, but the effect remains significantly positive. Comparatively, the effect of protected areas in Eastern China is the largest, while that in Central China is the least.

Although the number of protected areas in Eastern China is minimum, the effect on tourist arrivals is a maximum due to the impact of other factors. For instance, Eastern China has a developed economy, and the level of residents' income is relatively high so that they have strong

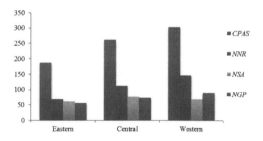

Figure 2. Numbers of protected areas as to three regions in China.

Table 3. Estimated results of domestic tourists arrivals (different effects in three regions in China).

Region	Model	CPAS	SCE	C	SERV	CITY	INFRA	CPGDP	R-square
Eastern	(6)	640.992***		−9685.068***	98.562***	8.313*	6093.927	390.265*	0.761
	(7)	541.507***	70.710*	−7501.867**	73.150***	7.204*	5726.002	827.155	0.715
Central	(8)	132.862*		−10502.159**	53.186*	15.146***	5807.690	504.944**	0.822
	(9)	116.815*	40.86	−8915.246**	47.287*	21.409***	4189.975	547.897**	0.804
Western	(10)	235.156**		−8302.497**	20.341	7.523**	10956.396***	45.293	0.750
	(11)	194.131**	20.91	−7682.044**	−13.197	4.626**	11648.892***	61.328	0.772

ability to travel [10]. Due to the high level of city modernisation, city tourism becomes an important kind of tourism product in the east and could form a product advantage combined with protected areas. As for the Central and Western regions, city tourism development is low, and the level of tourism service is deeply affected. In addition, Western China is sparcely populated, and human activities rarely have a negative impact on the protection of its protected areas. Thus, the quality of landscape and eco-environment here is excellent and widely recognizedin the field of domestic tourism.

4 CONCLUSIONS

With beautiful landscapes and an excellent eco-environment, protected areas are always important resources for tourism in China—some of them have even become world-famous destinations. However, the primary aim of constructing protected areas is to protect natural landscapes and eco-environment. Due to rapid development of tourism industry, the protection of resources and environment is under challenge, and a binary contradiction of tourism development and protection becomes prominent gradually. For a certain protected area, whether there is a massive influx of tourists and whether tourism would have a marked negative impact on protected areas or not all depends on the attractiveness of the protected areas to visitors. In other words, the effects of protected areas on tourist arrivals determine the negative impacts on protected areas.

This study confirms that provincial domestic tourist arrivals are significantly affected by the number of protected areas, and the coefficient is greater than that of scenic spots. The results of M-estimation shows that adding one protected area would increase the number of yearly domestic tourists by 2153.22–2841.06 thousand. Due to the differences of landscapes and their attractiveness, the effects on tourist arrivals differ as to the types of protected areas. National Geoparks have the highest effect on tourism, followed by National Scenic Areas and National Natural Reserves. At

present, visiting mountains and great rivers is still one of the main tourism preferences, and a biological landscape is not attracting a large number of tourists presently. But ecotourism is gradually flourishing in China, and the differences of effects caused by landscape will be changed with the augmentation of the ecological consciousness and the desire for eco-knowledge of the tourists.

Also, the regional differences of protected areas are evident in this study. That is, protected areas in each region have different tourism bearing pressures. Eastern China is under the greatest pressure while Central and Western regions have less pressure. Facing the pressure, on the one hand, we need to pay attention to the management of protected areas so as to ease the pressure on the protection of resources and environment and keep the balance of nature. On the other hand, we should divert tourist flows from the protected areas to other tourism spots, and strengthen effective tourism management in protected areas.

5 SUMMARY

About this paper, we used Microsoft word to edit it. The phone for communication is +86 18500237920, and E-mail address is:
zhangaiping89757@126.com.

ACKNOWLEDGEMENT

This study was supported by the National Natural Science Foundation of China (grant no. 41171435). Special thanks go to Mr. Zhao and Ms. Zhou, who helped to collect related data.

CORRESPONDING AUTHOR

Linsheng Zhong. Institute of Geographic Sciences and Natural Resources, Chinese Academy of Sciences (CAS), 11 Datun Road, Chaoyang District, Beijing 100101, China. Tel.: +86 13651112579. E-mail address: zhongls@igsnrr.ac.cn.

REFERENCES

[1] G.R. Zhang: A rational thinking on China's tourism development. *China Soft Science Magazine*. No. 2 (2011), p.16–33. (In Chinese).

[2] W.L. Gui, Z. Z Han: On the assessment of the impact of crisis events on the foreign exchange earnings of China's inbound tourism. *Tourism Tribune*. Vol. 25, No.2 (2010), p. 28–35. (In Chinese).

[3] L.S. Zhong, S.D. Zhao, B.H. Xiang: *Principle and method of ecotourism planning*. (2003). Beijing: Chemistry Industry Press. (In Chinese).

[4] Y.W. Su, H.L. Lin: Analysis of international tourist arrivals worldwide: The role of world heritage sites. *Tourism Management*. Vol.40, No.1 (2014), p. 46–58.

[5] C.H. Yang, H.L. Lin, C. Han: Analysis of international tourist arrivals in China: The role of World Heritage Sites. *Tourism Management*, Vol.31, No.6 (2010), p. 827–837.

[6] B. Baltagi: *Econometric Analysis of Panel Data*. (2005). USA: Wiley published.

[7] R.G. Wu: Robustness of the constrained LSE in terms of error distribution. *Journal of Xi'an University of Arts & Science*. Vol.11, No.2 (2008), p. 59–62. (In Chinese).

[8] R.Q. Zhang: Robust m-estimates for the partial linear models. *Acta Mathematics Application Sinca*. Vol.28, No.1 (2005), p. 151–157. (In Chinese).

[9] Y.L. Fang: Spatial and temporal analysis of Chinese provincial tourism economy. *Economic Geography*. Vol.32, No.8 (2012), p. 149–154. (In Chinese).

[10] R.J. Ao, Y.S. Wei: A study on the regional tourism resources and the unbalanced development of the tourism industry in China. *Journal of Finance and Economics*. Vol.32, No.3 (2006), p. 32–43. (In Chinese).

Advances in Future Manufacturing Engineering – Yang (Ed.)
© 2015 Taylor & Francis Group, London, ISBN 978-1-138-02817-3

The Data Acquisition System of the large-scale sliding bearing test rig design and research

Daqian Ren, Weizhong Zhang, Huanya Cao & Yao Zhang
Zhejiang Institute of Mechanical and Electrical Engineering, Hangzhou, China
The Sliding Bearing Technology Research Center of Zhejiang Province, Hangzhou, China

ABSTRACT: In order to test the static and dynamic properties of various types of sliding bearings, a large-scale sliding bearings test rig is developed. The mechanical system includes motor, transmission, shaft, support bearings and test bearings and other components. Data Acquisition System (DAQ) is consisted of sensors, pretreatment circuit, data acquisition cards, industrial PC and so on. Vibration, displacement, temperature and pressure signals are sampled by the data acquisition system. DAQ also can process the state information effectively. The sampled data is displayed in curve or chart form. The test results show that the system can meet the need of sliding bearing testing.

Keywords: sliding bearing; test rig; Data Acquisition System

1 INTRODUCTION

Large sliding bearings are often used in large-scale valuable equipment, such as large generators, large ship engines and turbines. The sliding bearing operating state affects the safe operation of the equipment. Since sliding bearings often work under high speed and high load operating conditions, lack of lubricating or serious wear can lead to failure. The critical damage of the machinery and equipment may cause severe safety accident. So the study is needed to determine the various factors affecting the performance of sliding bearings. The result of study will provide a theoretical basis for the optimum design of the sliding bearings and improve the reliability of bearing [1,2,3]. For the sake of experimental study, sliding bearing test rig is necessary.

The actual working conditions can be simulated on the test rig, such as speed, axial load, radial load, ambient temperature, lubrication. According to the result of test, the performance of bearing oil film can be effectively judged, and the design of sliding bearing can be improved.

2 THE PRINCIPAL OF BEARING OIL FILM

In the sliding bearing, there is a wedge gap between the shaft neck and bearing, filled with lubricant. When the shaft does not rotate, the shaft neck is submerged at the bottom of bearing hole. When the shaft starts rotating, due to the frictional force, the shaft neck will climb up along the opposite rotational direction until the frictional force cannot support. This is called semi-liquid friction. Since speed continues to rise, the shaft neck brings the lubricating oil into the wedge-shaped gap between the shaft neck and bearing. The accumulated oil on the gap is just the oil film, which left the shaft neck. When the pressure of oil film is balanced with the external load, the shaft neck will rotate stably without touching the bearing. Because of the oil film in the wedge gap, the axis of the shaft will slightly offset to one side. The oil pressure can be calculated by Eq.1, and The contribution of oil pressure and the position of shaft neck are shown on Figure 1.

$$P = S_0 \frac{\mu \omega l d}{\psi^2} \qquad (1)$$

where P—bearing load;
S_0—bearing load capacity coefficient and is also called Sommerfeld number;
μ—dynamic lubricant coefficient;
l—bearing width
d—shaft neck diameter;
ω—shaft neck rotational angular velocity;
ψ—relative gap
C—average gap, $C = R - r$
θ—deviation angle, e—eccentricity, h_{min}—minimum film thickness, $h_{min} = C - e = C(1 - \varepsilon)$.

The bearing load capacity coefficient S_0 is an important parameter to determine the working state of the sliding bearing. If the geometry similar bearings have same coefficients S_0, the bearings will have similar properties. When $S_0 > 1$, the rotor is called low-speed heavy-duty rotor. When $S_0 < 1$, the rotor is called high-speed light load one. S0 is the function of relative eccentricity $\varepsilon (\varepsilon = e/C)$ and the bearing width and diameter ratio l/d. The greater of the eccentricity or the greater of bearing width diameter ratio is, the greater of value S0 and the higher load capacity is[4].

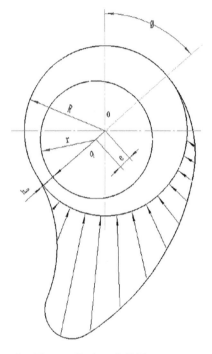

Figure 1. The contribution of oil film pressure.

3 THE STRUCTURE OF BEARING OIL FILM TEST RIG

The mechanical structure of the sliding bearing test rig is shown in Figure 2, which include motors, pulleys, unloading bearing set, shaft, coupling, front/rear bearing, front/rear bearing block, loading sets, stimulating devices, reducing vibration set, etc.. Mechanical transmission system is composed by inverter motor and coupling. The inverter controls the motor running, and drives shaft rotation through coupling. The tested bearing is mounted on the middle of the shaft. The load is applied directly to the test bearing. The support bearings are arranged at both ends of the shaft. This layout is suitable for the various static and dynamic test of sliding bearing. It is easy to align, install and eccentricity measure.

The control system of the sliding bearing test rig is consisted of industrial PC, PLC (Programmable Logic Controller) and hydraulic system. The working status of the test rig is displayed on PC. The control data can also be input and output through the program interface. The PC Sends control signal to the PLC, and control low-pressure pump, high pressure pumps, hydraulic loading cylinders, servo motors rapidly and precisely.

4 THE DATA ACQUISITION SYSTEM OF BEARING TEST RIG

In order to test the performance under various working condition, a complete Data Acquisition System (DAS) is needed. Since vibration signal are carrying rich information, the vibration signal DAS is often the most important part, as shown in Figure 3. Compared with temperature and pressure signal, vibration signal is the critical issue of bearing test rig designing.

For the convenient of process and display, the DAS is built based on an industrial PC. A data

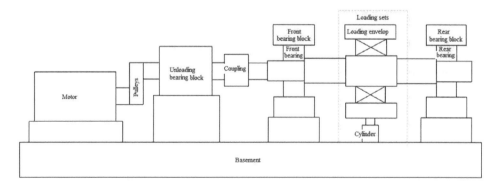

Figure 2. The mechanical structure of sliding bearing test rig.

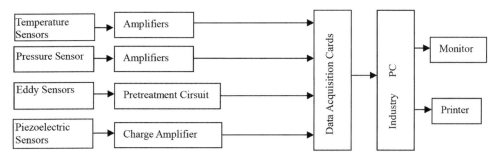

Figure 3.   The frame of data acquisition system.

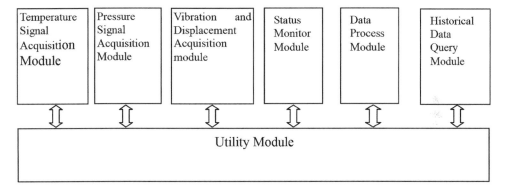

Figure 4.   Data acquisition system software architecture diagram.

acquisition card is inserted into a spare bus slot of the PC. The card will convert the analog signal into digital signal. The signal from temperature sensors, pressure sensors, eddy current sensors and piezo-electric transducer are all converted on the card. The digital signal is further processed and stored on the hardware and software of the PC. The eddy current sensor is used to measure the thickness of the oil film, and the piezoelectric sensor is applied on bearing vibration measurements.

The software on the industrial PC is assembled by utility module, temperature signal acquisition module, pressure signal acquisition module, vibration and displacement signal acquisition module, real-time status monitoring module, data processing module and historical data query module. The temperature signal acquisition module, pressure signal acquisition module, vibration and displacement signal acquisition module are closely related to hardware, including hardware drivers, as well as the unit for preliminary signal processing. The real-time status monitoring module will display the acquired signal. Data processing module is a collection of several signal processing algorithms, including spectral analysis, waterfall plot, and time-frequency analysis methods. By the historical

data query module the stored data can be inquired and display on the computer screen.

Eddy current sensor vibration signal acquisition needs pretreated and then input to the data acquisition card for A/D conversion. Preprocessing circuit AC and DC signals separately, the DC signal attenuation, AC signal amplification and noise filtering.

The vibration signal from the eddy current sensors should be pretreated and then be A/D converted on the data acquisition card. The AC component and DC component will be separated by the pretreating circuit. The AC component is amplified and the DC component is attenuated. The noise is filtered.

5   STATUS INFORMATION EXTRACTION

After signal acquisition, the data can be real-time processed. Some traditional method such as spectral analysis, shaft axis track analysis, and waterfall plot are available. There are also some modern signal process methods, such as Hilbert-Huang transform, wavelet analysis. The process methods can help engineer to analysis the test result [5, 6].

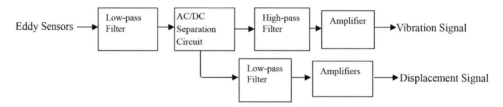

Figure 5. The frame of vibration pretreatment circuit.

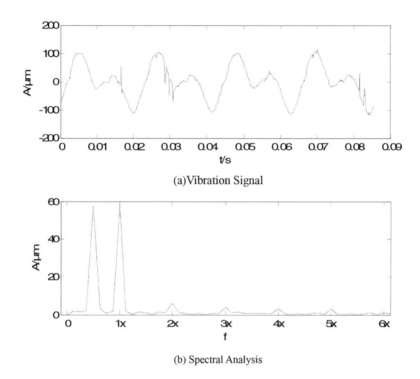

(a)Vibration Signal

(b) Spectral Analysis

Figure 6. Vibration signal spectral analysis.

As shown in Figure 6, 4 periods of vibration signal and some noise can be seen. After the spectral analysis, we can clearly see a harmonic frequency component and a half harmonic frequency component. The half harmonic frequency component prompted the oil whirl phenomenon.

Figure 7 shows the shaft axis track. We can see the movement track of shaft axis in the bearing hole. In order to obtain the axis track, two sensors are needed. The sensors are install on the X direction and Y direction of a cross-section. Vibration signals are collected by these 2 sensors, and display on the same graph. AS seen in Figure 7, the axis has a significant precession phenomenon. The graph can be a powerful tool to study the dynamic characteristics of the bearing oil film.

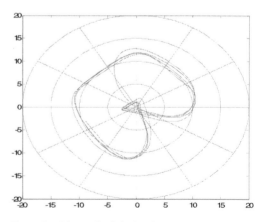

Figure 7. The track of shaft axis.

6 CONCLUSIONS

The design scheme of sliding bearing test rig is discussed in this paper. The vibration signal acquisition method is also studied. The actual test results show that the test system can meet the data collection requirements. Sliding bearing test rig provides a powerful tool for bearing design.

ACKNOWLEDGMENTS

The authors acknowledge the supports from the Specialized Research Fund for Post-doctor of Zhejiang Province, China under Grant No. BSH1301020. The supports of the Sliding Bearing Engineering Technology Research Center Fund of Zhejiang Province, China under Grant No. 2012E10028.

REFERENCES

[1] Huang Lei, Teng Qin, Sun Jun. Design of Measurement and Control System for Multi-functional Journal Bearing Test Bed [J]. Machine Design and Research, 2013, 29(3):92–96.

[2] Zhang wenming. Development of a measurement and control system of the sliding bearing test bench for the fuel pump[D]. Hefei University of Technology, 2013, 04.

[3] Hu Zhaoyang, Chang Shan. The structural design and the application of the large-scale radial plain bearing and thrust plain bearing platform. Journal of Qiqihar University, 2005, 21(2):79–82.

[4] Song Qiang, Sun Fengchun, Ma Jinkui. Research Situation of the Identification of the Dynamic Characteristic of Bearings and Oil Stability Analysis[J]. Lubrication Engineering, 2002(4):40–43.

[5] Ruqiang Yan, Robert X. Gao, Hilbert-Huang Transform-Based Vibration Signal Analysis for Machine Health Monitoring[J]. IEEE Transactions on Instrumentation and Measurement, 2006, 55(6):2320–2329.

[6] Zhong Haiquan. The Large-scale Low Speed Heavy Loading Radial Dynamic Bearing Research[J]. Journal of Southwest Petroleum Institute, 2003, 25(6):87–89.

Advances in Future Manufacturing Engineering – Yang (Ed.)

Research on the secure of embedded system database

Jun Chen

Experimental Center of Jiujiang University, Jiujiang, Jiangxi, China

ABSTRACT: Aiming at the security problems of the open source embedded database SQLite, the paper designs a set of feasible security mechanisms, based on the analysis on the commonly used security mechanism of database. The password authentication is combined with data encryption, by increasing the key creation, data encryption and decryption, page code and recording events and code—to achieve a more secure SQLite database. The test results show that: the system inherits the embedded database features at the same time, and the safety is improved.

Keywords: embedded system; database; security; encryption; SQLite

1 INTRODUCTION

With the rapid development of computer technology, the embedded system has become an important aspect of computer science—and is more and more widely used. The embedded database system is a database management system supporting mobile PC, or a particular computing mode, database system and operating system; it is application—specific and integrated together and runs on a variety of intelligent embedded devices or a mobile device. In mobile database systems it generally includes two parts: one, the host and the operation of a fixed position of a general database management system on top of it, the other, a mobile device and a small embedded database management system. As the mobile device resources are constrained, the embedded database management system and application systems integrated together—and the existence of it as the front of the whole application system, a its management data sets may be a copy of a general database management system data set on a back-end server or a subset. The two parts through the middle layer, such as a Mobile Support Station (MSS) or base station, are connected together. The embedded database system on mobile devices, because it involves database technology, distributed computing technology, multidisciplinary and mobile communications technology, has become a very active field of research and application.

SQLite is an open source, and has fast query, memory space and a higher storage efficiency characteristic, can be applied widely in enterprises. The security problem attracts high attention. Based on this, the database security mechanisms are analysed, and so will the design and implementation of the security mechanism of SQLite database.

2 INTRODUCED OF SQLITE DATABASE STRUCTURES

SQLite is an open source embedded database engine written in the C language. It is completely independent, and does not have any external dependence. The occupation of resources is very low in the embedded devices; only a few hundred K of memory is enough. It can support the Windows/Linux and other mainstream operating systems; it also has the TCL, PHP, Java program language combination, in order to provide an ODBC interface. Its processing speed cannot even be matched by the world famous open source database management system—Mysql, PostgreSQL.

SQLite supports the SQL92 standards along-with indexes, limit, triggering and view, support for atomic, consistent, independent and lasting (ACm) affairs. Internally, SQLite is composed of the SQL compiler, kernel, rear end and attachment of several components, SQLite is as shown in Figure 1.

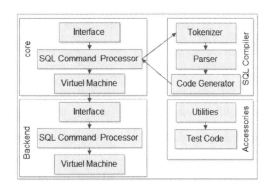

Figure 1. The internal structure of SQLite.

Through the use of virtual machines and Virtual Database Engine (VDB) it helps, make the debugging, modification and expansion of the kernel of SQLite become more convenient. All SQL statements are compiled into easy-to-read format and can be executed in the SQLite virtual machine assembly.

3 ANALYSIS ON THE SECURITY MECHANISM OF DATABASE

In order to make the embedded system meet the portability of the database itself, and the efficiency of data storage, and access speed, memory usage and other performance requirements, SQLite adopted different elements from the large database method of mechanism; at the same time it also brings in some potential safety hazard. The open source SQLite database does not provide an encryption mechanism of rigorous quality for the user, in the data whichcould have hidden safety dangers. Moreover, because the data storage mode of SQLite database is relatively simple, it currently does not have a specific storage tool. The user can use the text editor to view the data content. At the same time, the SQLite database does not provide a multi-user mechanism for everybody; the database on the lack of audit mechanism accordingly, backup and recovery of database in the data, usually just by hand to complete the copy and the safe use of the data is not safe.

4 THE DESIGN OF THE SECURITY MECHANISM OF SQLITE

4.1 *Password authentication*

SQLite database file is relatively simple; it is just plain text files. So access to the database is mainly dependent on the file access control. This paper adds the password authentication for the database file; only after the user provides the correct password, can he add, delete, modify or query the database file in the data operation. If the password is not correct, the user will not be able to perform any operation on the data file.

4.2 *Database encryption*

Database encryption has two kinds of database encryption modes:

1. The encryption functions will realise in the database management system; (2) to some fields in database applications value will be realized encryption; the database management system is already encrypted ciphertext. At the time of the

SQLite database development, designers have stayed well in advance for the related interface encryption on the database, to facilitate the use of the database staff further encryption operation on DBMS. It has realized the encryption setting to the SQLite database [1], and compared it with the previous version; its internal function and data structure have been further improved and enhanced.

4.3 *Audit mechanism*

This design uses a specific text fileand records simple important events of the system, such as opening and closing the database, designing and changing the password etc. Because the SQLite database system does not have this module design of user management, we need to therefore, develop a specific audit mechanism. On the basis of not changing the original export interface of database, the design uses a specific text file, to carry out a series of operations on the database system, opening and closing operation of the database, modifying and deleting the data operation.

4.4 *Backup and recovery mechanism*

According to the special way SQLite stores database content (single file to store the entire contents), we can use the file copy in the database backup and recovery.

5 THE REALISATION OF SQLITE ENCRYPTION MECHANISM

In this study, by modifying the SQLite source code in the DBMS class, to implement the log function password authentication, data encryption, simple and backup and restore function. The password authentication and data encryption are the combination of the two, for the reunification of the encryption mechanism—namely the database structure information and the stored data encryption. When accessing a database one must provide the password. If the password is not correct, then the database structure information cannot correctly parse, and therefore does not allow subsequent access.

5.1 *Encryption interfaces provided by SQLite*

In the open source SQLite code, have reserved the encryption interface, including 4 prototype of the function: in the sqlite3. ħ provides a sqlite3_key () and sqlite3_rekey () function prototype, the former is used to specify a database using the key, the latter is used for database reset key; in the file attach. C provide the interface function key associated with the database, namely the function sqlite3CodecGetKey () and sqlite3CodecAttach, the former is used to return the current key database, the latter is used to connect the key and page coding function and database. To realise SQLite database encryption function, there must be a specific implementation of key functions which are a few, but also provide the data encryption and decryption function, page code function and other necessary functions.

5.2 The selection of cryptographic algorithm

SQLite database saves data with page as a unit; the page size can be customized. The existing encryption algorithms are divided into block encryption algorithm and sequence encryption algorithm; packet encryption algorithm needs to be explicitly divided into fixed length blocks, and encrypted in block units; sequence encryption algorithm created to generate a key sequence, one by one byte (or one by one bit of plaintext encrypted). For SQLite, the packet encryption algorithm is also capable of encryption algorithm, based on sequence. But if the use of the packet encryption algorithm, and if the page size is not integer times the size of the block is encrypted, you need to fill data. So the encrypted page length will be greater than the original length of the page; all the data needs to adjust to the original database. If using the sequence encryption algorithm, it can avoid the problem of adjustment of the page.

5.3 Key

The user selected password length is generally not fixed—but also from the security point of view, if the direct use of the password input is as an encryption key, it may reveal the database in the application of key. Therefore, this paper provides a key derivation function, in order to transform the input password—then the result will transform as the database encryption key.

6 ENCRYPTION AND IMPLEMENTATION

A study on the source code based version of SQLite3. 6.15 in the paper, using the standard C to implement the SQLite security mechanism—through text file record, database open, close and encryption operations, to achieve the function of auditing; increase the export function, through the file copy of the database file backup and recovery and to use RC4 sequence to achieve the encryption mechanism of SQLite encryption algorithm. Code structure as shown in Figure 1 encryption mecha-

nism, wherein the arrow indicates the call relationship, function of sqlite3_key and sqlite3_rekey is derived from API function the final. The revised code on the platform of windows VC6.0 environment and the Linux platform under GCC environment are compiled, and dynamic link library for generating and static link library tested.

6.1 The test of encryption function

The author tested the modified SQLite database system. Test content includes: 1) the original application can run correctly in the new library; 2) database encryption functions correctly; 3) log function, backup and restore functions correctly. The test process is as follows. Write a SQLite3 application. The program code function is to open a database in test2. dB, create a database table (table), and insert the two records. The program code is as follows.

```
main(){
    sqlite3*db;
    char*zErrMsg = 0;
    sqlite3_open('test2 db',&db);
    if(db = = NULL)
    return -1;
    sqlite3_exec(db,"create    table    user(name
varchar(64) UNIQUE, age text);", 0, 0,&zErrMsg);
    sqlite3_exec(db,"insert into user values('Wang
H',''20');", 0, 0,&zErrMsg);
    sqlite3_exec(db,"insert into user values('Zhang
Y',''18');", 0, 0,&zErrMsg);
    sqlite3_free(zErrMsg);
    sqlite3_close(db);
}
```

To modify the above procedures, add a line of code after code SQLite3 _open ("test2.db", &db):
Sqlite3_key (DB ", 123456", 6);

It said the DB database encryption, password is 1234560, and the password length is 6.

Use Notepad to open the encrypted database file to see content. Database file content has become garbled, and achieved data confidentiality purposes.

Figure 2. Code structure of SQLite encryption mechanism.

6.2   *The test of decryption function*

Write a SQLite3 application; the function of the program is to read user-content after encrypted database test2. DB table, program code is as follows.

```
Int SelectTable (void*para, intn_column, char
**column_value, char** column_name);
voidmain(){
    sqlite3* db;
    char* zErrMsg = 0;
    sqlite3_open("test2. Db",&db);
    if(db == NULL)
    return;
    sqlite3_key(db,"123456", 6);
    sqlite3_exec( db,"select* from user", SelectTa-
ble, NULL, &zErrMsg);
    printf("errmsg is:% s", zErrMsg);
    sqlite3_free(zErrMsg);
    sqlite3_close(db);
}
    intSelectTable( void*para, intn_column, char**
column_value, char** column_name )
{
    printf("cloname=%s, colvalue=%s, cloname=%s,
colvalue = %s\n", *column_name, *column_value,
*(column_name+1), * (column_value+1) );
    printf("\n");
    return 0;
}
```

In which: Code sqlite3_key (DB, "123456", 6) specify the database encryption password, function. SelectTable is a callback function.

The program runs visibly: when provided the correct password, we can be obtained correctly before encryption of data.

7   CONCLUSIONS

The database is a centralised structure to store the data in the information system;it is the key to security. The paper designs the security mechanism for SQLite, based on the analysis of the security mechanism which is adopted on the current database. This paper explains in detail how to realise the process of encryption mechanism to modify the source of Sqlite code. The test results show that: the realisation of the SQLite system inherits the original system advantages, and security is improved at the same time.

REFERENCES

[1] Wu Taofeng, Zhang Yuqing. Database security review [J]. Computer engineering, 2006,32 (12): 85 –88.
[2] Zang Jinsong. Research and analysis of the security of database system [J]. Computer security, 2008 (7): 26 –30.
[3] Liao Shun He. Analysis and research on the encryption method [J]. Computer applications and software, 2008, 25 (11): 256–258.
[4] http://www.sqlite.org/SQLite. The official website, including a variety of information and source code.
[5] http://www.sqlite.com.cn/SQLite. Chinese website, including various Chinese data and application document.

Advances in Future Manufacturing Engineering – Yang (Ed.)
© 2015 Taylor & Francis Group, London, ISBN 978-1-138-02817-3

Study on encryption of embedded database system

Jun Chen
Experimental Center of Jiujiang University, Jiujiang, Jiangxi, China

ABSTRACT: Embedded systems have become a hot topic in today's IT industry. Security in embedded database draws more and more attention. This paper designs a set of relatively feasible security mechanisms. It maintains the basic characteristics of embedded database, through the use of encryption algorithm. The paper solves the hidden safety problems of database storage file, which will greatly improve the use safety.

Keywords: embedded system; embedded database system; SQLite; reduction; transplantation; encryption

1 INTRODUCTION

With the rapid development of computer technology, the embedded system has become an important aspect of computer science, and has been used more and more widely. As the shadow follows the form, embedded database shows more and more advantages, from the research stage to the wide application field. In many embedded operating systems, the Windows CE, and Linux operating systems are the most eye-catching. The development of database application based on the WINDOWS CE application platform is an important part of the development of palmtop computer applications. While SQLite is a data management of small, fast and minimization in the embedded database field which developers love, and now have a steady spiraling momentum.

After the embedded hardware design is finished, all kinds of functions depend on the software to be achieved. Value-addition greatly depends on the embedded equipment embedded software. On the same hardware platform, it can come to the platform according to the actual application of transplantation of embedded operating system and the corresponding software—and transplant the corresponding database, real-time storage and management of the data. This paper studies some of the embedded systems and embedded databases. On the basis of this, the Liod development board are cut and transplanted Linux and Windows CE—two operating systems and corresponding softwares based on graphical interface and database, and designed on the corresponding database management system (including: being based on the Liod development board, Linux operating system, Qt development platform, and SQLite embedded database management system design and implementation; based on the Liod development board and embedded Windows CE operating system, in the VS2005.NET EVC environment using ADOCE control and Structured Query Language ((SQL) design and implementation of the database management of the test system). In this paper, the application development, based on the Liod development board, has practical value; at the same time, for the development of other embedded applications, there are many worthy learning places from the aspect of operating system and database selection.

With the popularization of intelligent devices, embedded database security problems in the system are even more important. Through the use of encryption algorithm, we solve the hidden safety problems of the database storage file.

2 THE CHARACTERISTICS OF THE EMBEDDED DATABASE MANAGEMENT SYSTEM

The embedded database system is a database management system that supports mobile computing or a particular computing mode. Application of the database system and operating system is integrated together and runs on a variety of intelligent embedded devices or a mobile device. Because of the special limitation of application environment, the characteristics of the embedded system can be seen as an embedded database management system, a middle ware in the embedded system; it also must be restricted by various aspects of embedded system and the application of speed, resources and other factors. Of course, the essence of the embedded database management system is developed from the general database management system. It can also be hierarchical and the mesh or relational database even can be object-oriented. But we should be clear, that the embedded database management system and the general database management system in the operating environment, application field from many aspects is not the same, therefore, it cannot be simply an is the embedded database management system as the microfilm edition on the embedded device universal database management system. When compared to the conventional database management system, the embedded database management system has its own characteristics:

1. The data synchronization
   Since the mobile device resource has some constraints, the embedded database management system in general integrates together with the application system, and exists as the front-end of the whole application system,; its management data sets may be a copy of the data set in the back-end server in a subset. Embedded database replication model generally adopted some data (upload, download or mixed mode), mapping and database server, to meet the needs of people in any place, with any time access to any data demand. Because of the existence of data replication, in the system in each application front-end and back-end servers may need to synchronize all the necessary control process, and even some or all of the front-end application, which also need to synchronize data.
2. To simplify the backup, restoration
   Embedded database backup and recovery are different from a large DBMS management database, not simply just for independent service or a similar form to completed, according to some simplified way. The embedded database system transaction processing can be simple in the front-end—but in the entire application system, the characteristics of the mobile computing environment may need to combine the transaction control.
3. High security, high reliability, zero management
   The control access of database system is strictly on the embedded system. Many embedded devices have high mobility, portability and a non-fixed working environment. These also have brought the potential insecurity to the embedded mobile database. Another is that the personal privacy of certain data is very high, so in the collision prevention, magnetic interference, loss, theft of personal data security threats, there is the need to provide adequate security assurance. Because the embedded database management system is running on embedded devices, its operation basically does not need the participation of people. This is for the reliability and safety of the system put forward for higher requirements; to do so the system must be able to achieve zero management.
4. Independent platform
   In the field of the embedded system, the hardware platform and the embedded real-time operating system are of many kinds, and each different embedded application is running on hard, specific software platforms. The design of embedded database management system must make the operating system and the underlying support software interface as simple as possible, so that it can help the transplant of the system, and achieve an independent platform.
5. Scarce resources and the hardware is slow
   The running environment of embedded software is relatively poor—slow CPU and bus speed. The small capacity the RAM and ROM have a great influence on the design of the embedded database management system. To ensure that with the limited resources the system can operate normally, it is necessary to use the resources of the system in a seriously control manner; to try to improve the speed of the system, we must focus on more time-consuming operations by using the algorithm of elaborate design, so as to eliminate the bottleneck in the system performance.

3 KEY TECHNOLOGY OF DATABASE ENCRYPTION

3.1 *The selection of encryption standard*

Data encryption algorithm is generally a software mode for users. A variety of encryption algorithms come with differently defined interfaces. This paper uses encryption standard CryptoAPI V2.0.

3.2 *CryptoAPI*

As an application programming interface, CryptoAPI provides the function of information in data integrity and confidentiality, which are very helpful. CryptoAPI itself can not realise the encryption and decryption, or complete the function CSP. CSP exists independently of the underlying operating system; it internally contains the encryption and decryption function algorithm and the standard.

3.3 *The encryption process*

Encryption of data file, the process is as follows:
Data (plaintext) - >Access to CSP-> Build a hash table - >Generate a key - > Delete hash table > Encrypted data - > Generated encryption;
Data (ciphertext) - > Duild a hash table - > Derived key - >Delete hash table > Decrypt the data - > Data (plaintext).

4 REALIZATION OF DATA ENCRYPTION

The operation process firstly encrypts the data, then the encrypted data is written to the file; reading the data flow is firstly set on the data decryption, and then returns to the upper layer, in order to achieve data confidentiality.

4.1 Encryption algorithm

Figure 1 is the basic flow of encryption and decryption. The author achieves the encryption and decryption functions on the MinisQlite, and uses the PROV_RSA_FULL type and RSA, SHA algorithm. RSA is the realization of encryption and signature algorithm and its security relies on factorization; RC4 is a change in the length of key, byte oriented operation; the SHA algorithm is the Hash algorithm, to build a digital signature.

4.2 The realization of basic function

The following is the process of the realization of MiniSQlite encryption and decryption functions:

1. Data encryption
   When writing data to memory code encryption:
   CopyMemory (((LPBYTE) pBlock ->pvCryPt) +CRYPT_OFFSET, data, pBlock ->dwPageSize);
   Data = ((LPBYTE) pBlock->pvCrypt) + CRYPT_OFFSET;
   The encrypted data code:
   DwPageSize = = pBlock->dwPageSize;
   CryptEncrypt (pBlock->hReadKey, 0, TRUE, 0, ((LPBYTE) pBlock->pvCrypt) + CRYPTOFFSE T _, & DWPageSIZE, PBlock->dwCryptSize);
2. Data decryption
   Decryption of ciphertext data will be read into memory code:
   CopyMemory (((LPBYTE) PBlock ->pVCRYPT) +CRYPT`OFFSET, data, pBlock ->dwPageSize);
   PvTemp = data;
   Decrypt the data:
   DwPageSize = pBlock->dwCryptSize;

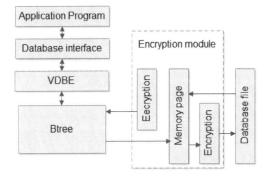

Figure 1.   The basic flow of miniSQlite encryption and decryption.

Figure 2.   Data before encryption.

```
00000420h: 02 E2 02 CC 02 BF 02 B0 02 A2 02 93 02 7C 02 69 ; .??????|.i
00000430h: 02 56 02 49 02 2F 02 18 02 C5 01 F3 01 E3 01 D6 ; .V.I./.....???
00000440h: 01 C4 01 B0 01 A0 01 83 01 70 01 48 01 35 01 28 ; .????p.H.5.+
00000450h: 01 15 00 FA 00 E2 00 CA 00 B4 00 9D 00 87 00 75 ; ...?????u
00000460h: 00 00 00 00 00 00 00 00 00 00 00 00 00 00 00 00 ; ...............
```

Figure 3.   Data behind decryption.

CryptDecrypt (pBlock->hReadKey, 0, TRUE, 0, (LPBYTE) data, & dwPageSize);

5 THE EXPERIMENTAL RESULTS

The results of data encryption and decryption is as shown in Figure 2 and Figure 3.

6 CONCLUSIONS

In the information system, the storage and use of all the data to be completed in the database are conducted through the database. The design of database safety is a critical security mechanism. The paper analyzes the security mechanism of the current database, and on the existing foundation, the embedded database is improved, which improves its safety as much as possible. Through the test, this paper completes the encryption, decryption of the database file; it ensures data security, but also enhances the security of the database.

REFERENCES

[1] Song Ling, Li Tao Shen, Chen Tuo. To ensure the safety of data communication with CryptoAPI in VC++ [J]. Computer applications and software, 2005 (07). Pp.29–31.
[2] Feng Wenfei, Wang Jiangquan. Study on the technology of repeat system based on Embedded Database [J]. Technology and application of digital, 2011 (08), p.143–144.
[3] Liu Haiyan, Yang Jiankang, Cai Hongliu. Analysis and design of the security mechanism of embedded database SQLite [J]. Academy of Armored Forces Engineering, 2009 (05), pp.68–71.
[4] Lu Xuan. Research and implementation of embedded database system SQLite [D]. Xi'an University of Electronic Science and Technology, 2009, pp.1–63.
[5] Tang Min, Song Jie. The principle and application of embedded database SQLite [J]. Computer knowledge and technology, 2008 (02): 18–21.

Multimedia applications

Advances in Future Manufacturing Engineering – Yang (Ed.)
© 2015 Taylor & Francis Group, London, ISBN 978-1-138-02817-3

On the strategies of foreign language teaching assisted by the network classes

Tianhao Wang & Xiyan Wang
Foreign Language Institute, Northeast Dianli University, Jilin, China

ABSTRACT: With the development of computer science, traditional teaching strategies are facing more and more challenges. New teaching strategies, assisted by network classes, are the first imperative. How to combine the traditional classroom teaching and the advanced network application, in order to improve the language ability of college students, is an important task for us to discuss here.

Keywords: teaching strategies; network classes; foreign language teaching

1 INTRODUCTION

With the ascendant trend in the number of college studentsr year by year, foreign language teaching is facing more and more pressure from society. On the one hand, the new teaching materials require the students to have a much higher standard of foreign language than ever before. On the other hand, the educational resources are so limited that they cannot meet the needs of the high-speed development of the subjects. There are many technical matters involved when teachers use the network teaching system. Interactive communication can't always satisfy both students and teachers. Some of these problems could be solved when we conduct more study on the network teaching systems and apply it in the classes. Through the cooperation of the teachers, the students and the network system, we will enrich the teaching materials and share the experience together; what's more, the comprehensive language application ability of the students will also improve.

2 CHARACTERISTICS OF THE NETWORK CLASSES

The network class is a new educational device, which is supported by computer science and the Internet. It can construct a student-centered curriculum environment. The advantages of the network classes are as followed:

2.1 Far distance between the teachers and the students

The activity of teaching and learning happens in different places and at different time. The medium is the network between the teachers and the students.

This characteristic is basic, because it is the answer to the questions on the differences between the network classes and traditional classes. Without the limitation of the place, students could have more choices on when and where to study.

2.2 Large amount of information

With the high speed development of the network technology, the Internet could offer the educational materials far more than any teacher, any book, and even any library could. This could enlarge the students' view in order to form the biggest classroom which contains the education of school, family and society. Students could choose the time, the content and the speed g to suit themselves. And they can choose whenever and wherever they want to communicate with each other. In a word, the Internet creates a convenient, fast and effective learning environment.

2.3 The virtual teaching and learning environment

For the division of the time and the place of education, the network class is a virtual environment. This could give rise to some limitations of the curriculum provision. For example, some experimental and training courses can't be studied in this way. Therefore, only by teaching on network classes we can't cultivate the high-quality creative personnel. [1]

3 ANALYSIS OF TEACHING AND LEARNING IN THE NETWORK ENVIRONMENT

Compared with the traditional classroom environment, the multimedia network classes could be learner-centric because it is varied and open. This environment could offer the learners the texts, pictures, icons, audio, video, animation and other information. What's more, the integrated resources could arouse the curiosity of the students and enforce their initiative. They could select, learn, generate or create new information according to their own interest, experience and needs. When network classes are formed, students could do their own work with more initiative. They will have more effective learning experiences. But every coin has two sides. Because it is a self-regulated learning system, whether the final result are good or bad depends on the students and the learning materials they choose. According to Hammond, the self-control of those who are learning is necessary and possible, in the course of learning, which could be done by themselves or with the help of the teachers. [2] Therefore, during the course, teachers should play an important role in this self-adjustment study. The teachers should design the teaching task, learning environment and courses carefully and effectively.

4 STRATEGIES OF FOREIGN LANGUAGE TEACHING ASSISTED BY THE NETWORK CLASSES

As to the foreign language classes, the network classes play an important role in the normal teaching process and the testing process. Based on the detailed situation in our Northeast Dianli University, there are English classes and Japanese classes which can be chosen by all the students. All the English classes are taught in the multi-media. Teachers should prepare the courseware of each lesson, including the audio, the video, the PPT and so on. Some Japanese lessons teach in this way— such as Japanese through video and speaking and interpretation courses.

Based on the teaching process, the teachers should arrange the teaching activities according to different potential standards of all the students. This is a contradiction all the time. The group education is effective in the efficiency of teaching, while it has some limitations in respecting the individual differences and variance in learning ability. In the new teaching environment system which is composed of the teachers, the students, the network and the teaching content maximises the favourable factors and minimises the unfavorable ones. The system offers a learning-centric environment, as

well as the strategies of self-learning, based on the diversity of students. [3]

Teachers should focus on the differences of the students. For example, the different standards of preparation for the classes, the learning interests, the learning habits, the potential abilities and so on.

Teachers should pay more attention to the quality than the quantity. Encouraging the students to cultivate themselves with proper strategies, the teachers should teach students to plan, supervise, check, evaluate, give feedback and make adjustments. In this course, students could make much progress in learning abilities and initiatives.

In student-centric classes, the teachers are the organizers and promoters to design the teaching process and the tasks for study. With the multimedia materials, teachers could organize the students as a whole, or in groups or individually, according to different tasks and students. Both in the traditional classes or the network classes, there are basic tasks and key tasks. The former is based on show and observation by the students, while the latter emphasises the generation and the structure of knowledge. Forasic tasks, we could have the teacher-centric classes. For the key tasks or creative tasks, we could have student-centric classes.

5 CONCLUSION

We can't totally do away with the strategies in the traditional teaching process. The above strategies are based on respecting the individual differences of the students and completing the task with initiative. It also gives some attention to the positive sides of group teaching, in order to complete some basic tasks. It optimises the different teaching materials so as to adjust the teacher-centric and the student-centric factors dynamically.

REFERENCES

[1] Lin Junfen (2010), Research on the Different Teaching Modes and Strategies with the Network Classes, [J] Educational Research, 2010 (05).
[2] Zhang Ning, Research on the Teaching Mode with the Internet Classes, [J] Journal of Adult Education College in Wu Lumuqi, 2007.5.
[3] Pan Xue & Ma Xinzhi & Wu Tongjun, *Research on the Self-regulation of Learning with the Network Classes*, [J] Educational Research, 2007.6.

Advances in Future Manufacturing Engineering – Yang (Ed.)
© 2015 Taylor & Francis Group, London, ISBN 978-1-138-02817-3

Application of network classes in Sports English

Tianhao Wang & Xiyan Wang
Foreign Language Institute, Northeast Dianli University, Jilin, China

ABSTRACT: With the development of computer science, the new teaching strategies assisted by network classes are popular in college education. The Sports English classes should be combined the traditional classroom teaching and the advanced network application, in order to improve the language ability of the college students of P.E. major. There are many advantages when teachers apply network classes in Sports English. However, there are some limitations which teachers should try to avoid.

Keywords: network classes; Sports English; teaching strategies

1 INTRODUCTION

With the development of the computer science and the Internet, network classes are becoming popular in college education. The application of network classes in English courses can improve the impact of teaching. At the same time it can strengthen the students" interest in English study.

Sports English is becoming popular after the Olympic Games in Beijing in 2008. As a tool of international sports communication, Sports English is widely used on the TV, in newspapers and the Internet. In college, Sports English is a required course in Physical Education. However, the application of network classes in Sports English for the college students of the PE major is rare. If we apply the network classes in Sports English, the students will show more interest and the impact of teaching will be much better.

2 CHARACTERISTICS OF THE NETWORK CLASSES

The network class is a new educational device, which is supported by computer science and the Internet. It can construct a student-centric curriculum environment. The advantages of the network classes are as follow:

2.1 *Large amount of information*

With the high speed development of network technology, the Internet could offer many educational materials than any teacher, any book, and even any library could. This could enlarge the scope of perception for students'vi in order to form the biggest classroom which would embrace the education of school, family and society. Students could choose the time, the content and the speed to to suit themselves. And they can choose when and where they want to communicate with others. In a word, the Internet creates a convenient, fast and effective learning environment.

2.2 *The virtual teaching and learning environment*

For the division of the time and the place of education the network class is a virtual environment. This could give rise to some limitations of the curriculum provision. For example, some experimental and training courses can't be studied in this way. Therefore, by only teaching on network classes we can't cultivate the top-quality creative personnel. [1]

3 CHARACTERISTICS OF SPORTS ENGLISH

Sports English becomes indispensable for daily life because of the popularity of sports with people. We can understand the development of international sports events, as well as gather more knowledge in this field.

3.1 *Abbreviations and compounds*

To be understood briefly and easily, Sports English has many abbreviations. Abbreviation is the short form of the initial word with the same meaning. It is quite common for many new words to be composed with the first letter of each word in a phrase. The name of the nations and associations are always abbreviations. For example, *CHN* is short for China. *RUS* is short for Russia. *GRE* is short for Greek.

Compound is a combination of two or more words which functions as a new word. It is widely used in Sports English. For example, *free kick*, *gold medal*, *goal keeper* and so on. [2]

3.2 *Loan words*

Many sports events are characteristic in some countries which are rare in other places. Many loan words are coming from languages with the similar pronunciation and grammar. For example, *Wushu, Taijiquan, Yoga* and so on.

3.3 *More declarative sentences and exclamatory sentences*

On the one hand, for the news or introduction of the sports, we find that there are more declarative sentences and exclamatory sentences than in other forms of passages. We can rarely read some questions or imperative sentences. On the other hand, considering the mood of the public who want to know the final results of the competitions, conciseness and briefness of the sentences is required.

4 STATUS ANALYSIS OF ENGLISH IN P.E. MAJOR

Students who major in Physical Education should pay more attention to the training lessons. Their English level is lower than other students' in the college. After the survey done by the author, it was found many students are not interested in English. They don't have a study goal like passing the CET4. The study materials for a P.E. major are much easier than other majors'. So is the level of difficulty of the final exams. When discussing the importance of the Sports English, they all consider it as a necessary tool to communicate with others and to get more knowledge of their major.

5 APPLICATION OF NETWORK CLASSES IN SPORTS ENGLISH

The network classes could offer a large amount of information such as through audio, the video, the text and the animation, which can arouse interest of learning English greatly. At the same time, the ability of listening, speaking, reading and writing is being improved in class. The input of multi-media equipment could save teachers a lot of labour. The class would be more effective than before. The network classes could provide the students with a lot of learning materials according to their own standard and interests. They can choose what to learn at any time without the limitation of places to do it.

Based on the teaching process, the teachers should arrange the teaching activities according to the different potential standards of all the students. The group education is effective in the efficiency of teaching, while it has some limitations in respecting the individual differences and variance in learning ability. The new teaching environment system, which is composed of the teachers, the students, the network and the teaching content, maximises the favourable factors and minimises the unfavourable ones. The system offers a learning-centric environment, as well as the strategies for self-learning, based on the diversity of students.

Teachers should focus on the differences of the students. For example, the different standards of preparation for the classes, the learning interests, the learning habits, the potential abilities and so on.

Teachers should pay more attention to quality than quantity. Encouraging the students to cultivate themselves with proper strategies, the teachers should teach the students to plan, supervise, check, evaluate, give feedback and make adjustments. In this course, students could make much progress in their learning abilities and initiatives.

In student-centric classes, the teachers are the organizers and promoters for designing the teaching process and the tasks for learning.. With the multimedia materials, teachers could organise the students as a whole, or in groups or individually according to different tasks and students.

However there are some limitations to which the teachers should pay much attention. Some teachers rely on the network classes. Too much information is transferred to the students,. but the students can't understand it all. Teachers should be activive and make the class student-centric. [3]

6 CONCLUSION

The application of network classes in the Sports English course could make the class lively and interesting. According to the students' English level, teachers should provide students with various forms of information in class. At the same time, teachers should pay much more attention to communication with the students. To organise the class effectively it requires the teachers to improve their professional standards.

REFERENCES

[1] Lin Junfen (2010), *Research on the Different Teaching Modes and Strategies with the Network Classes*, [J] Educational Research, 2010(05).

[2] Jiang Yang, *Analysis on the Characteristics of Sports English Vocabulary*, [J] Journal of Physical Education College in Ha Erbin, 2010.8.

[3] Zhang Yinghong, *Teachers' Leading Role in the Process of Multimedia English Teaching*, [J] Journal of Shenyang Physical Education Institute, 2005.10.

Advances in Future Manufacturing Engineering – Yang (Ed.)
© 2015 Taylor & Francis Group, London, ISBN 978-1-138-02817-3

A study on the oral test of college English in the multi-media environment

Tianhao Wang & Xiyan Wang

Foreign Language Institute, Northeast Dianli University, Jilin, China

ABSTRACT: The oral English ability is very important, and needs to be examined in the final English test every semester. However, the reliability and validity of the traditional oral test is lower in comparsion to the test for other abilities such as listening, writing and reading. In the mean time, the technology of multi-media is being widely used in college English classes. In such an environment the content and the forms of the final oral tests of college English can be enriched. In addition, the oral test could have a positive influence on oral English teaching.

Keywords: multi-media environment; oral English test; oral English teaching

1 INTRODUCTION

Nowadays in the final test of college English, the language ability of listening, reading, writing and translation can be all tested. However, the oral test is weak. There are some penalties in the traditional oral test. For example, the factors in different grades, different levels, and different psychological state can't be taken into consideration. In the mean time, multi-media technology is widely used in college English teaching. This thesis is based on the current multi-media environment. The form and content of the oral test should be varied. Based on the different oral English levels of the students, we should address the issue of reforming teaching English in college.

2 THEORETICAL FRAMEWORK

In 1972 Hymes put forward the definition of communicative competence. In his opinion, the communicative competence is the application of the potential knowledge and ability. A speaker who fulfills all four aspects is competent. The first one is formal possibility. It means something possible within a formal system is grammatical, cultural, or communicative. The second is feasibility in which the psycholinguistic factors are the predominant concern. When adding a cultural level, features of the body and the material environment must be considered as well. The third is the appropriateness in which he employs the tacit knowledge and the sense of relating to the contextual features. The last is the actual performance, in which occurrence can't be left out. If not, something may be possible, feasible and appropriate, but may never occur. In this theory, language acquisition not only includes the linguistic knowledge and skills, but also communicative competence, which is more important than the former.

A more recent survey of communicative competence by Bachman (1990) divides it into the broad headings of "organizational competence", which includes both grammatical and discourse (or textual) competence, and "pragmatic competence", which includes both sociolinguistic and illocutionary competence. [1] Strategic Competence is associated with the ability of the interlocutors in using communication strategies. [2]

3 FORMS OF ORAL ENGLISH TEST

Through the influence of communicative language teaching, it has been widely accepted that communicative competence should be the goal of language education. This is in contrast to previous views, in which grammatical competence was commonly given top priority. The details are as follows.

① Self-introduction: Students are required to make a brief self-introduction with the help of the multimedia equipment. The content could refer to family, hometown, hobbies and so on. With the help of the picture and video show, students can express themselves vividly and easily, to avoid forgetting words in case of stress.

② Reading: Students are asked to read the chosen material which is printed on a piece of paper, or some passages shown by the multimedia equipment. According to the different styles and individual pronunciation habits, students should try to imitate the English pronunciation.

③ Questions and answers: Students are asked to answer the questions raised by the teachers or according to some given topics. The topics, during the communication, should be relevant to their daily life, For example, *Do you like your major? What do you do at weekends?*

④ Retelling: After listening to some passages or watching some videos, students are asked to retell what he or she saw. After a few minutes of preparation, students can talk about the main idea or their impression of the materials. The student can make notes while listening or watching, which can make the student more confident.

⑤ Picture talk: Students would get some pictures or watch a video. After that they are asked to talk about the story that was featured in the picture or the video. They should talk about their own feelings after seeing the picture or video.

⑥ Conversation: There are two forms of conversations in the oral test. The first one is completed by the teacher and the student. As to a particular topic, which is familiar to the student, the teacher and the student can discuss the topic together. The second one is completed by two students. They could get the topic after they have prepared for a few minutes. They can have the discussion in the form of a conversation in front of the teacher. The teacher can give the score by holistic marking or impression marking.

⑦ Discussion: There are two forms of discussion in the oral test, between the teacher and the students or among the students. Students should draw a conclusion according to the materials or videos before the discussion. They should discuss the topic that they have concluded earlier. The interaction and the participation of each student would influence their score. In the other form of discussion, teachers should try to make the discussion student-centric. The performance of the students should be evaluated comprehensively.

⑧ Role Play: Students would get different role cards on which different roles are described. The video will be played to create the communicative atmosphere.

In the oral test, teachers should choose the proper forms for the students at different levels. Generally there are 4 semesters in the college English teaching. In the first semester, forms from 1 to 3 could be selected. In the second semester, forms 3 to 5 could be selected. In the third semester, forms 4 to 6 could be selected. In the fourth semester, forms 6 to 8 could be selected. [3]

4 TEACHING REFORM

Due to the different policies of the college entrance examinations in different places, so many freshmen haven't studied the listening part in middle school. Teachers will teach many students like this in a class. According to this situation, we should take the following measures to solve the problem:

① The level of listening and oral English is supplementary to each other. In the listening class, students should be asked to imitate the pronunciation of the dialogues and passages. In this way, their shortcoming in pronunciation and accent will show up in front of the teacher. The teacher could then try his best to help the student. The teacher could encourage students to listen to more English broadcasts such as BBC and VOA. During the course, the students could improve their listening ability as well as enlarge their vocabulary.

② In the intensive reading classes, students should read the passages. Teachers can correct their pronunciation and let them retell the content of the passage or recite some significant paragraphs.

③ Students are encouraged to participate in many English speaking contests. Giving a speech in front of many people is very necessary for a student. They should try to be more confident and active in the contests.

Teachers should also pay attention to the background information and the English culture of the English language. Students should be active in communicating in English and using the language properly. Teachers should give detailed explanations about nonverbal communication when the lecture refers to a different culture, communication strategies and body language.

5 CONCLUSION

By studying the oral test in the multimedia environment, we should try our best to set up an appropriate series of test theory, test forms, and evaluation standards, according to different situations in each university. The atmosphere for fluent English would be highly active. The impact of teaching would be better. Students would speak English fluently and properly. They would become aware of the differences between cultures and nations. The purpose of a college education is to address the needs of society.

REFERENCES

[1] Bachman, Lyle (1990). *Fundamental considerations in language testing*. [M] Oxford: Oxford University Press.
[2] Færch, C., & Kasper, G. (1983). *Strategies in Interlanguage communication*. [M] London: Longman.
[3] Lan Liwei & Zhang Qingdong, (2008) *Study on the Bachwash Effect of the Oral English Test*, [J] Journal Of Changchun University.

Advances in Future Manufacturing Engineering – Yang (Ed.)
© *2015 Taylor & Francis Group, London, ISBN 978-1-138-02817-3*

Teaching strategies of English listening assisted by multimedia

Xiyan Wang & Tianhao Wang
Foreign Language Institute, Northeast Dianli University, Jilin, China

ABSTRACT: On the basis of analysing the characteristics of multimedia applied in English teaching, this paper puts forward the new reform and the emphasis on the teaching strategies of English listening classes. The design and production of the English listening classes should be based on the levels of the students' English, so that multimedia can help improve their abilities of listening skills to solve more and more problems.

Keywords: multimedia; English listening teaching; teaching strategies

1 INTRODUCTION

With the development of globalization, the need for international communication is increasing as is the importance of the role of speaking and listening English. For the college students, in the English video course, their listening and speaking ability is improved greatly. According to the requirement of the college English course, the purpose of college English education is to cultivate the comprehensive application ability of the language, especially the listening and speaking ability. Nowadays the listening part of English course is short of requirement. As a result, the students can't improve their listening ability effectively in class. How to improve the effect of English listening teaching is becoming a hotly debated topic in teaching reform in this area. Since we can't change the schedule of the classes, the students should learn by themselves after class. However, the multimedia technology which is widely applied in the listening classes can solve the problem to some extent. This thesis is mainly about the teaching strategies in improving the impact of teaching.

2 CHARACTERISTICS OF MULTIMEDIA

Multimedia refers to content that uses a combination of different content forms. This contrasts with media that use only rudimentary computer displays such as text-only or traditional forms of printed or hand-produced material. Multimedia includes a combination of text, audio, still images, animation, video, or interactivity content forms. The characteristics of multimedia applied in English classes are as follows:

2.1 *To arouse great interests in learning English*

"According to the educational psychology, the key factor in learning is emotion control. The main factors of emotion control in foreign language learning are motivation, attitude and personality." [1] The traditional teaching methods are composed of the chalk, the blackboard, the tape recorder and books which are very boring. The students are always display a lack of interest in learning. However, with the help of multimedia, the class could be organized with greater animation by the PPT show, the audio and video materials which could stimulate the students' senses to enhance their perspective. They are keen on this new form of English teaching. They could select the information as they enjoy the process of learning more and more..

2.2 *To enrich the content of classes*

In the traditional classes, either the teaching part or the exercise part hardly manages to arouse the interests of the students. Because the various contents played by the multimedia, such as the audio, the video, the text, the animation and so on, these forms could display an animated class in front of the students. Teachers could choose the materials preferred by the students from the Internet. All these efforts could be improved by the multimedia. If the teachers could use some teaching software, the effect of the teaching will be better.

The multimedia applied in English teaching could save a lot of time in class. The different types of materials could also add the quantity of information in class, which could help raise the efficiency in teaching. The multimedia applied in English teaching can not only fulfill the task of teaching, but also help to promote the reform and regeneration in education. These English classes offer a proper environment for education for all-round development.

2.3 *To promote the teachers' comprehensive abilities*

Multimedia requires that the teachers not only be proficient in English but also be skillful in operating the computer. The methods of making the courseware are widely used in college English classes. In the summer training of the college English teachers, the lectures on this topic are warmly welcomed. In daily life, teachers should accumulate the proper pictures, the interesting videos and some funny jokes which are related to the classes. When making the courseware, the teachers should adopt a modest attitude, with the idea of learning from each other and to share with each other. [2]

3 TEACHING STRATEGIES OF ENGLISH LISTENING WITH MULTIMEDIA

3.1 *To apply the range of information into listening classes*

Making use of the multimedia classes, teachers should display as much information as possible to the students. The bright colours, the delightful images, the proper movement and the lively sound can help arouse the interests of the students. Once they are keen on English listening classes, the effect of teaching will improve. Multimedia is used to produce computer-based training courses and reference books such as encyclopedias and almanacs. The user could go through a series of presentations, texts about a particular topic, and associated illustrations in various information formats. Their favorite songs and films are the best items to choose to stimulate their enthusiasm.

3.2 *More background information*

Before listening, teachers should introduce the background information to the students. They can get the general idea about what is going to be listened to or talked about. This kind of communication is carried out on both between the teacher and students or among the students. With the discussion and interactions, students are well prepared for the next lot of listening materials. They can predict the stories in the video or the questions which will be asked according to the text. The background information could be shown in many forms—such as the pictures, the videos, the power point and so on. If the students can finish

the task by groups, the impact of teaching will be even better. [3]

3.3 *To predict actively*

Before listening, the background information is introduced. But that's far from enough to get a better accuracy of response. Students should be encouraged to predict the relative information as much as possible. Before the listening material is played, students should predict the topic, the key words and some important details in the material, by going through the choices marked A, B, C and D. Applying the background information as well as the requiremed responses, the students could pay greater attention to the target information, in order to finish the listening task confidently. However, the listening ability is varied amongst the students. At first teachers should explain many times how they should do the prediction properly. Then the students could follow the approach, and slowly they will become good at it.

3.4 *Student-centric classes*

Multimedia could help cultivate the process of thinking actively and creatively. It could turn abstract questions into more vivid ones. In the teaching activities, students should be the center of attention in the class. On the one hand, teachers could be the host or hostess in order to organize the class activities. The sequences and the difficulty level of the questions should be well designed before class begins. For the interesting parts, teachers should design more questions to promote a greater participation of the students in the class. Limited by many factors, students cannot answer the questions very well. At this point teachers can display the audio or the video materials on the screen to draw the attention of the students in a bigger way. On the other hand, teachers should design the questions and the activities according to the level of most students. If the materials are more complicated than needed, the students will lose interest in it. If the materials are much easier than required, their listening ability can't be improved anyway. Teachers should choose the proper materials and design the activities with more attention.

4 CONCLUSION

Multimedia is widely used in college English classes, especially in the listening and speaking classes. It improves the effect of teaching of the classes. To make good use of it is the purpose of all English teachers. After analysing the characteristics of the multimedia-assisted learning, the thesis is explored in the four teaching strategies. Because the time and space is limited here, more topics and details should be further discussed. More research could be carried out on other specific types of college English teaching.

REFERENCES

[1] Shu Dingfang, Zhuang Zhixiang: *Modern Foreign Language Teaching* [M]. Shanghai Foreign Language Education Press (1996), p. 46–50.

[2] Yang Yuchun, ZhangYang, *Analysis of Advantages and Disadvantages in Multimedia Applied in College English Teaching*, [J], Read and Write Periodical (2014).

[3] Xu Qiuling, *CALL Applied in English Listening Teaching*, [J] Journal of Shandong Youth Administrative Cadres College (2006).

Advances in Future Manufacturing Engineering – Yang (Ed.)
© 2015 Taylor & Francis Group, London, ISBN 978-1-138-02817-3

Inter-cultural communication competence training in multimedia-assisted College English teaching

Xiyan Wang & Tianhao Wang
Foreign Language Institute, Northeast Dianli University, Jilin, China

ABSTRACT: The inter-cultural communication competence training is one of the important goals of College English teaching. This thesis is explored in four effective teaching strategies, after introducing the inter-cultural communication competence and the characteristics of multimedia. College English teaching assisted by multimedia could be effective and efficient in cultivating inter-cultural communication competence.

Keywords: inter-cultural communication competence; multimedia; College English teaching

1   INTRODUCTION

Different people have different languages and cultures. The language reflects the different values and the concept of world of different people. With the rapid development of information technology, globalisation promotes communication between different countries and people. In order to reduce the gap between different languages and cultures, college students should learn to use a foreign language properly. To meet the requirements of the society, the inter-cultural communication competence must be cultivated. The College English course is the main platform where students can refer to cultural differences and linguistic application. All in all, to cultivate qualified professionals for inter-cultural communication there should be a direction in College English teaching.

2   INTERCULTURAL COMMUNICATION COMPETENCE

Language is part of culture. It is also a mirror of people's culture. We can understand the customs, the way of life as well as people's thinking patterns from the language. For example, westerners, especially Americans, worship Individualism. There are more than 100 words containing *self* as a prefix, such as self-esteem, self-evaluation and so on. In fact, the distillation of a certain culture can be reflected in the language.

Language and culture interact with each other constantly. Language is the vehicle of culture. It spreads and carries on the essence of each culture, generation after generation. And the development of culture promotes the development of language. The use of language should be based on the environment of a society and its culture. Therefore, language and culture amalgamate seamlessly with each other. The proper use of language should be based on a good understanding of people's culture. [1]

According to the Longman Dictionary of Language Teaching and Applied Linguistics, inter-cultural communication is an inter-displinary field of research that studies how people communicate and understand each other across group boundaries, or, discourse systems of many kinds, including national, geographical, linguistic, ethnic, occupation, class or gender-related boundaries and how such boundaries affect language use. [2] To cultivate the inter-cultural communication competence should be the important goal of the College English teaching. It is also a basic requirement of quality-oriented education.

3   CHARACTERISTICS OF MULTIMEDIA

Multimedia refers to content that uses a combination of different content forms. This contrasts with media that use only rudimentary computer displays, such as text-only or traditional forms of printed or hand-produced material. Multimedia includes a combination of text, audio, still images, animation, video, or interactivity content forms. The characteristics of multimedia applied in English classes are as follows:

First, it could arouse great interest in learning. In a traditional teaching environment, the students always lack of interest in learning. However, with the help of multimedia, the class could be organized to be more animated. Especially when students are asked to make a presentation, they can select the kind of information they wish to. They can also choose tdifferent topics and relevant information with the help of the Internet. Because of the various contents offered by multimedia (such as the audio, the video, the text, the animation and so on), these forms could create an animated class for the students. In this way, the content of the English classes can be enriched, which is the second characteristic of multimedia applied in English classes.

Third, the multimedia applied in English teaching could save a lot of time. The different types of materials available could add the quantity of information in class, which could help raise the efficiency of teaching.

4 EFFECTIVE TEACHING STRATEGIES IN ENGLISH CLASSES WITH MULTIMEDIA

4.1 *To introduce different standards of behaviour from different cultures*

Culture is the boundary of people's behaviour as well as the use of language. The key factor of communication is based on a good understanding of the language, the society and the culture of different people. To break the boundary of the cultural gap, students should understand the difference in habits, time consciousness, social norms and other aspects. For example, when you want to start talking with a friend, you may start with the question "*Have you eaten yet?*" If your friend is American, he will be puzzled. You should start the conversation with a question like "*It's a fine day today, isn't it?*" Your friend will understand what you mean.

Making use of multimedia, teachers should show relevant videos to the students. From the videos, the different behavioural habits and effective ways of communication can be understood easily. The real social environment and the lively sounds can help arouse the interest of the students.

4.2 *To compare the different values and thinking modes*

In the inter-cultural communication, speakers both have their own values and thinking modes. It is the root of contradictions and conflicts. For example, the modest attitude of a Chinese may be seen as hypocritical in the view of Americans. The hospitality of a Chinese may be seen as nosy in the American's opinion. There are a lot of differences in values and thinking modes of different cultures. College students are all familiar with the Chinese ways of thinking modes and values. [3]

To introduce the similarities and differences of two people, teachers could make use of different pictures, videos and animations. Multimedia could help the students understand the abstract values and modes in this way. The class teaching is efficient in helping the students use language properly.

4.3 *To analyse the cultural intention of vocabulary*

Vocabulary is the basis of communication, which carries the information of culture. The cultural gap is mainly reflected at the level of vocabulary. Therefore, to analyse the connotations of a word in different cultures is very significant. In the learning of some slang and jargon, the teachers should explain the different meanings and connotations in different cultures of a certain word. Take the word *dog* as an example. In the Chinese culture dog is the derogatory term with the connotation of being disgusting and contemptible such as *zou gou* and *luo shui gou*. However, in the culture of English, we have the phrases such as *a lucky dog, a gay dog*. We also know slang such as *Love me, love my dog* and *As a dog with two tails*. Teachers should analyse the differences of *gou* and *dog* in different cultures. Students could understand the word from different aspects and use this word in a proper way when communicating with foreigners.

4.4 *More emphasis on the non-verbal communication*

Non-verbal communication is the process of communication through sending and receiving wordless clues between people. It encompasses many aspects, such as the use of voice, touch, distance, and physical environment. It represents two-thirds of all communication. Culture plays an important role in non-verbal communication, and it is one aspect that influences how learning activities are organised. In this sense, learning is not dependent on verbal communication. It is non-verbal communication which serves as a primary means, not only organising interpersonal interactions, but also in conveying cultural values.

Some videos and films could help students understand non-verbal communication which, one can hardly learn in books. By watching real communications at work, students can easily understand what the non-verbal communication mean in a different cultures.

5 CONCLUSION

Multimedia is widely used in College English classes. To make good use of it could help to cultivate the inter-cultural communication competence of the college students. After introducing the idea of competence in inter-cultural communication and analysing the characteristics of the multimedia-assisted learning, the thesis is explored in the four effective teaching strategies. Because the time and space is limited here, more topics and details should be discussed further.

REFERENCES

[1] Sun Xiaohui, *Intercultural Communication and College English Teaching.* [J] Journal of Shanxi Normal University (Social Science) (2002).
[2] Jack C Richards, Richard Schmidt, *Longman Dictionary of Language Teaching and Applied Linguistics,* [M], Foreign Language Teaching and Research Press (2005).
[3] Zhang Jianfang, Peng Zhimin, *on Cultivating Students' Cross-cultural Communicative Competence in College English Teaching,* [J] Journal of Shangluo University (2007).

Advances in Future Manufacturing Engineering – Yang (Ed.)
© 2015 Taylor & Francis Group, London, ISBN 978-1-138-02817-3

English films in multimedia-assisted College English classes

Xiyan Wang & Tianhao Wang
Foreign Language Institute, Northeast Dianli University, Jilin, China

ABSTRACT: As multimedia is being widely used in College English teaching, there is a great need for all-round development of English. Films could be used in the English classes with multimedia. It is an effective way to promote the listening and speaking teaching reform. In this thesis the author discusses the advantages and the relevant teaching strategies.

Keywords: films; multimedia; College English

1 INTRODUCTION

With the indepth reform and opening up of global communication the fast development of the society requires high quality talent with all-round development. Nowadays the college students' speaking and listening ability are not satisfactory for this environment. English teaching is still focused on reading and writing. If we want to improve the comprehensive ability of the language, we should pay more attention to the speaking and listening classes. As information technology continues to develop, multimedia is being used widely in English education. Mnay different kinds of information can be sent through the multimedia, such as the audio, the video, the text, the pictures and so on. English films could also be shown in the speaking and listening classes. Making good use of English films can improve the impact of teaching English.. This thesis is focused on the advantages and the application of using English films, assisted by multimedia, in College English classes.

2 ADVANTAGES OF ENGLISH MOVIES USED IN CLASSES

2.1 *To create the real language environment*

According to the theory of *Language Acquisition*, S. D. Krashen put forward in 1970s, there are two ways to develop or learn a language. One is called acquisition. The other is called learning. Using *learning* means, a conscious process involving the study of explicit rules of language and monitoring one's performance, as is often typical of classroom learning in a foreign language context. Using *acquisition* refers to a nonconscious process of rule internalization resulting from exposure to comprehensible input when the learner's attention is on

meaning rather than form, as is more common in a second language context. [1]

Therefore, the language environment is very important in language acquisition. Especially, since the speaking and listening ability of English can't be improved if there's no language environment. Movies are stories from the real world. When watching the movies, students are transported to the movie environment. That is the acquisition.

English movies could help to improve the listening ability. They offer a range of interesting materials for the students to talk about. When watching movies, students could learn how they could apply words in real life, the different forms of pronunciation, the various rhymes and so on. Students could do well in speaking English, by learning through imitating, reciting, retelling and other forms.

2.2 *To arouse greater interest in learning English*

Interest is always the best teacher. The motivation of the learners contains attitude, which is the key to their study. For the students from the non-English major, their interests in English become less when they find the traditional English classes are boring. However, the films are vivid learning materials. The artige quantity of input. By the visual and aural stimulations, students will find it easy to understand the words.

2.3 *To study more about English culture*

Language can't exist without culture. The linguistic education should put more emphasis on culture. Films could show the students the background information of the country, the event, or the story which can hardly be discussed clearly in class. The lively teaching material could bring the students into the foreign country at a certain historical occasion. Understanding the film by the nonverbal information is very important. It is helpful for the students to learn about the culture.

For example, in the movie *Forrest Gump*, apart from the classic words, students will come across big events such as the Vietnam War, Watergate, and the Ping-pong Diplomacy which happened in America from 1950s to 1980s. They can also get to know more information about the American spirit, which Gump stands for, and the Lost Generation which Jenny stands for. [2]These classical movies could show the natural and the social environment of a country in a certain period vividly. Watching the movies in class is a good opportunity to enlarge their vision and knowledge.

3 EFFECTIVE TEACHING STRATEGIES

Multimedia equipment is the hardware of movie playing. The computer, the screen and the projector are required, as are the teaching strategies.

3.1 *To preview the films*

To preview the films doesn't mean to watch the films. What the author means here is to ask the students to preview the background information of the movies, such as the history, the geography, the characters, the customs, and the social events in the movie. If the students lack the necessary information, they can hardly understand the movie well. As to the film *Gone With the Wind*, the students should know something about the American Civil War.

Teachers are also required to do some preparation. On the one hand, some questions should be asked to enlighten the students before watching the movie. Either to help the students predict the content of the movie from the title of the film, or to talk about their feelings about the theme song, is an interesting warming-up activity. [3] On the other hand, teachers should choose the movies with enlightened themes, the clear language spoken, and the proper speed of words. Films such as *Lion King, Pride and Prejudice, the Brave Heart* are all good choices.

3.2 *To finish some tasks while watching*

Teachers should give some tasks to the students while they are watching the movie. To avoid turning the listening and speaking class into the film watching class, teachers should design the class carefully. The following are some suggestions:

Students should answer the questions about the important events that happened in the movie, after watching. As for some classic words or sentences—they should be asked to write them down. After watching the movie, they are asked to retell the story. In this way they could learn to make notes while watching. They could also practise stenography.

After watching they could write a composition about a certain theme reflected in the movie. In this way, their writing ability could be improved. For example, in the movie *Gone With the Wind*, the famous saying in the movie is that *'Tomorrow is another day'*. Students could write about their plans for their future life or their own thoughts about life.

Students could also have a discussion about their favourite actor or actress in the movie, or in their daily life to practise their spoken English. In this style of learning, students can improve their listening and speaking ability. Take the movie *the Pursuit of Happiness* as an example. The father tells his son, "You got a dream, you gotta protect it. People can't do something themselves, they wanna tell you you can't do it. If you want something, go and get it." When the movie comes to this period, students are all moved by the simple but significant words of the great father. They all have the answer to the question as to whether the father's suggestion will influence their life.

3.3 *Some matters needs attention*

When films are used in the speaking and listening classes, the teachers are required to design the class carefully. Some questions are needed. They should also pick some pieces out of the movie as teaching material if the time is limited. Considering the themes, the word speed, the vocabulary, the pronunciation and other aspects, the movies should be helpful in their English study. Teachers should share the good movies together.

4 CONCLUSION

Films could be used in multimedia-assisted English classes to improve the listening, speaking and writing ability of students. Students are interested in watching movies in class. The teachers should design the class carefully to help students make good use of the film. The effect of teaching will be better compared to traditional listening and speaking classes. Because of the limited time and space here, more details should be discussed in the future research.

REFERENCES

[1] Jack C. Richards, Richard Schmidt, *Longman Dictionary of Language Teaching and Applied Linguistics,* [M], Foreign Language Teaching and Research Press (2005).
[2] Zuo Yanlin, *Application of the English Films to Students of Non-Englisn Majors in the Listening and Speaking Classes,* [J], Movie Review (2008).
[3] Zou Shuli, *English Films as an Aid in the Teaching of Listening and Speaking in College English,* [J] Journal of Cangzhou Teachers' College (2010).

Advances in Future Manufacturing Engineering – Yang (Ed.)
© 2015 Taylor & Francis Group, London, ISBN 978-1-138-02817-3

Application of network in multimedia-assisted college English reading classes

Xiyan Wang & Tianhao Wang
Foreign Language Institute, Northeast Dianli University, Jilin, China

ABSTRACT: With the rapid development of network technology, multimedia-assisted English classes are common in colleges. Although reading ability is important, students find the reading classes uninteresting. To improve the impact of teaching and their reading ability, application of network technology is necessary. After introducing some proper teaching strategies, suggestions are put forward from the different perspectives of materials, teachers and students.

Keywords: network; multimedia; college English reading

1 CURRENT STATUS OF COLLEGE ENGLISH READING CLASSES

In college English teaching, the reading ability requires, from both teachers and students, the most attention, because people acquire the most knowledge by the means of reading. However, in the reading classes, students don't take it as seriously as required. Some students just want to learn the new words or grammar by reading. Some students just read because of the final exam. In the reading classes students are passive because of the lack of systematic reading skills or the reconstruction of the reading material. Some teachers regard the reading classes as the input of information in forms of paragraphs. They organize reading classes in the method of teaching grammar, vocabulary and translation. Students lack interest in these reading classes, although they know reading ability is very important.

In fact, reading is not only a language skill but also the key to improve other language skills such as listening, speaking, writing and translation. Not only can we monitor the impact of the teaching of vocabulary and grammar by reading, we can also learn many reading skills to improve reading efficiency. Reading can broaden people's horizons and open their mind. By reading students can get more background information to rise to a higher moral, cultural, or intellectual level.

2 ADVANTAGES OF APPLYING THE NETWORK IN READING CLASSES

Applying the network in the college English reading classes is a new educational device, which is assisted by multimedia and the Internet. It can construct a student-centric curriculum environment. The advantages of the network classes are as follows:

2.1 *Large amount of information to read*

The Internet is the largest library. By surfing the Internet students can search for and get the information that interests them. The college English classes assisted by the multimedia can also offer all kinds of information. Compared with the traditional printed books, most English learners can get vast amounts of useful information easily if they want to. Students could read the passages or news in English from different websites. They could also learn some useful reading skills and preview the teaching materials. By the network students could also finish some tests and receive the feedback quickly and clearly to check their reading efficiency. Making use of the multimedia, teachers could assign the students some supplementary materials. It could save a lot of time and labour for both the teachers and the students.

2.2 *Latest news and materials to read*

With the high speed development of the network technology, the Internet could offer the far more educational material than any teacher, any book, and even any library. What's more, it could offer the latest news and materials to read. Compared to this, the content is out of fashion of the reading materials at hand. The students can get the latest, or popular, information on the network quite easily. Students could get whatever they want by operating the computer and the mouse.

Because of the real-time, interaction and wide regional coverage of the network, students can read and study whenever and wherever they like. With no limitation of classes or books, all kindst of information can be found on the network including text, pictures, video and so on. In multimedia-assisted classes, teachers and students can share interesting and animated information together. [1]

3 APPLICATION OF NETWORK IN COLLEGE ENGLISH READING CLASSES

3.1 *Before the reading classes*

Teachers could assign the students some tasks that could be finished by the use of the network to preview the lessons. The assignment could be varied. Students could be asked to do some research about the theme or the background information of the lesson. Teachers could share some websites or appointed resources with the students. Or the students could read whatever they like according to their needs. During the course the students could prepare well for the new lesson as well as cultivate the ability to get useful information by themselves. Teachers could ask some students to do the presentations before the class to check their preview. Students could use power point to share their assignment. They could also use some texts, pictures, videos and animations to make their presentation more vividly.

3.2 *In the reading classes*

Teachers could organize the reading classes with the help of the network. They could have their own blogs or websites which arefamiliar to the students. When the teachers release some relevant content or some supplementary materials on these websites, students could read up the requirements and some other helpful passages. This could help the student understand the lesson better. During the classes, students could discuss and share their own opinions with each other.

In such student-centered classes, teachers are the organizers and promoters to design the teaching process and the task of learning. With multimedia materials, teachers could organise the students as a whole, or in groups or individually, according to different task and students No doubt that in the traditional classes or the network classes, there are basic tasks and key tasks. The former is based on the show and observation by the students, while the latter emphasises the generation and the construction of the knowledge. For the basic tasks, we could have teacher-centered classes. As to the key task or creative task, we could have student-centered classes. Therefore, teachers should organize the classes in different ways according to the lessons to be done.

3.3 *After the reading classes*

Studying the teaching materials is far from enough. To improve the reading ability of students, further reading is required after the reading classes. The assignment should be finished by reading and writing. Students are asked to read the relevant materials on the network and write the reports. They could also hold a speech competition every semester. In these activities, students could do well in reading, writing and speaking. Their English language ability could be improved from every aspect.

As to the assessment system of the reading classes, teachers should design a scientific system to evaluate the reading ability of the students. By the quantity of new words, the length of the sentences, the complexity of grammar and other aspects, teachers could measure the reading ability of the students. Teachers should give detailed explanations on reading skills, the difficulties in understanding English, as well as the new words and phrases. [2] The effect of teaching will be greatly improved.

4 MATTERS NEED ATTENTION

4.1 *As to the reading materials*

The reading materials should be relevant to the lessons of the teaching materials. The difficulty level of the materials should be varied according to the different abilities of the students. By searching through some useful and popular websites, students could save a lot of time and read efficiently. [3]

4.2 *As to the teachers*

Teachers should practice their computer skills in their spare time. They should change the educational philosophy by studying the new theories. In the multimedia classes, teachers are the hosts or organizers of the classes. They are not the projectionists of the power points or videos. They should communicate with the students all the time. Without the proper supervision of the teachers, the effect of teaching can't be improved.

4.3 *As to the students*

Students should correct their attitude towards the reading classes. The English classes in the college are different from the ones in middle school. They should adapt themselves to the new environment. Studying by the network doesn't mean they can do anything they like on the Internet. They should study according to the requirement of the teachers and difficulties of each unit.

5 CONCLUSION

Application the network in the multimedia-assisted college English reading classes is a new trend in teaching. It could make the reading classes actively interesting. It could also arouse the interests of the students in reading. However, we should realise the advantages and disadvantages of this clearly. According to different students and teaching materials, teachers should choose the proper teaching methods. The traditional teaching mode and the multimedia teaching modes should be applied together to promote the impact of teaching together if necessary.

REFERENCES

[1] Zhang Fuqiang & Zhang Xiangyang, *Discussion on the College English Reading Teaching in the IT Environment*, [J] Enterprisers, 2013(08)
[2] Zhang Dan, *A Study on Network-based Multimedia-assisted College English Reading Teaching and Learning*, [J] Theory Research,2011.8.
[3] Zhang Yalan, *A Study on the College English Reading Teaching Modes in the Multimedia Environment*, [J] Journal of Hunan University of Science and Engineering, 2014.3.

Advances in Future Manufacturing Engineering – Yang (Ed.)

A study on multimedia-assisted college English listening classes

Xiyan Wang & Tianhao Wang

Foreign Language Institute, Northeast Dianli University, Jilin, China

ABSTRACT: Based on the development of the IT technology and the reforms of college English teaching, there are some disadvantages in the traditional teaching of English listening. However, making use of the multimedia could improve the teaching effect of the listening classes. After analysing the advantages of the multimedia-assisted college English listening classes, the author gives some suggestions on effective teaching strategies.

Keywords: multimedia; listening; college English teaching

1 INTRODUCTION

With the rapid development of computer science and the teaching reforms in college, the multimedia-assisted college English has become widely spread in most colleges in recent years. In the multimedia environment, the network plays the role of a platform for the communication between teachers and students in listening classes. Multimedia provides many teaching resources and helps to create the ideal language environment. With the guidance of the teacher, the students complete the tasks given by the teacher in the listening class. It plays an important role in cultivating language skills and improving the teaching effect of the listening classes.

2 DISADVANTAGES OF TRADITIONAL ENGLISH LISTENING CLASSES

In the traditional English listening classes, there are several disadvantages as they have followed that influence for teaching effectively for a long time.

2.1 The dull instructional modes

The three steps in the listening classes seem dull. First, the students do the exercises in the books while listening to the material required. Second, the teacher provides the answers to the questions. Third, the listening material would be played again if necessary. Students get bored in the listening classes in college because of the tedious teaching strategies prevalent from their days in middle school. More than half of the class is occupied with the listening part and the teacher's explanation. The students' voice can't be heard, forget

about the communication between them. The students are passive in the traditional listening classes. An instructional mode like this will decrease the teaching effect.

2.2 Too much emphasis on the autonomic learning and less listening classes than required

On the one hand, in the English classes of middle school, the instructional emphasis is on the grammar and vocabulary, because of the stress of the entrance examination of the college. Many college students are from the provinces where the listening test doesn't count in the entrance examination of the college. [1] These policies result in the poor performance of the students in the college English listening classes. On the other hand, the number of the listening classes are less than required in most colleges. Therefore, most college English teachers ask the students to do a lot of listening exercises after class. Because of the poor listening ability and lack of interest, few students could do as the teachers required. This phenomenon leads to the unsatisfactory progress of collge students in listening classes.

2.3 The large scale of classes and the teacher-centered instructional mode

Because of the increased enrollment of the colleges, more and more students can now further their study after middle school. The scale of the English classes is larger than before. The number of the students in one class can go upto 90 sometimes. This phenomenon leads to the communication decreasing between the teacher and students in the listening class. In the traditional English classes the teachers are the center of the class. The students are passive. Some teaching activities can't be held because the time is so limited in classes and many students don't get the opportunity to attend. The students will lose interest in practicing listening and the listening ability can hardly be improved in the long run in such a situation.

3 ADVANTAGES OF THE MULTIMEDIA ASSISTED COLLEGE ENGLISH LISTENING CLASSES

Applying the network and multimedia in the college English listening classes is a new educational device. It can construct a student-centered curriculum environment. The advantages of the new type classes are as follows:

3.1 The combination of listening, speaking and viewing

By making full use of the multimedia equipment, the teaching material would be in various forms, such as the audio, the video, the text, the Power-Point, the animation and so on. These teaching forms could arouse the interest of the students. Students want to participate in the teaching activities. While playing the video, the combination of viewing and listening would stimulate the different parts of the brain to promote the potential of the students. It will improve the understanding and memory of the teaching material. Students would like to talk about what they have understood, after the input of the material has been absorbed. In this way, the combination of listening, speaking and viewing could be realized by making use of the multimedia.

3.2 To arouse the interest of the students

Interest is always the best teacher. The motivation of the learners contains the attitude, which is the key to their study. For the students from the non-English majors, their interests in English decreases when they find that traditional English classes are boring. However, the multimedia could provide the animated learning materials. The artistic quality and the animation from a large quantity of inputs could leave a deep impression on the students. [2] By the visual and aural stimulation students will find it easy to understand the conversations and the new words. The classroom climate will be comfortable and free.

3.3 More cultural input

The second language education should put more emphasis on culture. By making use of the multimedia, teachers could show the students the background information of the history, the geography, the lifestyle, the literature, and the traditions which can hardly be discussed at length and vividly in class. There are many cultural connotations in a language. Understanding the nonverbal information in a conversation or film is also very important. Learning the background information is helpful for the students to cultivate inter-cultural communication competence. In this way, the listening class turns into a lively class of inter-cultural introductions.

4 MATTERS NEED ATTENTION

4.1 As to the teachers

First, in the multimedia-assisted college English listening classes, teachers are the directors in transforming the passive role of the students in class into an active one. Teachers should encourage the students to practice listening in their daily life through all kinds of learning materials. Second, teachers should make full use of the multimedia properly. It means that teachers should keep the balance of the multimedia and the teachers' role in the listening classes. The multimedia could help teachers save the writing time on the blackboard and attract the attention of the students. However, some teachers spend much time on the various teaching modes, with less emphasis on the actual teaching effect. [3] Third, teachers should choose the proper listening materials for the students. The VOA and BBC are the best choices. There're some classic films and songs in English which could correct students' pronunciation.

4.2 As to the students

First, there are some students who didn't work on listening in middle school or in every class in college, because of the different policies of the college entrance exams. Because of the poor performance in listening, they always keep quiet in class. They should be encouraged to improve their listening ability from the basic level. Teachers should pay attention to these students from every class, and help them to be more confident in listening. Second, the students should correct their attitude towards listening classes. Some students just want to pass the final exams or the CET 4. They pay more attention on the listening skills of a certain exam. They neglect the listening ability in the daily communication. They should focus on cultivating listening ability and practicing listening skills together.

5 CONCLUSION

The importance of listening in college English is self-evident. Listening plays a key role in communicating, so it should arouse great attention amongst the teachers and students. This thesis focuses on the disadvantages of the traditional listening classes and the advantages of the multimedia-assisted listening classes. College teachers should pay attention to promoting the listening ability of college students by this new means. Teachers should take charge of the listening classes by making good use of the multimedia to improve the students' listening ability.

REFERENCES

[1] Liu Xiaoyan, *On the Application of Multi-media Technology to College English Listening Teaching*, [J] Overseas English, 2010.6.

[2] Zhang Cancan, *Multi-media's Research in the College English Listening Teaching*, [J] Journal of Hunan University of Science and Engineering, 2006.7.

[3] Xue Na, *Reflections on the Multimedia Assisted College English Listening Teaching*, [J] Journal of Language and Literature Studies, 2014(06).

Advances in Future Manufacturing Engineering – Yang (Ed.)
© 2015 Taylor & Francis Group, London, ISBN 978-1-138-02817-3

Popularity prediction based on ensemble learning in online forums

Fei Xiong, Yun Liu & Zhenjiang Zhang
School of Electronic and Information Engineering, Beijing Jiaotong University, Beijing, China
Key Laboratory of Communication and Information Systems, Beijing Municipal Commission of Education,
Beijing Jiaotong University, Beijing, China

Yingsi Zhao
School of Economics and Management, Beijing Jiaotong University, Beijing, China

ABSTRACT: We present a method to predict the popularity of posts in online forums based on ensemble learning. We extract content and the trend features from original posts, which are closely related to post popularity. The content features contain topic cluster similarity and key word frequency. A Support Vector Machine (SVM) is used to fuse these features to generate a predictive model of popular posts. Then, we use ensemble learning algorithm to process training data, and the original data are divided into several parts with overlapping. Experiment results prove that our method is efficient for online popular post prediction. The ensemble learning algorithm greatly improves the precision without loss of other performance.

Keywords: online virtual community; popular post prediction; feature fusion; ensemble learning

1   INTRODUCTION

Nowadays, Internet has become one of the most important ways to collect and share information for people. By web crawlers, one can easily collect plenty of data from the Internet, and the data support empirical analysis of Internet structure. With the development of high-performance processors and parallel computing technology, Internet data mining has become a research focus. In terms of time and space dependence among Internet data, one can predict undiscovered online phenomena.

Researchers explore Topic Detection and Tracking (TDT) to find out the document organization about a topic from multiple sources of information [1]. The main task of TDT is document clustering. The Single-pass algorithm inputs documents sequentially [2]. When a new document is input, the algorithm compares the similarity between it and other topic vectors, and the new document may be assigned to an existing topic or be treated as a new topic. In Ref. [3], the authors used the logarithmic maximum likelihood criterion to determine weighting values of terms, instead of the traditional method by frequency counting. Ref. [4] integrated the cosine distance in document vector space with time similarity, and more computing time was needed. In Ref. [5], the authors proposed a probability model of time discrimination based on temporary weight and feature selection,

and experiment validated the effectiveness of the model. Ref. [6] implemented hierarchical clustering to detect new topics based on user interest similarity.

At present, the studies of topic prediction concentrate on short-term trend prediction of online information diffusion. S. Jamali [7] predicted the content popularity on Digg website by combining user interest analysis with community structure detection. Ref. [8] acquired time series of posts in online forums, and conducted wavelet decomposition to separate high-frequency and low-frequency information. After document clustering, the authors used Auto Regressive Integrated Moving Average Model (ARIMA) to predict the growth trend in time series for each topic [9]. The back propagation neural network was also used to predict the evolutionary trend of topics, and the results proved that the performance of neural networks was higher than the ARIMA model [10].

Current research mainly aims at the prediction of short-term trend. These methods often do not have high effectiveness for long-term prediction. However, in actual situations, people always hope to quickly find out the potential popular topics to avoid online emergencies. Therefore, whether we are able to predict topic popularity according to early data still needs to be explored. In this paper, we collect post data from Tianya online forum. The Tianya community is one of the most well-known online forums in China, and a great many users publish posts and replies each day. We propose a method of post popularity prediction. We extract multiple features from original data, and use SVM to fuse these features. Then, we use ensemble learning to process the original data. Our method can detect potential popular posts at the early time. Without loss of generality, if the number of reads for a post is above 1000, we consider it as a popular post reasonably.

2 FEATURE EXTRACTION

Now we extract features related to the popularity of a post. These features can be divided into two aspects, i.e., content influence and trend influence.

1. Content influence. The popularity of a post closely correlates with its content. If a post belongs to a hot topic, it may absorb many users and have wide influence in online forums. Posts about a hot topic usually describe the same event. Therefore, users that are interested in a post may be attracted by other posts having the similar contents. The content influence is measured by the vector cosine similarity. Since texts written in Chinese are comprised of continuous terms without separate signs, we should split terms from continuous texts. For each post, we use the ICTCLAS tool [11] to implement term segmentation, so that a post is changed to a term vector. The formula $TF \times IDF$ is used to measure the weight of a term in a document. In the formula, TF means the number of times a term occurs in a document, while IDF characterizes how frequently the term occurs in different documents. Then, the similarity between two posts V_i and V_j is measured by the cosine value between the two term vectors, i.e. $\cos < V_i, V_j >$. In the training data, we use the K-Means [12] algorithm to cluster the posts based on their cosine similarity, so that posts that have similar contents are classified into the same topic. We calculate the popularity of those topics in training data, and rank the topics by their popularity from top to bottom. Thus, we can assign an ordered integer label for each topic, and a larger integer label means the topic is more popular. By now, we obtain the available topic list from training data. For a post in testing data, we calculate its cosine similarity with all these topics in training data, and assign it with the label of the corresponding topic that has the largest similarity with the post.

In addition, key words in the tile of a post make a summary of the whole content, and are of great importance to post attraction. Meanwhile, key words about a hot topic often burst out and grow explosively within a short time. We gain entity nouns in the title of each post for testing data, and calculate the number of entity nouns, the frequency of these nouns occurring in training data, the total number of posts in which these nouns occur, and the duration of these nouns occurring in training data.

In all, we obtain 1 topic feature and 4 key word features for the content influence.

2. Trend influence

The growth trend in the first few hours after a post is created may reflect its attraction. If a post is replied by many users at the early stage, it may strike a chord and absorb a great many participants. Even though some posts may demonstrate their attraction after a long period of time and few users concentrate on them at the beginning, the short-term trend can also be treated as a significant feature. The early growth trend can help to detect potential posts that correspond with the public interest. For each post, we calculate the number of its replies in the first T hours, the number of its participants in the first T hours, the total number of posts within the first T hours, and the average number of replies, the average number of participants for posts in the first T hours.

The diffusion of posts in online forums is closely related to user interactions. Therefore, whether there are plenty of active users affects the popularity of posts. Active users mean they participate in discussions in online forums, either publish a post or reply to other users. We define the total number of active users within an hour before the creation time of a post, and the total number of active users within a day before the creation time of the post.

Thus, the trend influence accounts for 6 features altogether.

3 FEATURE FUSION

In the section of feature extraction, we obtain 5 content features and 6 trend features. Now we use SVM to fuse these features to produce a predictive model of post popularity. The focus post is given by X_i, and the set of all posts is given by X_i. We define the set of all users as G. Since the features of a post are relevant with the content and creation time of the post, as well as the number of all users. Therefore, the 11 features are calculated from the post X_i and the set G. We express these features as an 11-dimensional vector $\bar{h}(X_i, G)$. Then, we use SVM to build a model. SVM is an optimal margin classifier that can integrate multiple features for classification. SVM uses linear mapping to find a hyperplane that can divide original data into two categories. SVM makes the two nearest samples that belong to different categories have the largest distance [13]. The output of SVM is defined as $y(X_i)$. $y(X_i) = 1$ means the post is popular, while $y(X_i) = -1$ means the post does not have attraction. SVM transfers the feature vector $\bar{h}(X_i,G)$ to one-dimensional real number. The mapping function that SVM uses is $\bar{w}^T \cdot \bar{h}(X_i,G) + b$, where $\bar{w}$ is an 11-dimensional weighting coefficient vector corresponding to the 11 features, and b is a constant. If the value after mapping satisfies $\bar{w}^T \cdot \bar{h}(X_i,G) + b \geq 0$, one gains $y(X_i) = 1$. Otherwise, if $\bar{w}^T \cdot \bar{h}(X_i,G) + b < 0$, then is $y(X_i) = -1$. For all training data, we extract 11 features of each post, and we already know the popularity of posts in training set. Then, we build a SVM model, and use the features and popularity of posts in training data to adapt the weighting coefficient vector $\bar{w}$ and constant b, making the SVM model optimal for training data. Assuming the training data contain m samples $\langle X_1, X_2, ... X_m \rangle$, the solution of SVM is equivalent to analyze the quadratic programming problem.

$$\min_{\bar{w},b} \frac{1}{2}\left\|\bar{w}\right\|^2 \tag{1}$$

Satisfying the following linear constraints

$$y(X_i)\left(\bar{w}^T \cdot \bar{h}(X_i,G) + b\right) \geq 1, i = 1, 2, ... m \tag{2}$$

For simplicity, we write $y(X_i)$ as y_i, and $\bar{h}(X_i,G)$ as $\bar{h}_i$. From Eqs. [1, 2], one can gain the Lagrangian function.

$$L(\alpha,\bar{w},b) = \frac{1}{2}\left\|\bar{w}\right\|^2 + \sum_{i=1}^{m} \alpha_i\left(1 - y_i(\bar{w}^T\bar{h}_i + b)\right) \tag{3}$$

After, some transformations $L(\alpha,\bar{w},b)$ can be transformed to the following equation.

$$L(\alpha,\bar{w},b) = \sum_{i=1}^{m} \alpha_i - \frac{1}{2} \sum_{i=1,j=1}^{m} y_i y_j \alpha_i \alpha_j \bar{h}_i^T \bar{h}_j \tag{4}$$

4   POST POPULARITY PREDICTION

From above processes, we calculate 11 features related to each post, and build a SVM model to fuse these features. Now we describe the overall procedures of our method for popularity prediction as follows.

1. We collect data from online forums, including user information, post title, creation time, etc. Then, redundant and spam posts are deleted. All the data are structured and stored in the database.

2. The whole data are divided into two parts according to post creation time, i.e., the training set and testing set. In the training data, we assume we are aware of the popularity of all posts.
3. For the training set and testing set, we calculate the 5 content features and 6 trend features.
4. We use the 11 features and popularity of posts in training data to train a SVM model. After training, we gain the weight value of each feature which represents the importance of features. Then, a predictive model for post popularity is produced.
5. We input multiply features of posts in testing data to the predictive model, and get the estimation of popularity for these posts. Therefore, we can obtain the potential popular posts.
6. We compare the prediction results of our model with actual data, and verify the performance of our model.

5   ENSEMBLE LEARNING

Now we use ensemble learning to improve the performance of our algorithm for popularity prediction. The bagging algorithm is used to process the training data. For a data set D that contains n samples, the bagging algorithm divides it into m subsets D_i, each of which contains n_i samples [14]. Obviously, the sizes of subsets should satisfy $n_i < n$. The subsets may overlap with common samples, and $\sum n_i \geq n$. We use each subset of training data to construct SVM, and use the trained SVM

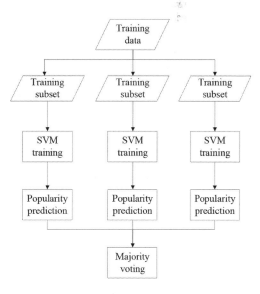

Figure 1.   Flow chart of the ensemble learning algorithm.

to predict the popularity of a post in testing data. Here, for a post, each SVM gives a prediction result, and therefore, we have several different prediction results. Then, we use the average or majority voting method to integrate these results, and achieve the final outcome. The flow chart of ensemble learning algorithm is illustrated in Figure 1. The overall procedures are listed as follows.

6 PERFORMANCE EVALUATION

Now we use actual data from Tianya to verify our algorithm. After preprocessing, we collect 8647 posts in total. We sort these posts by time. The first 4000 posts are used as training data, while the remaining posts are treated as testing data. In the testing data set, we assume the popularity of posts is unknown, then, we use our method to predict the popularity, and compare the prediction result with actual popularity. Since the measure of post popularity is the same in both training and testing data set, the specific parameter in popularity definition does not affect the performance of the method. We use accuracy, recall and precision to evaluate our method. If there are N posts in testing data, among which M posts are popular. For the method, L popular posts are predicted, among which C posts exist in the collection of M actual popular posts. Obviously, $C \leq M$ and $C \leq L$. Therefore, the recall of the method is given by C/M, and the precision is C/L. The accuracy refers the number of correct prediction results, including popular and non-popular posts, so the accuracy is given by $(2C + N - M - L)/N$.

Table 1 gives the recall, precision and accuracy of single SVM as a function of T. The parameter T is the inspection window to calculate trend features. For a larger T, most growth data of a post is calculated and the more appropriate trend is considered, so that the method has a higher performance. For a smaller T, the trend features may not adequately reflect the attraction of a post, and the performance especially the recall drops. However, in actual situations, to detect popular posts as early as possible can help to avoid and control online emergencies, and therefore, a smaller T is better from the practical application angle. In Table 1, our method has a high recall and precision, and especially the accuracy is larger than 90 percent. The recall is

Table 1. Recall, precision and accuracy versus T.

	Recall	Precision	Accuracy
$T = 2$	0.542	0.762	0.931
$T = 4$	0.668	0.786	0.961
$T = 6$	0.713	0.803	0.967

Table 2. Recall, precision and accuracy of ensemble learning.

	Recall	Precision	Accuracy
$T = 2$	0.538	0.827	0.954
$T = 4$	0.654	0.861	0.975

smaller than the precision for any T. The reason is that only an extremely small proportion of posts are popular and most posts are not attractive. Therefore, to maintain a high accuracy, the SVM classifies some popular posts into the non-popular group, and then, the recall is decreased. The precision only considers the popular posts that the method predicts, so it is not affected. Meanwhile, the precision and accuracy do not have clear correlation with the inspection window T, but the recall increases with the increase of T. It is noticed that even for a very small $T = 2$, more than 50 percent of popular posts are detected by the method.

Table 2 demonstrates the accuracy, recall and precision of ensemble learning method. We divide the original training data into three subsets by sampling and replacement, and each subset contains 2000 posts. Obviously, overlapping posts exist in the subsets. In Table 1, we find the accuracy has a small advance by ensemble learning method. Although the recall after ensemble learning decreases slightly, but the precision increases greatly, from 0.786 to 0.861 for $T = 4$. Therefore, the ensemble learning method can improve the prediction performance of the SVM.

7 CONCLUSIONS

In this paper, we proposed an algorithm of the post popularity prediction. We extracted features from post and user information, including content and trend features. We calculated the topic similarity, frequency of key words, and growth trend in the early stage. SVM was used to fuse these features to produce a predictive model, and its output indicated whether a post will become popular. Then, we used ensemble learning algorithm to process training data. Experiment results show our method has a high performance in recall, precision and accuracy. Through the process of ensemble learning, the precision is greatly improved without loss of other performance.

ACKNOWLEDGMENT

This work was supported by the Fundamental Research Funds for the Central Universities under Grant No. 2014 JBM018.

CORRESPONDING AUTHOR

Fei Xiong, Email: xiongf@bjtu.edu.cn, Mobile phone: +86-13811992620.

REFERENCES

[1] Pons-Porrata, R. Berlanga-Llavori and J. Ruiz-Shulcloper. Detecting events and topics by using temporal references. Advances in Artificial Intelligence Vol. 2527 (2007), p. 11.

[2] J.T. Tsai, T.K. Liu and J.H. Chou. Hybrid Taguchi-genetic algorithm for global numerical optimization. IEEE Trans on Evolutionary Computation Vol. 8 (2004), p. 365–377.

[3] Pouuque and R. Steinberger. Multilingual and cross-lingual news topic tracking. in: Proceedings of the 20th International Conference on Computational Linguistics, Geneva (2004), Switzerland.

[4] Y. Yang, T. Pierce and J. Carbonell. A study on retrospective and on-line event detection. in: Proceedings of SIGIR-98, Melbourne (1998), Australia.

[5] Q. He, K.Y. Chang, E.P. Lim and B. Arindam. Keep it simple with time: a reexamination of probabilistic topic detection models. IEEE Transactions on Pattern Analysis and Machine Intelligence Vol. 32 (2010), p. 1795–1808.

[6] M. Nakatsuji, M. Yoshida and T. Ishida. Detecting innovative topics based on user-interest ontology. Journal of Web Semantics Vol. 7 (2009), p. 107–120.

[7] S. Jamali and H. Rangwala. Digging Digg: Comment mining, popularity prediction, and social network analysis. International Conference on Web Information Systems and Mining, Shanghai (2009), 32–38.

[8] Z. Hong, Z. Bing and Z. Hua. Hot trend prediction of network forum topic based on wavelet multi-resolution analysis. Computer Technology and Development Vol. 19 (2009), p. 76–79.

[9] H. Cheng and Y. Liu. An online public forecast model based on time series. Journal of Internet Technology Vol. 9 (2008), p. 429–432.

[10] J.J. Cheng, Y. Liu, H. Cheng, Y.C. Zhang, X.M. Si and C.L. Zhang. Growth trends prediction of online forum topics based on artificial neural networks. Journal of Convergence Information Technology Vol. 6 (2011), p. 87–95.

[11] H.P. Zhang, H.K. Yu, D.Z. Xiong and Q. Liu. HHMM-based Chinese lexical analyzer ICTCLAS. in Proceedings 2nd SIGHAN Workshop on Chinese Language Processing, 2003.

[12] Arthur and S. Vassilvitskii. K-means++: the advantages of careful seeding. in Proceedings of the Eighteenth Annual ACM-SIAM symposium on Discrete Algorithms, 1027–1035, 2007.

[13] J.A.K. Suykens and J. Vandewalle. Least squares Support Vector Machine classifiers. Neural Processing Letters, Vol. 9 (1999), p. 293–300.

[14] Information on http://en.wikipedia.org/wiki/Bootstrap_aggregating.

Advances in Future Manufacturing Engineering – Yang (Ed.)
© 2015 Taylor & Francis Group, London, ISBN 978-1-138-02817-3

Blind persons' online e-commerce learning influence factors analysis, based on social media

Wanqiong Tao & Chunhua Ju
Zhejiang Gongshang University, Hangzhou, Zhejiang, China

ABSTRACT: With the rapid development of e-commerce, the way of learning and using its related knowledge has drawn great attention. In recent years, the development of social media has created a favourable platform for the study of related knowledge on e-commerce. And this behavior on a special group—the blind has become a focus of the study. This article takes a social network site as an example. It stands on the issue from the blind user's point of view and is also based on relevant theory to explore the impacting factors of the blind user's online e-commerce studying on a social contact website. Experiments draw the conclusion that blind user's experience on a social contact website, as well as web information resources and website services has a positive impact for online e-commerce studying. However, the user experience has the maximum impact on online e-commerce learning behavior.

Keywords: social media; blind user; electronic commerce; online learning; social networking sites

1 INTRODUCTION

In 2013, Chinese e-commerce transactions reached 10 trillion yuan, a 26.8% increase over last year's figure. The development of e-commerce is overwhelming, the statistics from Taobao show that by 2017 the electricity industries could face a shortage of up to 4.457 million professionals. It is very urgent to cultivate e-commerce professionals. Its traditional teaching mode is only one, and teaching content updates are slow, so it is difficult to keep pace with the development of electronic commerce [1]. Social media has developed vigorously on the basis of the Internet's enormous influence. Now, in daily life, people prefer to acquire information and knowledge spread by social media from the Internet. Therefore, e-commerce online learning, based on social media, has become an important way to train e-commerce professionals.

Foreign research on social media pre-dates that of our country; the related research on electronic commerce online learning in the Chinese academic circle is developing step by step. At present, a lot of academic research focus is on the electronic commerce online learning, based on the computer network. Based on related theory research, for blind users, this paper uses social websites as examples to collect data and using spss19.0 tools to analyse data and exploring the factors which influence online e-commerce learning behaviour of the blind social website users.

2 THE OVERVIEW OF EXISTING E-COMMERCE LEARNING MODE

Electronic commerce, commonly known as E-commerce or eCommerce, is about trading in products or services using the computer networks, such as the Internet. The real e-commerce covers a great deal of knowledge, including management, computer technology and other various aspects of knowledge. So the requirements of enterprises for electronic commerce professionals are to generate more professionals, harder training of professionals as well as longer training cycles. [2].

2.1 *Traditional e-commerce learning mode*

The traditional e-commerce model mainly includes two aspects—e-commerce theory and e-commerce experiment. The e-commerce student's traditional theory learning style sees teachers as the core that mainly solves the problem of "what". The experimental study is heavily dependent on training simulation software, and plays an important role in the study of e-commerce; its purpose is to make things easy to understand and for students to remember the theoretical knowledge of electronic commerce. However, in modern society, e-commerce is developing at an amazing speed. So, simulation software is very difficult to keep pace with. In this case, there is a large gap between the effect of the student's practice and expectations—thus its significance for improving the students' skills is lost.

2.2 E-commerce learning mode based on network education

The concept of an e-commerce education model is that students can regulate, with recognition, motivation and behaviour according to their own needs, preferences and so on in the network [3]. This kind of learning mode includes two main characteristics —the interactivity in the learning process, and learning resources sharing.

At present, there are three main types of electronic commerce learning modes which are based on network education—these are linear learning mode, select learning mode and independent interactive learning mode.

Linear learning mode: Teaching content is transmitted mainly by video playback and information through real-time transmission, and teaching both sides to communicate directly on the network in real time; this model is usually applied based on Local Area Network (LAN).

Select learning mode: Through the teaching information resource in the network system, students, based on their own needs, can choose to watch videos or teaching courseware; it is the main mode of network teaching over a period of time.

Independent interactive learning model: The whole teaching system is parallel and open; there are only the organizer and guidance without an information controller in this model. But, it truly embodies the network spirit, therefore, it is the development direction of network teaching of the future. However, it is difficult to implement, unless there are powerful search engines, abundant network information resources, faster network speed, perfect organisation management and certain legal guarantees, learners with a strong independent consciousness, and above all, a revolution in teaching the ideas and learning concept.

The above three styles of network learning modes are the general learning pattern of electronic commerce in the network environment presently as well as in the future. The electronic commerce courses are strongly comprehensive, timely and practical . The student who is learning E-commerce not only has the knowledge of e-commerce, but can also use other knowledge for practical operations. Through experiencing the operation of various trading platforms, they can ultimately improve their abilities.

Electronic commerce online learning based on social media: Social media is the social interaction among people in which they create, share or exchange information and ideas in virtual communities and networks. It is a platform or tool which allows people to share ideas, opinions, views and experiences with others. Now, social media mainly includes search engine, Weibo, micro letter, social networking sites, q&a sites, BBS and so on.

2.3 The electronic commerce online learning based on social media

It has some advantages such as interactivity, sharing, at any time and practicality. Students and teachers can discuss problems more easily. Furthermore, they also can analyse new knowledge to achieve their better learning targets [4]. Learners can share useful e-commerce learning resources; teachers can also share good learning resources and students can obtain resources by using social media so as to be able to share. Since the transmission and update of information is rapid, learners can get the latest and the most cutting-edge e-commerce development information. At the same time, the study of e-commerce is no longer restricted by time and space because of social media. As long as there is network, computer and mobile device, we can study everywhere and anytime and that would improve the utilisation rate of the students' time greatly [5]. In addition, learners can obtain practical opportunities like online shopping, etc. Those practical operations help them strengthen the theoretical knowledge and enhance the learning effect.

3 AN EMPIRICAL ANALYSIS OF THE ELECTRONIC COMMERCE ONLINE LEARNING OF THE BLIND, BASED ON SOCIAL NETWORKING SITES AS AN EXAMPLE

With the rapid development of the Web2.0, online electronic commerce learning, based on social media has been overwhelming. Currently, Weibo and social networking sites have played an important role in the online electronic commerce learning. This paper takes the social websites as an example to analyse the online e-commerce learning behaviour of the blind.

The theory is based on the influencing factor of social networking sites on the electronic commerce online learning behaviour of the blind. Deng and Bao [6] built a model that analysed the influence factors of user interactive learning behaviour based on the social networking site environment and user characteristics. They concluded two main motivations—the external motivation and internal motivation. External motivation includes SNS technology environment, the SNS service method and SNS information resources. Internal motivation includes SNS user knowledge accumulation and self-worth ascension.

This paper analysed the influencing factors from three aspects, including personal qualities, relationships and network environment; it further subdivided the personal qualities and network environment factors. Personal traits included user experience, perceived usefulness and user preferences; the SNS network environment included site quality, information resource, service functions, and incentive mechanism. Then, this paper established a corresponding index system.

In this paper, the basic theories are from the following three aspects:

1. Wilson put forward the viewpoint of three factors including personality, interpersonal relationship and environmental factors;
2. Deng [6] concludes the influence factors of network user's information behaviour in his book—*Web Users Information Behavior Research.* as three aspects: The users' own factors, information and information factors, and social and natural environment factors.
3. The eight-factor theory summarised by Paisley, includes a formal organization, society, reference group, political system, cultural system, psychological factors, formal information systems, law and economy; this theory puts particular emphasis on environment. The focus of the three-factor theory summarised by A. Bouazza is individual traits, while the focus of the three-factor theory summarised by C. L. Mick and G. N. Lindsey is individual traits and interpersonal relationship.

Model building. Based on the above theories, this paper integrates the interactive study behaviour of social web users, e-commerce knowledge contents learning and the special requests of users who are blind on the Internet, and then makes a appropriate adjustment on the influencing factors of the e-commerce online study behaviour of the blind. The definitions of all the influencing factors and measurement indexes are illustrated in Table 1.

1. The user experience, measured by the length of time and frequency of using social networking sites for the blind user.
2. Effective awareness, measured by whether social networking sites can help the blind access electronic business knowledge, whether the blind think social networking siteshave a promoting effect on the development of electronic commerce, and the users' personal feeling about the importance of e-commerce in today's society.
3. Physical conditions, measured by the amount of time the blind users can continue surfing the Internet.
4. Interpersonal relationships, refers to the blind user's number of friends, friends; behaviour activity, the number of interaction with friends, and friend's status and logging on to the social networking site.
5. Information resources, refers to relevant information number and quality of e-commerce, as well as the relevance, accessibility and representation of it.
6. Functional service, including the function of comprehensiveness, ease of use.
7. Configuration quality refers to the design of screen reading software and keyboard.

Table 1. The influencing factors of blind person to social network sites.

Variable	The dimension	Dimensions of content
The user experience	UE1	The length of using
	UE2	The frequency of using
Effective perception	EP1	Whether access to electronic business knowledge
	EP2	Whether to promote the development of electronic commerce
	EP3	Feel the importance of electronic commerce
Physical condition	PC1	The body would be accepting of surfing the Internet over time
Interpersonal relationships	IRS1	Number of friends
	IRS2	Friends activity
	IR3	The number of interactions with close friends
	IR4	Friend's status and login
Information resources	IR1	Information quality
	IR2	The number of e-commerce knowledge information
	IR3	The representation of information resources
Functional service	FS1	Ease of use
	FS2	Comprehensiveness
Configure the quality	CQ1	Read the screen software
	CQ2	The design of the keyboard
The blind electronic commerce online learning	BEL1	Update, evaluation, sharing, forward or group discussion
	BEL2	Active learning

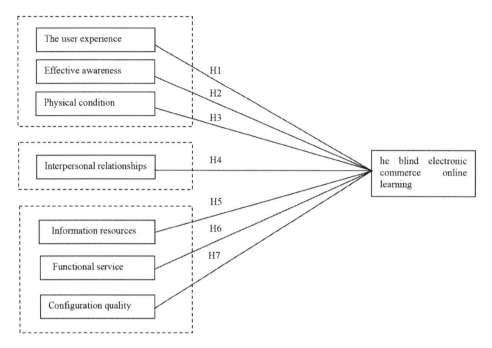

Figure 1. The model and hypothesis of the influence factor of blind users' e-commerce online learning behavior on social networking sites.

8. The blind electronic commerce online learning, refers to the behaviour of original, forwarding, comments, reading, and group participation of social networking site users and the degree of taking the initiative to proceed.

According to the analysis of the above several dimensions, a research model and hypothesis of the influence factor of blind user's e-commerce online learning behaviour on social networking sites is built, as shown in Figure 1.

Assume the following:

H1: The blind user's experience in social networking sites has a positive effect on electronic commerce online learning behaviour;

H2: The blind user's Effective perception in social networking sites has a positive effect on electronic commerce online learning behaviour;

H3: The blind user's physical conditionon social networking sites has a positive effect on electronic commerce online learning behaviour;

H4: The blind user's interpersonal relationships on social networking sites has a positive effect on the electronic commerce online learning behaviour;

H5: Social networking website information resources have a positive effect on the blind e-commerce online learning behaviour;

Table 2. The reliability of latent variables.

Research factors	Alpha
The user experience	0.829
Effective awareness	0.813
Physical condition	0.818
Interpersonal relationships	0.819
Information resources	0.834
Functional service	0.837
Configuration quality	0.812
The blind electronic commerce online learning	0.845

H6: Social networking site service completion degree has a positive effect on the blind e-commerce online learning behaviour;

H7: The Internet configuration quality has a positive effect on the blind e-commerce online learning behaviour.

In this study, we use the questionnaire survey method, according to the user's experience, effective perception, physical condition, interpersonal relationship, information resources, functional service, configuration quality and the blind electronic commerce online learning; a total of 8 dimensions and 19 indexes to design the corresponding questionnaire were used as a measure.

Table 3. Regression coefficient and significance test results.

Coefficient model		Thestandardized coefficients		Standardized coefficient		
		B	Standard error	A trial version	t	Sig.
1	(constant)	1.226	0.274		4.469	0.000
	The user experience	0.301	0.072	0.287	4.149	0.000
	Effective awareness	−0.081	0.074	−0.100	−1.098	0.273
	Physical condition	0.075	0.056	0.104	1.338	0.182
	Interpersonal relationships	−0.004	0.060	−0.005	−0.063	0.950
	Information resources	0.152	0.067	0.155	2.270	0.024
	Functional service	0.166	0.067	0.163	2.469	0.014
	Configuration quality	0.045	0.078	0.054	0.585	0.559

a. Dependent variable: the blind electronic commerce online learning.

Study questionnaires were issued for the Internet Blind.

1. The data collection
   This investigation by looking for online QQ group of the blind conducted the survey for the blind. A total of 256 questionnaires were received, 245 valid questionnaires, 11 copies of the questionnaire to be invalid (respondent never used social networking websites).
2. Data analysis and results
   1. The reliability analysis. Reliability refers to consistency or stability of the measurement result. This study used Spss 19.0 for reliability analysis. The overall reliability of Alpha index is 0.845, is greater than 0.7, so reliability is high. Each latent variable reliability is shown in Table 2. Cronbach's Alpha of all factors are greater than 0.7, so the measurement has good reliability.
   2. Regression analysis and hypothesis testing results. Using spss19.0 to proceed regression analysis, the regression coefficient and significance test results of the blind persons' electronic commerce online learning and other factors are shown in Table 3.

According to the above perceptions, only the user experience, information resources and functional service are significantly higher, so only three hypothesis are true:

H1: The blind users' experience in social networking sites has a positive effect on the electronic commerce online learning behaviour;
H5: Social networking website information resources have a positive effect on the blind e-commerce online learning behaviour;
H6: Social networking site service complete degree has a positive effect on the blind e-commerce online learning behaviour;

And by the above-mentioned B value of the user experience of is 0.301, so the user experience on the influence of the electronic commerce online learning is the largest.

4 SUMMARY

This study aims to explore the influencing factors of the e-commerce online learning of the blind based on social networking sites, and the degree of these factors influencing the electronic commerce online learning for the blind. At first, we put forward some possible influencing factors according to the experts in the related field of the theoretical research, combined with the characteristics of blind people to get to the Internet, establishing model and then we designed the questionnaire. At the end we analyzed the data form and collected the questionnaires. Proceeding regression analysis is high reliability of the questionnaire; it is concluded that social networking sites for the blind user experience, pros or cons of social networking site information resources and the degree of perfection of the social networking site functional service have a positive impact on the e-commerce online learning behaviour of the blind, and that the user experience on the influence of the electronic commerce online learning is the largest.

ACKNOWLEDGMENTS

This research is supported by The National Key Technology R&D Program of China (Grant 2012BAI34B01-5), Major Project for Science and Technology Plan of Hangzhou (No. 20122511A26), Natural Science Foundation of Zhejiang Province (Grant LY14F020002), This research is supported by the Contemporary Business and Trade Research

Center of Zhejiang Gongshang University which is the Key Research Institutes of Social Sciences and Humanities Ministry of Education (Grant 14SMXY04YB).

REFERENCES

[1] Hu Tao and Yao Shun. Electronic commerce teaching method reform and practice [J]. *China's university teaching*, 2012,10: p67–70.

[2] Lin Yuejuan. E-commerce professional students' practical ability training of blended learning management research [D]. Guangzhou University, 2013.

[3] Zhang Weiyuan. Network development of teaching model theory construction and applied [J]. *Modern Distance Education Research*, 2012, 19: p7–23.

[4] Michael Thomas, Howard Thomas. Using new social media and Web 2.0 technologies in business school teaching and learning [J]. *Journal of Management Development*, 2012, Vol. 31 (4): pp.358–367.

[5] Yingxia Cao, Paul Hong. Antecedents and consequences of social media utilization in college teaching: a proposed model with mixed-methods investigation [J]. *On the Horizon*, 2011, Vol. 19 (4): p297–306.

[6] Deng ShengLi, Bao Wei. Factors affecting the empirical study on the social networking site user interaction learning [J]. *The intelligence theory and practice*, 2012, 03: p57–61.

Advances in Future Manufacturing Engineering – Yang (Ed.)
© 2015 Taylor & Francis Group, London, ISBN 978-1-138-02817-3

The application of virtual simulation technology to information technology training

Qingyu Xie & Heng Qin
State Grid of China Technology College, Jinan, Shandong, China

ABSTRACT: In today's society, information technology develops very rapidly along with the unceasing progresses of technology, and information technology training has already become one of the important parts in the current enterprise training. In the information technology training, virtual simulation technology can well meet the high requirements of practical trainings for the number and configuration of terminal equipment, and can also play an important role in the information technology training such as system security, database, and office software. In this paper, the development of virtual simulation technology and its current application to information technology training are mainly analyzed.

Keywords: virtual simulation technology; information technology; technical training; application methods

1 INTRODUCTION

In the 21st century, the modern information technology represented by computer technology and Internet technology develop fiercely and rapidly in the fields such as computer and small phones. But the subsequent influence is that the information technology in all industries is comprehensively upgraded. In the information technology training, virtual simulation technology can well meet the high requirements of practical trainings for the number and configuration of terminal equipment, and can also play an important role in the information technology training such as system security, database, and office software. In this paper, the development of virtual simulation technology and its current application to the information technology training are mainly analyzed, and the application, the advantages and disadvantages of virtual simulation technology in the practice of information technology training are discussed. Finally the future application to the information technology training in China is prospected.

2 THE DEVELOPMENT OF VIRTUAL SIMULATION TECHNOLOGY

Along with the rapid development of information technology and the wide application of computer, virtual simulation technology gets a breakthrough and is extensively applied to all fields. Today's computer virtual simulation technology means that a virtual environment is available for users through a certain way of apparent perception using computer technology and its hardware. Computer virtual simulation technology combines computer graphics technology, simulation technology, artificial intelligence technology, computer sensor technology, computer display technology, and network technology together to process the actual conditions in all industries using a three-dimensional digital model including computer graphics [1].

In a broad sense, computer virtual simulation technology means that the hardware and software system demanded by users is simulated with software to commonly work with computer entity system [2]. It is mainly divided as follows: the simulation for computer operating systems; the simulation for computer running environment. At present, the terminals of most computers are Windows system and Linux system, and therefore the terminal system is simulated mainly by using Vmware series software in the information technology training. The most significant feature of computer virtual simulation technology is that the operator can take advantage of the natural existing way and virtual 3d environment to exchange and operate, so that the previous model of people to perceive the environment only through personal experience is changed. In other words, people can indirectly perceive and know the environment through a virtual environment.

3 THE APPLICATION OF VIRTUAL SIMULATION TECHNOLOGY TO INFORMATION TECHNOLOGY TRAINING

In the information technology training, virtual simulation technology is mainly used in information security research and computer information technology teaching.

3.1 The application of virtual simulation technology to information security research

Resource sharing and information sharing are realized in the network environment because of the development of computer network technology. But information security attacks feature destructivity and uncontrollability, so the application of virtual simulation technology for information security protection is necessary [4].

1. Mathematical analytical model
   Mathematical analytical model is to abstract information security research as a mathematical process and simulate the changes of the objects in various environments and circumstances using a set of mathematical formulas. A set of differential equations and discrete recursion expression are the most commonly used.
2. Monte Carlo simulation
   In Monte Carlo simulation, a more accurate mathematical model is not established, but the status change of all hosts is determined through probability statistics method and the whole process is divided into many successive time slices. A Monte Carlo simulation is a simulation mechanism using the discrete time, so it cannot cope with the complex, dynamic network behaviors.
3. Emulation system
   Emulation system is a small-scale real network system and makes multiple hosts linked together using the network, and one of them is used for intensively managing software and network performance observation analysis tools, etc. Emulation system can implement the network traffic simulation in different scenarios and the validation and analysis under a different network condition.

3.2 The application of virtual simulation technology to computer information technology teaching

At present, the development of information technology has influenced all aspects of people's life; the Internet also has entered into all families; microcomputer has become one of necessities of the new-generation families. The development of smartphone promotes information technology to complete the permeation from macroscopic objects to microcosmic objects. Information technology training has become a fashionable trend, and the application of virtual simulation technology to the teaching in information technology training can help better complete the training. Vmware Workstation and Vmware ESX Server are the common local virtual software and server version software

in the process of information technology training. The local virtual software and server version software can virtualize the vast majority of Windows and Linux operating systems.

1. The application method of Vmware Workstation
   In an information machine room, the terminal of the computer used by every student needs to install VM virtual software and operating system image. In the concrete operation process, a virtual computer for the training can be set up through Vmware Workstation in teacher's computer, in which necessary hardware parameters and corresponding operating system as well as the initial system configuration and the database software operation are necessary. The installed image is shared using the Local Area Network (LAN) to every student's machine terminal, and finally learners load the system image at the student terminal and then can virtually operate it using VM virtual software on the student machine [5].
2. The application method of Vmware ESX Server
   Vmware ESX Server software system should be installed in a computer Server and then a virtual machine is established as Vcenter management machine; after planning and processing IP address and hardware, users can log in Vcenter management machine to create and duplicate the virtual machine for the training, so learners can finally connect and operate the virtual machine through the student remote desktop [6].

4 THE ADVANTAGES AND DISADVANTAGES OF APPLYING VIRTUAL SIMULATION TECHNOLOGY TO INFORMATION TECHNOLOGY TRAINING

Virtual simulation technology has not been applied to information technology training for a not very long time, and therefore there are still many disadvantages. However, it possesses absolute advantages to let virtual simulation technology continuously develop in information technology training. In the following, the advantages and disadvantages of applying the current virtual simulation technology to information technology training are analyzed mainly by studying the application of virtual simulation technology to information technology.

4.1 The disadvantages of applying virtual simulation technology to information technology training

1. High cost
   Vmware Workstation virtual machine should be required in each student's machine, but this

has high requirements for the hardware memory of the machine and also needs a software cost. If the hardware condition of Vmware ESX Server in information room terminal is the same as that of Vmware Workstation virtual machine, the main cost is on the Server and also very high [7].

2. Virtual simulation technology is not fully recognized by learners

   Virtual simulation technology has not been input into the actual information technology training for a not very long time, and therefore it is not fully recognized by learners, and they will not adapt to the use of virtual simulation technology in the beginning and accordingly produce psychological resistance.

4.2 The absolute advantages of virtual simulation technology to information technology training

1. Reducing expenses

   In spite of the high cost of virtual simulation technology, one or even more operable virtual machine systems can be created using virtual technology for each student under the conditions of limited funds and insufficient student terminals. Thus, practice methods are added for learners in information technology courses, and also users are needless to worry about the uncontrollable destruction to physical environment or ontology information if a virtual machine operating system is applied.

2. Helpful for implementation of information technology training

   Virtual machine is easy to create and stable to operate, and also it can create the same hardware, the same configuration, the same software, and the same operating system for each student. For example, the same Windows Server virtual machine and Red Hat Linux virtual machine are created, so that learners can install system, operate database, and make exercises in the system at the initial environment with the same virtual terminal, so as to let learners deepen the understanding of virtual simulation technology in the practice.

5 THE PROSPECT FOR THE APPLICATION OF VIRTUAL SIMULATION TECHNOLOGY TO INFORMATION TECHNOLOGY TRAINING

Information technology is a new development form of computer technology and has been toward diversified development currently. Moreover, information technology has become an important course of the nine-year compulsory education in China. The specialty has been very popular in recent years, and it is learned by a large number of students. All these mean that the future work and life will be fully filled with information technology.

6 CONCLUSION

According to the above introduction of virtual simulation technology and the application of information technology training, the integration of virtual simulation technology in the information technology training is an inevitable developing trend of information technology in China. And also sharing and virtualization will become new labels in the information age.

REFERENCES

[1] Liang Peng, Pengyu-hua, Xu Bing-yin, The application of correlation method in cable fault location using pulse reflection [J]. Shandong University (Engineering Science), 2004, 34(1): 89–91.

[2] Desheng Xing. The Application of Virtual Simulation Technology in the Electronic Technology Teaching [J]. Journal of Shandong Institute of Agricultural Engineering, 2014, 8 (11): 1186–1188.

[3] Songxiang Li. Research on the Application of Virtual Simulation Technology in Electronic Technology Course Teaching [J]. Journal of Hunan Normal University, 2012, 34 (1): 120–121+125.

[4] Yang Liu. Discussion on the Application of Virtual Technology in the Information Technology Training [J]. Journal of Electronic Technology and Software Engineering: Software Applications, 2010, 7 (22): 111–112.

[5] Jiangang Liao. The Application of Computer Virtual Simulation Technology in the Numerical Control Teaching [J]. Journal of Teaching Practice, 2011, 4 (20): 178+180+185.

[6] Yaxin Zeng. The Application Practice of Information Technology in Enterprise Technical Staff Training [J]. Journal of Hebei Normal University, 2010, 8 (11): 249–251.

[7] Zhihong Sun. The Exploration and Practice of the Training Pattern Combining Task with Cases—the Application of Teachers' Information Technology Training for Example [J]. Journal of Shandong Normal University, 2011, 4 (20): 178+180+185.

[8] Hong Chen, Yuewu Wang, Jiwu Jing, Xiaofeng Niu. The Application of Simulation Technology in Information Security Research [J]. The Collection of the Papers at the National Computer Security Academic Exchange, 2010, 8 (11): 378–379.

Biochemistry and medicine

Advances in Future Manufacturing Engineering – Yang (Ed.)
© 2015 Taylor & Francis Group, London, ISBN 978-1-138-02817-3

Automated identification of Fetal Heart Rate acceleration combined with fetal movement

Xiaodong Li, Yaosheng Lu & Zhaoxia Chen
Department of Electronic Engineering, College of Information Science and Technology, Jinan University, Guangzhou, China

ABSTRACT: The Fetal Heart Rate (FHR) acceleration is one of the most important parameters in FHR analysis. Traditional acceleration identification is based on FHR baseline estimation. A novel automated acceleration identification algorithm combined with fetal movement was proposed in our study. In this method, acceleration identification does not rely on the baseline. To assess the performance of this algorithm, a comparative study was made against an existing acceleration identification algorithm of a fetal monitor's computerized analysis system and the visual counting by an experienced obstetrician. The proposed algorithm achieved sensitivity above 80% and Positive Predictive Value (PPV) of 99% for 125 FHR tracings, showing that it has a good performance. Meanwhile the algorithm would perform better if the fetal movements are detected exactly and the coupling of acceleration and fetal movement is good. Overall the design solution that accelerations can be identified independent of the baseline has been proven feasible.

Keywords: acceleration; automated identification; fetal movement

1 INTRODUCTION

Electronic Fetal Heart Rate (FHR) monitoring has become a widely used way for fetal surveillance in order to assess fetal condition. Analysis of the FHR patterns defined by the characteristics of baseline, variability, accelerations and decelerations is of significance for understanding the status of fetus [1].

As for FHR acceleration, it is critical to antenatal Non-Stress Test (NST), which aims to assess the "reactive or non-reactive" fetus according to whether or not the acceleration criteria were met. A normal NST necessitates the presence of two or more accelerations during a 20-min period. In most cases, a normal NST is predictive of a good perinatal outcome for one week [2].

Many researchers have studied and proposed different automated FHR acceleration identification algorithms. Mantel et al. [3] detected accelerations in two phases: basic criteria for detection of accelerations are introduced in phase 1 and signal loss is dealt with in phase 2. In Andersson's method, if the calculated baseline is considered unreliable for certain FHR segment, no accelerations are detected in that segment [4]. In addition, automated acceleration identification function is one of the basic modules of computerized FHR analysis systems, such as System 8000/8002 [5], 2CTG [6] and SisPorto 2.0 [7].

In general, the currently-used acceleration identification algorithms are on the basis of FHR baseline. Once the baseline has been estimated, accelerations can be recognized consequently. However, it brings a circular problem. Baseline is by definition determined after accelerations and decelerations exclusion [1], but accelerations and decelerations are identified depending on baseline estimation in practice. Therefore, acceleration identification previous baseline recognition should be a solution.

The relationship between FHR accelerations and fetal movements has been studied for several decades [8–11]. Wheeler et al. [9] thought that fetal body movements were associated with a brief tachycardia. According to DiPietro et al.'s study, fetal movement and FHR became more integrated with advancing gestation, manifesting as the increased coupling and the decreased latency between fetal movements and FHR changes [11].

In our study, an automated FHR acceleration identification algorithm was proposed to bring a solution to the circular problem. The major difference between the proposed algorithm and others is that fetal movements are taken into consideration and acceleration identification is no longer relied on baseline estimation. Therefore, this approach is complied with the definitions of baseline and acceleration.

2 MATERIALS AND METHOD

Signal acquisition and preprocessing. One hundred and twenty-five antepartum FHR tracings, which were obtained clinically from the People's Hospital of Baiyun District, Guangzhou by SFR618 fetal monitors made in Sunray Medical Apparatus Co, Ltd., were chosen according to following criteria: (1) good quality with <10% signal loss and (2) the duration of >20 minutes. The average duration of these tracings were 22.65 ± 3.84 (mean ± SD) minutes. During signal acquisition, the patients were asked to press a handle event marker whenever fetal movements occur.

However, due to the signal acquisition problem, there are many missing points existing in original FHR tracings. Meanwhile, FHR changed to a much bigger or smaller value abruptly are regarded as artifacts and needed to be removed. Then linear interpolation is used to fill the gaps of tracings. It is worth mentioning that fetal movement signals are treated as well as FHR in order to keep time consistence.

Base value determination. Although an FHR baseline is not a necessity in our method, a temporary reference line or incompetent baseline is needed as a reference for subsequent acceleration identification. It is notable that the temporary baseline mentioned herein is not a true FHR baseline and cannot replace the baseline in FHR analysis.

After preprocessing, the FHR tracings are classified into three groups according to the gross fetal movements. If the fetal movement rate is more than 60 per hour, the FHR tracing is categorized as Group 1; if the rate is less than 40 per hour, the FHR tracing is categorized as Group 3. And the remaining tracings belong to Group 2.

For a FHR tracing, the base value is determined by a histogram analysis of the stable FHR segment defined as a group of five adjacent samples whose differences are less than 10 beats per minute (bpm). All FHR values of stable segments are rounded to increments of 5 bpm, of which the frequency distribution is calculated. If the number of FHR value in histogram is more than or equal to 8, 8 FHR values with eight biggest frequency are selected as options for the tracings in Group 1, and 5 FHR values are selected for that in Group 2, and 3 FHR values are chosen for that in Group 3. If the histogram does not contain more than 3 FHR values, all the values are regarded as options, whichever group the FHR tracing belongs to. According to the clinical findings, we found that the more fetal movements are, the more accelerations are [12], and consequentially, the larger the frequencies of FHR values in acceleration segments may be. This finding is noteworthy because the stable segment is not equivalent to baseline segment. Therefore, the number of optional FHR base value differs based on the number of fetal movement. Among all the options, the smallest FHR value is confirmed as the base value of tracing. This is done to make the base value as small as possible to avoid the corresponding temporary baseline crossing the FHR accelerations in next section.

Fetal movement-coupled acceleration identification. In the guideline issued by the National Institute of Child Health and Human Development (NICHD), acceleration is defined as an abrupt FHR increase above the baseline with an acme of ≥15 bpm and duration of ≥15 seconds [1].

In this section, the base value computed above is used to establish a temporary baseline to recognize accelerations. The temporary baseline is a straight line and its initial value is equivalent to the base value.

For the whole FHR tracing, it is split into several 10-minute segments. However, a 10-minute FHR segment is extracted every five minutes to make sure that there is an overlap between two adjacent segments to avoid an acceleration being separated. For each FHR segment, a threshold for the longest duration of the acceleration is preset by the corresponding gross fetal movement linked with that segment. If the fetal movement rate for 10-minute segment is more than 60 per hour, the threshold is set to be 6 minutes. And the threshold is set to be 2 minutes when fetal movement rate is less than 30. Otherwise, this threshold will be 4 minutes. The reason for this is that accelerations appear constantly if fetal movement becomes frequent.

For each 10-minute segment, suspicious accelerations are recognized firstly based on the definition by temporary baseline. Then accelerations would be confirmed and identified if they satisfy the following requirements: (1) the duration of less than the preset threshold and (2) the proportion that the biggest 30% FHR values account for is more than 0.3.

For the suspicious accelerations which are not identified as accelerations, the temporary baseline will move up 1 bpm and accelerations will

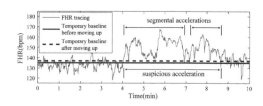

Figure 1. Suspicious accelerations segmented from a big suspicious acceleration due to an adjusted temporary baseline.

be recognized anew. If necessary, repeat this step until all the accelerations that meet requirements are found or no acceleration is recognized. If a suspicious acceleration is divided into several segmental accelerations during this period as shown in Figure 1, all segmental accelerations should be confirmed and processed independently.

It is important that the temporary baseline should be reset to the base value when the steps for an FHR segment end and those for the next segment are about to start.

In our study, if the acceleration occurs within 2 seconds of fetal movement, it would be confirmed ultimately and regarded as the fetal movement-coupled acceleration which is the outcome of the proposed algorithm. Although it is proven that there is an association between fetal movement and acceleration, the detailed time relationship is not yet known. Zhao et al. [13] found that FHR accelerations coincide with or preceded fetal movements nearly all the time, while DiPietro et al. [11] gave an opposite conclusion that accelerations lag fetal movements. Therefore, the accelerations that occur within 2 seconds of the fetal movement occurrence are regarded as coupled in our algorithm.

3 RESULTS AND DISCUSSION

The automated acceleration identification algorithm was tested with 125 FHR tracings in this study, and Figure 2 illustrates how this algorithm works to recognize accelerations. The identification processing takes place in three phases: (1) the base value of a FHR tracing is determined based on the histogram analysis, (2) suspicious accelerations are found in accord with definition and requirements and (3) accelerations are regarded as the fetal movement-coupled if they couple with fetal movement.

In order to compare this algorithm with the traditional algorithm, the acceleration identification algorithm [14] from Sunray's computerized analysis system installed in fetal monitor was selected. In this selected algorithm, accelerations are detected based on the definition with the premise that an FHR baseline is given. As shown in Figure 3, the accelerations identified by our algorithm (marked as Alg1 thereinafter) were marked as ACC 1 and the outcome of the selected algorithm (marked as Alg2) were marked as ACC 2. The result of Alg1 is as good as that of Alg2 if the coupling of acceleration and fetal movement is good as shown in Figure 3(a). However, the accuracy of Alg1 is significantly affected by the fetal movement. Even if the acceleration accords with the definition and requirements, it would not be found out on the condition that no fetal movement occurs. As such, an acceleration was overlooked in Figure 3(b). However, this approach is considered as reasonable in essence. In the case of accelerations, one would rather want to fail to detect some than over detect false accelerations.

For the purpose of assessing the performance of acceleration identification algorithms quantitatively, an obstetric expert with more than thirty years of experience was invited to count the accelerations visually. By taking the result of expert as the gold standard, the sensitivity and Positive Predictive Value (PPV) for two acceleration identification algorithms are shown in Table 1. Sensitivity is the ratio of accurate automated identified accelerations to the total accelerations identified by the expert, while PPV is calculated as the ratio of accurate automated accelerations detected to the total accelerations by the algorithm. Overall Alg1 is slightly inferior to Alg2, and this result is predictable because FHR baseline plays a significant role in practical acceleration identification. Without precise baselines, the sensitivity achieved

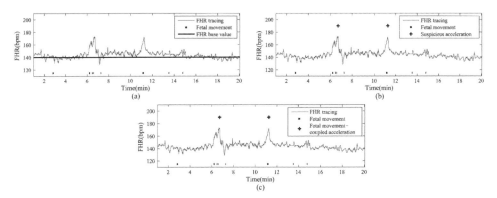

Figure 2. Sequence of signal processing to identify the accelerations: (a) the base value determined, (b) suspicious accelerations found based on definition and requirements, and (c) acceleration confirmed by fetal movement.

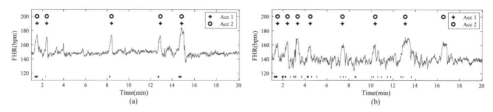

Figure 3. Acceleration identification by the proposed algorithm and traditional algorithm in the condition that: (a) all the accelerations relate with fetal movement, and (b) not all the accelerations relate with fetal movement.

Table 1. The sensitivity and PPV for two automated acceleration identification algorithms.

	Alg1	Alg2
Sensitivity (%)	80.34	90.45
PPV (%)	98.96	95.83

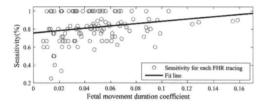

Figure 4. Sensitivities for the FHR tracings.

by our proposed algorithm is still more than 80%. In addition, it has higher PPV indicating that the automated acceleration identified by this method is more likely to be accurate.

In 125 FHR tracings, 632 accelerations identified by the expert were confirmed to have relationship with fetal movement, accounting for 88.76% of total accelerations. That also means that 90.51% of fetal movement-coupled accelerations were detected by our algorithm, and this value is equal to the sensitivity of the traditional algorithm.

To assess further the effect of fetal movement on the proposed acceleration identification algorithm, the sensitivity for each FHR tracing achieved by this algorithm was calculated respectively. The results are shown in Figure 4, in which the scattered circles represent the sensitivities, with specific fetal movement duration coefficient indicating the frequency and duration of fetal movements, which was the time ratio of fetal movement part to a whole FM tracing. A linear curve is obtained by the least squares method, which fitted these circles. The results suggested that our algorithm may perform more reliably with greater fetal movement duration coefficient, which means that more FHR accelerations are likely to couple with fetal movement.

4 CONCLUSIONS

An automated FHR acceleration identification algorithm combined with fetal movement was proposed in our work. This approach offered a novel solution to acceleration identification, for acceleration identification no longer depend only on FHR baseline. In this algorithm, the base value or the temporary baseline of FHR tracing is determined firstly and then accelerations are found based on the definition and the requirements and confirmed by fetal movement finally. Experimental results indicate that our proposal achieved a good performance with a sensitivity of 80% and a PPV of 99%. Although it had slightly worse availability than the existing traditional algorithm, the outcomes were acceptable because no precise baseline was provided. Admittedly there was room for improvement in our experiment. Due to the limitations in terms of the medical devices, handle event markers which are used to record fetal movements artificially still serve in many hospitals. Therefore, fetal movement recognition is potentially subjective. In this study, antepartum signals were selected to insure the veracity of the fetal movement as much as possible because a manual indication of the presence of fetal movement may be inaccurate in the latter stages of delivery.

In conclusion, a high rate of coupling of acceleration and fetal movement contributes to a high acceleration recognition rate in our algorithm. Therefore, this algorithm will perform better if the fetal movements are detected exactly. Therefore, we will look for a new fetal movement recognition method to further improve the availability of the proposed algorithm.

On the other hand, this acceleration identification algorithm has the model significance to baseline estimation algorithm. Although a number of FHR baseline estimation algorithms have been proposed, they are still not used clinically at a large scale [15]. Many existing methods are not satisfactory when the FHR tracings are irregular with frequent and dramatic fluctuations for the difficulties in distinguishing baseline segments from non-baseline segments. Therefore, the proposed algorithm obviously helps with the problem, as accelerations are part of the non-baseline segments. It is also of value to apply this acceleration identification algorithm to baseline estimation, and it will be our further study.

ACKNOWLEDGMENTS

We are grateful to Ruyi Qin, a specialist obstetrician from the People's Hospital of Baiyun District, Guangzhou, for her help in reading the FHR tracings and detecting the accelerations. This work was supported by the Science and Technology Planning Project of Guangdong Province (No. 2012B091000053) and Guangzhou city (No. 201207YH017).

REFERENCES

[1] G.A. Macones, G.D. Hankins, C.Y. Spong, J. Hauth and T. Moore: The 2008 National Institute of Child Health and Human Development Workshop Report on Electronic Fetal Monitoring: Update on Definitions, Interpretation, and Research Guidelines. J. Obstet. Gynecol. Neonatal Nurs. Vol. 37 (2008), pp. 510–515.

[2] R. Liston, D. Sawchuck and D. Young: for the Society of Obstetrics and Gynaecologists of Canada, and the British Columbia Perinatal Health Program. Fetal health surveillance: antepartum and intrapartum consensus guideline. J. Obstet. Gynaecol. Can. Vol. 29 (2007), pp. S3–56.

[3] R. Mantel, H.P. Van Geijn, F.J.M. Caron, J.M. Swartjes, E.E. Van Woerden and H.W. Jongswa: Computer analysis of antepartum fetal heart rate: 2. Detection of accelerations and decelerations. Int. J. Biomed. Comput. Vol. 25 (1990), pp. 273–286.

[4] S. Andersson: Acceleration and deceleration and baseline estimation[D]. Chalmers University of Technology, Goteborg, 2011.

[5] G.S. Dawes, M. Moulden and C.W.G. Redman: System 8000: computerized antenatal FHR analysis. J. Perinat. Med. Vol. 19 (1991), p. 47–51.

[6] D. Arduini, G. Rizzo, G. Piana, A. Bonalumi, P. Brambilla and C. Romanini: "Computerized analysis of fetal heart rate. I. Description of the system (2CTG)", J. Matern. Fetal Invest. Vol. 3 (1993), p. 159–163.

[7] D. Ayres-de-Campos, J. Bernardes, A. Garrido, J. Marques-de-Sa and L. Pereira-Leite: Pereira-Leite L (2000) Sisporto 2.0: a program for automated analysis of cardiotocograms. J. Matern. Fetal Neonatal Med. Vol. 9 (2000), p. 311–318.

[8] S. Aladjem, A. Feria, J. Rest and J. Stojanovic: Fetal Heart Rate Responses to Fetal Movements. Bjog: Int. J. Obstet. Gynaecol. Vol. 84 (1977), p. 487–491.

[9] T. Wheeler, G. Gennser, R. Lindvall and A.J. Murrills: Changes In The Fetal Heart Rate Associated With Fetal Breathing and Fetal Movement. BJOG: Int. J. Obstet. Gynaecol. Vol. 87 (1980), p. 1068–1079.

[10] I. Baser, T.R.B. Johnson and L.L. Paine: Coupling of Fetal Movement and Fetal Heart Rate Accelerations as an Indicator of Fetal Health. Obstet. Gynecol. Vol. 80 (1992), p. 62–66.

[11] J.A. DiPietro, D.M. Hodgson, K.A. Costigan, S.C. Hilton and T.R. Development of fetal movement—fetal heart rate coupling from 20 weeks through term. Johnson: Early Hum. Dev. Vol. 44 (1996), p. 139–151.

[12] L.M. Jansson, J.A. DiPietro, M. Velez, A. Elko, E. Williams, L. Milio, K. O'Grady and H. Jones: Fetal neurobehavioral effects of exposure to methadone or buprenorphine. Neurotoxicol. Teratol. Vol. 33 (2011), p. 240–243.

[13] H. Zhao and R.T. Wakai: Simultaneity of foetal heart rate acceleration and foetal trunk movement determined by foetal magnetocardiogram actocardiography. Phys. Med. Biol. Vol. 47 (2002), p. 839–846.

[14] S. Wei: Algorithms and Systemization of Automatic Cardiotocography analysis[D]. Jinan University, Guangzhou, 2013.

[15] B.N. Krupa, M.A.M. Ali and E. Zahedi, Computerized Fetal Heart Rate Baseline Estimation Based on Number and Continuity of Occurrences. in: IFMBE Proceedings Vol. 21 (2008), p. 162–165.

Advances in Future Manufacturing Engineering – Yang (Ed.)
© 2015 Taylor & Francis Group, London, ISBN 978-1-138-02817-3

Study on evaluation of sampling inspection plan for cigarette trademark paper

Wei He, Li Cheng & Zhekun Li
College of Mechanical and Electrical Engineering, Kunming University of Science and Technology, Kunming, China

Geyi Liu
Qujing Cigarette Factory, HongYunHongHe Tobacco Co. Ltd., Qujing, China

ABSTRACT: In order to effectively evaluate the science and rationality of sampling inspection plans for cigarette trademark paper, a binomial distribution-based method for calculating the characteristic values of sampling is proposed, by analyzing the sampling inspection plans' characteristics of cigarette trademark paper. The characteristic values of three sampling inspection plans, such as, acceptance probability, risk quality, etc. can be calculated and analysed with Excel. The results show that plan C (Testing sample size 100 from the overall volume of a batch of products, when the number of nonconforming products X ≤ 3, determines to receive the batch of products—or judging to reject) is the best for the enterprise.

Keywords: sampling inspection plans; cigarette trademark paper; binomial distribution; acceptance probability

1 INTRODUCTION

Sampling inspection is the application of the statistical sampling method in product quality inspection. The sampling inspection can not only guarantee the quality of products, but also improve economic benefit, therefore, it is widely used in domestic industrial production [1]. A scientific and rational sampling inspection plan has been the subject of concern of cigarette manufacturers and material producers also, to develop scientific sampling plans and reasonably control sampling risk, we must first calculate and analyse the characteristic value of the sampling plan [2]. At present, when enterprises make a sampling plan for cigarette trademark paper, if using the national sampling standard (GB/T2828-2003), tone can find the corresponding appendix obtaining solution sampling characteristic value [3]; if using custom plans, there is no related literature to provide the calculation method of characteristic values of sampling inspection plan for cigarette trademark paper; moreso, there can be no scientific comparison of different plans. Therefore, research on the calculation method of characteristic of sampling inspection plan for cigarette trademark paper, aims to make a comparative study of the different sampling plans, thereby evaluating the science and rationality of sampling plans.

Sampling inspection is done through the sample to judge the entire issue; it is inevitable to that two two types of errors are made. Type 1 error: the qualified products approved wrongly and are assumed to have failed the test—leading to the rejection of the entire batch, so that producers suffere losses—this is called producer risk; the risk ratio is represented by α; Type 2 error: the unqualified products wrongly are wrongly assumed to qualify the test; the consumer thus suffers losses—this is called consumer risk; the risk ratio is represented by β [1]. An effective inspection must first ensure that α is not too large, and secondly in the α controlled conditions to ensure that β should be as little as possible—that is, the test power $(1-\beta)$ is greater, the discriminating ability of the test better.

2 THE PROCESS OF SAMPLING INSPECTION PLAN FOR CIGARETTE TRADEMARK PAPER

Through the investigation, the author learned that the bulk purchase of cigarette materials are usually larger; because of testing costs and other factors the sampling quantity is usually small, therefore industrial enterprises—choose the binomial distribution [4] method for calculating the acceptance probability of the sampling inspection plan for cigarette materials [2]. Namely, the unqualified number X may be found in cigarette materials sampling inspection obeys:

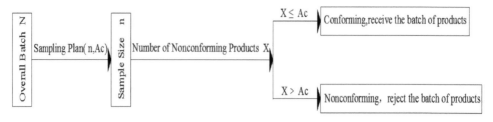

Figure 1. The sampling process of cigarette trademark paper.

$$P(X = x) = C_n^x p^x (1-p)^x \qquad (1)$$

Among them, P is the probability of occurrence, n represents the sample size, p represents batch of nonconforming rate, X represents the number of defective products in the sample.

With sample size n and receiving number Ac these two parameters to represent a sampling plan, being expressed as (n, Ac). Its meaning is: testing sample size n from the overall volume N of a batch of product, when the number of nonconforming products $X \le Ac$, determining to receive the batch of products, or judging to reject, the sampling process is shown in Figure 1. Using Pa(p) for expressing the acceptance probability of the sampling plan when the nonconforming rate is p. By the nature of discrete random variables and addition theorem of probability, we could calculateapproximately that:

$$Pa(p) = P(X \le Ac) = \sum_{X=0}^{Ac} P(X) = \sum_{X=0}^{Ac} C_n^X p^X (1-p)^{n-X}$$
$$\qquad (2)$$

3 ANALYSIS AND COMPARISON OF THREE DIFFERENT SAMPLING INSPECTION PLANS FOR CIGARETTE TRADEMARK PAPER

Qujing Cigarette Factory needed to compare three different kinds of sampling inspection plans: A (48,1), B (80,2), C (100,3), for cigarette trademark paper. Qujing Cigarette Factory has its own acceptance criteria. When the nonconforming probability of cigarette paper trademark is $p_0 \le 0.010$, it is a high level of quality; when the nonconforming probability is $p_1 \ge 0.065$, it is unacceptably poor quality. In this paper, the author made a statistical sampling about several indicators of stock cigarette paper trademarks from January to September 2014 in Qujing Cigarette Factory; statistical results are shown in Table 1.

Sampling inspection plan A (48,1), that is, when it is $p_0 = 0.010$, the acceptance probability

can be calculated by BINOM.DIST, which is in Excel formula function. We get $Pa(p_0) = 0.9166$, the rejection probability is 0.0834—that is the producer risk α. When it is $p_1 = 0.065$, we can also get $Pa(p_1) = 0.1722$—that is the consumer risk.

For a specified high quality, the producer requires a high probability of being accepted—this high quality is called the Producer Risk Quality (PRQ); for a specified low quality, the user requires a low probability of being accepted—this low quality is called the Consumer Risk Quality (CRQ) [5]. The Standard [6] sampling plan study usually specifies: α is 0.05, β is 0.01, and is then based on this research to study other sample characteristics. Therefore, PRQ is the nonconforming rate and the corresponding acceptance probability is 0.95, CRQ is the nonconforming rate and the corresponding acceptance probability is 0.1. Resolution OR is usually used to quantitatively measure the discrimination of a sampling plan, that is, the integrated ability of sampling plan which will distinguishgood or bad from abatch. The smaller the OR value, the higher discernment of the plan—indicating that nonconforming rate increases once and acceptance probability decreases rapidly. The OR expression is as follows [7]:

$$OR = \frac{p_{0.10}}{p_{0.95}} \qquad (3)$$

where: $p_{0.10}$ represents the level of quality corresponds to acceptance probability 0.10;

$p_{0.95}$ represents the level of quality which corresponds to the acceptance probability 0.95.

That is, in this paper, OR = CRQ/PRQ.

For sampling inspection plan A, when the nonconforming rate varies from 0 to 1, acceptance probability can be calculated by BINOM.DIST, which is in Excel formula function. By calculating, we can get—PRQ is 0.007, and CRQ is 0.078 in plan A, that is, resolution OR = 0.078/0.007 = 11.14. Similarly, we can get the acceptance probability and other property values of plan B and plan C at p_0 and p_1.

Table 1. Testing results of several indicators of cigarette paper trademarks.

Testing items	Specified index	Testing results						
Fluorescent whiteness (%)	<1.0	0.5	0.3	0.8	0.1	0.2	0.2	...
Thickness (mm)	0.10–0.13	0.10	009	0.11	0.10	0.12	0.11	...
Quantitative (g/m²)	105 ± 4.2	103.79	103.65	105.71	103.95	106.77	105.89	...

Table 2. Comparison of characteristic values of different sampling plans.

Plans	Pa (P_0)	Pa (P_1)	α	β	PRQ	CRQ	Testing power	OR
Plan A	0.9166	0.1722	0.0834	0.1722	0.007	0.078	0.8278	11.14
Plan B	0.9534	0.1001	0.0466	0.1001	0.010	0.064	0.8999	6.4
Plan C	0.9816	0.1039	0.0184	0.1039	0.013	0.065	0.8961	5.0

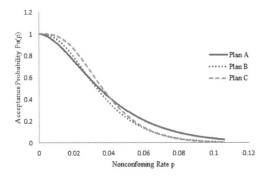

Figure 2. OC curves of the three plans.

The sampling characteristic values of plan A, plan B and plan C are shown in Table 2.

By calculating the sampling characteristics of the plans, OC curves of the three plans are available [8], as shown in Figure 2.

4 CONCLUSIONS

From the Table 2, we find that as to the higher quality level of $p_0 = 0.010$ that the cigarette factory argues, the acceptance probability of plan A is the lowest, followed by plan B; the acceptance probability of plan C is the highest. This suggests that as to the higher level of quality, hamartia probability of plan A is the highest, namely the producer has the greatest risk, the hamartia probability of plan C is the lowest, namely the producer has the minimum risk. As to the lower quality level of $p_1 = 0.065$ that the cigarette factory argues, the acceptance probability of plan A is higher than plan B and plan C, but the acceptance probability of plan A is higher than the normal consumer risk of 0.1 and the acceptance probability of plan B and plan C is almost equal to the consumer risk. Therefore, plan A can't effectively protect the user, but plan B and plan C can. The discriminant ability of plan B and plan C is similar, and is higher than plan A. As to the resolution OR, plan C is the minimum, so the discrimination of plan C is the highest.

From the Figure 2, we find that the acceptance probability decreases with the increase of the nonconforming rate. When the nonconforming rate varies from 0 to 0.04, the acceptance probability of plan C is the highest; when the nonconforming rate is more than 0.04, the acceptance probability of plan A is the highest.

In summary, for the cigarette factory, the plan C is more scientific and reasonable, but the sample of plan C is the maximum, therefore, the inspection cost of plan C is relatively the highest.

ACKNOWLEDGEMENTS

This work was sponsored by Yunnan Cigarette Company Research Foundation Project (No. 2012GY08), China and Yunnan Natural Science Foundation Project (No. 2013FZ024), China.

CORRESPONDING AUTHOR

The corresponding author of this paper is Li Cheng, his email is chengli_1009@126.com.

REFERENCES

[1] Zhuang Changling. Venture analysis of sampling inspection [J], Journal of Shaoxing University, 2002, 22(2), pp, 10–12.

[2] Du Fangqi, Qian Cuizhu, Tang Defang, Wang Hui. Calculation and application of operating characteristic values of acceptance sampling for cigarette materia l[J], Tobacco Science & Technology, 2013(4), pp. 11–13.

[3] GB/T the 2828–2003 Sampling procedures for inspection by attributes [S].

[4] Fan Xiaodong, Sun Lei. Calculation methods for probability of acceptance with sampling plan for inspection by attributes [J], Journal of Bohai University, 2005, 26(2), pp. 10–12.

[5] GB/T 13393–2008 Guide to acceptance sampling inspection [S].

[6] Fan Shuhai, Cai Hong, Wang Yuqian, Amanda Elizabeth, Shao Jianfeng. Intercept-based performance evaluation of operation characteristic curve in product quality inspection [J], Industrial Engineering Journal, 2011, 14(5), pp. 65–68.

[7] Cheng Wanying, Xu Yi, Hao XueYing. Evaluation of sampling plan by OC curve [J]. Technology Foundation of National Defence, 2008(5), pp. 26–29.

[8] Zhuang Changling. Design method and principle of a standard sampling inspection plan by counting [J], Automobile Science and Technonogy, 2002(5), pp. 43–45.

Advances in Future Manufacturing Engineering – Yang (Ed.)
© 2015 Taylor & Francis Group, London, ISBN 978-1-138-02817-3

Oral ginsenosides-Rb1 supplementation ameliorates physical fatigue

Bo Qi & Xuejun Li
Department of Physical Education, Central South University, Changsha, Hunan, P.R. China

ABSTRACT: The active components of Asian ginseng are considered to be ginsenosides. The aim of this study was to investigate the effects of ginsenosides-Rbl (GRb1) supplementation on physical fatigue, using the forced swimming test in mice. The mice were randomly divided into four groups—a control group and three GRb1 supplementation groups. The GRb1 supplementation groups received different doses of GRb1 (25, 50 and 100 mg/kg), while the control group received physiological saline. Each treatment was continued daily for 28 days. After 28 days, the forced swimming test and some biochemical parameters were measured. The data showed that GRb1 supplementation could extend the exhaustive swimming time of mice, as well as decrease the blood lactate and serum urea nitrogen contents and increase the hemoglobin, blood glucose, liver glycogen and muscle glycogen contents. The results indicating an alleviating effect of oral GRb1 supplementation on physical fatigue. In addition, according to acute toxicity studies, GRb1 were considered to be a practically non-toxic substance.

Keywords: ginsenosides-Rb1; alleviating effect; physical fatigue; forced swimming test

1 INTRODUCTION

The root of *Panax ginseng* C.A. Meyer (Family *Araliaceae*, Asian ginseng, which is mainly produced in northeast China and Korea) has been used extensively in traditional Chinese medicine for more than 2,000 years [1]. The pharmacological effects of Asian ginseng have been demonstrated in the central nervous system, the cardiovascular system, as well as the endocrine and immune systems [2]. The active components of Asian ginseng are considered to be ginsenosides, a group of steroidal saponins [3]. There are two major classes of ginsenosides—namely, the derivatives of either protopanaxatriol (Rg1, Rg2, Re and Rf) or protopanaxadiol (Rb1, Rb2, Rc and Rd). They possess four trans-ring rigid steroid skeletons with a modified side-chain at C20, whereas estradiol has no side-chain [4]. Pharmacological actions of these ginsenosides are different, and some of them even showed theopposite effect. In these ginsenosides, Rb1 makes up 0.37–0.5% of ginseng extracts, and it has been reported to exhibit multiple pharmacological activities, such as anti-oxidant, anti-bacterial, anti-cancer, protection against liver damage and promoting neurogenesis effects [5]. Unfortunately, the effects of ginsenosides-Rb1(GRb1) on physical fatigue have never been investigated. Therefore, the aim of this study was to investigate the effects of oral GRb1 supplementation on physical fatigue, using the forced swimming test in mice.

2 MATERIALS AND METHODS

Chemicals and reagents. Ginsenosides-Rb1 (it were isolated from the the root of Panax ginseng C.A. Meyer, chemical purification > 96.2%) was purchased from the Fanke Pharmaceutical Co. (Shanghai, China). The detection kits for blood lactate, Serum Urea Nitrogen (SUN), liver glycogen and muscle glycogen were purchased from Jiancheng Institute of Biotechnology (Nanjing, China). The detection kits for blood glucose were purchased from (Biosino Biotechnology Co., Ltd (Beijing, China). Hemoglobin Enzyme-Linked Immunosorbent Assay (ELISA) kit was purchased from Yueyan Biological Technology Co., Ltd (Shanghai, China).

Selection of animals and care. Male Kunming mice (6–8 weeks old, weighing 18–22 g) were provided by the Center of Experimental Animal of Hunan Province (Changsha, China). The mice were kept in stainless steel cages and maintained under standard laboratory conditions of temperature (21 ± 2 °C), relative humidity (55 ± 10%), 12 h light: Dark cycle (lights on at 07:00), standard rodent diet and water *ad libitum*. All experimental protocols described in this study followed the Institutional Guidelines of Hunan Province and were approved by Ethics Commission of Central South University (Changsha, China).

Grouping of animals. After one week of accommodation, the mice were randomly divided into the following four groups (24 mice in each group)

based on their body weight, a control group and three GRb1 supplementation (GRb1–25, GRb1–50 and GRb1–100) groups. The mice of control group received an oral administration of physiological saline in a volume of 2.0 mL, and the mice of three GRb1 supplementation received different doses of GRb1 (25, 50 and 100 mg/kg body weight, respectively). Each treatment was continued daily for 28 days. After 28 days, the effects of oral GRb1 supplementation on physical fatiguewas assessed by the forced swimming test.

Forced swimming test. After the final oral administration of GRb1, the mice were allowed to rest for 30 min. Then, eight mice were taken out from each group for the forced swimming test. The procedure used was described previously [6]. The forced swimming test was carried out in an acrylic plastic tanks ($50 \times 50 \times 40$ cm), filled with water to a depth of 30 cm and maintained at a temperature of 25 ± 1 °C. The mice had a load attached (6% body weight) to their tails to reduce the swimming time. Exhaustion was determined by observing loss of coordinated movements and failure to return to the surface within 10 s.

Assay of biochemical parameters. After the final oral administration of GRb1, the mice were allowed to rest for 30 min. Then, eight mice were taken out from each group for blood lactate analyses. The mice were forced to swim for 30 min after weight loading (2% body weight), and blood was collected from the tail vein before and after swimming [7]. Blood lactate contents were tested following the recommended procedures provided by the kits.

After the final oral administration of GRb1, the mice were allowed to rest for 30 min. Then, the remaining eight mice were taken out from each group for blood glucose, SUN, hemoglobin, liver glycogen and muscle glycogen analyses. The mice were forced to swim for 90 min without loads [7]. Immediately after the swimming exercise, the mice were anesthetized, using ethyl ether, then blood samples were collected from the abdominal aorta for the determination of glucose, SUN and hemoglobin. Next, the liver and hindlimb skeletal muscles were carefully removed and rinsed in ice cold physiological saline solution (0.9% NaCl), blotted dry and stored at -80 °C until analysis for determination of liver glycogen and muscle glycogen. Blood glucose, SUN, hemoglobin, liver glycogen and muscle glycogen contents were tested following the recommended procedures provided by the kits.

Acute toxicology test. The acute toxicology test in the mice was performed according to the method of Li et al. [8]. Male mice were divided into four groups (8 mice in each group), a control group and three test groups. The test was performed by using increasing oral doses of GRb1 (250, 500, and 1000 mg/kg body weight) to different test groups. The Control

group received s physiological saline. The mice were allowed a diet *ad libitum*, and were all kept under regular observation for 48 h, for any mortality or behavioural changes (irritation, restlessness, respiratory distress, abnormal locomotion and catalepsy).

Statistical analysis. Statistical analysis were performed using SPSS software (version 15.0; SPSS Inc., Chicago, IL, USA). The data was presented as the mean $\pm$ SD. The significance of the mean difference between the control group and each supplementation group was determined by Student's t-test. $p < 0.05$ were regarded as having statistical significance.

3 RESULTS AND DISCUSSION

Acute toxicology test of GRb1. The various observations showed the normal behaviour of the mice. No toxic effects were observed at a higher dose of 1000 mg/kg. Hence, there were no lethal effects in any of the groups during the experimental period. GRb1 were considered to be a practically non-toxic substance.

Effects of GRb1 on exhaustive swimming time of mice. As shown in Figure 1, the exhaustive swimming time in the GRb1–25, GRb1–50 and GRb1–100 groups were significantly longer compared with that in the control group ($p < 0.05$). Forced swimming test is commonly used as anti-fatigue and exercise endurance tests. Other methods of forced exercise such as the motor driven treadmill or a wheel can cause animal injury and may not be routinely acceptable [6]. In the current study, GRb1 supplementation could significantly have extended exhaustive swimming time, which demonstrated that GRb1 had anti-fatigue effects and could elevate the exercise tolerance.

Effects of GRb1 on blood lactate of mice. As shown in Figure 2, before swimming, the blood lactate contents were not significantly different between groups ($p > 0.05$). After swimming, the blood lactate contents in the GRb1–25, GRb1–50 and GRb1–100 groups were significantly lower compared to that in the control group ($p < 0.05$). Blood lactate is the glycolysis product of carbohydrates under an anaerobic condition. The increased lactate level further reduces pH value, which induces various biochemical and physiological side effects, including glycolysis and phosphofructokinase and calcium ion release, through muscular contraction, that are harmful for the body's performance [9]. Blood lactate accumulation has been considered a major inducer of physical fatigue. In the current study, GRb1 supplementation could significantly decrease the blood lactate contents after the swimming exercise, which demonstrated that GRb1 could effectively retard and lower the blood lactate produced, and postpone the appearance of physical fatigue.

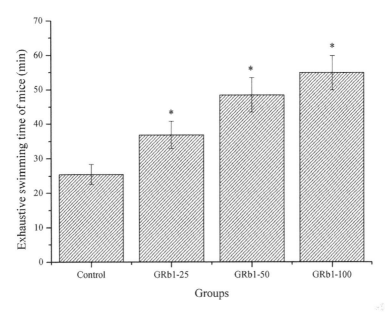

Figure 1. Effects of GRb1 on exhaustive swimming time of mice. *p < 0.05 when compared to the control group.

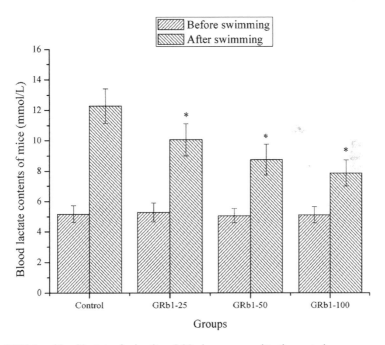

Figure 2. Effects of GRb1 on blood lactate of mice. *p < 0.05 when compared to the control group.

Effects of GRb1 on serum urea nitrogen of mice. As shown in Figure 3, the Serum Urea Nitrogen (SUN) contents in the GRb1–25, GRb1–50 and GRb1–100 groups were significantly lower compared to that in the control group (p < 0.05).

SUN—the product of energy metabolism when moving, is a sensitive index used to evaluate the bearing capability when human bodies suffer from a physical load. In other words, the worse the body is adapted to exercise tolerance, the more

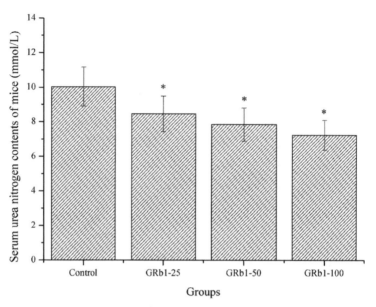

Figure 3.   Effects of GRb1 on serum urea nitrogen of mice. *p < 0.05 when compared to the control group.

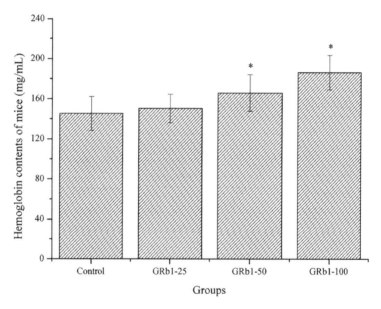

Figure 4.   Effects of GRb1 on hemoglobin contents of mice. *p < 0.05 when compared to the control group.

significantly the SUN level increases [7]. Therefore, SUN is another index of fatigue status. In the current study, GRb1 supplementation could significantly decrease SUN contents after the swimming exercise, which demonstrated that GRb1 could reduce the decomposition of nitrogenous substances in the body and improved endurance capacity during exercise.

Effects of GRb1 on hemoglobin of mice. As shown in Figure 4, the hemoglobin contents in the GRb1–50 and GRb1–100 groups were significantly higher compared to that in the control group (p < 0.05). The hemoglobin contents in the GRb1–25 group were also higher but not significantly (p > 0.05) so. Hemoglobin is one of the indicators that reflect the degree of recovery from fatigue after exercise. Its main function is to serve as the carrier for the erythrocyte, to transport oxygen and carbon dioxide [10]. Studies have demonstrated that a higher level of hemoglobin is helpful to ameliorate fatigue [11]. In the current study, GRb1 supplementation could significantly increase serum hemoglobin contents after

the swimming exercise. It could be considered that increasing the hemoglobin contents may be another pathway of GRb1 alleviating physical fatigue.

Effects of GRb1 on blood glucose of mice. As shown in Figure 5, the blood glucose contents in the GRb1–25, GRb1–50 and GRb1–100 groups were

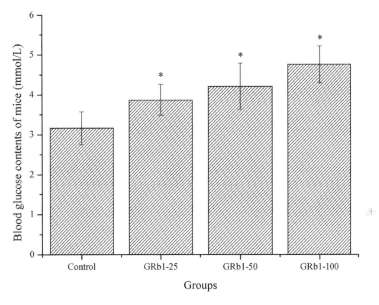

Figure 5.   Effects of GRb1 on blood glucose contents of mice. *$p < 0.05$ when compared to the control group.

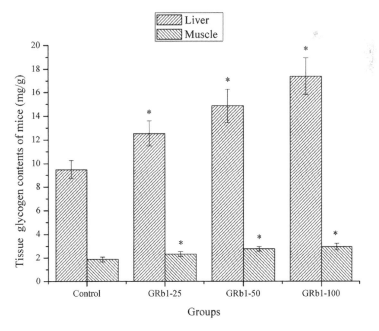

Figure 6.   Effects of GRb1 on liver glycogen and muscle glycogen of mice. *$p < 0.05$ when compared to the control group.

significantly higher compared to that in the control group (p < 0.05). Homeostasis of blood glucose is important for the prolongation of endurance exercise. Previously study hase pointed out that continuous exercises often leads to hypoglycemia and can suppress the active functioning of the brain. Therefore, the amount of blood glucose can illustrate the speed and degree of fatigue development [9]. In the current study, GRb1 supplementation could significantly increase blood glucose contents after the swimming exercise, which demonstrated that the blood glucose regulating ability of G-Rb1 was certainly related to improvements in exercise tolerance and resistance to physical fatigue.

Effects of GRb1 on liver glycogen and muscle glycogen of mice. As shown in Figure 6, the liver glycogen and muscle glycogen contents in the GRb1–25, GRb1–50 and GRb1–100 groups were significantly higher compared to that in the control group (p < 0.05). It was known that endurance capacity of body was markedly decreased if the energy was exhausted. As glycogen was the important resource of energy during exercise, the increasing of glycogen stored in liver and muscle is advantageous to enhancing the endurance of the exercise. So liver glycogen and muscle glycogen are sensitive parameters related to physical fatigue [12]. In the current study, GRb1 supplementation could significantly increase liver glycogen and muscle glycogen contents after swimming exercise, which demonstrated that the anti-fatigue effects of GRb1 might be related to the improvement in the metabolic control of exercise and the activation of energy metabolism.

4 SUMMARY

The present study showed that oral GRb1 supplementation could extend exhaustive swimming time of mice, as well as decrease the blood lactate and serum urea nitrogen contents and increase the hemoglobin, blood glucose, liver glycogen and muscle glycogen contents. The results indicated an alleviating effect of oral GRb1 supplementation on physical fatigue. In addition, according to acute toxicity studies, GRb1 was considered to be a practically non-toxic substance. Further study is needed to elucidate the exact mechanism of anti-fatigue effects of GRb1.

CORRESPONDING AUTHOR

Xuejun Li, Emai: xuejunznLi@163.com, Mobile phone:+86-18973110102.

REFERENCES

[1] Y.S. Chang, E.K. Seo, C. Gyllenhaal and K.I. Block: Panax ginseng: a role in cancer therapy?.Integr. Cancer Ther Vol. 2 (2003), p. 13–33.

[2] H.Y. Shin, H.J. Jeong, J.A. Hyo, S.H. Hong, J.Y. Um, T.Y. Shin, S.J. Kwon, S.Y. Jee, B.I. Seo, S.S. Shin, D.C. Yang and H.M. Kim: The effect of Panax ginseng on forced immobility time & immune function in mice. Indian J. Med. Res Vol. 124 (2006), p. 199–206.

[3] A.S. Attele, Y.P. Zhou, J.T. Xie, J.A. Wu, L. Zhang, L. Dey, W. Pugh, P.A. Rue, K.S. Polonsky and C.S. Yuan: Antidiabetic effects of Panax ginseng berry extract and the identification of an effective component. Diabetes Vol. 51 (2002), p. 1851–1858.

[4] R.Y. Chan, W.F. Chen, A. Dong, D. Guo and M.S. Wong: Estrogen-like activity of ginsenoside Rg1 derived from Panax notoginseng. J. Clin. Endocrinol. Metab Vol. 87 (2002), p. 3691–3695.

[5] Y.L. Hou, Y.H. Tsai, Y.H. Lin and J.C. Chao: Ginseng extract and ginsenoside Rb1 attenuate carbon tetrachloride-induced liver fibrosis in rats. BMC Complement. Altern. Med Vol. 14 (2014), p. 415–416.

[6] B. Qi, L. Zhang, Z.Q. Zhang, J.Q. Ouyang and H. Huang: Effects of ginsenosides-Rb1 on exercise-induced oxidative stress in forced swimming mice. Phcog. Mag Vol. 10 (2014), p. 458–463.

[7] Y.X. Xu and J.J. Zhang: Evaluation of anti-fatigue activity of total saponins of Radix notoginseng. Indian J. Med. Res Vol. 137 (2013), p. 151–155.

[8] F. Li, Y. Zhang and Z. Zhong: Antihyperglycemic effect of ganoderma lucidum polysaccharides on streptozotocin-induced diabetic mice. Int. J. Mol. Sci Vol. 12 (2011), p. 6135–6145.

[9] B.B. Wang, F. Yan and L.M. Cai: Anti-fatigue properties of icariin from Epimedium brevicornum. Biomed.Res Vol. 25 (2014), p. 297–302.

[10] J.C. Wang and K.K. Wang: Fatigue-alleviating effect of polysaccharides from Cyclocarya paliurus (Batal) Iljinskaja in mice. Afr. J. Microbiol. Res Vol. 6 (2012), p. 5243–5248.

[11] M.P.V. Prasad and F. Khanum: Antifatigue Activity of Ethanolic Extract of Ocimum sanctum in Rats. Res. J. Med. Plant Vol. 6 (2012), p. 37–46.

[12] L.N. Shan and Y.X. Shi: Effects of polysaccharides from gynostemma pentaphyllum (thunb.), makino on physical fatigue. Afr. J. Tradit. Complemen.t Altern. Med Vol. 11 (2014), p. 112–117.

Advances in Future Manufacturing Engineering – Yang (Ed.)
© 2015 Taylor & Francis Group, London, ISBN 978-1-138-02817-3

Prediction of lactic acid from *Lactobacillus plantarum* LAB-5 in mulberry fruit

Jian Zhang, Kaixuan Wang, Jun Wang, Jiahui Ma, Guoxia Sun, Yandong Zhang & Fuan Wu
Jiangsu University of Science and Technology, Zhenjiang, P.R. China

ABSTRACT: Mulberry fruit, a rich source of raw food protein, has been long valued for its good taste, nutritional value, and biological activities. However, the high sugar content in mulberry also has limited its application as a popular food. Here we report how to utilize an isolated lactic acid bacterium named LAB-5 to convert sugars to lactic acid from fermented mulberry fruit; its bioconversion characteristics and process dynamics were also investigated. The results indicated that the sugar content of the mulberry fruit was significantly decreased during the fermentation process by *Lactobacillus plantarum* LAB-5. The optimum bioconversion conditions were as follows: temperature of 40°C, duration of 6 d, and 4% (v/v) as the amount of inoculum. The segment models for the biomass were $Y = 3.89 \times 10^8 - 3.89 \times 10^8/(1 + (x/1.27)^{7.23})$ (0–2 d) and $Y = 1.29 \times 10^8 - 1.15 \times 10^8/(1 + \exp(x - 4.45/0.14))$ (3.5–5.5 d); the segment models for lactic acid were $Y = 16.40 - 14.94/(1 + (x/3.69)^{1.46})$ (0–4.5 d) and $Y = 13.70 \times \exp(-\exp(-2.10 \times (x - 3.90)))$ (4.5–7 d); and the models for substrate consumption were as follows: glucose $Y = 15.90 + 111.08/(1 + 10^{0.3x - 0.38})$ (0–5 d) and fructose $Y = 19.95 + 21.12/(1 + 10^{0.90x - 4.47})$ (4.5–8 d).

Keywords: mulberry fruit; lactic acid bacteria; dynamics; bioconversion

1 INTRODUCTION

The mulberry tree belongs to the genus *MorusalbaL* in the family Moraceae; its fruits, which generally mature in late spring and early summer, have wrinkled skin and a sour-sweet taste and is juicy. The fruit contains nutrient materials such as glucose, fructose, sucrose, proteins and amino acid, and other functional compounds like vitamins, rutin, anthocyanin (especially cyanidin) and resveratrol, which benefit human health[1]. Mulberry fruit has a delicious taste, and the functional active elements in it can help the body prevent cancer and resist mutation, strengthen the body's immunity, protect the kidneys and liver, improve looks and reduce skin aging, promote blood corpuscles to grow, and reduce blood sugar and blood lipids[2,3]. In 1993, the Ministry of Health of the People's Republic of China listed mulberry fruit as an edible and medicinal agricultural product[4]. Therefore, mulberry fruit is not only a traditional Chinese medicine, but has also become a more popular food because of its abundant nutritional ingredients and being highly active physiologically.

However, it is known that, Chinese traditional medicine (TCM) considers the mulberry a cold fruit and that should not be eaten in large quantities[5]. Moreover, the mature mulberry fruit is rich in sugars (9%–12%), in contradiction with the modern concepts of a diet with low sugar content or without sugars[6]. In addition, mulberry fruit easily rots and can not be stored for long at room temperature[7]. These factors limit its large-scale consumption. Since ancient times, people have been keenly interested in the agricultural processing of mulberry fruit. Products such as dried mulberry fruit, mulberry juice, mulberry wine, mulberry ice cream, mulberry cake, and mulberry cookies, extend the storage life of mulberry fruit and increase added value to some extent[8]. Nevertheless, with the further improvement of the requirements for safe and health food, the existing forms of mulberry fruit products are relatively simple, and its high-sugar-content problem has not changed. Therefore, new products with high added value and good adaptability need to be developed.

Lactic acid bacteria have been widely used in the modern food industry to change the flavour and nutritional content of food[9,10,11]. These lactic acid bacteria are edible probiotics that are beneficial to human health. Lactic acid bacteria are added to a variety of dairy-based products and fruits to develop new-types of drinks and foods[12,13]. Essentially, lactic acid bacteria perform a biotransformation process achieved by consuming sugars to produce lactic acid. The fermentation with vegetable juices or fruit juices as the raw materials, and lactic acid bacteria as the fermentation strains, can bestow a lactic acid flavour on the original juice flavour and increase the nutritional value, from the beneficial amino acids produced[13]. As this process has not been tested with mulberry fruit, a novel idea that lactic acid bacteria could be inoculated into mulberry fruit to decrease the sugar content was conceived and attempted. Drawing lessons from the biotransformation process performed by the lactic acid bacteria, the assumption was made that converting the sugars in the mulberry fruit to lactic acid would not only effectively reduce the sugar content in mulberry fruit, but also improve its flavour, strengthen its nutritional contents and further increase the added value of new products and the scope of its applications.

Here we show through a variety of experiments using pure mulberry juice as a raw material and edible lactic acid bacteria as bioconversion strains to convert sugar to lactic acid. The bioconversion conditions, the dynamic process of sugar conversion to lactic acid, and dynamic models of bacterial growth and biological transformation were also discussed.

2 MATERIALS AND METHODS

2.1 Materials and reagents

Mulberry fruits at the commercially mature stage were purchased from a mulberry farm (Zhenjiang, China) in May 2012; the mulberry variety was Zhongsang 5801. The fruit was carefully selected in terms of shape and its ripeness needed to be homogeneous this was then mixed with a blender. A juicer was used to repeatedly extract the juice to increase the yield. Finally, the juice was collected and stored at 4°C prior for experimentation.

The lactic acid bacteria were isolated from commercially available yogurts, and stored in 50% (v/v) glycerin.

To prepare 1 L of MRS culture medium[14], the following ingredients were combined: peptone 5.0 g, yeast extract 5.0 g, beef extract 10.0 g, ammonium citrate dibasic 2.0 g, sodium acetate 5.0 g, dipotassium phosphate 2.0 g, manganese sulfate 0.25 g, magnesium sulfate 0.58 g, glucose 20.0 g, Tween80 1.0 mL, and agar 15.0 g, added in that order. The pH was adjusted to 5.6, and the solution was sterilised for storage.

Standard lactic acid was purchased from Sigma-Aldrich Co. (USA), Wahaha pure water was purchased from Hangzhou Wahaha Group Co., Ltd (China), and glucose, fructose, sucrose, ammonium citrate dibasic, sodium acetate, dipotassium phosphate, manganese sulfate, inositol, hexamethyldisilazane, trimethylchlorosilane, cyclohexane, and Tween 80 were purchased from the China National Pharmaceutical Group Industry Corporation Ltd, all of which were AR grade.

2.2 The screening methods for the lactic acid bacteria

The specific methods for isolating the lactic acid bacteria have been described[15], and the strains were named LAB-1, LAB-2, LAB-3, etc. These lactic acid bacteria were inoculated into 5.0 mL of mulberry juice and cultivated at 37°C for 4 d. After this period of bioconversion, the lactic acid content was determined by HPLC and the strain that produced the largest yield was chosen as the experimental strain for PCR.

The methods for isolating the 16S rDNA and PCR amplification have been described[16].

Samples were delivered to Shanghai Shenggong Biological Co., Ltd for sequencing.

2.3 The methods of lactic acid bacteria bioconversion on mulberry juice

Aliquots of 5.0 mL of mulberry juice were placed in 15 mL test tubs, sterilised by pasteurisation, inoculated with lactic acid bacteria (2.5×10^6 CFUs/mL), that had been counted using the plate count method, and an aerobically cultured by sealing the tubes.

2.4 HPLC analysis of lactic acid

A HPLC system equipped with a UV detector operated at 210 nm was used to analyse the lactic acid contents in the mulberry bioconversion broth. The chromatographic column was a Kromasil C_{18}, 5 μm × 4.6 mm × 250 mm; the temperature of the column was 30 °C; the mobile phase was a solution of 0.02 mol/L disodium hydrogen phosphate with a pH of 3.6 (vacuum filtered with 0.45 μm membrane before using) at a flow rate of 1.0 mL/min. The injection volume of each sample was 20 μL, and five injections were performed to attain the average values and variances.

A standard curve was drawn, with the horizontal axis being the concentrations of the lactic acid standards and the vertical axis being the averages of the peak values of the lactic acid standards[17], with Y = 733600.546X–42808.433, R^2 = 0.999.

After bioconversion, the fluid supernatant of the bioconversion broth was obtained by centrifuging it for 5 minutes at 4000 r/min, moderately diluting it, and filtering it through a 0.45 μm filter membrane. The content of lactic acid was determined by HPLC of 20 μL samples.

2.5 GC analysis of sugars

GC was performed with a HP-PLOT column (30.0 m × 32.0 μm); the temperature of the injection port was 240 °C, and that of the detector and column oven were respectively 270 °C and 180 °C.

The flow of H_2 was 50 mL/min with high purity N_2 as the carrier, and that of air was 400 mL/min. The programmed temperature increase condition was from 180°C to 250 °C at a rate of 10 °C/min.

Standard curves of glucose, fructose and sucrose were drawn, with the horizontal axis being the concentrations of the three standard sugars and the vertical axis being the averages of the peaks of the three sugars[18]. That is:

$Y_{Glu} = 17.000X + 0.172$, $R^2 = 0.9990$;
$Y_{Fru} = 5.966X + 0.256$, $R^2 = 0.9990$;
$Y_{Suc} = 5.471X + 0.636$, $R^2 = 0.9980$.

After bioconversion, 1.0 mL of the supernatant of the bioconversion broth was obtained by centrifuging it for 5 minutes at 4000 r/min. The supernatants were placed in 0.5 mL of inosite solvent, and freeze-dried to powder. The powder was placed into tubes and dissolved in 1.0 mL of dimethylsulfoxide, cleared sonication for 15 minutes and placed into which 0.8 mL of hexamethyldisilazane and 0.4 mL of trimethylchlorosilane were added prior to incubating the tubes for 3 h at room temperature with oscillatory rotation. Finally, the reaction solvent was extracted by adding 2.0 mL of cyclohexane, oscillating until the solution was uniform, and let it stand for a few minutes. An aliquot of 1 μL of the supernatant was taken for GC to determine the sugar content.

2.6 Bioconversion dynamic experiments

Aliquots of 5.0 mL of mulberry juice were placed into test tubes, which were sterilised by pasteurisation, inoculated with the lactic acid bacteria isolated from the yogurt in inoculation volumes of 1%, 2%, 3%, 4%, and 5% (v/v) and then cultured for 4 d (days) at 35°C to allow fermentation. Finally, the lactic acid content in the tube tests was determined by HPLC analysis to evaluate the effect of the inoculation volumes on the bioconversion of mulberry juice by the lactic acid bacteria.

A bioconversion period of 2, 3, 4, 5, and 6 d and an inoculation volume of 4% (v/v) were chosen to determinate the effect of time on the bioconversion of mulberry juice by the lactic acid bacteria.

Similarly, the bioconversion tests were conducted at temperatures of 25, 30, 35, 40, and 45 °C with a fixed bioconversion time of 4 d and an inoculation volume of 4% (v/v) to determinate the effect of temperature on the bioconversion of mulberry juice by the lactic acid bacteria.

Bioconversion experiments were conducted after the three tests above were completed. An aliquot of 5.0 mL of mulberry juice was placed into 48 test tubes, which were sterilised by pasteurisation, inoculated with an inoculation volume of 4% (v/v) of lactic acid bacteria, and cultivated at 40°C. Samples were obtained from three tubes every 12 h for 8 days. The sugar contents were determined as the method described in section 2.5 and the lactic acid content were determined as the method described in section 2.4; at the same time, the number of living bacteria in each sample was counted by the plate count method.

2.7 Establishment of bioconversion dynamic models

The processes of lactic acid bacteria growth in mulberry juice, product formation and substrate consumption were nonlinearly fitted using Origin 8.6 software, establishing dynamic models of the growth and metabolism of lactic acid bacteria[19,20].

Three experimental models of lactic acid bacteria growth and metabolism were chosen after a thorough research literature search. The equations are as follows:

Boltzmann model:

$$Y = A_2 + \frac{A_1 - A_2}{1 + \exp\left(\dfrac{x - x_0}{dx}\right)} \tag{1}$$

Logistic model:

$$Y = A_2 + \frac{A_1 - A_2}{1 + \left(\dfrac{x}{x_0}\right)^p} \tag{2}$$

Gompertz model:

$$Y = a \times \exp(-\exp(-k \times (x - x_c))) \tag{3}$$

in which A_1, A_2, dx, x_0, a, k and x_c are all constant[21,22,23].

In the process of the bioconversion on mulberry fruit, the lactic acid bacteria mainly consume sugars for their metabolism, producing lactic acid. After a literature review, the DoseResp experimental model was chosen to describe the process of sugar consumption[24].

The DoseResp model equation is:

$$Y = A_1 + \frac{A_2 - A_1}{1 + 10^{(\log x_0 - x)*p}} \tag{4}$$

in which A_1, A_2, x_0 and p are constants.

The model evaluation criterion that was adopted was the goodness of fit $R^2 \geq 0.85$, so that the larger the R^2 is, the better the model is; F-statistics is another indicator of the significance of the equations fitted, whose significant level (P > F) was 0.05. Using nonlinear model fitting, the model with the largest R^2 and significant level was chosen.

3 RESULTS AND DISCUSSION

3.1 *Identification of the isolated strain*

The results of strain screening experiments revealed that the strain with the highest yield was the lactic acid bacteria named LAB-5. After isolation, cultivation, PCR amplification of the 16S rDNA using universal primers (Fig. 1) and gene sequencing, the gene fragment of the LAB-5 lactic acid bacteria was found to contain 1466 bp, in accordance with the apparent size of the PCR product in the gel. The results obtained from the alignment of this product with sequences on the NBCI website using BLAST showed that the 16S rDNA sequence of LAB-5 was highly similar to that of *Lactobacillus plantarum* and their homology reached 99%. Therefore, the results suggest that LAB-5 belongs to the genus *Lactobacillus plantarum*.

3.2 *The optimal bioconversion conditions of LAB-5*

The results of inoculation volume experiments for LAB-5 fermenting mulberry juice were shown in Figure 2A, demonstrating that with the increase of

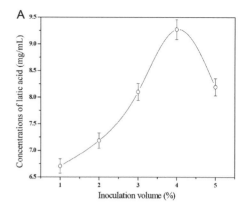

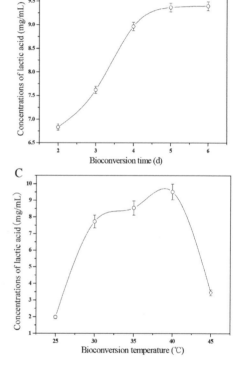

Figure 2. Effects of the inoculation volume (A), bioconversion time (B) and temperature (C) on LAB-5 bioconversion. Aliquots of 5.0 mL of mulberry juice were placed into test tubes, which were sterilised by pasteursation, then inoculated with the lactic acid bacteria isolated from the yogurt in inoculation volumes of 1%, 2%, 3%, 4% and 5% (v/v) and then cultured for 4 d at the 35 °C to allow fermentation. Finally, the lactic acid content in the cultures was determined by HPLC analysis to evaluate the effect of the inoculation volumes on the bioconversion of mulberry juice by LAB-5. The experiment was repeated by respectively changing bioconversion time and temperature to 2, 3, 4, 5, and 6 d and 25, 30, 35, 40, and 45 °C to determine the effects of bioconversion time and temperature on bioconversion of mulberry juice by LAB-5.

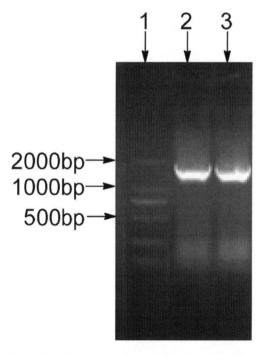

Figure 1. Gel electrophoresis of the 16S rDNA PCR products from LAB-5 lactic acid bacteria. Lane 1 contains DNA molecular weight standards (Maker DL 2000), and both Lane 2 and Lane 3 contain LAB-5 PCR products.

the strain inoculation volume, the concentrations of lactic acid first increased then declined, and that the optimal inoculation volume was 4% (v/v). When the inoculation volume was lower, the yield of the relevant enzyme for bacterial metabolism was limited and the bioconversion effect was poor; while when the inoculation volume was higher, the number of bacteria increased rapidly, leading to a rapid drop in pH and a shortage of dissolved oxygen and nutrients in the later stages, so the bioconversion effects were also undesirable.

The effect of the bioconversion time on LAB-5 fermenting mulberry juice was shown in Figure 2B, demonstrating that with an increase in bioconversion time, the concentrations of lactic acid gradually rose and eventually reached a steady state, and that the optimal bioconversion time was 6 d. Lactic acid was the metabolic product of the bacteria, whose yield increased with the bacterial growth. However, when the nutrients in the culture were depleted, the yield slowly reached a plateau.

The results of the bioconversion temperature on the LAB-5 fermenting mulberry juice were shown in Figure 2C, revealing that with an increase in bioconversion temperature, the concentration of lactic acid significantly increased then decreased. The maximal concentration was achieved when the temperature was 40 °C. When the temperature was low, all types of physiological activities of the bacteria were inhibited, leading to the poor cellular metabolic ability. When the temperature was high, the bacterial enzyme activities were inhibited, the bacteria grew slowly and the metabolic rate declined.

3.3   Bioconversion characteristics of mulberry fruit by LAB-5

In the bioconversion process, the lactic acid bacteria consume the sugars as carbon sources to produce lactic acid as a metabolic product. Glucose, fructose and sucrose are the prevalent sugars in mulberry fruit[25]. However, the sugars in the mulberry fruit used in this experiment are mainly glucose and fructose, and there is no sucrose. Figure 3 shows the changes of the contents of glucose, fructose and lactic acid in the process of bioconversion using mulberry fruit, and demonstrates that with an increase in bioconversion time, the glucose content reduced gradually to be stabilised, did the fructose content. The qualitative values of the glucose, fructose and lactic acid in the mulberry juice before and after bioconversion ranged respectively from 93.34 g/L, 50.27 g/L, 0.00 g/L to 16.82 g/L, 19.53 g/L, 12.18 g/L, from which we can see that the effect of bioconversion by LAB-5 on mulberry fruit is highly significant.

It has been reported that bacteria cultivated in the presence of several sugars grow by preferentially

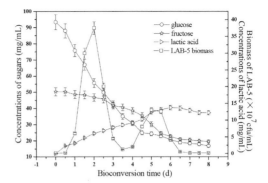

Figure 3.   Dynamic changes in the contents of glucose, fructose, lactic acid and the biomass of the LAB-5 lactic acid bacteria throughout the 8 d process of batch bioconversion of mulberry fruit. An aliquot of 5.0 mL of mulberry juice was placed into 48 test tubes that were sterilized by pasteurization, inoculated with an inoculation volume of 4% (v/v) of lactic acid bacteria, and cultivated at 40 °C. Samples were obtained from three tubes every 12 h for 8 days. The contents of sugars and lactic acid were respectively determined by GC and HPLC.

metabolizing glucose via the sugar phosphotransferase system (PTS)[26,27]. The functions of the PTS include regulating the transcription of a wide variety of inducible proteins dependent on cAMP, as well as regulating the levels of the endogenous inducible permeases (inducer exclusion)[26,27]. In the presence of glucose, not only is the expression level of the inducible transport proteins dramatically reduced. due to decreased transcription, but the activity of the non-PTS sugar permeases expressed at a basal level are inhibited by binding unphosphorylated IIA^{Glc}[26]. Therefore, the inference is that in this experiment, LAB-5 first utilised the glucose in mulberry fruit to reproduce until the content of glucose reached a low level, and then fructose was consumed as a second carbon source. In connection with Figure 3, the reason why the contents of the sugars continually decrease is that the entire bioconversion process occurred in a static culture because anaerobic cultivation was required, leading to an uneven distribution of LAB-5 bacteria. The result of this phenomenon is that the content of glucose used in the early stage is quickly reduced, while the fructose content significantly declines in the later stages.

In the course of the dynamic experiments described in section 2.6, not only were the sugar and lactic acid concentrations determined, but the LAB-5 biomass in the samples was also quantified. The growth curve of LAB-5 is shown in Figure 3, clearly showing that there are two peaks in the curve. The two peaks indicate that with the sugars in the culture consumed LAB-5 used two different metabolic pathways for growth. Due to glucose and fructose being the two main carbon sources in the mulberry, the conclusion was that the two peaks from left to right of the biomass in Figure 3 respectively represented the cell growth using glucose and fructose as the carbon sources.

3.4 Bioconversion dynamics models of mulberry fruit by LAB-5

The significance of the bioconversion dynamics models is to establish relationships among the thalli, matrixes and products in the process of bioconversion and to explore their inherent laws. The aim of building dynamics models is to control the bioconversion process according to practical demands.

The process of bacterial growth in batch culture can be modeled with four different phases: lag phase, exponential or log phase, stationary phase and death phase, which are connected together to form a batch bioconversion growth curve. According to Figure 3, the growth curve of LAB-5 tends be an "S"-curve. Combined with experimental models, the typical "S"-curve models are the Logistic model, the Gompertz model and the Boltzmann model.

The biomass was fitted using the fitting equations presented in Table 1, showing that the different models have different goodness of fit values but are all significant at $\alpha = 0.05$, which suggests that all models fitted the cell growth well at the 95% confidence level. In the first stage, because the goodness of the three models fit (R^2) is larger than 0.99, R^2 was not chosen as the standard of the optimal model. Therefore, the Akaike information criterion (AIC)[28] was introduced as a standard for confirming the fitness of the models. Moreover, generally speaking, the smaller the AIC value, the higher the degree of fitting. The calculations suggested that in the first stage, the AIC values of the Boltzmann model, the Logistic model and the Gompertz model are respectively 135.09, 129.00 and 139.34, showing that the AIC value of the Logistic model is the smallest. In the second stage, the R^2 of 0.9904 for the Boltzmann model is the largest. Therefore, the Logistic model was chosen for the first stage and the Boltzmann model was chosen for the second stage, for which the equations are respectively $Y = 3.89 \times 10^8 - 3.89 \times 10^8/(1 + (x/1.27)^{7.23})$ and $Y = 1.29 \times 10^8 - 1.15 \times 10^8/(1 + \exp(x - 4.45/0.14))$. Connected with the experiment data, a fragmented fitting curve is shown in Figure 4A and Figure 4B,

Table 1. Fitting results of the Boltzmann model, the logistic model and the Gompertz model for the LAB-5 biomass and lactic acid production.

Model	The first stage (0–2 d for LAB-5 biomass/ 0.5–4.5 d for lactic acid production)			The second stage (3.5–5.5 d LAB-5 biomass/ 4.5–6.5 d for lactic acid production)		
	Fitting equations	R^2	p value	Fitting equations	R^2	p value
LAB-5 biomass						
Boltzmann	$Y = 3.80 \times 10^8 - \dfrac{3.81 \times 10^8}{1 + \exp\left(\dfrac{x - 1.28}{0.17}\right)}$	0.9994	0.0041	$Y = 1.29 \times 10^8 - \dfrac{1.15 \times 10^8}{1 + \exp\left(\dfrac{x - 4.45}{0.14}\right)}$	0.9904	0.0400
Logistic	$Y = 3.89 \times 10^8 - \dfrac{3.89 \times 10^8}{1 + \left(\dfrac{x}{1.27}\right)^{7.23}}$	0.9998	0.0022	$Y = 1.29 \times 10^8 - \dfrac{1.14 \times 10^8}{1 + \left(\dfrac{x}{4.46}\right)^{32.39}}$	0.9884	0.0440
Gompertz	$Y = 3.89 \times 10^8 \times \exp(-\exp(-3.95 \times (x - 1.16)))$	0.9993	0.0003	$Y = 1.35 \times 10^8 \times \exp(-\exp(-2.93 \times (x - 4.23)))$	0.9527	0.0110
Lactic acid production						
Boltzmann	$Y = 14.31 - \dfrac{33.55}{1 + \exp\left(\dfrac{x - 1.17}{2.94}\right)}$	0.9921	0.0001	$Y = 13.73 - \dfrac{425.77}{1 + \exp\left(\dfrac{x - 2}{0.52}\right)}$	0.9666	0.0135
Logistic	$Y = 16.40 - \dfrac{14.94}{1 + \left(\dfrac{x}{3.69}\right)^{1.46}}$	0.9930	0.0001	$Y = 13.77 - \dfrac{25.98}{1 + \left(\dfrac{x}{3.73}\right)^{9.8}}$	0.9659	0.0136
Gompertz	$Y = 12.04 \times \exp(-\exp(-0.54 \times (x - 1.44)))$	0.9929	0.0001	$Y = 13.70 \times \exp(-\exp(-2.10 \times (x - 3.90)))$	0.9831	0.0001

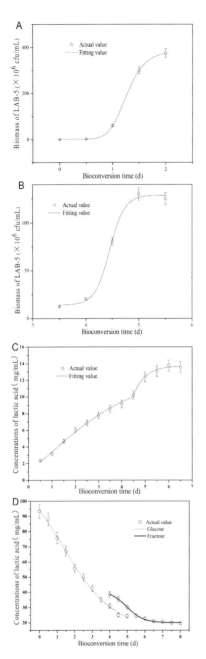

A

B

C

D

Figure 4. Fitting curves for the LAB-5 biomass, lactic acid production and glucose & fructose consumption in the diauxic stages. A is the fitting curve of the Logistic model for the first stage (0–2 d) of the LAB-5 biomass; B is the fitting curve of the Boltzmann model for the second stage (3.5–5.5 d) of the LAB-5 biomass; C is the fitting curve for lactic acid production in the diauxic stages: Logistic model for the first stage (0–4.5 d), Gompertz model for the second stage (4.5–6.5 d); D is the fitting curves for glucose (0–5 d) and fructose (4–8 d) consumption using the DoseResp model in the diauxic stages.

which confirmed that Lab-5 growth is similar to that of other bacteria, with the growth curve for fermentation of mulberry fruit being the "S"-curve.

The generation process of metabolic product occurs in relationship with cell growth. The relationship between a metabolic product generation and its cell growth can be divided into three categories, that is—when product generation and cell growth are coupled, and when product generation and cell growth are partially coupled and when there is no relationship between product generation and cell growth. In the bioconversion on mulberry fruit, LAB-5 consumed the sugars to produce lactic acid and the generation of lactic acid had a direct relationship with the growth of LAB-5, coupling the product generation and the cell growth[29].

Because there are two types of carbon sources in mulberry fruit, the generation of lactic acid is divided into two phases—one, in which glucose is used as the carbon source, and the other, in which fructose is used as the carbon source. Therefore, the fitted models of lactic acid production are segmented into two equations, similar to the cell growth curve models, in which the breakpoint was chosen as the 4.5th d to relate it to the bacterial growth curve. Similar to the models of cells growth, the models of product generation that are frequently used are the Logistic model, the Gompertz model and the Boltzmann model. The equations and fitted effects of lactic acid production by the three models were presented in Table. 1, which showed that all of models fitted could account for the generation of lactic acid with a significance level of $\alpha = 0.05$ and a 95% confidence interval. In the first stage of lactic acid generation, similar to the first stage of the bacterial growth, all of the R^2 values are larger than 0.99, which shows that the R^2 value is not an accurate indicator of the optimal model; therefore, the AIC value was introduced. Calculations suggested that in the first stage, the AIC values of the Boltzmann model, the Logistic model and the Gompertz model were respectively -29.12, -31.92 and -17.48, and thus the AIC value of the Logistic model is the smallest. In the second stage, the R^2 of 0.9930 for the Gompertz model is the largest. Therefore, the Logistic model was chosen for the first stage and the Gompertz model was chosen for the second stage, of which the equations are respectively $Y = 16.40 - 14.94/(1+(x/3.69)^{1.46})$ and $Y = 13.70 \times \exp(-\exp(-2.10 \times (x - 3.90)))$. Moreover, the fitting curves of lactic acid production were created in Figure 4C by connecting the experiment data, showing that with an increase in the bioconversion time, the lactic acid accumulation constantly increased. When glucose was depleted, lactic acid production decreased, as did the biomass. When LAB-5 fermented on mulberry fruit, using fructose as a carbon source, thalli growth speed increased gradually, while, due to the limitation of the fructose content, the yield of lactic acid tended to be stable after 6 d bioconversion.

In the process of bioconversion of mulberry fruit, the main energy substrates for LAB-5 metabolism are glucose and fructose, whose consumption decreased the sugar content, thus achieving the aim of reducing the sugar concentration in the mulberry juice. Because the cell biomass reached the maximal value by the 4th d (Fig. 3), and considering that the thalli are metabolic in the log phase and the death phase, the 4.5th d was chosen as the breakpoint. The bioconversion models were fitted, using the glucose consumption dynamics from 0 d to 5 d and the fructose consumption dynamics from 4 d to 8 d. The fitting results were provided in Table 2, showing that that the R^2 values of the glucose model and the fructose model were respectively 0.9966 and 0.9957, and that both of the fitting effects were significant at the level of $\alpha = 0.01$— which indicated that both the models described the processes of the matrixes consumption with a 99% confidence interval. Therefore, the equations for glucose and fructose were respectively $Y = 15.90 + 111.08/(1 + 10^{0.3x-0.38})$ and $Y = 19.95 + (21.12/1 + 10^{0.90x-4.47})$. The fitting curves for the two sugars were shown in Figure 4D, which demonstrated that when the substrate concentrations were too low, the bioconversion by LAB-5 ceased.

The bioconversion dynamics models reflect the relationship between the cells or its components and its environment. The aim of modeling is to guide practice; if models are built but not converted to practice, modeling loses its purpose[30]. According to the batch bioconversion models above, after 6 d of bioconversion, the amount of free glucose and fructose is insufficient for the production of lactic acid, which tends to take supremacy, and the growth of LAB-5 enters the death phase. Therefore, in continuous bioconversion, the 6th d was chosen as the optimal time to feed supplement; moreover, fresh bioconversion broth not only provides the carbon sources that the cells grow to need, but also improves the microenvironment conditions, such as the pH. In industrialised bioconversion processes, dynamics models provide guidance

and allow adjustments for practice. In sum, future experiments may confirm the dynamics models regarding continuous bioconversion, which will be applied in production as soon as possible.

4 SUMMARY

The lactic acid bacterial isolate LAB-5 was chosen from the strain screening assay because it significantly decreased the sugar content in mulberry juice by converting sugars to lactic acid. The contents of glucose and fructose in mulberry juice were respectively reduced by 82% and 61% by the bioconversion of LAB-5. The result of gene sequencing showed that the LAB-5 strain belongs to the genus *Lactobacillus plantarum*. The optimal bioconversion conditions for LAB-5 in mulberry fruit juice were determined to be 6 d at 40 °C with an inoculation volume of 4% (v/v).

Because there are two carbon sources, glucose and fructose, which are consumed in different stages by LAB-5, the bioconversion model of its cell growth is a fragmented equation, that is, $Y = 3.89 \times 10^8 - 3.89 \times 10^8/(1 + (x/1.27)^{7.23})$ (0–2 d) and $Y = 1.29 \times 10^8 - 1.15 \times 10^8/(1 + \exp(x - 4.45/0.14))$ (3.5–5.5 d). Lactic acid is the metabolic product of bioconversion by LAB-5, as well as a key factor of this experiment. Because lactic acid generation and cell growth are in a coupled relationship, the fitting model of lactic acid generation is also a fragmented equation, that is, $Y = 16.40 - 14.94/(1 + (x/3.69)^{1.46})$ (0–4.5 d) and $Y = 13.70 \times \exp(-\exp(-2.10 \times (x - 3.90)))$ (4.5–7 d). The equations of substrate consumption fitted by the DoseResp model are $Y = 15.90 + 111.08/(1 + 10^{0.3x-0.38})$ (0–5 d) (glucose), and $Y = 19.95 + 21.12/(1 + 10^{0.90x-4.47})$ (4–8 d) (fructose).

ACKNOWLEDGEMENTS

This work was financially supported by the Qing Lan Project of Jiangsu Province, the Science and Technology Support Programs of Jiangsu Province (BE2012365, BE2010419), the Modern Agroindustry Technology Research System of China (CARS-22), and the Research Project of Jiangsu University of Science and Technology (35211002). The authors have declared no conflict of interest.

Table 2. Fitting results of the DoseResp model for glucose and fructose consumption.

Sugars	Fitting equations	R^2	p value
Glucose (0–5 d)	$Y = 15.90 + \dfrac{111.08}{1 + 10^{0.3x-0.38}}$	0.9966	0.0001
Fructose (4–8 d)	$Y = 19.95 + \dfrac{21.12}{1 + 10^{0.90x-4.47}}$	0.9957	0.0001

REFERENCES

[1] Sezai, E., & Emine, O. Chemical composition of white (*Morus alba*), red (*Morus rubra*) and black (*Morus nigra*) mulberry fruits. *Food Chemistry*, 2007,103(4), 1380–1384.

[2] Chen, C.C., Liu, L.K., Hsu, J.D., Huang, H.P., Yang, M.Y., & Wang, C.J. Mulberry extract inhibits the development of atherosclerosis in cholesterol-fed rabbits. *Food Chemistry,* 2005, 91(4), 601–607.

[3] Chen, Z., Zhu, C.H., & Han, Z.Q. Effects of aqueous chlorine dioxide treatment on nutritional components and shelf-life of mulberry fruit (*Morus alba* L.). *J Biosci Bioeng,* 2011,111(6), 675–681

[4] Wang, P., Zhang, Y.X., & Liu, D.H. Research advancement of the nutrition value of mulberry and its functional components. *Food and Nutrient In China,* 2008,8(3), 57–59.

[5] Liang, G.Q., Wu, J.J., Qi, G.J., Li, Q., & Lu, F. Nutritional value and health function of mulberry wine. *Modern Food Science & Technology,* 2011,27(4), 457–460.

[6] Chen,Z.Y.,Wu,J.J.,Chen,W.D.,Xiao,G.S.,Liu,X.M. Xu, Y.J., & Li, S.F. Analysis of sugar content and acid content in mulberry juice. *China Sericulture,* 2002, 23(4), 18–19.

[7] He, X.M., Liao, S.T., & Liu, J.P. Research progress on nutrient and funcitional components and pharmacology of mulberry. *ACTA Sericologica Sinica,* 2004,30(4), 390–393.

[8] Li, D.X., & Chen, Q.X. Study progress of functional ingredient and development in mulberry. *Chinese Agricultural Science Bulletin,* 2009,25(24), 293–297.

[9] Claudia, N.B., Valeria, A.C., Cintia, W.R., Jorge, E.S., Mario, E.L., & Jorge, A.T. Biotransformation of halogenated 2′-deoxyribosides by immobilized lactic acid bacteria. *Journal of Molecular Catalysis B: Enzymatic,* 2012,79, 49–53.

[10] Frédéric, L., & Luc, D.V. Lactic acid bacteria as functional starter cultures for the food fermentation industry. *Trends in Food Science & Technology,* 2004,15(2), 67–78.

[11] Cátia, M.P., Cidália, P., Hernández-Mendoza, A., & Xavier, M.F. Review on fermented plant materials as carriers and sources of potentially probiotic lactic acid bacteria – With an emphasis on table olives. *Trends in Food Science & Technology,2012,* 26(1), 31–42.

[12] Chen, Y.M., Shih, T.W., Chiu, C., Pan, T.M., & Tsay, T. Y. Effects of lactic acid bacteria-fermented soy milk on melanogenesis in B16F0 melanocytes. *Journal of Functional Foods,2012, Avaiable online 21.*

[13] Raffaella, D.C., Rossana, C., Maria, D.A., & Marco, G. Exploitation of vegetables and fruits through lactic acid fermentation. *Food Microbiology,2013, 33*(1), 1–10.

[14] Ulrich, S., & Jéssica, V.V. Inhibition of Penicilliumnordicum in MRS medium by lactic acid bacteria isolated from foods. *Food Control, 2010,21*(2), 107–111.

[15] Zou, X., & Yi, X. Isolation of the Iactic acid-forming bacteria from the yoghurt and the analysis of the ratio to bacillus and COCCUS. *Food Science and Technology,2008, 33*(3), 118–122.

[16] Kye Man, C., Renukaradhya, K.M., Shah Md., A.I., Woo Jin, L., Younghong, S., Jongmin, K., Myoung Geun, Y., Ji Joong, C., & Han Dae, Y. Novel multiplex PCR for the detection of lactic acid bacteria during kimchi fermentation. *Mol Cell Probes, 2009,23*(2), 90–94.

[17] Rodrigo, S., Ana Cecília, P.R., Cristiano Augusto, B., Adriana Dillenburg, M., José Teixeira, F., & Helena Teixeira, G. Validation of a HPLC method for simultaneous determination of main organic acids in fruits and juices. *Food Chemistry, 2012,135*(1), 150–154.

[18] Bystrom, L.M., Lewis, B.A., Brown, D.L., Rodriguez, E., & Obendorf, R.L. (2008). Characterisation of phenolics by LC-UV/Vis, LC-MS/MS and sugars by GC in *Melicoccus bijugatus* Jacq. 'Montgomery' fruits. *Food Chemistry, 111*(4), 1017–1024.

[19] Annadurai, G., Rajesh Babu, S., & Srinivasamoorthy, V.R. Development of mathematical models (Logistic, Gompertz and Richards models) describing the growth pattern of Pseudomonas putida. *Bioprocess Engineering,2000, 23*(6), 607–612.

[20] Mikael, H., Junchen, L., Noriaki, O., & Simon, S. Descriptive and Predictive Growth Curves in Energy System Analysis. *Nonrenewable Resources,2011, 20*(2), 103–116.

[21] Jhony, T.T., Weber da, S.R., & Gilmar de, A.G. Mathematical modeling of microbial growth in milk. *Ciênc. Tecnol,* 2011,*31*(4), 891–896.

[22] Daniel, E.W., John, A.P., & Micheal, R.L. Analysis of the logistic function model: derivation and applications specific to batch cultured microorganisms. *Bioresour Technol,* 2003, 86(2), 157–164.

[23] Banani, R.C., Runu, C., & Utpal, R.C. (2007). Validity of modified Gompertz and Logistic models in predicting cell growth of Pediococcus acidilactici H during the production of bacteriocin pediocin AcH. *Journal of Food Engineering, 80*(4), 1171–1175.

[24] Lu, Y.Z., Mu, H.L., & Li, H.N. An analysis of present situation and future trend about the energy consumption of Chinese Agriculture Sector. *Procedia Environmental Sciences,* 2011, *11,* 1400–1406.

[25] Wu, Z.F., & Weng, P.F. The Nutrient Compositions of Mulberry and Its Functionality. *Journal of Chinese Institute of Food Science and Technology,* 2005, 5(3), 102–107.

[26] Lan, G., & Kaback, H.R. Glucose/Sugar Transport in Bacteria. *Encyclopedia of Biological Chemistry,* 2004, 204–207.

[27] John, T. Lactic acid bacteria: model systems for in vivo studies of sugar transport and metabolism in gram-positive organisms. *Biochimie,* 1988,*70*(3), 325–336.

[28] Akaike, H. A new look at the statistical model identification. *Automatic Control, IEEE Transactions on,* 1974, *19*(6), 716–223.

[29] Lin, Q., Li, Y.D., Sun, X.B., & Li, L.C. Fermentation dynamics of cherry wine. *China Brewing,* 2009,7, 65–68.

[30] Dai, Z.K., Yin, Y.L., & Ruan, Z. Microbial fermentation kinetic model and its parameters estimation by software. *Computers and Applied Chemistry,* 2011, *28*(4), 437–440.

517

Advances in Future Manufacturing Engineering – Yang (Ed.)
© 2015 Taylor & Francis Group, London, ISBN 978-1-138-02817-3

Study of the effect of using CAF process for concentrating sludge

Xiaotao Guan & Fengping Hu
School of Civil Engineering and Architecture, East China Jiaotong University, Nanchang, Jiangxi, China

ABSTRACT: The CAF process is used for concentrating the surplus low concentration activated sludge We changed the characteristics of sludge by adding the sludge conditioner in this test to achieve a good concentration effect. Test results showed: the best dosage appears when the dosage of cationic polymer flocculant is 1 kg/tDS, the SS is the lowest value and the sludge recovery reaches the highest of 92% to 94%; the best dosage appears when the surfactant's dosage is 0.2 kg/tDS, the SS is the lowest value and the sludge recovery reaches the highest of 92%.

Keywords: CAF; concentrate; sludge; factors

1 INTRODUCTION

CAF (Cavitation Air Flotation) is a patented product (patent number US00524600A) of a Hydrocal environmental company from USA, it is a sewage treatment equipment with characteristics of low investment, high efficiency, low processing cost and good efficiency, which is widely adopted and promoted at present. CAF has been widely used in the pre-treatment of industrial wastewater and urban sewage. In order to develop the CAF and promote the research work of CAF, in terms of sludge treatment, the CAF process is used for concentrating activated sludge by changing the characteristics of sludge in this test. It can be concluded from the test that: the dosing point and dosage of floccu-lant and surfactant have a significant impact on the CAF sludge concentrating process.

2 PROCESS

The process diagram of the CAF sludge concentrating process is shown in Figure 1.

3 THE IMPACT OF FLOCCULANT'S DOSAGE ON CONCENTRATION EFFECT

This test is mainly concentrated on studying the impact of flocculant FO4440SH on mixture's

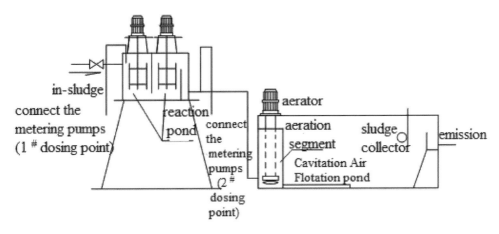

Figure 1.   Process diagram of the CAF sludge concentrating process.

CAF concentrating process. The in-sludge amount of CAF is about 5 m³/h, the sludge concentration is 3.6~4 g/L, the solid load is about 230 kgMLSS/m² • d, the flocculant is put at the inlet of reaction pond (1# dosing point in Fig. 1).

1227 is adopted as surfactant, when the surfactant's dosing points are selected at the inlet or outlet of reaction pond (2# dosing point in Fig. 1), the impact of dosage of flocculant FO4440SH on the CAF sludge concentrating process is shown in Figure 2.

a. The impact on the outlet of water SS, (b) the impact on sludge recovery (c) the impact on concentrated sludge's rate of solid content

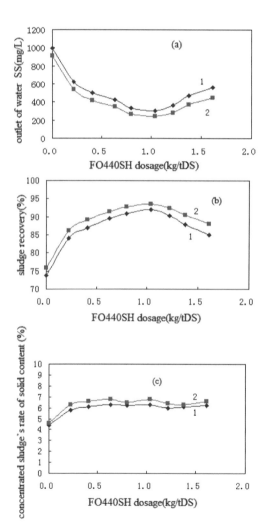

Figure 2. Impact of dosage of flocculant FO4440SH on the CAF sludge concentrating process.

1: The surfactant is put at the inlet of reaction pond; 2: The surfactant is put at the outlet of the reaction pond;

It can be concluded from Figure 2 that: when the dosing point is located at the outlet of the reaction pond, the surfactant's concentration effect is better than that at the inlet of the reaction pond—therefore, the surfactant should be put at the inlet of the reaction pond. (2) With the addition of cationic polymer flocculant, the SS in concentrated sludge gradually decreases and the sludge recovery gradually increases, the best dosage appears when the addition of cationic polymer flocculant is 1 kg/tDS, the SS is the lowest value and the sludge recovery reaches the highest of 92% to 94%, after that, with the addition of cationic polymer flocculant, the SS in concentrated sludge gradually increases and the sludge recovery gradually decreases, the change of dosage of cationic polymer flocculant has little impact on the concentrated sludge's rate of solid content, which is generally 6% to 8%.

4   THE IMPACT OF SURFACTANT'S DOSAGE ON CONCENTRATION EFFECT

This test is mainly concentrated on studying the impact of surfactant's dosage on mixture's CAF concentrating process. The in-sludge amount of CAF is about 5 m³/h, the sludge concentration is 3.6 ~ 4g/L, the flocculant's dosage is 1.0 kg FO4440SH/tDS, FO4440SH is adopted as flocculant, 1227 is adopted as surfactant, the flocculant is put at the inlet of the reaction pond (1# dosing point in Fig. 1), the surfactant's dosing point is selected at the outlet of the reaction pond (2# dosing point in Fig. 1), the impact of surfactant's dosage on the CAF sludge concentrating process is shown in Figure 3.

The impact on the outlet of water SS, (b) the impact on sludge recovery (c) the impact on concentrated sludge's rate of solid content.

It can be concluded from Figure 3 that: with the addition of surfactant, the SS in concentrated sludge gradually decreases and the sludge recovery gradually increases, the best dosage appears when the surfactant's dosage is 0.2 kg/tDS, the SS is the lowest value and the sludge recovery reaches the highest of 92%, after that, with the addition of surfactant, the SS in concentrated sludge and sludge recovery gradually tend to balance. The increase of surfactant's dosage can improve the concentrated sludge's rate of solid content, the surfactant can't be added in great amounts, otherwise, it will affect the concentration effect.

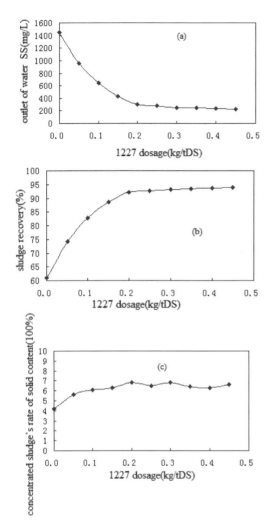

Figure 3. Impact of dosage of surfactant 1227 on the CAF sludge concentrating process.

5 CONCLUSION

It can be concluded from the test that the CAF process can be used for concentrating activated sludge when changing the sludge's characteristics by adding sludge conditioner; one must especially pay attention to the conditioner's dosing point and dosage. The best dosage of surfactant is 0.2 kg/tDS, and it should be put at the outlet of the reaction pond; the best dosage of cationic polymer flocculant is 1 kg/tDS. This process's lowest outlet of water SS can be 230–250 mg/L, the highest sludge recovery can be 92%–94%, the concentrated sludge's rate of solid content is generally 6%–8%.

CORRESPONDING AUTHOR

About the Author: Guan Xiaotao (1968–), female, master's degree, an associate professor. Corresponding author (responsible author).

REFERENCES

[1] Shang Xingying, etc.. Application of CAF in Chemical Area's Sewage Treatment Plant [J]. Northern Environment, 2011, 23 (11), 188–190.
[2] Shen Caiqin, etc.. The Treatment of Oily Wastewater by Using CAF Equipment [J]. Shanghai Environmental Sciences, 2001, 20 (10), 506–507.
[3] Ning Ping, etc.. Applications of CAF—Coagulation Sedimentation Process in the Treatment of Papermaking Wastewater [J]. Environmental Engineering, 1998, 23 (4), 7–9.

Advances in Future Manufacturing Engineering – Yang (Ed.)
© 2015 Taylor & Francis Group, London, ISBN 978-1-138-02817-3

Study on a SIRS epidemic model with constant input and vaccination treatment

Xingyang Ye
School of Science, Jimei University, Xiamen, China

ABSTRACT: In this paper, we consider an SIRS model with variable size population and vaccination treatment. Our model analysis shows that the disease—free equilibrium is globally asymptotically stable if the basic reproduction number is less than unity while the endemic equilibrium exists, and it is globally asymptotically stable whenever the basic reproduction number is greater than unity. Numerical simulations are carried out to illustrate the analytical results. Results show that vaccination and treatment has a great effect on infectious disease prevention.

Keywords: SIRS model; basic reproduction number; equilibrium point; stability

1 INTRODUCTION

Many diseases are transmitted by viruses. Viral levels often determine the ability of transmission for some diseases such as malaria, where the infectivity depends on parasite or viral loads in infected hosts or vectors [1,2]. Consequently, there is the need for proactive steps towards controlling the spread of infectious diseases, particularly those for which both vaccine and cure are available.

Mathematical models have become important tools in analysing the spread and control of infectious diseases. Some examples on the use of the mathematical model for the analysis of treatment and control of infectious disease can be found in [3,4]. A SIR model with variable size population and an optimal control problem, subject to the model with vaccination and treatment as controls, was considered in [4]. However, the individuals with vaccination and treatment may not keep permanent immunity, and some individuals would lose the immunity. Hence, we consider a SIRS model with variable size population and vaccination treatment to improve the work made in [4].

The paper is organised as follows. In section 2, we present the SIRS model to be investigated. In section 3, we carry out local and global stability analysis on the model equilibria. Section 4 presents different computer simulations of the system. Finally, the biological significance of our analytical and numerical findings are discussed.

2 MODEL FORMULATION

Let $S = S(t)$, $I = I(t)$, $R = R(t)$ be the number of susceptible individuals, infectious individuals, removed individuals at time t, respectively. The epidemic model considered here is as follows:

$$\begin{cases} \dot{S} = b + \sigma R - dS - u_1 S - \beta SI, \\ \dot{I} = \beta SI - (d + \alpha + u_2) I, \\ \dot{R} = u_2 I + u_1 S - (d + \sigma) R. \end{cases} \tag{1}$$

In the above model, $b > 0$ is the recruitment rate, $\sigma > 0$ is the rate constant of losing immunity, $\beta > 0$ is the disease transmission rate, $d > 0$ is the natural death rate, $u_1 > 0$ is the proportion of the susceptible that is vaccinated per unit time, $u_2 > 0$ is the proportion of the infectives that is treated per unit time, and $\alpha > 0$ is the disease-induced death rate. Let $N = N(t)$ be the total populations size at time t. Due to $N = S + I + R$, from Eq. 1 we obtain

$$\dot{N} = b - Nd - \alpha I. \tag{2}$$

Moreover, under the dynamics described by Eqs. 1 and 2, the region

$$\Omega = \left\{ x = (S, I, N) \in R_+^3 \middle| S \geq 0, I \geq 0, S + I \leq N \leq \frac{b}{d} \right\},$$

is positively invariant. Hence, the system is both mathematically and epidemiologically well-posed. Therefore, for the initial starting point $x \in R_+^3$, the trajectory lies in Ω. Thus, we can restrict our analysis to the region Ω.

3 STABILITY ANALYSIS

Since R can always be obtained by the equation $R = N - S - I$, we will use the system below in place of Eq. 1 in our analysis:

$$
\begin{cases}
\dot{S} = b + \sigma(N - S - I) - dS - u_1 S - \beta SI, \\
\dot{I} = \beta SI - (d + \alpha + u_2)I, \\
\dot{N} = b - Nd - \alpha I.
\end{cases}
\tag{3}
$$

3.1 Equilibrium

Let

$$
R_0 = \frac{b\beta(d + \sigma)}{d(d + \alpha + u_2)(d + \sigma + u_1)}.
$$

R_0 will be called basic reproduction number, which is the number of secondary infectious cases produced by an infectious individual during his or her effective infectious period, when introduced midst a susceptible population. By simple calculation, the following result holds for Eq. 3.

Theorem 1 For Eq. 3, there always exists a disease-free equilibrium $\varepsilon_0 = ((\sigma + d)b/(\sigma + d + u_1)d,\ 0,\ b/d)$, and there is also a unique endemic equilibrium solution $\varepsilon_1 = (S^*, I^*, N^*)$ if and only if $R_0 > 1$, where

$$
S^* = \frac{d + \alpha + u_2}{\beta},
$$

$$
I^* = (R_0 - 1)\frac{d(d + \sigma + u_1)(d + \alpha + u_2)}{\beta[(d + \alpha)\sigma + d(d + \alpha + u_2)]},
$$

$$
N^* = \frac{b\beta(\sigma - \alpha) + (d + \alpha + u_2)(b\beta + d + u_1 + \sigma)}{\beta[(d + \alpha)\sigma + d(d + \alpha + u_2)]}.
$$

3.2 Local stability of the equilibria

Theorem 2 The disease-free equilibrium is locally asymptotically stable if $R_0 < 1$, and it is unstable if $R_0 > 1$.

Proof. The Jacobian matrix J of the Eq. 3 is

$$
J = \begin{pmatrix}
-d - u_1 - \sigma - \beta I & -\beta S - \sigma & \sigma \\
\beta I & \beta S - (d + \alpha + u_2) & 0 \\
0 & -\alpha & -d
\end{pmatrix}.
$$

Evaluating matrix J at the disease free equilibrium gives

$$
J_0 = \begin{pmatrix}
-d - u_1 - \sigma & -\dfrac{\beta(\sigma + d)b}{(\sigma + d + u_1)d} - \sigma & \sigma \\
0 & \dfrac{\beta(\sigma + d)b}{(\sigma + d + u_1)d} - (d + \alpha + u_2) & 0 \\
0 & -\alpha & -d
\end{pmatrix}.
$$

The matrix J_0 has eigenvalues

$$
\lambda_1 = -d < 0,\quad \lambda_2 = -(d + \sigma + u_1) < 0,
$$

$$
\lambda_3 = \frac{\beta(\sigma + d)b}{(\sigma + d + u_1)d} - (d + \alpha + u_2).
$$

For local asymptotic stability, we require $\lambda_3 < 0$, which is equivalent to $R_0 < 1$. Thus, the disease-free equilibrium ε_0 is locally asymptotically stable if $R_0 < 1$, and unstable if $R_0 > 1$.

Theorem 3 The endemic equilibrium is locally asymptotically stable if $R_0 > 1$.

Proof. The Jacobian matrix J evaluated at the endemic equilibrium gives

$$
J_1 =
$$

$$
\begin{pmatrix}
-(\Re_0 - 1)\dfrac{d(d + \sigma + u_1)(d + \alpha + u_2)}{(d + \alpha)\sigma + d(d + \alpha + u_2)} & -(d + \alpha + \sigma + u_2) & \sigma \\
\quad -(d + \sigma + u_1) & & \\
(\Re_0 - 1)\dfrac{d(d + \sigma + u_1)(d + \alpha + u_2)}{(d + \alpha)\sigma + d(d + \alpha + u_2)} & 0 & 0 \\
0 & -\alpha & -d
\end{pmatrix}.
$$

The characteristic equation of J_1 is

$$
\lambda^3 + a_1\lambda^2 + a_2\lambda + a_3 = 0,
\tag{4}
$$

where

$$
a_1 = m + (2d + \sigma + u_1),
$$

$$
a_2 = (d + \alpha + \sigma + u_2)m + \big[m + (d + \sigma + u_1)\big]d,
$$

$$
a_3 = \big[(d + \alpha + \sigma + u_2)d + \sigma\alpha\big]m,
$$

$$
m = (R_0 - 1)\frac{d(d + \sigma + u_1)(d + \alpha + u_2)}{(d + \alpha)\sigma + d(d + \alpha + u_2)}.
$$

If $R_0 > 1$, then we have $a_1 > 0$, and

524

$$a_1a_2 - a_3 = (m+2d+\sigma+u_1)[(d+\alpha+\sigma+u_2)m$$
$$+(m+d+\alpha+\sigma+u_1)d]$$
$$-m[(d+\alpha+\sigma+u_2)d+\sigma\alpha]$$
$$>m(m+2d+\sigma+u_1)(d+\alpha+\sigma+u_2)$$
$$-m[d(d+\alpha+\sigma+u_2)+\sigma\alpha]$$
$$=m[(m+2d+\sigma+u_1)(d+\alpha+\sigma+u_2)$$
$$-d(d+\alpha+\sigma+u_2)-\sigma\alpha]$$
$$=m[(d+\alpha+\sigma+u_2)(m+d+\sigma+u_1)-\sigma\alpha]$$
$$>0.$$

Therefore, the Routh-Hurwitz stability conditions are satisfied. It follows that the endemic equilibrium ε_1 is locally asymptotically stable.

3.3 Global stability

Theorem 4 If $R_0 \leq 1$, then the disease free equilibrium ε_0 is globally asymptotically stable on Ω.

Proof Considering the Lyapunov function $V(S,I,N): R^3 \to R^+$ defined as

$$V(S,I,N) = \omega I, \omega \geq 0,$$

Differentiating $V(S,I,N)$ with respect to time yields

$$\dot{V} = \omega \dot{I}, \tag{5}$$

Note that $S \leq (d+\sigma)b/((d+\sigma+u_1)d)$, then substituting the system Eq. 3 and using $\omega = 1/(d+\alpha+u_2)$, we obtain

$$\dot{V} = \omega \dot{I} = \omega[\beta S - (d+\alpha+u_2)]I$$
$$\leq \omega\left[\frac{\beta(d+\sigma)b}{(d+\sigma+u_1)d} - (d+\alpha+u_2)\right]I$$
$$= \omega(d+\alpha+u_2)\left[\frac{\beta(d+\sigma)b}{(d+\sigma+u_1)(d+\alpha+u_2)d} - 1\right]I$$
$$= (R_0 - 1)I \leq 0.$$

It is important to note that $\dot{V} = 0$ only when $I = 0$. However, substituting $I = 0$ into the equations for S and N in Eq. 3 shows that $S \to (d+\sigma)b/(d+\sigma+u_1)d$ and $N \to b/d$ as $t \to \infty$. Therefore, the maximum invariant set in $\{(S,I,N) \in \Omega \,|\, \dot{V} \leq 0\}$ is the singleton set $\{\varepsilon_0\}$. Hence, the global stability of ε_0 when $R_0 \leq 1$ follows from LaSalle's invariance principle (see [5]).

Theorem 5 If $R_0 > 1$, then the endemic equilibrium ε_1 is globally asymptotically stable on Ω.

Proof. Let us consider the Lyapunov function

$$V = \omega_1(N-N^*)^2 + \omega_2(N-N^*)(S-S^*)$$
$$+ \omega_3\left[I - I^* - I^*\ln\left(\frac{I}{I^*}\right)\right], \omega_1 \geq 0, \omega_2 \geq 0, \omega_3 \geq 0.$$

Then the derivative of V along the solution curve of the system Eq. 3 yields

$$\dot{V} = 2\omega_1(N-N^*)\dot{N} + \omega_2(S-S^*)\dot{N}$$
$$+ \omega_2(N-N^*)\dot{S} + \omega_3\left(1-\frac{I^*}{I}\right)\dot{I}$$
$$= \left[2\omega_1(N-N^*) + \omega_2(S-S^*)\right](b-dN-\alpha I)$$
$$+ \omega_2(N-N^*)\left[b+\sigma(N-S-I)-dS-u_1S-\beta SI\right]$$
$$+ \omega_3\left(1-\frac{I^*}{I}\right)\left[\beta SI - (d+\alpha+u_2)I\right].$$

$$\tag{6}$$

We recall that, at the endemic equilibrium, we have

$$\begin{cases} b = \beta S^*I^* + dS^* + u_1S^* - \sigma(N^*-I^*-S^*) = dN^* + \alpha I^*, \\ \beta S^*I^* = (d+\alpha+u_2)I^* = b + \sigma(N^*-I^*-S^*) - dS^* - u_1S^*. \end{cases}$$

$$\tag{7}$$

Using the equilibrium condition Eq. 7 above, Eq. 6 becomes

$$\dot{V} = \left[2\omega_1(N-N^*) + \omega_2(S-S^*)\right](dN^* + \alpha I^* - dN - \alpha I)$$
$$+ \omega_2(N-N^*)\left[\beta S^*I^* + dS^* + u_1S^* - \sigma(N^*-I^*-S^*)\right.$$
$$\left. + \sigma(N-S-I) - dS - u_1S - \beta SI\right]$$
$$+ \frac{\omega_3}{I}(I-I^*)(\beta SI - \beta S^*I)$$
$$= -2\omega_1 d(N-N^*)^2 - 2\omega_1\alpha(N-N^*)(I-I^*)$$
$$- 2\omega_2 d(S-S^*)(N-N^*) - 2\omega_2\alpha(S-S^*)(I-I^*)$$
$$+ \omega_2(N-N^*)\left[\beta(S^*I^* - S^*I + S^*I - SI)\right.$$
$$\left. + (d+u_1)(S-S^*) + \sigma(N-N^*) + \sigma(I^*-I) + \sigma(S^*-S)\right]$$
$$+ \beta\omega_3(I-I^*)(S-S^*)$$
$$= -2\omega_1 d(N-N^*)^2 - 2\omega_1\alpha(N-N^*)(I-I^*)$$
$$- 2\omega_2 d(S-S^*)(N-N^*) - 2\omega_2\alpha(S-S^*)(I-I^*)$$
$$+ \omega_2(N-N^*)\left[(\beta S^* + \sigma)(I^*-I)\right.$$
$$\left. + (\beta I + d + u_1 + \sigma)(S^*-S) + \sigma(N-N^*)\right]$$
$$+ \beta\omega_3(I-I^*)(S-S^*)$$
$$= (\sigma\omega_2 - 2\omega_1 d)(N-N^*)^2$$
$$- \left[2\omega_1\alpha + \omega_2(\beta S^* + \sigma)\right](N-N^*)(I-I^*)$$
$$- \left[2\omega_2 d + \omega_2(\beta I + d + u_1 + \sigma)\right](N-N^*)(S-S^*)$$
$$+ (\beta\omega_3 - 2\omega_1\alpha)(S-S^*)(I-I^*)$$
$$\leq 0,$$

525

where we set $\omega_2 = 1$, $\omega_1 = \sigma/2d$ and $\omega_3 = 2d/\beta$. We note that $\dot{V} = 0$ holds only at ε_1. By Lasalle's invariant principle [5], every solution to the Eq. 3, with the initial conditions in Ω, approaches ε_1 as $t \to \infty$ if $R_0 > 1$. Hence, the endemic equilibrium ε_1 is globally asymptotically stable in Ω if $R_0 > 1$.

4 SIMULATION RESULTS

In this section we present a computer simulation of some solutions of the system (3). From the practical point of view, numerical solutions are very important besides analytical study.

$$b = 0.05,\ \sigma = 0.04,\ d = 0.04,\ u_1 = 0.2,$$
$$u_2 = 0.5,\ \alpha = 0.08,\ \beta = 0.75,\ (S(0), I(0), N(0))$$
$$= (0.95, 0.05, 1.0).$$

Then $\varepsilon_0 = (0.357, 0, 1.25)$, $R_0 = 0.432 < 1$. Therefore, by Theorem 4, ε_0 is globally asymptotically stable in the first octant. Figure 1 shows that $S(t)$ and $N(t)$ approach to its steady-state value while $I(t)$ approaches zero as time progresses, the disease dies out.

Now we take the parameters of the system as

$$b = 0.08,\ \sigma = 0.2,\ d = 0.02,\ u_1 = 0.2,\ u_2 = 0.15,$$
$$\alpha = 0.1,\ \beta = 0.75,\ (S(0), I(0), N(0))$$
$$= (0.95, 0.05, 1.0).$$

Then $\varepsilon_1 = (0.360, 0.496, 1.521)$ and $R_0 = 5.815 > 1$. Therefore, by Theorem 5, the endemic equilibrium ε_1 is globally asymptotically stable in the interior of the first octant. Figure 2 shows that all three components, $S(t), I(t)$ and $N(t)$ approach to their steady-state values, the disease becomes endemic.

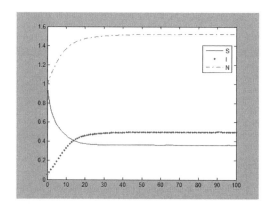

Figure 2.   When $R_0 < 1$, the disease becomes endemic.

5 CONCLUDING REMARKS

In this paper we have carried out the global analysis of a realistic SIRS model. In terms of the basic reproduction number R_0, our main results indicate that when $R_0 < 1$, the disease-free equilibrium is globally stable. When $R_0 > 1$, the endemic equilibrium exists and is globally stable. Therefore, we can adopt vaccination and treatment to decrease R_0 to control the spread of disease.

ACKNOWLEDGMENTS

The research was supported by the Science Foundation of Jimei University, China (No. ZQ2013005 and ZC2013021), Foundation (Class B) of Fujian Educational Committee, China (No. FB2013005).

REFERENCES

[1] R.M. Anderson, R.M. May. Infectious Diseases of Humans[M]. Oxford University Press, Oxford, (1991).
[2] L. Esteva, C. Vargas. Analysis of a dengue disease transmission model[J]. Math. Biosci. Vol. 150, (1998), p.131–151.
[3] H. Gaff, E. Schaefer. Optimal control applied to vaccination and treatment strategies for various epidemiological models[J]. Mathematical Biosciences and Engineering. Vol. 6 (2009), p.469–492.
[4] T.T. Yusuf, F. Benysh. Optimal control of vaccination and treatment for an SIR epidemiological model[J]. World Journal of Modelling and Simulation. Vol. 8 (2012), p.194–204.
[5] J.P. Lasalle. The stability of dynamical systems[M]. Philadelphia: SIAM, (1976).

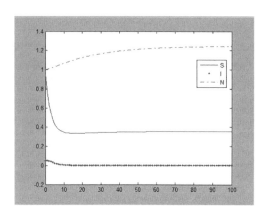

Figure 1.   When $R_0 < 1$, the disease dies out.

Author index

Printed and bound by CPI Group (UK) Ltd, Croydon, CR0 4YY

30/10/2024

01781067-0001